# AS/A Level course structure

This book has been written to support students studying for AQA AS Physics and for students in their first year of studying for AQA A Level Physics. It covers the AS sections from the specification, the content of which will also be examined at A Level. The sections covered are shown in the contents list, which also shows you the page numbers for the main topics within each section. There is also an index at the back to help you find what you are looking for. If you are studying for AS Physics, you will only need to know the content in the blue box.

**AS exam**

**Year 1 content**

1 Measurements and their errors
2 Particles and radiation
3 Waves
4 Mechanics and energy
5 Electricity

**A level exam**

**Year 2 content**

6 Further mechanics and thermal physics
7 Fields
8 Nuclear physics

Plus one option from the following:

- Astrophysics
- Medical physics
- Engineering physics
- Turning points in physics
- Electronics

A Level exams will cover content from Year 1 and Year 2 and will be at a higher demand. You will also carry out practical activities throughout your course.

# Contents


How to use this book iv
Kerboodle vii
Skills for starting AS and A Level Physics x

## Section 1
## Particles and radiation 2

**1 Matter and radiation** 4
1.1 Inside the atom 4
1.2 Stable and unstable nuclei 6
1.3 Photons 8
1.4 Particles and antiparticles 10
1.5 Particle interactions 13
Practice questions 16

**2 Quarks and leptons** 18
2.1 The particle zoo 18
2.2 Particle sorting 20
2.3 Leptons at work 22
2.4 Quarks and antiquarks 24
2.5 Conservation rules 26
Practice questions 28

**3 Quantum phenomena** 30
3.1 The photoelectric effect 30
3.2 More about photoelectricity 32
3.3 Collisions of electrons with atoms 34
3.4 Energy levels in atoms 36
3.5 Energy levels and spectra 39
3.6 Wave–particle duality 41
Practice questions 44
**Section 1 summary** 46
**End of Section 1 questions** 48

## Section 2
## Waves and optics 50

**4 Waves** 52
4.1 Waves and vibrations 52
4.2 Measuring waves 54
4.3 Wave properties 1 56
4.4 Wave properties 2 58
4.5 Stationary and progressive waves 60
4.6 More about stationary waves on strings 62
4.7 Using an oscilloscope 64
Practice questions 66

**5 Optics** 68
5.1 Refraction of light 68
5.2 More about refraction 70
5.3 Total internal reflection 73
5.4 Double slit interference 76
5.5 More about interference 79
5.6 Diffraction 82
5.7 The diffraction grating 85
Practice questions 88
**Section 2 summary** 90
**End of Section 2 questions** 92

## Section 3
## Mechanics and materials 94

**6 Forces in equilibrium** 96
6.1 Vectors and scalars 96
6.2 Balanced forces 100
6.3 The principle of moments 103
6.4 More on moments 105
6.5 Stability 107
6.6 Equilibrium rules 110
6.7 Statics calculations 114
Practice questions 116

**7 On the move** 118
7.1 Speed and velocity 118
7.2 Acceleration 120
7.3 Motion along a straight line at constant acceleration 122
7.4 Free fall 125
7.5 Motion graphs 128
7.6 More calculations on motion along a straight line 130

7.7 Projectile motion 1 132
7.8 Projectile motion 2 134
Practice questions 136

**8 Newton's laws of motion** 138
8.1 Force and acceleration 138
8.2 Using $F = ma$ 141
8.3 Terminal speed 144
8.4 On the road 146
8.5 Vehicle safety 149
Practice questions 152

**9 Force and momentum** 154
9.1 Momentum and impulse 154
9.2 Impact forces 158
9.3 Conservation of momentum 161
9.4 Elastic and inelastic collisions 164
9.5 Explosions 166
Practice questions 168

**10 Work, energy, and power** 170
10.1 Work and energy 170
10.2 Kinetic energy and potential energy 173
10.3 Power 175
10.4 Energy and efficiency 180
Practice questions 182

**11 Materials** 184
11.1 Density 184
11.2 Springs 186
11.3 Deformation of solids 189
11.4 More about stress and strain 192
Practice questions 194
**Section 3 summary** 196
**End of Section 3 questions** 198

**Section 4**
**Electricity** 200
**12 Electric current** 202
12.1 Current and charge 202
12.2 Potential difference and power 204
12.3 Resistance 206
12.4 Components and their characteristics 209
Practice questions 212

**13 DC circuits** 214
13.1 Circuit rules 214
13.2 More about resistance 217
13.3 Electromotive force and internal resistance 220
13.4 More about circuit calculations 223
13.5 The potential divider 226
Practice questions 228
**Section 4 summary** 230
**End of Section 4 questions** 232
**Further practice questions** 234

**Section 5**
**Skills in AS physics** 242
**14 Practical work in physics** 243
14.1 Moving on from GCSE 243
14.2 Making careful measurements 245
14.3 Everyday physics instruments 247
14.4 Analysis and evaluation 249

**15 About practical assessment** 252
15.1 Assessment outline 252
15.2 Direct assessment 254
15.3 Indirect assessment 256

**16 More on mathematical skills** 258
16.1 Data handling 258
16.2 Trigonometry 260
16.3 More about algebra 262
16.4 Straight line graphs 265
16.5 More on graphs 267
16.6 Graphs, gradients, and areas 269

**Reference data and equations** 272
**Answers** 274
**Glossary** 284
**Index** 287
**Acknowledgements** 292

# How to use this book

## Learning objectives

- → At the beginning of each topic, there is a list of learning objectives.
- → These are matched to the specification and allow you to monitor your progress.
- → A specification reference is also included.

*Specification reference: 3.1.1*

This book contains many different features. Each feature is designed to support and develop the skills you will need for your examinations, as well as foster and stimulate your interest in physics.

Terms that you will need to be able to define and understand are highlighted in **bold orange text**. You can look these words up in the glossary.

Sometimes a word appears in **bold**. These are words that are useful to know but are not used on the specification. They therefore do not have to be learnt for examination purposes.

### Synoptic link

These highlight how the sections relate to each other. Linking different areas of physics together becomes increasingly important, as many exam questions (particularly at A Level) will require you to bring together your knowledge from different areas.

There are also links to the maths section to support the development of these skills.

### Application features

These features contain important and interesting applications of physics in order to emphasise how scientists and engineers have used their scientific knowledge and understanding to develop new applications and technologies. There are also application features to develop your maths skills, with the icon $\sqrt{x}$, and to develop your practical skills, with the icon.

### Study tips

Study tips contain prompts to help you with your revision. They can also support the development of your practical skills (with the practical symbol) and your mathematical skills (with the math symbol $\sqrt{x}$).

### Extension features

These features contain material that is beyond the specification designed to stretch and provide you with a broader knowledge and understanding and lead the way into the types of thinking and areas you might study in further education. As such, neither the detail nor the depth of questioning will be required for the examinations. But this book is about more than getting through the examinations.

1 Extension and application features have questions that link the material with concepts that are covered in the specification. Short answers are inverted at the bottom of the feature, whilst longer answers can be found in the answers section at the back of the book.

### Hint

Hint features give other information or ways of thinking about a concept to support your understanding. They can also relate to practical or mathematical skills and use the symbols and $\sqrt{x}$.

### Summary questions

1 These are short questions that test your understanding of the topic and allow you to apply the knowledge and skills you have acquired. The questions are ramped in order of difficulty.

2 $\sqrt{x}$ Questions that will test and develop your mathematical and practical skills are labelled with the mathematical symbol ($\sqrt{x}$) and the practical symbol ().

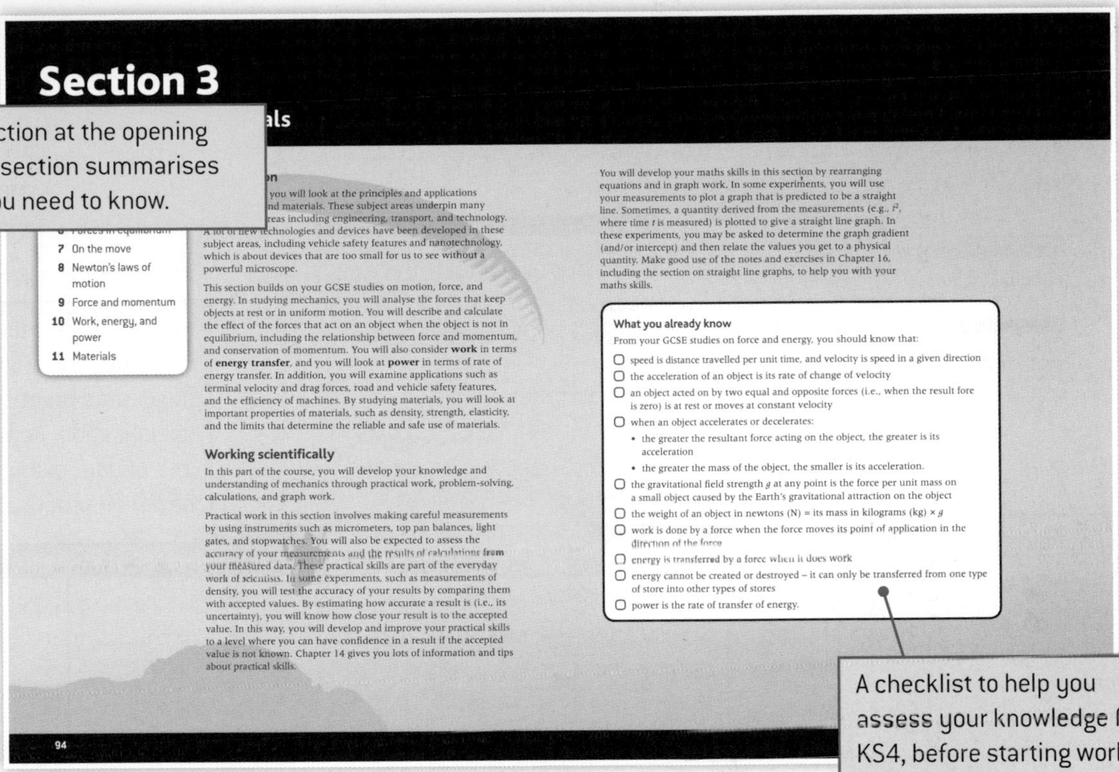

Introduction at the opening of each section summarises what you need to know.

# Section 3

...als

...n

... you will look at the principles and applications ...nd materials. These subject areas underpin many ...reas including engineering, transport, and technology. A lot of new technologies and devices have been developed in these subject areas, including vehicle safety features and nanotechnology, which is about devices that are too small for us to see without a powerful microscope.


7 On the move
8 Newton's laws of motion
9 Force and momentum
10 Work, energy, and power
11 Materials


This section builds on your GCSE studies on motion, force, and energy. In studying mechanics, you will analyse the forces that keep objects at rest or in uniform motion. You will describe and calculate the effect of the forces that act on an object when the object is not in equilibrium, including the relationship between force and momentum, and conservation of momentum. You will also consider **work** in terms of **energy transfer**, and you will look at **power** in terms of rate of energy transfer. In addition, you will examine applications such as terminal velocity and drag forces, road and vehicle safety features, and the efficiency of machines. By studying materials, you will look at important properties of materials, such as density, strength, elasticity, and the limits that determine the reliable and safe use of materials.

### Working scientifically

In this part of the course, you will develop your knowledge and understanding of mechanics through practical work, problem-solving, calculations, and graph work.

Practical work in this section involves making careful measurements by using instruments such as micrometers, top pan balances, light gates, and stopwatches. You will also be expected to assess the accuracy of your measurements and the results of calculations from your measured data. These practical skills are part of the everyday work of scientists. In some experiments, such as measurements of density, you will test the accuracy of your results by comparing them with accepted values. By estimating how accurate a result is (i.e., its uncertainty), you will know how close your result is to the accepted value. In this way, you will develop and improve your practical skills to a level where you can have confidence in a result if the accepted value is not known. Chapter 14 gives you lots of information and tips about practical skills.

You will develop your maths skills in this section by rearranging equations and in graph work. In some experiments, you will use your measurements to plot a graph that is predicted to be a straight line. Sometimes, a quantity derived from the measurements (e.g., $t^2$, where time $t$ is measured) is plotted to give a straight line graph. In these experiments, you may be asked to determine the graph gradient (and/or intercept) and then relate the values you get to a physical quantity. Make good use of the notes and exercises in Chapter 16, including the section on straight line graphs, to help you with your maths skills.

**What you already know**

From your GCSE studies on force and energy, you should know that:

- ☐ speed is distance travelled per unit time, and velocity is speed in a given direction
- ☐ the acceleration of an object is its rate of change of velocity
- ☐ an object acted on by two equal and opposite forces (i.e., when the result fore is zero) is at rest or moves at constant velocity
- ☐ when an object accelerates or decelerates:
  - the greater the resultant force acting on the object, the greater is its acceleration
  - the greater the mass of the object, the smaller is its acceleration.
- ☐ the gravitational field strength $g$ at any point is the force per unit mass on a small object caused by the Earth's gravitational attraction on the object
- ☐ the weight of an object in newtons (N) = its mass in kilograms (kg) × $g$
- ☐ work is done by a force when the force moves its point of application in the direction of the force
- ☐ energy is transferred by a force when it does work
- ☐ energy cannot be created or destroyed – it can only be transferred from one type of store into other types of stores
- ☐ power is the rate of transfer of energy.

94

A checklist to help you assess your knowledge from KS4, before starting work on the section.

Visual summaries of each section show how some of the key concepts of that section interlink with other sections.

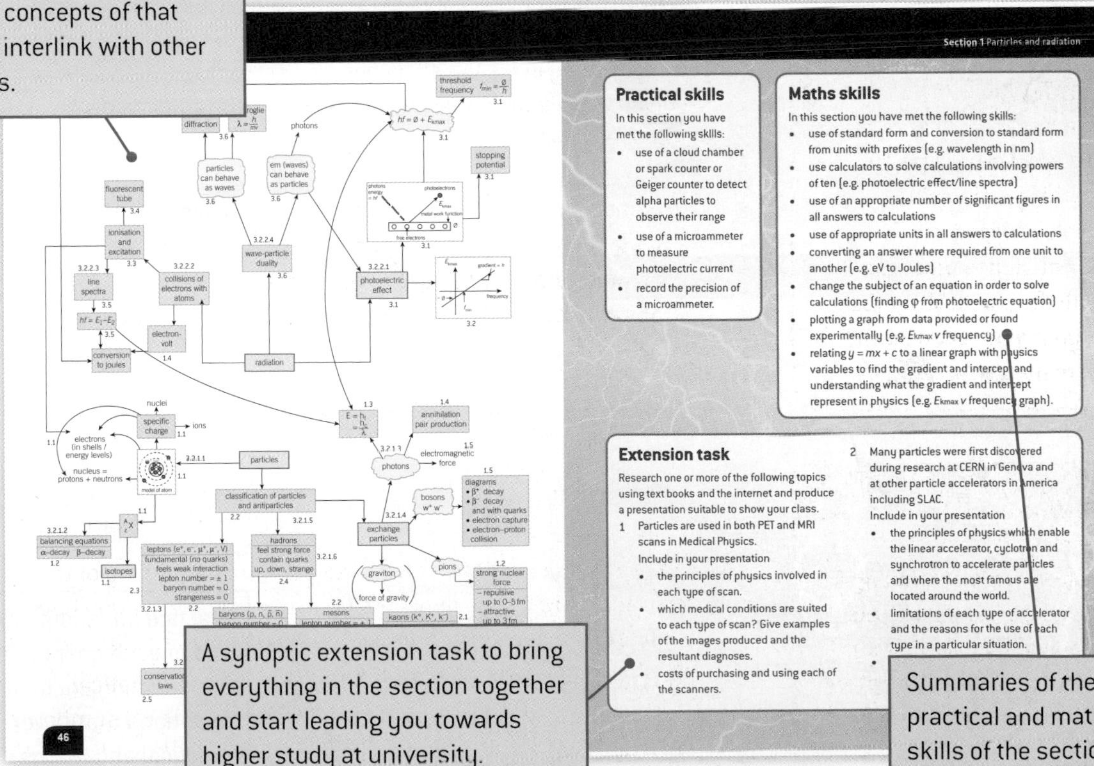

Section 1 Particles and radiation

### Practical skills

In this section you have met the following skills:

- use of a cloud chamber or spark counter or Geiger counter to detect alpha particles to observe their range
- use of a microammeter to measure photoelectric current
- record the precision of a microammeter.

### Maths skills

In this section you have met the following skills:

- use of standard form and conversion to standard form from units with prefixes (e.g. wavelength in nm)
- use calculators to solve calculations involving powers of ten (e.g. photoelectric effect/line spectra)
- use of an appropriate number of significant figures in all answers to calculations
- use of appropriate units in all answers to calculations
- converting an answer where required from one unit to another (e.g. eV to Joules)
- change the subject of an equation in order to solve calculations (finding φ from photoelectric equation)
- plotting a graph from data provided or found experimentally (e.g. $E_{kmax}$ v frequency)
- relating $y = mx + c$ to a linear graph with physics variables to find the gradient and intercept and understanding what the gradient and intercept represent in physics (e.g. $E_{kmax}$ v frequency graph).

### Extension task

Research one or more of the following topics using text books and the internet and produce a presentation suitable to show your class.

1 Particles are used in both PET and MRI scans in Medical Physics.
Include in your presentation
- the principles of physics involved in each type of scan.
- which medical conditions are suited to each type of scan? Give examples of the images produced and the resultant diagnoses.
- costs of purchasing and using each of the scanners.

2 Many particles were first discovered during research at CERN in Geneva and at other particle accelerators in America including SLAC.
Include in your presentation
- the principles of physics which enable the linear accelerator, cyclotron and synchrotron to accelerate particles and where the most famous are located around the world.
- limitations of each type of accelerator and the reasons for the use of each type in a particular situation.

46

A synoptic extension task to bring everything in the section together and start leading you towards higher study at university.

Summaries of the key practical and maths skills of the section.

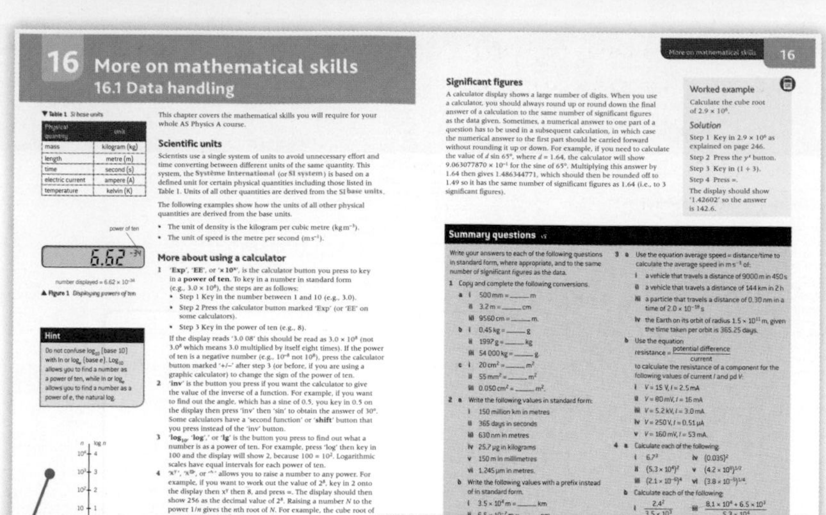

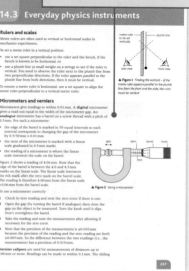

Practical skills section with questions for each suggested practical on the specification. Remember, 15% of your exam will be based on practical skills.

Mathematical section to support and develop your mathematical skills required for your course. Remember, 40% of your exam will involve mathematical skills.

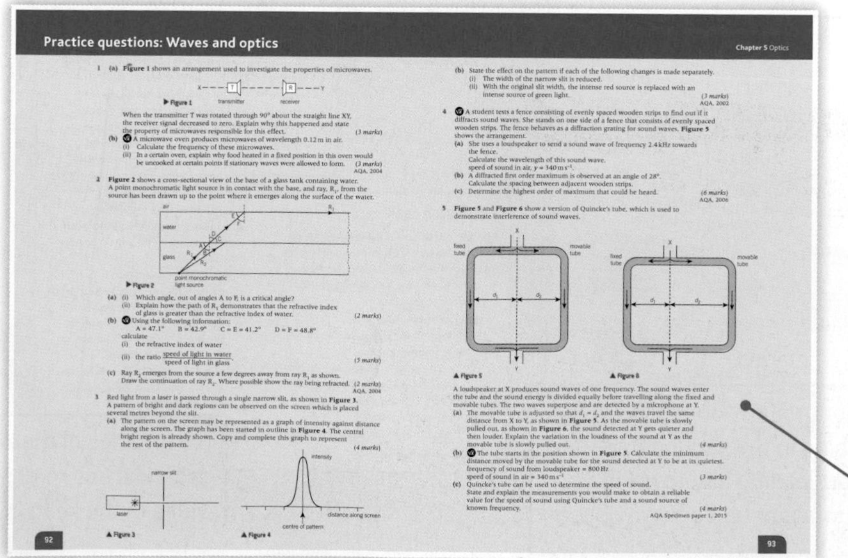

Practice questions at the end of each chapter and each section, including questions that cover practical and maths skills.

# Kerboodle

This book is supported by next generation Kerboodle, offering unrivalled digital support for independent study, differentiation, assessment, and the new practical endorsement.

If your school subscribes to Kerboodle, you will also find a wealth of additional resources to help you with your studies and with revision:

- Study guides
- Maths skills boosters and calculation worksheets
- On your marks activities to help you achieve your best
- Practicals and follow up activities to support the practical endorsement
- Interactive objective tests that give question-by-question feedback
- Animations and revision podcasts
- Self-assessment checklists.

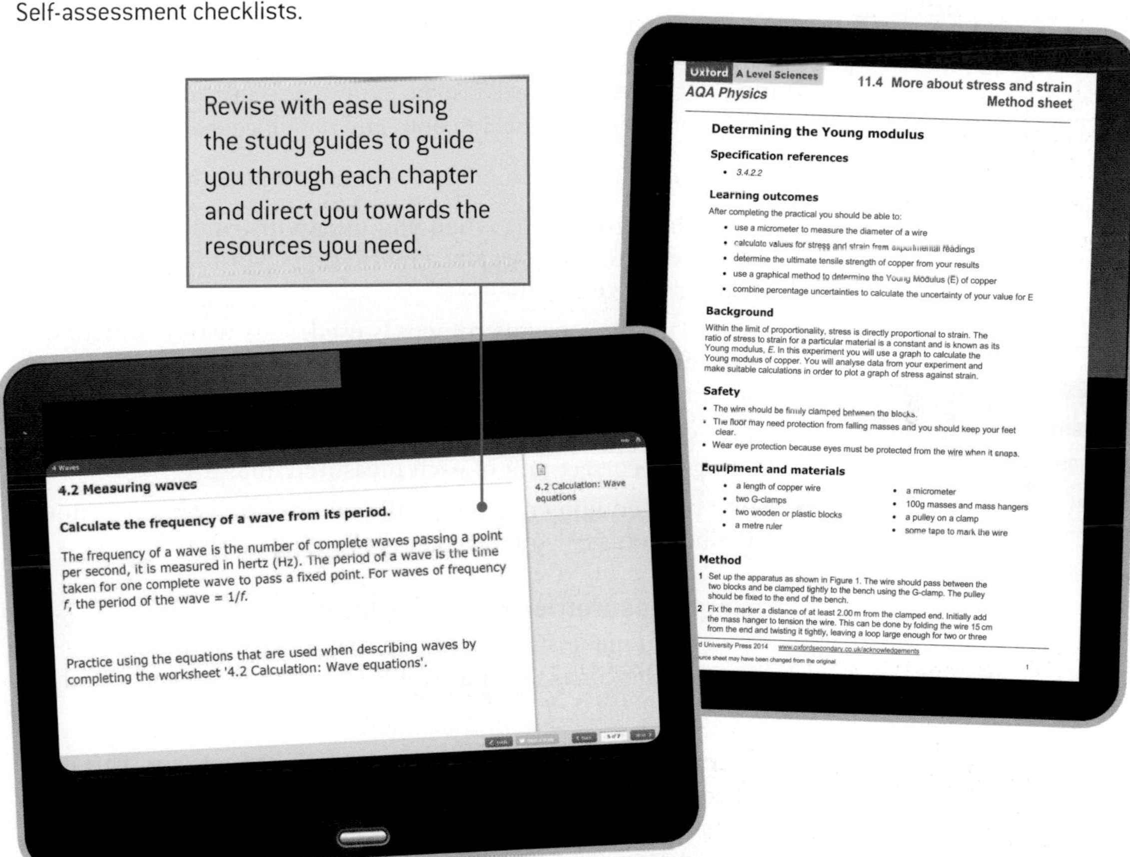


If you are a teacher reading this, Kerboodle also has plenty of further assessment resources, answers to the questions in the book, and a digital markbook along with full teacher support for practicals and the worksheets, which include suggestions on how to support and stretch your students. All of the resources that you need are pulled together into teacher guides that suggest a route through each chapter.

## 1. Using a calculator

Practice makes perfect when it comes to using a calculator. For AS and A level physics, you need no more than a scientific calculator. You need to make sure you master the technicalities of using a scientific calculator as early as possible in your physics course. At this stage, you should be able to use a calculator to add, subtract, multiply, divide, find squares and square roots and calculate sines, cosines and tangents of angles. But remember when using a calculator, it's all too easy to make a mistake, for example pressing the wrong button. So when carrying out a calculation using a calculator, check the answer by making an order-of-magnitude **estimate** of the answer in your head.

**Maths link** 

Further important calculator functions are described in 16.1.

## 2. Making measurements

You should know at this stage how to make measurements using basic equipment such as metre rules, protractors, stopwatches, thermometers, balances (for weighing an object) and ammeters and voltmeters. During the course, you will also be expected to use equipment such as micrometers, verniers, oscilloscopes and data loggers.

**Practical link**

The use of these items is described for reference in 14.3.

Here are some useful reminders about making measurements:

- check the zero reading when you use an instrument to make a measurement, for example, a metre ruler worn away at one end might give a zero error
- when a multi-range instrument is used, start with the highest range and switch to a lower range if the reading is too small to measure accurately
- make sure you record all your measurements in a logical order, stating the correct unit of each measured quantity
- don't pack equipment away until you are sure you have enough measurements or you have checked unexpected readings

**Practical link**

See 14.2 for anomalous measurements.

## 3. Using measurements in calculations

whenever you make a record of a measurement, you must always note the correct unit as well as the numerical value of the measurements.

**The scientific system of units** is called the S.I. system. This is described in more detail in 16.1 and 16.3. The base units of the S.I. system you need to remember are listed below. All other units are derived from the S.I. base units.

1. the metre (m) is the S.I. unit of length. Note also that 1 m = 100 cm = 1000 mm.
2. the kilogram (kg) is the S.I. unit of mass. Note that 1 kg = 1000 grams.
3. the second (s) is the S.I. unit of time.
4. the ampere (A) is the S.I. unit of current.

**Powers of ten and numerical prefixes** are used to avoid unwieldy numerical values. For example:

- $1\,000\,000 = 10^6$ which is 10 raised to the power 6 (usually stated as '10 to the 6').
- $0.000\,000\,1 = 10^{-7}$ which is 10 raised to the power −7 (usually stated as '10 to the −7').

Prefixes are used with units as abbreviations for powers of ten. For example, a distance of 1 kilometre may be written as 1000 m or $10^3$ m or 1 km. The most common prefixes are shown in Table 1.

▼ **Table 1** *Prefixes*

| Prefix | pico- | nano- | micro- | milli- | kilo- | mega- | giga- | tera- |
|---|---|---|---|---|---|---|---|---|
| Value | $10^{-12}$ | $10^{-9}$ | $10^{-6}$ | $10^{-3}$ | $10^{3}$ | $10^{6}$ | $10^{9}$ | $10^{12}$ |
| Prefix symbol | p | n | μ | m | k | M | G | T |

Note that the cubic centimetre ($cm^3$) and the gram (g) are in common use and are therefore allowed as exceptions to the prefixes shown.

**Standard form** is usually used for numerical values smaller than 0.001 or larger than 1000.

- The numerical value is written as a number between 1 and 10 multiplied by the appropriate power of ten. For example,

  $64\,000\,\text{m} = 6.4 \times 10^4\,\text{m}$

  $0.000\,005\,1\,\text{s} = 5.1 \times 10^{-6}\,\text{s}$

- A prefix may be used instead of some or all of the powers of ten. For example:

  $35\,000\,\text{m} = 35 \times 10^3\,\text{m} = 35\,\text{km}$

  $0.000\,000\,59\,\text{m} = 5.9 \times 10^{-7}\,\text{m} = 590\,\text{nm}$

To convert a number to standard form, count how many places the decimal point must be moved to make the number between 1 and 10. The number of places moved is the power of ten that must accompany the number between 1 and 10. Moving the decimal place:

- to the left gives a positive power of ten (for example, $64\,000 = 6.4 \times 10^4$)
- to the right gives a negative power of ten (for example, $0.000\,005\,1 = 5.1 \times 10^{-6}$)

## 4. Using the right-angled triangle

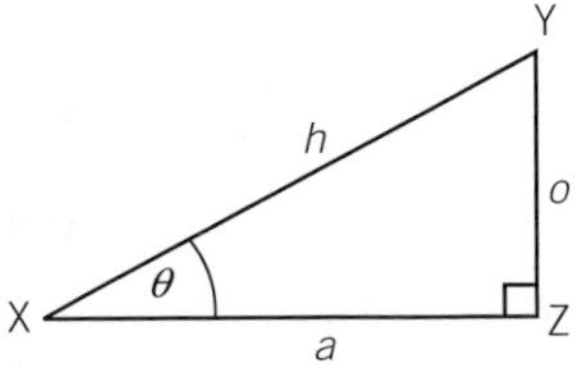


▲ **Figure 1** *The right-angled triangle*

**The sine, cosine and tangent of an angle** are defined from the right-angled triangle. Figure 1 on the previous page shows a right-angled triangle XYZ in which side XY is the hypotenuse (i.e. the side opposite the right angle), side YZ is opposite angle θ and side XZ is adjacent to angle θ.

$$\sin\theta = \frac{YZ}{XY} = \frac{o}{h}$$

$$\cos\theta = \frac{XZ}{XY} = \frac{a}{h}$$

$$\tan\theta = \frac{YZ}{XZ} = \frac{o}{a}$$

where o = YZ , the side opposite angle θ

h = XY, the hypotenuse

and a = XZ , the side adjacent to angle θ

To remember these formulas, recall SOHCAHTOA.

**Pythagoras' theorem** states that for any right angle triangle,

**the square of the hypotenuse = the sum of the squares of the other two sides**

Applying Pythagoras' theorem to the right angle triangle XYZ in Figure 1 gives:

$$\mathbf{(XY)^2 = (XZ)^2 + (YZ)^2}$$

## 5. Using equations

**Symbols** are used in equations and formulas to represent physical variables. In your GCSE course, you may have used equations with words instead of symbols to represent physical variables. For example, you will have met the equation 'distance moved = speed × time'. Perhaps you were introduced to the same equation in the symbolic form '$s = v\,t$' where $s$ is the symbol used to represent distance, $v$ is the symbol used to represent speed and $t$ is the symbol used to represent time. The equation in symbolic form is easier to use because the rules of algebra are more easily applied to it than to a word equation. In addition, writing words in equations wastes valuable time. However, you need to remember the agreed symbol for each physical quantity.

**Equations** often need to be rearranged. This can be confusing if you don't learn the following basic rules at an early stage:

1 Learn to read an equation properly. For example, the equation $v = 3\,t + 2$ is not the same as the equation $v = 3\,(\,t + 2\,)$. If you forget the brackets when you use the second equation to calculate $v$ when $t = 1$, then you will get $v = 5$ instead of the correct answer $v = 9$. The first equation tells you to multiply $t$ by 3 then add 2. The second equation tells you to add $t$ and 2 then multiply the sum by 3.

2 Learn to rearrange an equation properly. In simple terms, always make the same change to both sides of an equation.

For example, to make $t$ the subject of the equation $v = 3t + 2$

***Step 1:*** Subtract 2 from both sides of the equation so $v - 2 = 3t + 2 - 2 = 3t$

***Step 2:*** The equation is now $v - 2 = 3t$ and can be written $3t = v - 2$

***Step 3:*** Divide both sides of the equation by 3 so $\frac{3t}{3} = \frac{v-2}{3}$

***Step 4:*** Cancel 3 on the top and the bottom of the left hand side to finish with $t = \frac{v-2}{3}$

**To use an equation as part of a calculation:**

- start by making the quantity to be calculated the subject of the equation
- write the equation out with the numerical values in place of the symbols
- carry out the calculation and make sure you give the answer with the correct unit

AND remember to check your answer by making an order-of-magnitude estimate.

Unless the equation is simple (for example, $V = IR$), don't insert numerical values then rearrange the equation. Rearrange then insert the numerical values as you are less likely to make an error if the numbers are inserted later in the process rather than earlier.

**Maths link**

See 16.1 to 16.3 for more about data handling, trigonometry and algebra.

# Section 1

## Particles and radiation

### Chapters in this section:

1 Matter and radiation

2 Quarks and leptons

3 Quantum phenomena

### Introduction

Particle physics is at the frontiers of physics. It is about the fundamental properties of matter, radiation, and energy. In this section, you will gain awareness of the ongoing development of new ideas in physics and of the application of in-depth knowledge to well-established concepts.

Experiments in particle physics have supported the idea that protons and neutrons consist of smaller particles called **quarks**. They have also shown that electrons are fundamental particles belonging to a small family of particles called **leptons**. Quarks and leptons are the building blocks of matter. The key developments in this journey into the atom are highlighted in the first two chapters of this section. Investigations into the effect of light on metals have revealed that light consists of wave packets of electromagnetic radiation called **photons**. A photon is a quantum, or the least amount, of electromagnetic radiation. The theory that energy is in these lumps (quanta) instead of being evenly spread makes up part of **quantum theory**. It has led to the discovery of **energy levels** in the atom, **light spectra**, and **wave–particle duality**. In this section, we will be looking at all of these subject areas and at the experimental evidence that led to the ideas and theories in them. In addition, we will be looking at matter and **antimatter** and their properties, including annihilation, in which matter and antimatter interact and turn into radiation energy, and **pair production**, in which the opposite happens. We will also be studying how physicists detected uncharged leptons called **neutrinos**. These are so elusive that billions of them pass through us every second without our knowledge! Many questions have arisen because of the work being done by physicists across the world, for example, what exactly is the mechanism that turns mass into radiation energy in annihilation? We will look at some of these questions in this section.

### Working scientifically

In this part of the course, you will meet big ideas and find out how scientists discovered and confirmed them. You will learn how to observe and detect particles by using a cloud chamber or a spark counter – these instruments are the forerunners of the massive detectors being used at international nuclear research laboratories such as CERN in Geneva in Switzerland.

The maths skills you will need are mainly about handling numbers in standard form, carrying out scientific calculations, and rearranging equations. Make good use of the notes and exercises in Chapter 16 to help you improve your maths skills.

The notes on straight line graphs in Chapter 16 will also help you in Chapter 3 when we look at the photoelectric effect. This is the emission of electrons from metal when light is shone on the metal's surface. The puzzling results from these metal experiments were not explained until Albert Einstein formed a revolutionary new theory about light called the photon theory. His theory predicted that a graph of the kinetic energy of the emitted electrons versus the frequency of the incident light should give a straight line – and that's exactly what was found! As you progress through the course, you will carry out many experiments where straight line graphs are used to test an equation about a theory.

## What you already know

From your GCSE studies on radioactivity, you should know that:

- ☐ every atom has a positively charged nucleus at its centre that is surrounded by negatively charged particles called electrons
- ☐ the nucleus of the atom is composed of protons and neutrons, which have approximately the same mass
- ☐ protons and electrons have fixed equal and opposite amounts of charge, and neutrons are uncharged
- ☐ an uncharged atom has equal amounts of electrons and protons
- ☐ radioactive substances emit radiation because the nuclei of their atoms are unstable
- ☐ when the radiation from radioactive substances passes through other substances, it can ionise them
- ☐ there are three types of radiation from radioactive substances:
    - alpha radiation, which is stopped by paper and is strongly ionising
    - beta radiation, which is stopped by 2–3 mm of metal and is less ionising than alpha radiation
    - gamma radiation, which is very penetrating and is weakly ionising
- ☐ the half-life of a radioactive substance is the time it takes for the number of atoms of that substance to decrease by half.

# 1 Matter and radiation
## 1.1 Inside the atom

### Learning objectives:
- → Describe what is inside an atom.
- → Explain the term isotope.
- → Represent different atoms.

*Specification reference: 3.2.1.1*

### Study tip
Don't mix up 'n' words – nucleus, neutron, nucleon, nuclide!

▲ **Figure 1** *Atoms seen using a scanning tunnelling microscope (STM)*

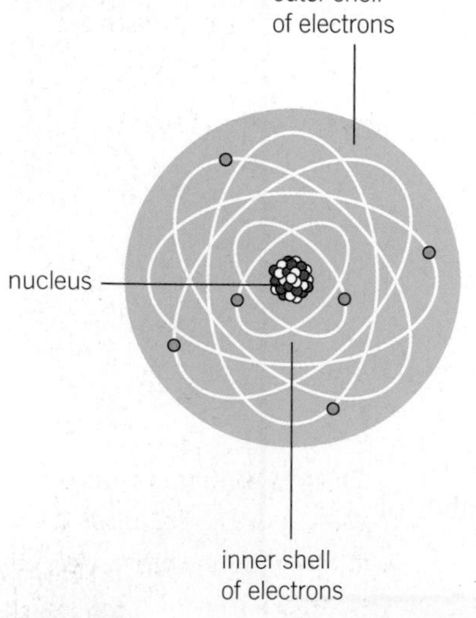


▲ **Figure 2** *Inside the atom (not to scale)*

## The structure of the atom

Atoms are so small (less than a millionth of a millimetre in diameter) that we need to use an electron microscope to see images of them. Although we cannot see inside them, we know, from Rutherford's alpha-scattering investigations, that every atom contains

- a positively charged nucleus composed of protons and neutrons
- electrons that surround the nucleus.

We use the word **nucleon** for a proton or a neutron in the nucleus.

Each electron has a negative charge. Because the nucleus is positively charged, the electrons are held in the atom by the electrostatic force of attraction between them and the nucleus. Rutherford's investigations showed that the nucleus contains most of the mass of the atom and its diameter is of the order of 0.000 01 times the diameter of a typical atom.

Table 1 shows the charge and the mass of the proton, the neutron, and the electron in SI units (coulomb for charge and kilogram for mass) and relative to the charge and mass of the proton. Notice that:

1. The electron has a much smaller mass than the proton or the neutron.
2. The proton and the neutron have almost equal mass.
3. The electron has equal and opposite charge to the proton. The neutron is uncharged.

▼ **Table 1** *Inside the atom*

| | Charge / C | Charge relative to proton | Mass / kg | Mass relative to proton |
|---|---|---|---|---|
| proton | $+1.60 \times 10^{-19}$ | 1 | $1.67 \times 10^{-27}$ | 1 |
| neutron | 0 | 0 | $1.67 \times 10^{-27}$ | 1 |
| electron | $-1.60 \times 10^{-19}$ | −1 | $9.11 \times 10^{-31}$ | 0.0005 |

This means that an uncharged atom has equal numbers of protons and electrons. An uncharged atom becomes an ion if it gains or loses electrons.

## Isotopes

Every atom of a given element has the same number of protons as any other atom of the same element. The **proton number** is also called the **atomic number** (symbol $Z$) of the element. For example:

- $Z = 6$ for carbon because every carbon atom has six protons in its nucleus.
- $Z = 92$ for uranium because every uranium atom has 92 protons in its nucleus.

The atoms of an element can have different numbers of neutrons. Atoms of the same element with different numbers of neutrons are called **isotopes**.

For example, the most abundant isotope of natural uranium contains 146 neutrons and the next most abundant contains 143 neutrons.

**Isotopes are atoms with the same number of protons and different numbers of neutrons.**

The total number of protons and neutrons in an atom is called the **nucleon number** (symbol $A$) or sometimes the **mass number** of the atom. This is because it is almost numerically equal to the mass of the atom in relative units (where the mass of a proton or neutron is approximately 1). A nucleon is a neutron or a proton in the nucleus.

We label the isotopes of an element according to their atomic number $Z$, their mass number $A$, and the chemical symbol of the element. Figure 3 shows how we do this. Notice that:

- $Z$ is at the bottom left of the element symbol and gives the number of protons in the nucleus.
- $A$ is at the top left of the element symbol and gives the number of protons and neutrons in the nucleus.
- The number of neutrons in the nucleus = $A - Z$.

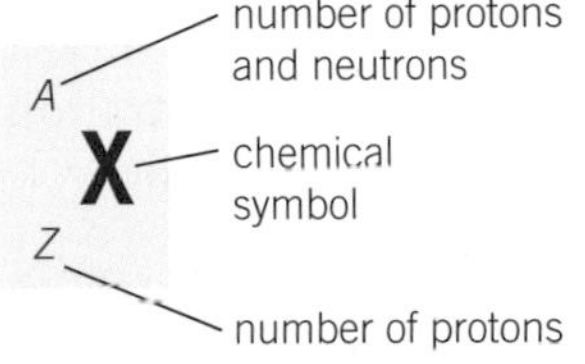


Example: the symbol for the uranium isotope with 92 protons and 146 neutrons is

$^{238}_{92}\mathrm{U}$ (or sometimes U-238)

▲ **Figure 3** *Isotope notation*

Each type of nucleus is called a **nuclide** and is labelled using the isotope notation. For example, a nuclide of the carbon isotope $^{12}_{6}\mathrm{C}$ has two fewer neutrons and two fewer protons than a nuclide of the oxygen isotope $^{16}_{8}\mathrm{O}$.

## Specific charge

The **specific charge** of a charged particle is defined as its charge divided by its mass. We can calculate the specific charge of a charged particle if we know the charge and the mass of the particle. For example:

A nucleus of $^{1}_{1}\mathrm{H}$ has a charge of $1.60 \times 10^{-19}$ C and a mass of $1.67 \times 10^{-27}$ kg. Its specific charge is therefore $9.58 \times 10^{7}$ C kg$^{-1}$.

The electron has a charge of $-1.60 \times 10^{-19}$ C and a mass of $9.11 \times 10^{-31}$ kg. Its specific charge is therefore $1.76 \times 10^{11}$ C kg$^{-1}$. Note that the electron has the largest specific charge of any particle.

An ion of the magnesium isotope $^{24}_{12}\mathrm{Mg}$ has a charge of $+3.2 \times 10^{-19}$ C and a mass of $3.98 \times 10^{-26}$ kg (ignoring the mass of the electrons). Its specific charge is therefore $8.04 \times 10^{6}$ C kg$^{-1}$.

## Summary questions

**You will need to use data from the data sheet on pages 262–265 to answer some of the questions below.**

**1** **a** State the number of protons and the number of neutrons in a nucleus of

**i** $^{12}_{6}\mathrm{C}$ **ii** $^{16}_{8}\mathrm{O}$

**iii** $^{235}_{92}\mathrm{U}$ **iv** $^{24}_{11}\mathrm{Na}$

**v** $^{63}_{29}\mathrm{Cu}$.

**b** Which of the above nuclei has

**i** the smallest specific charge?

**ii** the largest specific charge?

**2** Name the part of an atom which

**a** has zero charge

**b** has the largest specific charge

**c** when removed, leaves a different isotope of the element.

**3** A $^{63}_{29}\mathrm{Cu}$ atom loses two electrons. For the ion formed,

**a** calculate its charge in C

**b** state the number of nucleons it contains

**c** calculate its specific charge in C kg$^{-1}$.

**4** **a** Calculate the mass of an ion with a specific charge of $1.20 \times 10^{7}$ C kg$^{-1}$ and a negative charge of $3.2 \times 10^{-19}$ C.

**b** The ion has eight protons in its nucleus. Calculate its number of neutrons and electrons.

# 1.2 Stable and unstable nuclei

## Learning objectives:

- → State what keeps the protons and neutrons in a nucleus together.
- → Explain why some nuclei are stable and others unstable.
- → Describe what happens when an unstable nucleus emits an alpha particle or a beta minus particle.

*Specification reference: 3.2.1.2*

## The strong nuclear force

A stable isotope has nuclei that do not disintegrate, so there must be a force holding them together. We call this force the **strong nuclear force** because it overcomes the electrostatic force of repulsion between the protons in the nucleus and keeps the protons and neutrons together.

Some further important points about the strong nuclear force are:

- Its range is no more than about 3–4 femtometres (fm), where 1 fm = $10^{-15}$ m = 0.000 000 000 000 001 m. This range is about the same as the diameter of a small nucleus. In comparison, the electrostatic force between two charged particles has an infinite range (although it decreases as the range increases).
- It has the same effect between two protons as it does between two neutrons or a proton and a neutron.
- It is an attractive force from 3–4 fm down to about 0.5 fm. At separations smaller than this, it is a repulsive force that acts to prevent neutrons and protons being pushed into each other.

Figure 1 shows how the strong nuclear force varies with separation between two protons or neutrons. Notice that the equilibrium separation is where the force curve crosses the *x*-axis.

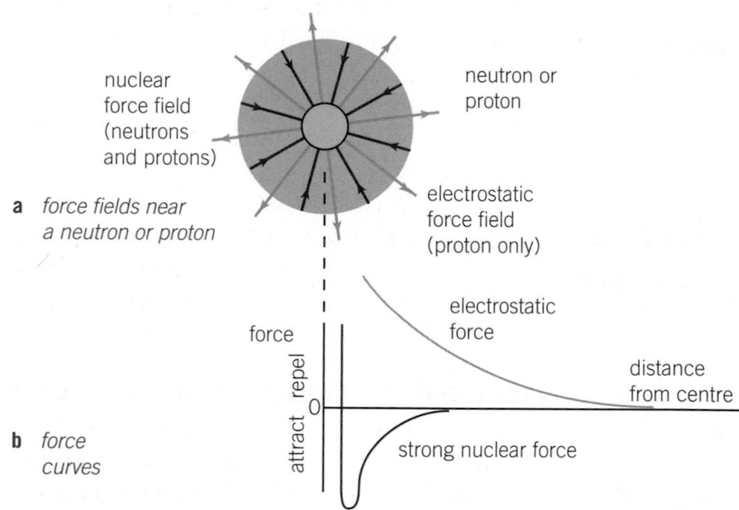


▲ **Figure 1** *The strong nuclear force*

the nucleus emits an α particle and forms a new nucleus

α particle

proton

neutron

▲ **Figure 2** *Alpha particle emission (not to scale)*

## Radioactive decay

Naturally occurring radioactive isotopes release three types of radiation.

1 **Alpha radiation** consists of alpha particles which each comprise two protons and two neutrons. The symbol for an alpha particle is ${}^{4}_{2}\alpha$ because its proton number is 2 and its mass number is 4. Figure 2 shows what happens to an unstable nucleus of an element X when it emits an alpha particle. Its nucleon number *A* decreases by 4 and its atomic number *Z* decreases by 2. As a result of the change, the product nucleus belongs to a different element Y.

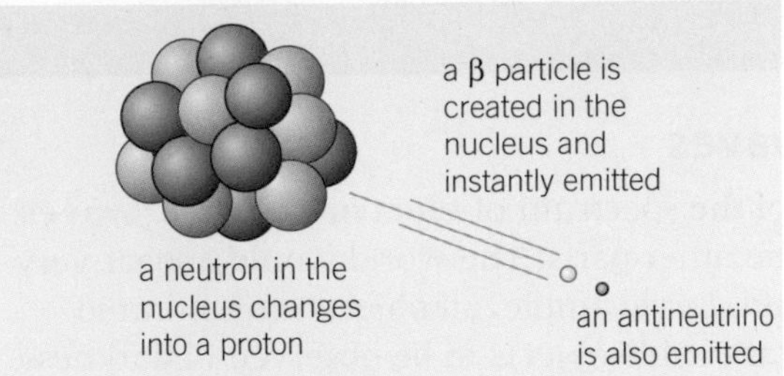
a β particle is
created in the
nucleus and
instantly emitted
a neutron in the
nucleus changes
into a proton
an antineutrino
is also emitted

▲ **Figure 3** *Beta particle emission (not to scale)*

We can represent this change by means of the equation below:

$$^{A}_{Z}X \rightarrow {}^{A-4}_{Z-2}Y + {}^{4}_{2}\alpha$$

2 **Beta radiation** consists of fast-moving electrons. The symbol for an electron as a beta particle is $^{0}_{-1}\beta$ (or $\beta^-$) because its charge is equal and opposite to that of the proton and its mass is much smaller than the proton's mass.

Figure 3 shows what happens to an unstable nucleus of an element X when it emits a $\beta^-$ particle. This happens as a result of a neutron in the nucleus changing into a proton. The beta particle is created when the change happens and is emitted instantly. In addition, an antiparticle with no charge, called an antineutrino (symbol $\bar{\nu}$), is emitted. You will learn more about antiparticles in Topic 1.4, and neutrinos in Topic 1.5. Because a neutron changes into a proton in the nucleus, the atomic number increases by 1 but the nucleon number stays the same. As a result of the change, the product nucleus belongs to a different element Y. This type of change happens to nuclei that have too many neutrons.

We can represent this change by means of the equation below:

$$^{A}_{Z}X \rightarrow {}^{A}_{Z+1}Y + {}^{0}_{-1}\beta + \bar{\nu}$$

3 **Gamma radiation** (symbol γ) is electromagnetic radiation emitted by an unstable nucleus. It can pass through thick metal plates. It has no mass and no charge. It is emitted by a nucleus with too much energy, following an alpha or beta emission.

### Journey into the atom (part 1): A very elusive particle!

When the energy spectrum of beta particles was first measured, it was found that beta particles were released with kinetic energies up to a maximum that depended on the isotope. The scientists at the time were puzzled why the energy of the beta particles varied up to a maximum, when each unstable nucleus lost a certain amount of energy in the process. Either energy was not conserved in the change or some of it was carried away by mystery particles, which they called **neutrinos** and **antineutrinos**. This hypothesis was unproven for over 20 years until antineutrinos were detected. Antineutrinos were detected as a result of their interaction with cadmium nuclei in a large tank of water. This was installed next to a nuclear reactor as a controllable source of these very elusive particles. Now we know that billions of these elusive particles from the Sun sweep through our bodies every second without interacting!

## Summary questions

1 Which force, the strong nuclear force or the electrostatic force,

a does not affect a neutron

b has a limited range

c holds the nucleons in a nucleus

d tends to make a nucleus unstable?

2 Complete the following radioactive decay equations:

a $^{229}_{90}Th \rightarrow Ra + \alpha$

b $^{65}_{28}Ni \rightarrow Cu + \beta + \bar{\nu}$.

3 A bismuth $^{213}_{83}Bi$ nucleus emits a beta particle then an alpha particle then another beta particle before it becomes a nucleus X.

a Show that X is a bismuth isotope.

b i Determine the nucleon number of X.

ii How many protons and how many neutrons are in the nucleus just after it emits the alpha particle?

4 The neutrino hypothesis was put forward to explain beta decay.

a Explain the term *hypothesis*.

b i State one property of the neutrino.

ii Name two objects that produce neutrinos.

# 1.3 Photons

## Learning objectives:

- → Recall what is meant by a photon.
- → Calculate the energy of a photon.
- → Estimate how many photons a light source emits every second.

*Specification reference: 3.2.1.3*

## Electromagnetic waves

Light is just a small part of the spectrum of **electromagnetic waves**. Our eyes cannot detect the other parts. The world would appear very different to us if they could. For example, all objects emit infrared radiation. Infrared cameras enable objects to be observed in darkness.

In a vacuum, all electromagnetic waves travel at the speed of light, $c$, which is $3.00 \times 10^8\ \mathrm{m\,s^{-1}}$. As you know from GCSE, the wavelength $\lambda$ of **electromagnetic radiation** of frequency $f$ in a vacuum is given by the equation

$$\lambda = \frac{c}{f}$$

Note that we often express light wavelengths in nanometres (nm), where $1\ \mathrm{nm} = 0.000\,000\,001\ \mathrm{m} = 10^{-9}\ \mathrm{m}$.

The main parts of the electromagnetic spectrum are listed in Table 1.

▼ **Table 1** *The main parts of the electromagnetic spectrum*

| Type | radio | microwave | infrared | visible | ultraviolet | X-rays | gamma rays |
|---|---|---|---|---|---|---|---|
| Wavelength range | >0.1 m | 0.1 m to 1 mm | 1 mm to 700 nm | 700 nm to 400 nm | 400 nm to 1 nm | 10 nm to 0.001 nm | <1 nm |

As shown in Figure 1, an electromagnetic wave consists of an electric wave and a magnetic wave which travel together and vibrate

- at right angles to each other and to the direction in which they are travelling
- in phase with each other. As you can see the two waves reach a peak together so they are in step. When waves do this we say they are in phase.

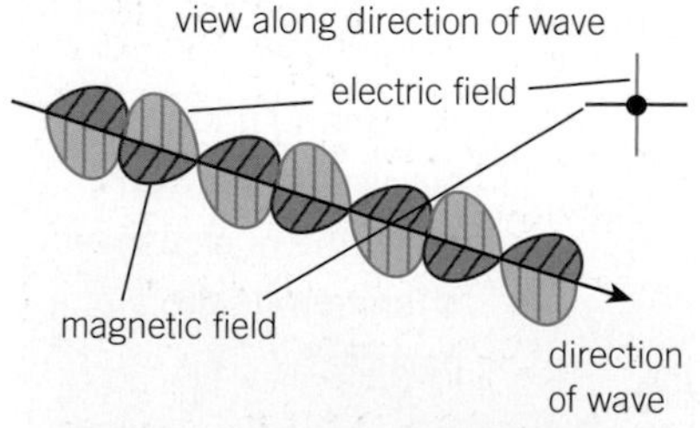


▲ **Figure 1** *Electromagnetic waves*

## Photons

Electromagnetic waves are emitted by a charged particle when it loses energy. This can happen when

- a fast-moving electron is stopped (for example, in an X-ray tube) or slows down or changes direction
- an electron in a shell of an atom moves to a different shell of lower energy.

Electromagnetic waves are emitted as short bursts of waves, each burst leaving the source in a different direction. Each burst is a packet of electromagnetic waves and is referred to as a **photon**. The photon theory was established by Einstein in 1905, when he used his ideas to explain the **photoelectric effect**. This is the emission of electrons from a metal surface when light is directed at the surface.

### Synoptic link

You will meet the photoelectric effect and the photon theory in more detail in Topic 3.1, The photoelectric effect.

Einstein imagined photons to be like *flying needles,* and he assumed that the energy $E$ of a photon depends on its frequency $f$ in accordance with the equation

$$\textbf{photon energy } E = hf$$

where $h$ is a constant referred to as the Planck constant. The value of $h$ is $6.63 \times 10^{-34}$ J s.

## Worked example

Calculate the frequency and the energy of a photon of wavelength 590 nm.

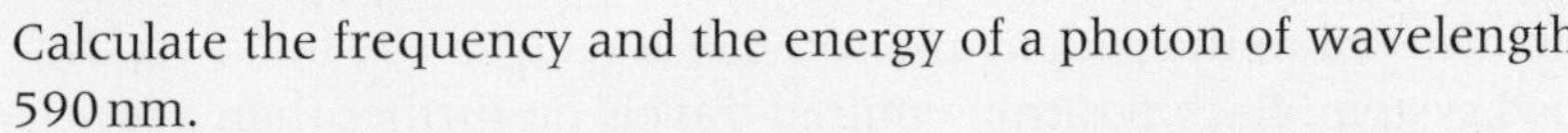

$h = 6.63 \times 10^{-34}$ **J s,** $c = 3.00 \times 10^{8}$ **m s**$^{-1}$

### *Solution*

To calculate the frequency, use $f = \dfrac{c}{\lambda} = \dfrac{3.00 \times 10^{8}}{590 \times 10^{-9}} = 5.08 \times 10^{14}$ Hz.

To calculate the energy of a photon of this wavelength, we use $E = hf$.

$E = hf = 6.63 \times 10^{-34} \times 5.08 \times 10^{14} = 3.37 \times 10^{-19}$ J.

### Laser power

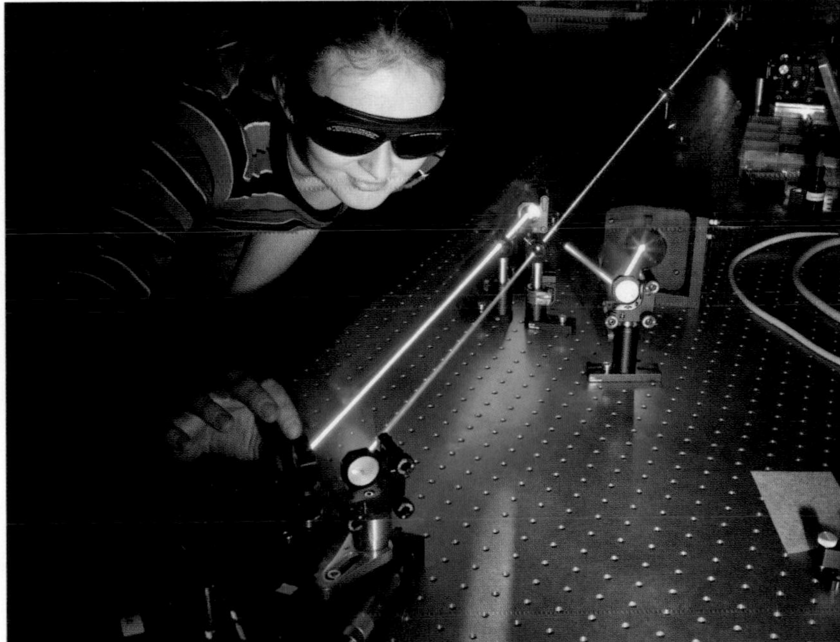

▲ **Figure 2** *A laser at work*

A laser beam consists of photons of the same frequency. The power of a laser beam is the energy per second transferred by the photons. For a beam consisting of photons of frequency $f$,

$$\textbf{the power of the beam} = nhf$$

where $n$ is the number of photons in the beam passing a fixed point each second. This is because each photon has energy $hf$. Therefore, if $n$ photons pass a fixed point each second, the energy per second (or power) is $nhf$.

## Summary questions

$c = 3.00 \times 10^{8}$ m s$^{-1}$

$h = 6.63 \times 10^{-34}$ J s

1 a List the main parts of the electromagnetic spectrum in order of increasing wavelength.

  b Calculate the frequency of

   i light of wavelength 590 nm

   ii radio waves of wavelength 200 m.

2 With the aid of a suitable diagram, explain what is meant by an electromagnetic wave.

3 Light from a certain light source has a wavelength of 430 nm. Calculate

  a the frequency of light of this wavelength

  b the energy of a photon of this wavelength.

4 a Calculate the frequency and energy of a photon of wavelength 635 nm.

  b A laser emits light of wavelength 635 nm in a beam of power 1.5 mW. Calculate the number of photons emitted by the laser each second.

# 1.4 Particles and antiparticles

**Learning objectives:**

- → Define antimatter.
- → Describe what happens when a particle and its antiparticle meet.
- → Discuss whether anti-atoms are possible.

*Specification reference: 3.2.1.3*

**Brain imaging**

PET scanners are used in hospitals for brain imaging of, for example, stroke patients. The T stands for tomography, which is the name for the electronic and mechanical system used to perform the scan.

**Q:** What type of radiation is detected in a PET scanner?

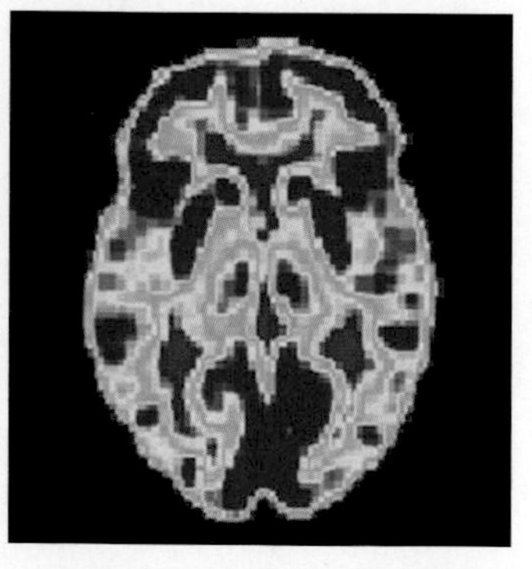

A Gamma radiation

## Antimatter

When **antimatter** and matter particles (i.e., the particles that make up everything in our universe) meet, they destroy each other and radiation is released. We make use of this effect in a positron emitting tomography (PET) hospital scanner. The P in PET stands for the **positron**, which is the antiparticle of the electron. When a PET scanner is used for a brain scan, a positron-emitting isotope is administered to the patient and some of it reaches the brain via the blood system. Each positron emitted travels no further than a few millimetres before it meets an electron and they annihilate each other. Two gamma photons, produced as a result, are sensed by detectors linked to computers. Gradually, an image is built up from the detector signals of where the positron-emitting nuclei are inside the brain.

Positron emission takes place when a proton changes into a neutron in an unstable nucleus with too many protons. The positron (symbol $^{0}_{+1}\beta$ or $\beta^{+}$) is the antiparticle of the electron, so it carries a positive charge. In addition, a neutrino (symbol $\nu$), which is uncharged, is emitted.

$$^{A}_{Z}\mathbf{X} \rightarrow {}^{A}_{Z-1}\mathbf{Y} + {}^{0}_{+1}\beta + \nu$$

Positron-emitting isotopes do not occur naturally. They are manufactured by placing a stable isotope, in liquid or solid form, in the path of a beam of protons. Some of the nuclei in the substance absorb extra protons and become unstable positron-emitters.

Antimatter was predicted in 1928 by the English physicist Paul Dirac, before the first antiparticle, the positron, was discovered. More than 20 years earlier, Einstein had shown that the mass of a particle increases the faster it travels, and that his famous equation $E = mc^2$ related the energy supplied to the particle to its increase in mass. More significantly, Einstein said that the mass of a particle when it is stationary, its rest mass ($m_0$), corresponds to **rest energy** $m_0c^2$ locked up as mass. He showed that rest energy must be included in the conservation of energy. Dirac predicted the existence of antimatter particles (or **antiparticles**) that would unlock rest energy, whenever a particle and a corresponding antiparticle meet and annihilate each other.

Dirac's theory of antiparticles predicted that for every type of particle there is a corresponding antiparticle that:

- annihilates the particle and itself if they meet, converting their total mass into photons
- has exactly the same rest mass as the particle
- has exactly opposite charge to the particle if the particle has a charge.

In addition to the annihilation process described above, Dirac predicted the opposite process of **pair production**. In this process, Dirac predicted that a photon with sufficient energy passing near a nucleus or an electron can suddenly change into a particle–antiparticle pair, which would then separate from each other. Figure 1 shows both of these processes.

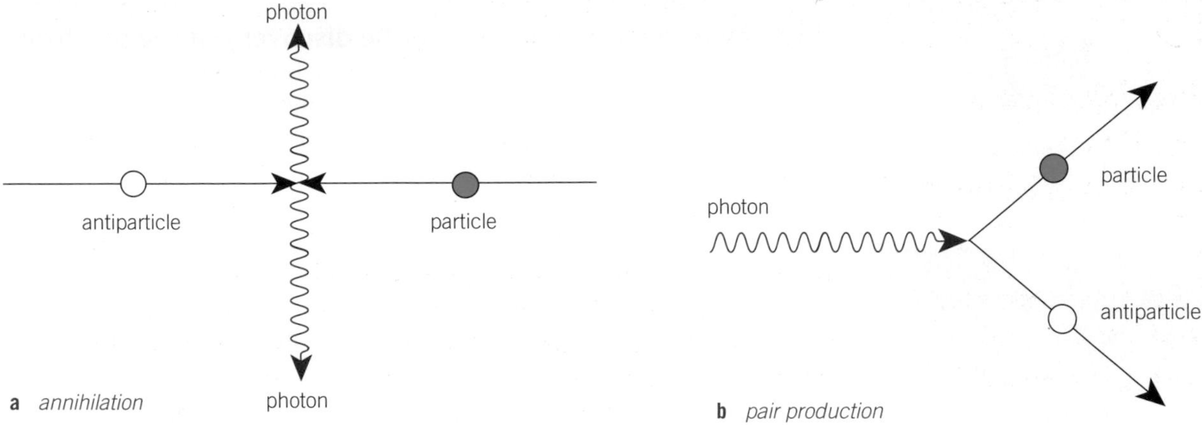


▲ **Figure 1** *Particles and antiparticles*

## Particles, antiparticles, and $E = mc^2$

The energy of a particle or antiparticle is often expressed in millions of electron volts (MeV), where **1 MeV = 1.60 × 10$^{-13}$ J**. One electron volt is defined as the energy transferred when an electron is moved through a potential difference of 1 volt. Given the rest mass of a particle or antiparticle, its rest energy in MeV can be calculated using $E = mc^2$. However, you won't need to do this type of calculation in this topic, as the rest energies of different particles are listed on pages 262–265 of this book.

**Annihilation** occurs when a particle and a corresponding antiparticle meet and their mass is converted into radiation energy. Two photons are produced in this process (as a single photon cannot ensure a total momentum of zero after the collision). Therefore, the minimum energy of each photon, $hf_{min}$, is given by equating the energy of the two photons, $2hf_{min}$, to the rest energy of the particle and of the antiparticle (i.e. $2hf_{min} = 2E_0$, where $E_0$ is the rest energy of the particle).

**Minimum energy of each photon produced, $hf_{min} = E_0$**

In **pair production**, a photon creates a particle and a corresponding antiparticle, and vanishes in the process. For a particle and antiparticle, each of rest energy $E_0$, we can calculate the minimum energy and minimum frequency $f_{min}$ that the photon must have to produce this particle–antiparticle pair. Remember, $c$ is the speed of light in a vacuum ($3.00 \times 10^8\ \mathrm{m\,s^{-1}}$).

**Minimum energy of photon needed = $hf_{min} = 2E_0$**

For example, the electron has a rest energy of 0.511 MeV. Therefore, for pair production of an electron and a positron from a photon:

**Minimum energy of a photon = 2 × 0.511 MeV = 1.022 MeV**

**= 1.64 × 10$^{-13}$ J**

A photon with less energy could not therefore create a positron and an electron.

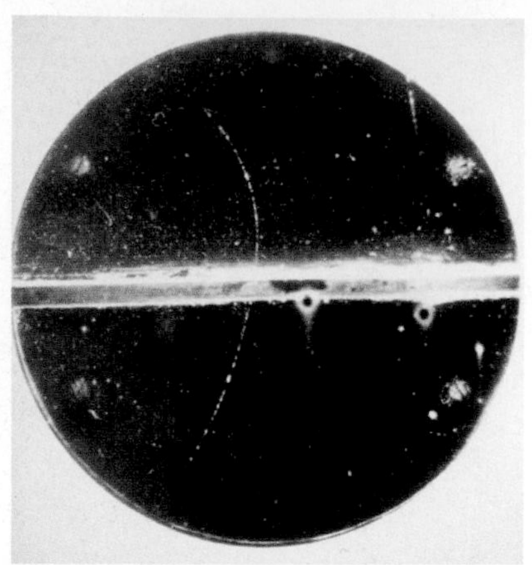

▲ **Figure 2** *The discovery of the positron*

## Journey into the atom (part 2): The discovery of the positron

We can see the path of alpha and beta particles using a cloud chamber. This is a small transparent container containing air saturated with vapour and made very cold. The same conditions exist high in the atmosphere. Ionising particles leave a visible trail of liquid droplets when they pass through the air – just like a jet plane does when it passes high overhead on a clear day. In 1932, the American physicist Carl Anderson was using a cloud chamber and a camera to photograph trails produced by cosmic rays. He decided to see if the particles could pass through a lead plate in the chamber. With a magnetic field applied to the chamber, he knew the trail of a charged particle would bend in the field.

- A positive particle would be deflected by the magnetic field in the opposite direction to a negative particle travelling in the same direction.
- The slower it went, the more it would bend.

If a particle went through the plate, he thought it would be slowed down so its trail would bend more afterwards. Imagine his surprise when he discovered a beta particle that slowed down but bent in the opposite direction to all the other beta trails he had photographed. He had made a momentous discovery – a positron, the first antiparticle to be detected (Figure 2).

## Summary questions

**1 MeV = $1.60 \times 10^{-13}$ J**

**1** a The rest energy of a proton is $1.501 \times 10^{-10}$ J. Calculate its rest energy in MeV.

  **b** Show that a photon must have a minimum energy of 1876 MeV to create a proton–antiproton pair.

**2** Explain why a photon of energy 2 MeV could produce an electron–positron pair but not a proton–antiproton pair.

**3** The rest energy of an electron is 0.511 MeV.

  **a** State the minimum energy in J of each photon created when a positron and an electron annihilate each other.

  **b** A positron created in a cloud chamber in an experiment has 0.158 MeV of kinetic energy. It collides with an electron at rest, creating two photons of equal energies as a result of annihilation.

    **i** Calculate the total energy of the positron and the electron.

    **ii** Show that the energy of each photon is 0.590 MeV.

**4** A positron can be produced by pair production or by positron emission from a proton-rich nucleus.

  **a** Describe the changes that take place in a proton-rich nucleus when it emits a positron.

  **b** State two ways in which pair production of a positron and an electron differs from positron emission.

# 1.5 Particle interactions

## The electromagnetic force

What is a force? We can measure forces using newtonmeters and force sensors. We know that when a single force acts on an object, it changes the momentum of the object. The **momentum** of an object is its mass multiplied by its velocity.

When two objects interact, they exert equal and opposite forces on each other. Momentum is transferred between the objects by these forces, if no other forces act on them. For example, if two protons approach each other, they repel each other and move away from each other. The American physicist Richard Feynman worked out in detail how this happens. He said the electromagnetic force between two charged objects is due to the exchange of **virtual photons**. He described them as virtual because we can't detect them directly. If we intercepted them, for example, by using a detector, we would stop the force acting. The interaction is represented by the diagram shown in Figure 1. This is a simplified version of what is known as a **Feynman diagram**. Note that the lines do not represent the paths of the particles. The virtual photon exchanged between the two protons is represented by a wave.

### An interaction model

Imagine two people on skateboards facing each other and one of them throws a ball to the other. The thrower recoils when the ball leaves his or her hands; the other person recoils when he or she catches the ball. Figure 2 shows the idea. The ball transfers momentum from the thrower to the catcher so they repel each other. The exchange of a virtual photon between two like-charged particles has the same effect – except we can't detect the virtual photon. But the ball analogy won't work for two oppositely charged particles – it would need to be changed to a boomerang, so the thrower and the catcher attract each other!

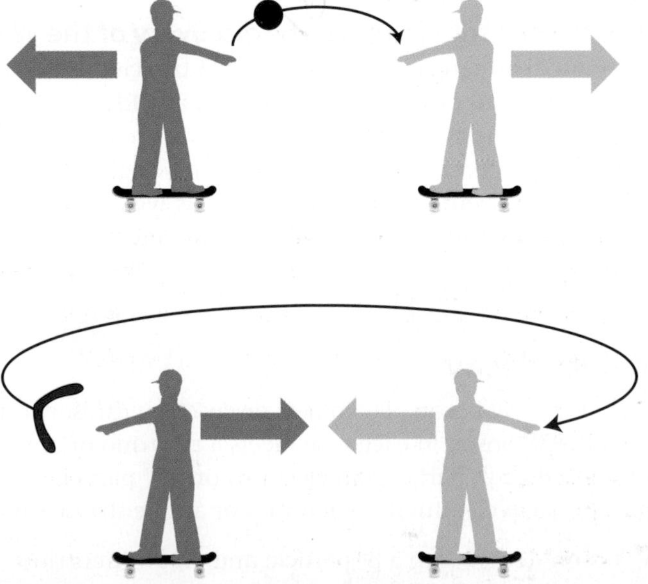

▲ **Figure 2** *Exchange analogies*

## Learning objectives:

- → Describe what is meant by an interaction.
- → Name different types of interaction.
- → Explain what makes charged particles attract or repel each other.
- → Describe an exchange particle.

*Specification reference: 3.2.1.4*

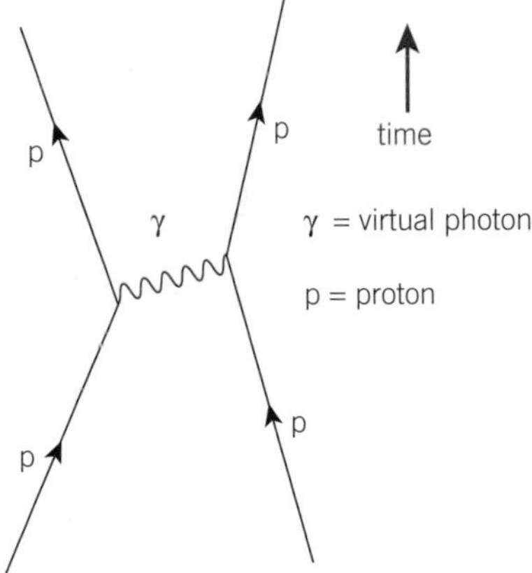


▲ **Figure 1** *Diagram for the electromagnetic force between two protons*

## The weak nuclear force

The strong nuclear force holds the neutrons and protons in a nucleus together. But it doesn't cause a neutron to change into a proton in $\beta^-$ decay, or a proton to change into a neutron in $\beta^+$ decay. These changes can't be due to the electromagnetic force, as the neutron is uncharged. There must be a different force at work in the nucleus causing these changes. It must be weaker than the strong nuclear force, otherwise it would affect stable nuclei – we refer to it as the **weak nuclear force**.

In both $\beta^-$ decay and $\beta^+$ decay, a new particle and a new antiparticle are created in each type of decay – but they're not a corresponding particle–antiparticle pair, as one is an electron or a positron and the other is a neutrino or an antineutrino.

Neutrinos and antineutrinos hardly interact with other particles, but such interactions do sometimes happen. For example:

- A neutrino can interact with a neutron and make it change into a proton. A $\beta^-$ particle (an electron) is created and emitted as a result of the change.
- An antineutrino can interact with a proton and make it change into a neutron. A $\beta^+$ particle (a positron) is created and emitted as a result of the change.

These interactions are due to the exchange of particles referred to as **W bosons**. Unlike photons, these exchange particles:

- have a non-zero rest mass
- have a very short range of no more than about 0.001 fm
- are positively charged (the $W^+$ boson) or negatively charged (the $W^-$ boson).

The diagram for each of these changes is shown in Figure 3. Notice that the total charge at the end is the same as at the start (i.e., charge is conserved).

### Journey into the atom (part 3): The discovery of the W boson

W bosons were first detected directly in 1983 by scientists using the 2 km diameter Super Proton Synchrotron at CERN in Geneva. Protons and antiprotons at very high energies were made to collide and annihilate each other. At sufficiently high energies, these annihilation events produce W bosons as well as photons. The β particles from the W boson decays were detected exactly as predicted.

## Back to beta decay

So what role does a W boson play in beta decay? You can see from Figure 3 that the W boson in each case meets a neutrino or antineutrino, changing them into a $\beta^-$ particle (an electron) or a $\beta^+$ particle (a positron), respectively. But if no neutrino or antineutrino is present:

- the $W^-$ boson decays into a $\beta^-$ particle and an antineutrino
- the $W^+$ boson decays into a $\beta^+$ particle and a neutrino.

p
$\beta^-$
$W^-$
p = proton
n = neutron
n
ν

**a** *A neutron–neutrino interaction*

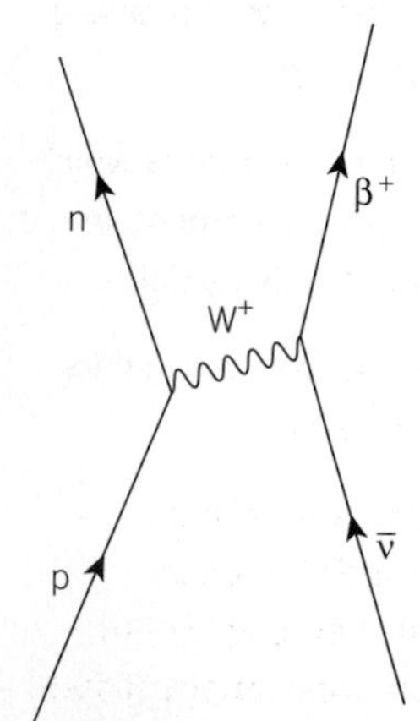


**b** *A proton–antineutrino interaction*

▲ **Figure 3** *The weak interaction*

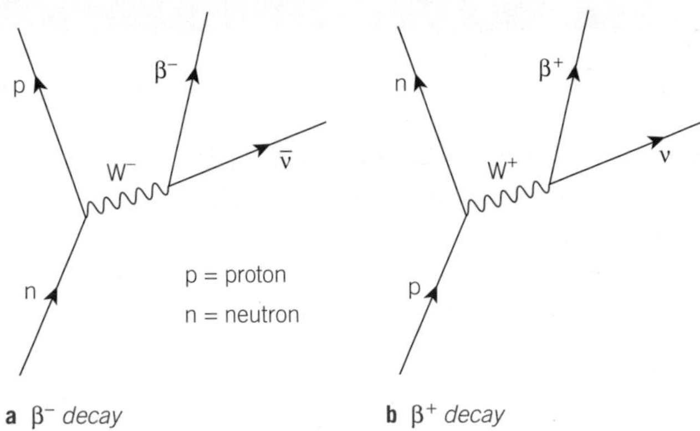


▲ **Figure 4** *W bosons in beta decay*

Figure 4 shows the diagram for each of these decay processes. Notice that charge is conserved in both processes.

## Electron capture

Sometimes a proton in a proton-rich nucleus turns into a neutron as a result of interacting through the weak interaction with an inner-shell electron from outside the nucleus (**electron capture**). Figure 5 shows the diagram for this process. Notice that the $W^+$ boson changes the electron into a neutrino.

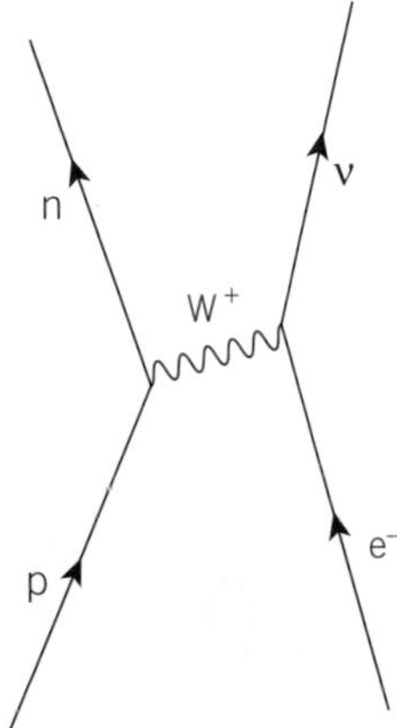


▲ **Figure 5** *Electron capture*

The same change can happen when a proton and an electron collide at very high speed. In addition, for an electron with sufficient energy, the overall change could also occur as a $W^-$ exchange from the electron to the proton.

### Force carriers

The photon and the W boson are known as force carriers because they are exchanged when the electromagnetic force and the weak nuclear force act respectively. You will soon meet the pion, which is known to be the exchange particle of the strong nuclear force. But what about gravity? Scientists think the carrier of the force of gravity is the graviton – but it has yet to be observed.

## Study tip

Scientists think there are four fundamental interactions in nature:

1 the force of gravitational attraction between any two objects due to their mass
2 the electromagnetic force which acts between objects due to their electric charge, and explains all electrical and magnetic effects
3 the strong nuclear force which holds the protons and neutrons together in stable nuclei
4 the weak nuclear force which causes β-decay.

Only the gravitational and electromagnetic forces are observed in everyday life but you will learn more about all four of these forces as you progress through your physics course.

## Summary questions

$c = 3.00 \times 10^8\ \text{m s}^{-1}$

1 Sketch the diagram for the electromagnetic interaction between
 **a** two protons **b** a proton and an electron **c** two electrons.
2 Sketch the diagram for **a** $\beta^+$ decay **b** $\beta^-$ decay.
3 **a** Sketch the diagram for the interaction between **i** a neutron and a neutrino, **ii** a proton and an antineutrino.
 **b** State the approximate range of the W boson and estimate its lifetime, given it cannot travel faster than the speed of light.
4 **a** State three differences between a W boson and a virtual photon.
 **b** In both Figure 4b and Figure 5, a proton changes into a neutron. Describe how these two processes differ.

# Practice questions: Chapter 1

1 **(a)** Copy and complete the table below: *(3 marks)*

| | Particle or antiparticle | Charge / proton charge |
|---|---|---|
| antiproton | antiparticle | |
| neutrino | | |
| neutron | | 0 |
| positron | | |

**(b)** The tracks of a positron and an electron created by pair production in a magnetic field curve in opposite directions as shown above.

(i) Why do they curve in opposite directions?

(ii) Both particles spiral inwards. What can you deduce from this observation about their kinetic energy?

(iii) In **Figure 4** of Topic 1.4, the track is a typical beta track. Explain how Carl Anderson deduced from the photograph that the track was created by a positron rather than an electron travelling in the opposite direction. *(5 marks)*

2 **(a)** $^{229}_{90}$Th is a neutral atom of thorium. How many protons, neutrons, and electrons does it contain? *(2 marks)*

**(b)** $^{Y}_{X}$Th is a neutral atom of a different isotope of thorium which contains *Z* electrons. Give possible values for *X*, *Y*, and *Z*. *(3 marks)*

AQA, 2001

3 An atom of argon $^{37}_{18}$Ar is ionised by the removal of two orbiting electrons.

**(a)** How many protons and neutrons are there in this ion? *(2 marks)*

**(b)** What is the charge, in C, of this ion? *(2 marks)*

**(c)** Which constituent particle of this ion has

(i) a zero charge per unit mass ratio?

(ii) the largest charge per unit mass ratio? *(2 marks)*

**(d)** Calculate the percentage of the total mass of this ion that is accounted for by the mass of its electrons. *(3 marks)*

AQA, 2002

4 **(a)** A neutral atom contains 28 nucleons.
Write down a possible number of protons, neutrons, and electrons contained in the atom. *(2 marks)*

**(b)** A certain isotope of uranium may split into a caesium nucleus, a rubidium nucleus, and four neutrons in the following process:

$$^{236}_{92}\text{U} \rightarrow {}^{137}_{55}\text{Cs} + {}^{X}_{37}\text{Rb} + 4\,^{1}_{0}\text{n}$$

(i) Explain what is meant by isotopes.

(ii) How many neutrons are there in the $^{137}_{55}$Cs nucleus?

(iii) Calculate the ratio $\frac{\text{charge}}{\text{mass}}$, in $\text{C kg}^{-1}$, for the $^{236}_{92}$U nucleus.

(iv) Determine the value of *X* for the rubidium nucleus. *(4 marks)*

AQA, 2003

5 An α particle is the same as a nucleus of helium, $^{4}_{2}$He.
The equation $^{229}_{90}\text{Th} \rightarrow {}^{X}_{Y}\text{Ra} + \alpha$
represents the decay of a thorium isotope by the emission of an α particle.
Determine:

**(a)** the values of *X* and *Y*, shown in the equation *(2 marks)*

**(b)** the ratio $\frac{\text{mass of } ^{X}_{Y}\text{Ra nucleus}}{\text{mass of } \alpha \text{ particle}}$ *(1 mark)*

AQA, 2005

6 **(a)** (i) Describe an α particle and state its properties.
(ii) $^{218}_{85}At$ is an isotope of the element astatine (At) which decays into an isotope of bismuth (Bi) by emitting an α particle. Write down the equation to represent this reaction. *(5 marks)*

**(b)** (i) State what happens when an unstable nucleus decays by emitting a $\beta^-$ particle.
(ii) Write down and complete the following equation, showing how an isotope of molybdenum decays into an isotope of technetium:

$$^{...}_{42}Mo \rightarrow {}^{99}_{...}Tc + \beta^- + ...$$ *(5 marks)*

7 In a radioactive decay of a nucleus, a $\beta^+$ particle is emitted followed by a γ photon of wavelength 8.30 $10^{-13}$ m.

**(a)** √x (i) State the rest mass, in kg, of the $\beta^+$ particle.
(ii) Calculate the energy of the γ photon.
(iii) Determine the energy of the γ photon in MeV. *(6 marks)*

**(b)** Name the fundamental interaction or force responsible for $\beta^+$ decay. *(1 mark)*

AQA, 2004

8 **(a)** (i) State the name of the antiparticle of a positron.
(ii) Describe what happens when a positron and its antiparticle meet. *(3 marks)*

**(b)** √x Calculate the minimum amount of energy, in J, released as radiation energy when a particle of rest energy 0.51 MeV meets its corresponding antiparticle. *(2 marks)*

AQA, 2005

9 In a particle accelerator a proton and an antiproton, travelling at the same speed, undergo a head-on collision and produce subatomic particles.

**(a)** The total kinetic energy of the two particles just before the collision is $3.2 \times 10^{-10}$ J.
(i) What happens to the proton and antiproton during the collision?
(ii) State why the total energy after the collision is more than $3.2 \times 10^{-10}$ J. *(2 marks)*

**(b)** In a second experiment the total kinetic energy of the colliding proton and antiproton is greater than $3.2 \times 10^{-10}$ J.
State *two* possible differences this could make to the subatomic particles produced. *(2 marks)*

AQA, 2001

10 An electron may interact with a proton in the following way:
$e^- + p \rightarrow n + \nu_e$.
Name the fundamental force responsible for this interaction. *(1 mark)*

AQA, 2003

11 **(a)** Give an example of an exchange particle other than a $W^+$ or $W^-$ particle, and state the fundamental force involved when it is produced. *(2 marks)*
**(b)** State what roles exchange particles can play in an interaction. *(2 marks)*

AQA, 2006

12 Describe what happens in pair production and give *one* example of this process. *(3 marks)*

AQA, 2005

# 2 Quarks and leptons

## 2.1 The particle zoo

**Learning objectives:**

- → Explain how we can find new particles.
- → State whether we can predict new particles.
- → Describe strange particles.

*Specification reference: 3.2.1.5*

### Practical link

Two Geiger counters can be used to detect the presence of cosmic radiation. Cosmic rays are probably present if the two counters click simultaneously.

### Space invaders

Cosmic rays are high-energy particles that travel through space from the stars, including the Sun. When cosmic rays enter the Earth's atmosphere, they create new short-lived particles and antiparticles, as well as photons. Most physicists thought, when cosmic rays were first discovered, that they were from terrestrial radioactive substances. This theory was disproved when the physicist and amateur balloonist, Victor Hess, found the ionising effect of the rays was significantly greater at 5000 m than at ground level.

Further investigations showed that most cosmic rays were fast-moving protons or small nuclei. They collide with gas atoms in the atmosphere, creating showers of particles and antiparticles that can be detected at ground level. By using cloud chambers and other detectors, new types of short-lived particles and antiparticles were discovered, including:

- the **muon** or heavy electron (symbol $\mu$), a negatively charged particle with a rest mass over 200 times the rest mass of the electron
- the **pion** or $\pi$ meson, a particle which can be positively charged ($\pi^+$), negatively charged ($\pi^-$), or neutral ($\pi^0$), and has a rest mass greater than a muon but less than a proton
- the **kaon** or K meson, which also can be positively charged ($K^+$), negatively charged ($K^-$), or neutral ($K^0$), and has a rest mass greater than a pion but still less than a proton.

▲ **Figure 1** *Creation and decay of a pion. The short spiral track is a $\pi^+$ meson created when an antiproton from the bottom edge annihilates a proton. The $\pi^+$ meson decays into an antimuon that spirals and decays into a positron.*

### Journey into the atom (part 4): An unusual prediction

Before the above particles were first detected, the Japanese physicist Hideki Yukawa predicted the existence of exchange particles for the strong nuclear force between nucleons. He thought they would have a range of no more than about $10^{-15}$ m and he calculated that their mass would be between the electron and the proton mass. He called them **mesons**, because the predicted mass was somewhere in the middle between the electron mass and the proton mass.

A year later, a cloud chamber photograph obtained by Carl Anderson showed an unusual track that could have been produced by such a particle. However, the track length of 40 mm indicated that it lasted much longer than a strongly interacting particle should. Further investigations showed this unexpected particle to be a heavy electron, now referred to as a **muon** but not classed as a meson. It decays through the weak interaction.

Yukawa's meson was discovered some years later by the British physicist Cecil Powell from tiny microscopic tracks found in photographic emulsion exposed to cosmic rays at high altitude. Powell called these particles $\pi$ **mesons** or **pions** – Yukawa's prediction was correct!

## A strange puzzle

Less than a year after Powell's discovery, further cloud chamber photographs revealed the existence of short-lived particles we now refer to as kaons. Like pions, kaons are produced in twos through the strong interaction, when protons moving at high speed crash into nuclei and they each travel far beyond the nucleus in which they originate before they decay. However, the decay of kaons took longer than expected and included pions as the product. This means that kaons must decay via the weak interaction. These and other properties of kaons led to them being called **strange particles** (see Topic 2.4).

All the new particles listed above can also be created using accelerators in which protons collide head-on with other protons at high speed. The kinetic energy of the protons is converted into mass in the creation of these new particles. So these new particles could be studied under controlled conditions by teams of physicists using accelerators to create them.

The rest masses, charge (if they were charged), and lifetimes of the new particles were measured. Their antiparticles, including the **antimuon**, were detected. Their decay modes were worked out:

- A kaon can decay into pions, or a muon and an antineutrino, or an antimuon and a neutrino.
- A charged pion can decay into a muon and an antineutrino, or an antimuon and a neutrino. A $\pi^0$ meson decays into high-energy photons.
- A muon decays into an electron and an antineutrino. An antimuon decays into a positron and a neutrino.
- Decays always obey the conservation rules for energy, momentum, and charge (see Topic 2.5).

### About accelerators

A TV tube in a traditional TV is an accelerator because it accelerates electrons inside the tube through a potential difference of about 5000 V. The electrons form a beam that hits the inside of the TV screen. Magnetic fields produced by electromagnets deflect the beam so it scans the screen to create an image. LCD (liquid crystal display) TVs work on a different principle.

- The longest accelerator in the world is the Stanford linear accelerator in California. It accelerates electrons over a distance of 3 km through a potential difference of 50 000 million volts. The energy of an electron accelerated through this pd is 50 000 MeV (= 50 GeV). When the electrons collide with a target, they can create lots of particle–antiparticle pairs.
- The biggest accelerator in the world, the Large Hadron Collider at CERN near Geneva, is designed to accelerate charged particles to energies of more than 7000 GeV. Unlike a linear accelerator, this accelerator is a 27 km circumference ring constructed in a circular tunnel below the ground. It is used by physicists from many countries to find out more about the fundamental nature of matter and radiation. In 2012, an uncharged particle called the Higgs boson was discovered at CERN. It had been predicted in 1964 as the cause of the rest mass of W bosons.

▲ **Figure 2** *A particle accelerator*

### Summary questions

1 a List the following particles in order of increasing rest mass:
the electron, the $K^0$ meson, the muon, the neutron, the $\pi^+$ meson, the proton.

b Which of the above particles are uncharged?

2 State whether the strong or the weak interaction acts when

a a muon decays

b a kaon decays

c pions are produced.

3 a Which creates the longer trail in a cloud chamber, a muon or a $\pi^+$ meson?

b In terms of their properties, state one similarity and one difference between a muon and a pion.

4 A high-energy collision between two protons creates pions and kaons which then decay.

a What happens to

i the kaons

ii the pions?

b What particles are ultimately formed by the above decays?

# 2.2 Particle sorting

## Learning objectives:

- → Identify different classifications of particles.
- → Recognise hadrons.
- → Recognise leptons.

*Specification reference: 3.2.1.5*

## Classifying particles and antiparticles

The new particle discoveries raised the important question – how do all these new particles and antiparticles fit in with each other and with protons, neutrons, and electrons? They are created through high-energy interactions and, apart from the neutrino, they decay into other particles and antiparticles. Also, charged pions were often produced in pairs – leading to the conclusion that the $\pi^+$ meson and the $\pi^-$ meson are a particle–antiparticle pair. The same conclusion was reached for charged kaons. So is there an underlying pattern?

All the particles we have discussed so far are listed in Table 1 together with some of their properties. Can you see any links between the different particles?

▼ **Table 1** *Particles and their properties*

| Particle and symbol | Charge / proton charge | Antiparticle and symbol | Charge / proton charge | Rest energy / MeV | Interaction |
|---|---|---|---|---|---|
| proton p | +1 | antiproton $\bar{p}$ | −1 | 938 | strong, weak, electromagnetic |
| neutron n | 0 | antineutron $\bar{n}$ | 0 | 939 | strong, weak |
| electron $e^-$ | −1 | positron $e^+$ | +1 | 0.511 | weak, electromagnetic |
| neutrino $\nu$ | 0 | antineutrino $\bar{\nu}$ | 0 | 0 | weak |
| muon $\mu^-$ | −1 | antimuon $\mu^+$ | +1 | 106 | weak, electromagnetic |
| pions $\pi^+, \pi^0, \pi^-$ | +1, 0, −1 | $\pi^+$ for a $\pi^-$<br>$\pi^-$ for a $\pi^+$<br>$\pi^0$ for a $\pi^0$ | −1, 0, +1 | 140, 135, 140 | strong, electromagnetic ($\pi^+, \pi^-$) |
| kaons $K^+, K^0, K^-$ | +1, 0, −1 | see Topic 2.4, Quarks and antiquarks | −1, 0, +1 | 494, 498, 494 | strong, electromagnetic ($K^+, K^-$) |

Notice the symbol for an antiproton, an antineutron, or an antineutrino is the particle symbol with a bar above it, except for the positron and the antimuon which have their own symbols. The charged pions are antiparticles of each other.

How would you classify these particles? In fact, we can divide them into two groups, called **hadrons** and **leptons**, according to whether or not they interact through the strong interaction.

- Hadrons are particles and antiparticles that can interact through the strong interaction (e.g., protons, neutrons, π mesons, K mesons).
- Leptons are particles and antiparticles that do not interact through the strong interaction (e.g., electrons, muons, neutrinos).

1 **Leptons interact through the weak interaction, the gravitational interaction, and through the electromagnetic interaction (if charged).**

2 **Hadrons can interact through all four fundamental interactions. They interact through the strong interaction and through the electromagnetic interaction if charged.**

### Study tip

lepton – light and weak
hadron – strong and heavy

### Synoptic link

You will meet baryons, mesons, and the smaller particles that make them up in more detail in Topic 2.4, Quarks and antiquarks.

**Apart from the proton, which is stable, hadrons tend to decay through the weak interaction.**

## Energy matters

The Large Hadron Collider is a ring-shaped accelerator that boosts the kinetic energy of the charged particles in the ring at several places round it. Fixed magnets all the way round the ring bend the path of the particles to keep them in the ring. When they collide with other particles:

- The total energy of the particles and antiparticles before the collision = their rest energy + their kinetic energy.
- The total energy of the new particles and antiparticles after the collision = their rest energy + their kinetic energy.

Using conservation of energy,

$$\text{the rest energy of the products} = \text{total energy before} - \text{the kinetic energy of the products}$$

For example, if a proton and an antiproton each with kinetic energy of 2 GeV collide, their total energy before the collision is 6 GeV (as each particle has 2 GeV of kinetic energy and approximately 1 GeV of rest energy – see Table 1). The result could be a range of products, as long as their total energy does not exceed 6 GeV, provided the conservation rules (e.g., charge) are obeyed.

*Note:*

1 GeV = 1000 MeV

### Baryons and mesons

When kaons are created, short-lived particles with greater rest masses than protons may also be produced. These particles are created through the strong interaction, so they are hadrons. However, in comparison with kaons, they decay into protons as well as into pions (whereas protons are never among the decay products of kaons). Consequently, we can divide hadrons into two groups:

1. **Baryons** are protons and all other hadrons (including neutrons) that decay into protons, either directly or indirectly.
2. **Mesons** are hadrons that do *not* include protons in their decay products. In other words, kaons and pions are not baryons.

Baryons and mesons are composed of smaller particles called **quarks** and **antiquarks**.

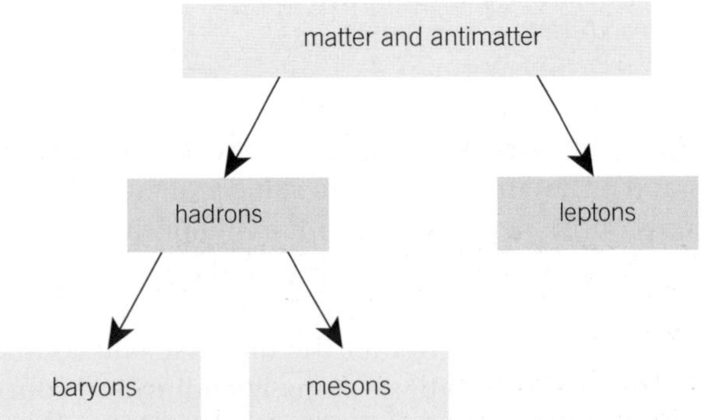


▲ **Figure 2** *Particle sorts*

▲ **Figure 1** *The Large Hadron Collider*

## Summary questions

1. a What is
   i a hadron
   ii a lepton?
   b State whether each of the following particles or antiparticles is a baryon, a meson, or a lepton:
   i an electron
   ii a neutron
   iii a kaon.
2. State one similarity and one difference between
   a a muon and a $\pi^-$ meson
   b a $\pi^+$ meson and a $K^+$ meson
   c a $K^0$ meson and a neutron.
3. Complete the following sentences:
   a A particle that is not a hadron is a ________.
   b Hadrons are divided into ________ and ______.
   c _____ do not include protons in their decay products.
4. A $K^+$ meson can decay into three charged mesons.
   a Complete the following equation for this decay:
   $K^+ \rightarrow \quad + \pi^+ +$
   b Use the rest energy values in Table 1 to calculate the maximum kinetic energy of the pions, assuming the kaon is at rest before it decays.

# 2.3 Leptons at work

**Learning objectives:**

- → Consider whether leptons are elementary.
- → Distinguish between different types of neutrinos.
- → Evaluate the importance of lepton numbers.

*Specification reference: 3.2.1.5*

## Lepton collisions

The universe would be very dull if all its particles were leptons. Neutrinos interact very little, muons are very short-lived, and the electrons would repel each other. However, leptons and antileptons can interact to produce hadrons. Figure 1 shows two jets of hadrons produced from a single electron–positron annihilation event. This event produces a **quark** and a corresponding **antiquark**, which move away in opposite directions, producing a shower of hadrons in each direction.

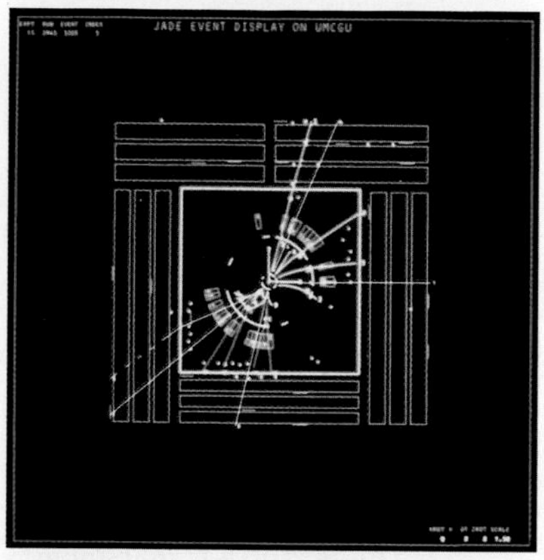


▲ **Figure 1** *A two-jet event after an electron–positron collision*

## Neutrino types

Neutrinos travel almost as fast as light, billions of them sweeping through the Earth from space every second with almost no interaction. They are produced in much smaller numbers when particles in accelerators collide.

Further research on neutrinos showed that the neutrinos and antineutrinos produced in beta decays were different from those produced by muon decays. In effect, neutrinos and antineutrinos from muon and antimuon decays create only muons and no electrons when they interact with protons and neutrons. If there were only one type of neutrino and antineutrino, equal numbers of electrons and muons would be produced. We use the symbol $\nu_\mu$ for the **muon neutrino** and $\nu_e$ for the **electron neutrino** (and similarly for the two types of antineutrinos).

**Synoptic link**

You have met neutrinos in Topic 1.2, Stable and unstable nuclei.

**Synoptic link**

You met neutrino–neutron and antineutrino–proton interactions in Topic 1.5, Particle interactions.

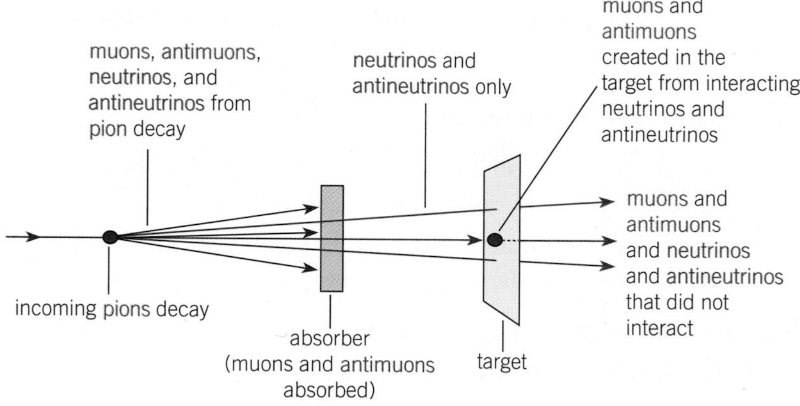


▲ **Figure 2** *Muon and electron neutrinos*

**Journey into the atom (part 5): The future of the universe?**

Neutrinos and antineutrinos are the subject of on-going research. The rest mass of the neutrino is about a millionth of the electron's rest mass. If this rest mass can be measured accurately, physicists will be able to work out how much neutrinos contribute to the mass of the universe and whether the universe will continue to expand forever (Big Yawn) or stop expanding and contract (Big Crunch).

## Lepton rules

Leptons can change into other leptons through the weak interaction and can be produced or annihilated in particle–antiparticle interactions. But all the experiments on them indicate that they don't break down into non-leptons. They appear to be fundamental.

1 **In an interaction between a lepton and a hadron, a neutrino or antineutrino** can change into or from a corresponding charged lepton. An electron neutrino can interact with a neutron to produce a proton and an electron:

$$\nu_e + n \rightarrow p + e^-$$

However, even though charge is conserved, an electron neutrino and a neutron could not change into an antiproton and a positron:

$$\nu_e + n \nrightarrow \bar{p} + e^+$$

By assigning +1 to a lepton, –1 to an antilepton, and 0 for any non-lepton, we can see that the first equation has a lepton number of +1 before the change and a lepton number of +1 after the change. This equation is permitted. The second equation has a lepton number of +1 before the change and a lepton number of –1 after the change. This change is not permitted.

2 **In muon decay**, the muon changes into a muon neutrino. In addition, an electron is created to conserve charge and a corresponding antineutrino is created to conserve lepton number. For example:

$$\mu^- \rightarrow e^- + \bar{\nu}_e + \nu_\mu$$

However, a muon cannot decay into a muon antineutrino, an electron, and an electron antineutrino even though charge is conserved.

$$\mu^- \nrightarrow e^- + \bar{\nu}_e + \bar{\nu}_\mu$$

The first equation has a lepton number of +1 before the change and a lepton number of +1 (= +1 – 1 + 1) after the change. This equation is permitted.

The second equation has a lepton number of +1 before the change and a lepton number of – 1 (= +1 – 1 – 1) after the change. This change is not permitted.

**The lepton number is conserved in any change.**

### *A weak puzzle*

Consider the following possible change for a muon decay:

$$\mu^- \xrightarrow{?} e^- + \nu_e + \bar{\nu}_\mu$$

It obeys the lepton conservation law and conservation of charge but it is never seen. The reason is that the muon can only change into a muon neutrino. A muon can't change into a muon antineutrino, and an electron can only be created with an electron antineutrino. So we always need to apply the rule below separately to electrons and muons.

**The lepton number is +1 for any lepton, –1 for any antilepton, and 0 for any non-lepton.**

## Summary questions

1 State one similarity and one difference between

a a positron and a muon

b an electron neutrino and a muon neutrino.

2 a What type of interaction occurs when

i two protons collide and create pions

ii a beta particle is emitted from an unstable nucleus

iii a muon decays?

b The muon decays into an electron, a neutrino, and an antineutrino.

i Complete the equation below representing this decay:

$\mu^- \rightarrow e^- \quad + \quad +$

ii Use the rest energy values in Table 1 on page 20 to calculate how much energy, in MeV, is removed by the neutrino and the antineutrino, if the muon and the electron have no kinetic energy.

3 a What is the charge of

i a muon neutrino

ii an antimuon

iii a positron

iv an electron antineutrino?

b What is the lepton number of each of the above leptons?

4 State whether or not each of the following reactions is permitted, giving a reason if it is not permitted:

a $\nu_e + p \rightarrow n + e^-$

b $\bar{\nu}_e + p \rightarrow n + e^+$

c $\nu_e + p \rightarrow n + e^+$

d $\bar{\nu}_e + p \rightarrow n + \mu^+$.

# 2.4 Quarks and antiquarks

## Learning objectives:

- → Define strange particles.
- → Define quarks and explain how we know they exist.
- → Explain the quark changes that happen in beta decay.
- → Explain why it could be said that there are no antimesons.

*Specification reference: 3.2.1.6; 3.2.1.7*

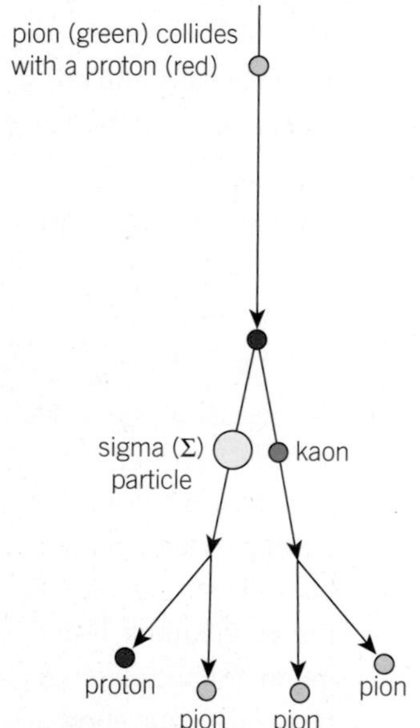


▲ **Figure 1** *Strange particles*

▼ **Table 1** *Some reactions that were predicted*

| | | |
|---|---|---|
| 1 | $\pi^- + p \rightarrow K^+ + \Sigma^-$ | observed |
| 2 | $\pi^+ + n \rightarrow K^+ + \Sigma^0$ | observed |
| 3 | $\pi^- + n \rightarrow K^0 + \Sigma^-$ | observed |
| 4 | $\pi^- + n \rightarrow K^- + \Sigma^0$ | *not* observed |

### Study tip

You may be asked to work out the charge or strangeness of any baryon or antibaryon from its quark composition.

## Strangeness

When kaons were first discovered they were called V particles because the cloud chamber photographs often showed V-shaped tracks. They were called strange particles after investigations showed that the V tracks decay into pions only, or into pions and protons. Although these strange particles all decay through the weak interaction:

1. those that decay into pions only were referred to as kaons
2. the others, such as the sigma (Σ) particle in Figure 1, were found to:
   - have different rest masses which were always greater than the proton's rest mass
   - decay either in sequence or directly into protons and pions.
3. strange particles are created in twos.

### Journey into the atom (part 6): Predictions and observations

Accelerators were used to create these other particles and to study their properties. The results showed that some reactions that were predicted were never observed (Table 1).

All the observed reactions conserve charge. To explain why certain reactions were not observed, a **strangeness number** S was introduced for each particle and antiparticle (starting with +1 for the $K^+$ meson) so that strangeness is always conserved in strong interactions. Non-strange particles (i.e., the proton, the neutron, pions, leptons) were assigned zero strangeness. The strangeness numbers for the other strange particles and antiparticles can then be deduced from the observed reactions.

In reaction 1, $S = -1$ for $\Sigma^-$ so that the total strangeness is zero before and after the reaction.

In reaction 2, $S = -1$ for $\Sigma^0$ so that the total strangeness is zero before and after the reaction.

In reaction 3, $S = +1$ for $K^0$ so that the total strangeness is zero before and after the reaction.

But reaction 4 is not observed so the $K^-$ meson does not have $S = +1$.

The above interactions are all strong interactions. Observations that strangeness is not conserved when strange particles decay leads to the conclusion that strangeness is not conserved in a weak interaction.

**Strangeness is always conserved in a strong interaction, whereas strangeness can change by 0, +1, or −1 in weak interactions.**

## The quark model

The properties of the hadrons, such as charge, strangeness, and rest mass can be explained by assuming they are composed of smaller particles known as quarks and antiquarks. For the hadrons studied in this course, three different types of quarks and their corresponding antiquarks are necessary. These three types are referred to as up, down, and strange quarks, and are denoted by the symbols u, d, and s, respectively. The properties of these three quarks are shown in Table 1.

▼ **Table 2** *Quark properties*

| | Quarks | | | Antiquarks | | |
|---|---|---|---|---|---|---|
| | up u | down d | strange s | up $\bar{u}$ | down $\bar{d}$ | strange $\bar{s}$ |
| charge *Q* | $+\frac{2}{3}$ | $-\frac{1}{3}$ | $-\frac{1}{3}$ | $-\frac{2}{3}$ | $+\frac{1}{3}$ | $+\frac{1}{3}$ |
| strangeness *S* | 0 | 0 | −1 | 0 | 0 | +1 |
| baryon number *B* | $+\frac{1}{3}$ | $+\frac{1}{3}$ | $+\frac{1}{3}$ | $-\frac{1}{3}$ | $-\frac{1}{3}$ | $-\frac{1}{3}$ |

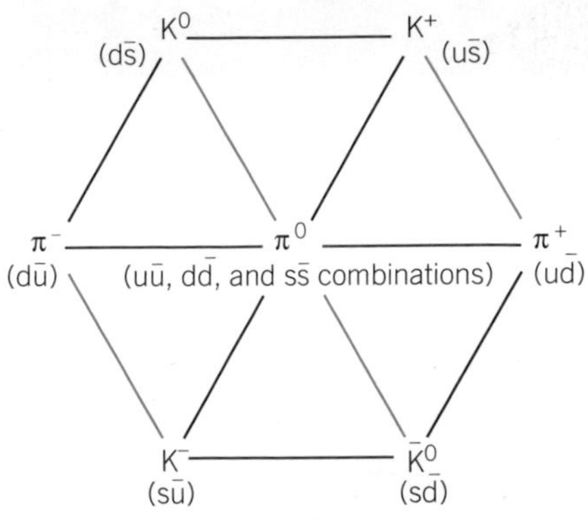


▲ **Figure 2** *Quark combinations for the mesons*

## Quark combinations

The rules for combining quarks to form baryons and mesons are astonishingly simple.

**Mesons** are hadrons, each consisting of a quark and an antiquark. Figure 2 shows all nine different quark–antiquark combinations and the meson in each case. Notice that:

- a $\pi^0$ meson can be any quark–corresponding antiquark combination
- each pair of charged mesons is a particle–antiparticle pair
- there are two uncharged kaons, the $K^0$ meson and the $\bar{K}^0$ meson
- the antiparticle of any meson is a quark–antiquark pair and therefore another meson.

**Baryons** and **antibaryons** are hadrons that consist of three quarks for a baryon or three antiquarks for an antibaryon.

- A proton is the uud combination.
- A neutron is the udd combination.
- An antiproton is the $\bar{u}\bar{u}\bar{d}$ combination.
- The Σ particle is a baryon containing a strange quark.

The proton is the only stable baryon. A free neutron decays into a proton, releasing an electron and an electron antineutrino, as in $\beta^-$ decay.

## Quarks and beta decay

In $\beta^-$ decay, a neutron in a neutron-rich nucleus changes into a proton, releasing an electron and an electron antineutrino. In quark terms, a down quark changes to an up quark. The diagram for this change is shown in Figure 3a.

In $\beta^+$ decay, a proton in a proton-rich nucleus changes into a neutron, releasing a positron and an electron neutrino. In quark terms, an up quark changes to a down quark. The diagram for this change is shown in Figure 3b.

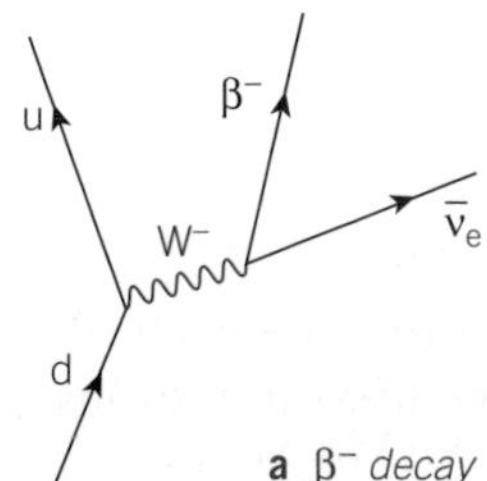


**a** *$\beta^-$ decay*

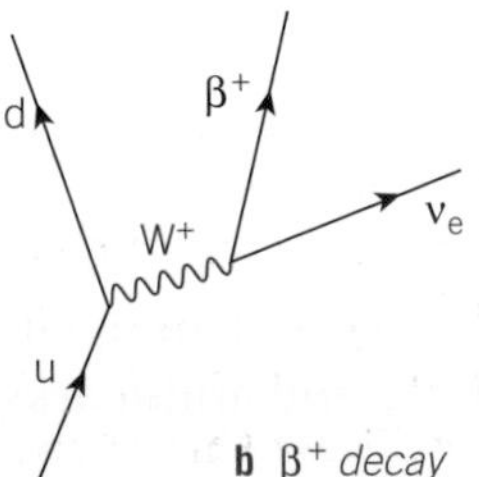


**b** *$\beta^+$ decay*

▲ **Figure 3** *Quark changes in beta decay*

## Summary questions

**1** Determine the quark composition of each of the following hadrons, given its strangeness:

**a** a $\pi^0$ meson (S = 0)

**b** an antiproton (S = 0)

**c** a $K^-$ meson (S = −1)

**d** a $\Sigma^0$ baryon (S = −1).

**2** In $\beta^+$ decay, a positron and an electron neutrino are emitted when a proton in the nucleus changes into a neutron.

**a** In terms of quarks, draw a diagram to represent this change.

**b** Describe the changes that are represented in the diagram.

**3 a** A $\Sigma^-$ particle has a strangeness of −1. Show that it is composed of a strange quark and two down quarks.

**b** A $K^+$ meson is composed of a strange antiquark and an up quark. A $\Sigma^-$ baryon is composed of a strange quark and two down quarks. They can be created in the reaction: $\pi^- + p \rightarrow K^+ + \Sigma^-$

Describe this reaction in terms of quarks and antiquarks.

# 2.5 Conservation rules

## Learning objectives:

- → State the conservation rules for particle interactions.
- → Explain what is sometimes conserved.
- → Explain what is never conserved.

*Specification reference: 3.2.1.6; 3.2.1.7*

## Particles and properties

Particles and antiparticles possess energy – they may be charged or uncharged, they may have non-zero strangeness, and they may not be stable. They obey certain **conservation rules** when they interact. Some rules apply to all interactions and decays and some do not.

1. **Conservation of energy and conservation of charge** apply to all changes in science, not just to all particle and antiparticle interactions and decays. Remember conservation of energy includes the rest energy of the particles (see Topic 2.2, Particle sorting).
2. **Conservation rules used only for particle and antiparticle interactions and decays** are essentially particle-counting rules, based on what reactions are observed and what reactions are not observed. So far, we have used the following conservation rules (in addition to energy and charge):

   **a** **Conservation of lepton numbers**. In any change, the total lepton number for each lepton branch before the change is equal to the total lepton number for that branch after the change.

   For example, when an antimuon decays, it changes into a positron, an electron neutrino, and a muon antineutrino, as shown below:

   $$\mu^+ \rightarrow e^+ + \bar{\nu}_\mu + \nu_e$$

   The properties of each particle or antiparticle in this change are shown in Table 1. We can see from this table that the lepton numbers for each type of lepton are conserved, as in all changes involving leptons.

   **b** **Conservation of strangeness**. In any strong interaction, strangeness is always conserved.

   For example, a $K^0$ meson and a strange baryon called the neutral lambda particle (symbol $\Lambda^0$) are produced when a $\pi^-$ meson collides with a proton, as shown below:

   $$\pi^- + p \rightarrow K^0 + \Lambda^0$$

   The properties of each particle involved in this change are shown in Table 2. We can see from this table that strangeness is conserved, just as in all changes involving the strong interaction.

▼ **Table 1** *The decay of the antimuon (interaction is weak)*

| | $\mu^+$ | $e^+$ | $\nu_e$ | $\bar{\nu}_\mu$ |
|---|---|---|---|---|
| charge | +1 | +1 | 0 | 0 |
| muon lepton number | −1 | 0 | 0 | −1 |
| electron lepton number | 0 | −1 | +1 | 0 |

▼ **Table 2** *A proton–$\pi^-$ meson collision (interaction is strong)*

| | $\pi^-$ | p | $K^0$ | $\Lambda^0$ |
|---|---|---|---|---|
| charge | −1 | +1 | 0 | 0 |
| strangeness | 0 | 0 | +1 | −1 |

**Study tip**

Strangeness is not necessarily conserved when the weak interaction is involved. In weak interactions, strangeness can be unchanged or changed by ±1.

## Are baryons and mesons conserved?

Table 3 shows some reactions that are observed and some that are not observed.They all obey the above conservation laws so why are some not observed?

Let's consider the first reaction in terms of quarks (Figure 1):

$$uud + \bar{u}\bar{u}\bar{d} \rightarrow u\bar{d} + \bar{u}d$$

We can see that an up quark and an up antiquark annihilate each other and the other quarks and antiquarks rearrange to form mesons.

▼ **Table 3** *Some reactions conserve baryons and mesons*

| | |
|---|---|
| $p + \bar{p} \rightarrow \pi^+ + \pi^-$ | observed |
| $p + \bar{p} \rightarrow p + \pi^-$ | not observed |
| $p + p \rightarrow p + p + p + \bar{p}$ | observed |
| $p + \bar{p} \rightarrow \bar{p} + \pi^+$ | not observed |

This reaction shows that we can keep track of the baryons and antibaryons, if we assign a **baryon number** of:

- +1 to any baryon and −1 to any antibaryon
- 0 to any meson or lepton.

In effect, the above rules amount to assigning a baryon number of $+\frac{1}{3}$ to any quark, $-\frac{1}{3}$ to any antiquark, and 0 to any lepton.

So in the first reaction, the baryon number is 0 (= +1 − 1) for the left-hand side and 0 for the right-hand side. In other words, the baryon number is conserved. Prove for yourself that this rule applies to the third reaction but not to the second or fourth reactions.

**In any reaction, the total baryon number is conserved.**

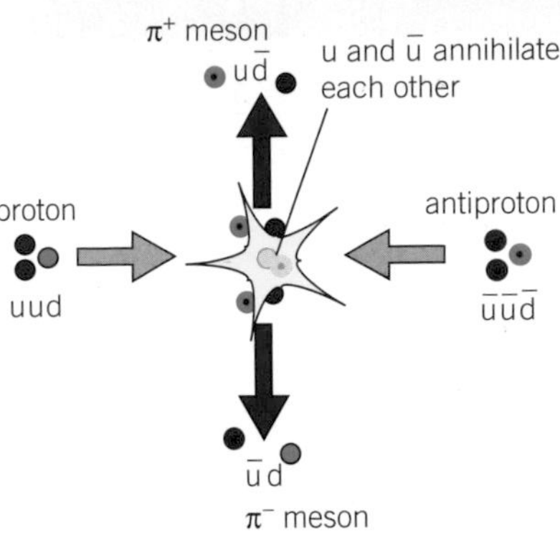


▲ **Figure 1** *Using the quark model*

## Journey into the atom (part 7): Big questions

Is there a deeper structure of matter that is still to be discovered? Radiation released in the Big Bang produced quarks and leptons in equal numbers to their antiparticles. So why haven't they all annihilated to turn back into radiation and why is the universe made of matter, not antimatter?

The study of particle physics raises lots of interesting questions! Scientists now know there are six different quarks and six different leptons. The properties and behaviour of matter on the smallest possible scale affects what happens on the largest scale (i.e., the universe). Perhaps the Large Hadron Collider and the discovery of the Higgs boson (see Topic 2.1) at CERN will provide some of the answers.

▲ **Figure 2** *Hubble Ultra Deep Field*

## Summary questions

**1** The following reaction was observed when two protons collided head-on at high speed:

$p + p \rightarrow p + n + \pi^+$

**a i** Show that charge is conserved in this reaction.

**ii** What other quantity is conserved in this reaction?

**b** Use the rest energy values in Table 1 on page 20 to show that the two protons must have at least 141 MeV of kinetic energy to create the $\pi^+$ meson.

**c i** State the quark compositions of a proton and of a neutron.

**ii** The $\pi^+$ meson has a quark compositions of $u\bar{d}$. Describe the above reaction in terms of quarks.

**2 a** State the baryon number of

**i** an up quark **ii** an antiproton **iii** a muon.

**b** State the lepton number of

**i** a neutrino **ii** an antimuon **iii** a neutron.

**3** The omega minus particle ($\Omega^-$) is a baryon with a strangeness of −3 and a charge of −1.

**a** Determine its quark composition.

**b** It decays into a $\pi^-$ meson and a baryon X composed of two strange quarks and an up quark.

**i** Determine the charge and the strangeness of X from its quark composition.

**ii** What type of interaction occurs in this change?

**iii** Describe the quark changes that take place.

# Practice questions: Chapter 2

**1** **(a)** Kaons were originally called strange because they are produced in *one* type of interaction and they decay through a different type of interaction.
(i) In which type of interaction are they produced?
(ii) Which type of interaction causes them to decay?
(iii) In which type of interaction is strangeness always conserved? *(3 marks)*

**(b)** (i) Describe the three-quark model of hadrons.
(ii) The three-quark model originated from a theory that explained the properties of baryons and mesons. Show that the three-quark model explains why there are just four charged mesons. *(6 marks)*

**2** From the following list of particles

$p \quad \bar{n} \quad \nu_e \quad e^+ \quad \mu^- \quad \pi^0$

identify all the examples of
(i) hadrons (ii) leptons
(iii) antiparticles (iv) charged particles. *(4 marks)*

AQA, 2006

**3** **(a)** (i) Write down the particles in the following list that may be affected by the weak interaction:

positron neutron photon neutrino positive pion

(ii) Write down the particles in the following list that may be affected by the electromagnetic force:
electron antineutrino proton neutral pion negative muon *(4 marks)*

**(b)** A positive muon may decay in the following way:

$\mu^+ \rightarrow e^+ + \nu_e + \bar{\nu}_\mu$

(i) Exchange each particle for its corresponding antiparticle and complete the equation to show how a negative muon may decay:

$\mu^- \rightarrow$

(ii) Give *one* difference and *one* similarity between a negative muon and an electron. *(3 marks)*

AQA, 2003

**4** **(a)** State the quark composition of a neutron. *(2 marks)*
**(b)** Which of the following terms can be used to describe a neutron?
antiparticle baryon fundamental particle hadron lepton meson *(2 marks)*

AQA, 2004

**5** **(a)** A particle is made up of an up antiquark and a down quark.
(i) Name the classification of particles that has this type of structure.
(ii) Find the charge on the particle.
(iii) State the baryon number of the particle. *(3 marks)*

**(b)** A suggested decay for the positive muon ($\mu^+$) is

$\mu^+ \rightarrow e^+ + \nu_e$

Showing your reasoning clearly, deduce whether this decay satisfies the conservation rules that relate to baryon number, lepton number, and charge. *(3 marks)*

AQA, 2003

**6** A sigma plus particle, $\Sigma^+$, is a baryon.
(i) How many quarks does the $\Sigma^+$ contain?
(ii) If *one* of these quarks is an s quark, by what interaction will it decay?
(iii) Which baryon will the $\Sigma^+$ eventually decay into? *(3 marks)*

AQA, 2005

**7** Some subatomic particles are classified as *hadrons*.

**(a)** What distinguishes a hadron from other subatomic particles? *(1 mark)*

**(b)** Hadrons fall into two subgroups.
Name each subgroup and describe the general structure of each. *(3 marks)*

**(c)** The following equation represents an event in which a positive muon collides with a neutron to produce a proton and an antineutrino:

$$n + \mu^+ \rightarrow p + \bar{\nu}_\mu$$

Show that this equation obeys the conservation laws of charge, lepton number, and baryon number. *(3 marks)*

AQA, 2004

**8** A negative pion ($\pi^-$) is a meson with a charge of −1e.
State and explain the structure of the $\pi^-$ in terms of the up and down quarks. *(3 marks)*

AQA, 2002

**9** The following is an incomplete equation for the decay of a free neutron:

$$^1_0n \rightarrow {}^1_1p + {}^{\ 0}_{-1}e + \ldots$$

**(a)** Complete the equation by writing down the symbol for the missing particle. *(2 marks)*

**(b)** Use the principles of conservation of charge, baryon number, and lepton number to demonstrate that decay is possible. *(3 marks)*

**(c)** The following reaction can take place when two protons meet head-on, provided the two colliding protons have sufficient kinetic energy:

$$p + p \rightarrow p + p + \bar{p} + p$$

If the two colliding protons each have the same amount of energy, calculate the minimum kinetic energy, in MeV, each must have for the reaction to proceed. *(2 marks)*

AQA, 2005

**10 (a)** (i) What class of particle is represented by the combination of three antiquarks, $\bar{q}\bar{q}\bar{q}$?

(ii) Name a hadron that has an antiparticle identical to itself. *(3 marks)*

**(b)** The kaon $K^+$ has a strangeness of +1.

(i) Give its quark composition

(ii) The $K^+$ may decay via the process

$$K^+ \rightarrow \pi^+ + \pi^0$$

State the interaction responsible for this decay.

(iii) The $K^+$ may also decay via the process

$$K^+ \rightarrow \mu^+ + \nu_\mu$$

Change each particle of this equation to its corresponding antiparticle in order to complete an allowed decay process for the negative kaon $K^-$:

$$K^- \rightarrow$$

(iv) Into what class of particle can both the $\mu^+$ and the $\nu_\mu$ be placed?

(v) State *one* difference between a positive muon and a positron, $e^+$. *(6 marks)*

AQA, 2002

**11** The following equation represents the collision of a neutral kaon with a proton, resulting in the production of a neutron and a positive pion:

$$K^0 + p \rightarrow n + \pi^+$$

**(a)** Show that this collision obeys *three* conservation laws in addition to energy and momentum. *(3 marks)*

**(b)** The neutral kaon has a strangeness of +1.
Write down the quark structure of the following particles:

$K^0$ $\quad \pi^+ \quad$ p *(4 marks)*

AQA, 2005

# 3 Quantum phenomena

## 3.1 The photoelectric effect

### Learning objectives:

- Explain the photoelectric effect.
- Define a photon.
- Discuss how the photon model was established.

*Specification reference: 3.2.2.1*

### Practical link

**Observing the photoelectric effect**

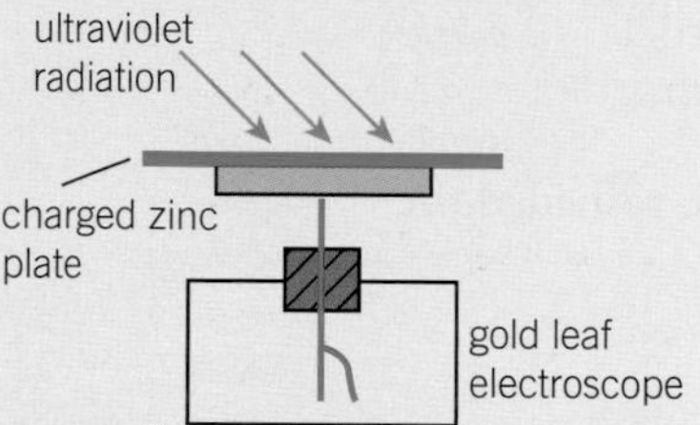


Ultraviolet radiation from a UV lamp is directed at the surface of a zinc plate placed on the cap of a gold leaf electroscope. This device is a very sensitive detector of charge. When it is charged, the thin gold leaf of the electroscope rises – it is repelled from the metal stem, because they both have the same type of charge.

If the electroscope is charged negatively, the leaf rises and stays in position. However, if ultraviolet light is directed at the zinc plate, the leaf gradually falls. The leaf falls because conduction electrons at the zinc surface leave the zinc surface when ultraviolet light is directed at it. The emitted electrons are referred to as **photoelectrons**.

If the electroscope is charged positively, the leaf rises and stays in position, regardless of whether or not ultraviolet light is directed at the zinc plate.

## The discovery of the photoelectric effect

A metal contains conduction electrons, which move about freely inside the metal. These electrons collide with each other and with the positive ions of the metal. When Heinrich Hertz discovered how to produce and detect radio waves, he found that the sparks produced in his spark gap detector when radio waves were being transmitted were stronger when ultraviolet radiation was directed at the spark gap. Further investigations on the effect of electromagnetic radiation on metals showed that electrons are emitted from the surface of a metal when electromagnetic radiation above a certain frequency was directed at the metal. This effect is known as the **photoelectric effect**.

### Puzzling problems

The following observations were made about the photoelectric effect after Hertz's discovery. These observations were a major problem because they could not be explained using the idea that light is a wave.

1. Photoelectric emission of electrons from a metal surface does *not* take place if the frequency of the incident electromagnetic radiation is below a certain value known as the **threshold frequency**. This minimum frequency depends on the type of metal. This means that the **wavelength** of the incident light must be less than a *maximum* value equal to the speed of light divided by the threshold frequency since $\lambda = \frac{c}{f}$.
2. The number of electrons emitted per second is proportional to the intensity of the incident radiation, provided the frequency is greater than the threshold frequency. However, if the frequency of the incident radiation is less than the threshold frequency, no photoelectric emission from that metal surface can take place, no matter how intense the incident radiation is.
3. Photoelectric emission occurs without delay as soon as the incident radiation is directed at the surface, provided the frequency of the radiation exceeds the threshold frequency, and regardless of intensity.

The wave theory of light cannot explain either the existence of a threshold frequency or why photoelectric emission occurs without delay. According to wave theory, each conduction electron at the surface of a metal should gain some energy from the incoming waves, regardless of how many waves arrive each second.

## Einstein's explanation of the photoelectric effect

The photon theory of light was put forward by Einstein in 1905 to explain the photoelectric effect. Einstein assumed that light is composed of wavepackets or **photons**, each of energy equal to $hf$, where $f$ is the frequency of the light and $h$ is the Planck constant. The accepted value for $h$ is $6.63 \times 10^{-34}$ J s.

**Energy of a photon = $hf$**

For electromagnetic waves of wavelength $\lambda$, the energy of each photon $E = hf = \frac{hc}{\lambda}$, where $c$ is the speed of the electromagnetic waves.

To explain the photoelectric effect, Einstein said that:

- When light is incident on a metal surface, an electron at the surface absorbs a *single* photon from the incident light and therefore gains energy equal to $hf$, where $hf$ is the energy of a light photon.
- An electron can leave the metal surface if the energy gained from a single photon exceeds the **work function** $\phi$ of the metal. This is the minimum energy needed by an electron to escape from the metal surface. Excess energy gained by the photoelectron becomes its kinetic energy.

The maximum kinetic energy of an emitted electron is therefore

$$E_{\text{Kmax}} = hf - \phi$$

Rearranging this equation gives

$$hf = E_{\text{Kmax}} + \phi$$

Emission can take place from a metal surface provided $E_{\text{Kmax}} > 0$ or $hf > \phi$, so the threshold frequency of the metal is $f_{\text{min}} = \dfrac{\phi}{h}$

### *Stopping potential*

Electrons that escape from the metal plate can be attracted back to it by giving the plate a sufficient positive charge. The minimum potential needed to stop photoelectric emission is called the **stopping potential $V_S$**. At this potential, the maximum kinetic energy of the emitted electron is reduced to zero because each emitted electron must do extra work equal to $e \times V_S$ to leave the metal surface. Hence its maximum kinetic energy is equal to $e \times V_S$.

Conclusive experimental evidence for Einstein's photon theory was obtained by Robert Millikan. Millikan measured the stopping potential for a range of metals using light of different frequencies. His results fitted Einstein's photoelectric equation very closely. After these results were checked and evaluated by other physicists through peer review, the scientific community accepted that light consists of photons.

**Synoptic link**

You have met Einstein's ideas about photons in Topic 1.3, Photons.

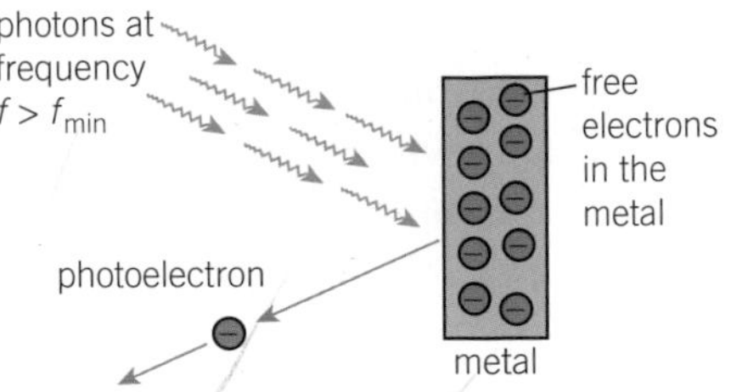


▲ **Figure 1** *Explaining the photoelectric effect – one electron absorbs one photon*

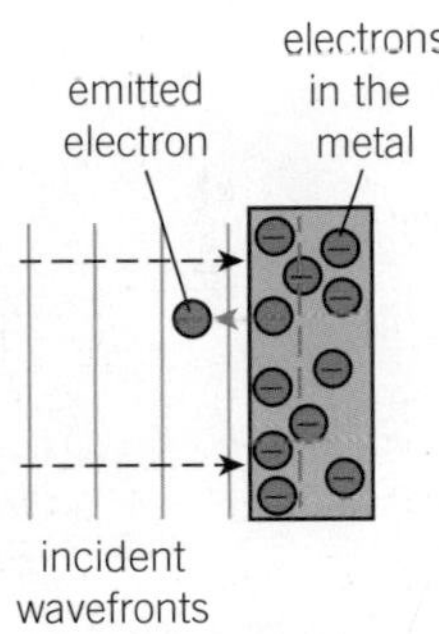


▲ **Figure 2** *Each electron in the metal can only absorb some of energy from each incident wavefront if light is a wave, so emission would take longer – this model is incorrect*

## Summary questions

**$h = 6.63 \times 10^{-34}$ J s, $c = 3.00 \times 10^{8}$ m s$^{-1}$**

**1 a** What is meant by photoelectric emission from a metal surface?

**b** Explain why photoelectric emission from a metal surface only takes place if the frequency of the incident radiation is greater than a certain value.

**2 a** Calculate the frequency and energy of a photon of wavelength

i 450 nm

ii 1500 nm.

**b** A metal surface at zero potential emits electrons from its surface if light of wavelength 450 nm is directed at it. However, electrons are not emitted when light of wavelength 650 nm is used. Explain these observations.

**3** The work function of a certain metal plate is $1.1 \times 10^{-19}$ J. Calculate

**a** the threshold frequency of incident radiation for this metal

**b** the maximum kinetic energy of photoelectrons emitted from this plate when light of wavelength 520 nm is directed at the metal surface.

**4** Light of wavelength 635 nm is directed at a metal plate at zero potential. Electrons are emitted from the plate with a maximum kinetic energy of $1.5 \times 10^{-19}$ J. Calculate

**a** the energy of a photon of this wavelength

**b** the work function of the metal

**c** the threshold frequency of electromagnetic radiation incident on this metal.

# 3.2 More about photoelectricity

## Learning objectives:

- → Explain why Einstein's photon model was revolutionary.
- → Define a quantum.
- → Explain why an electron can't absorb several photons to escape from a metal.

*Specification reference: 3.2.2.1*

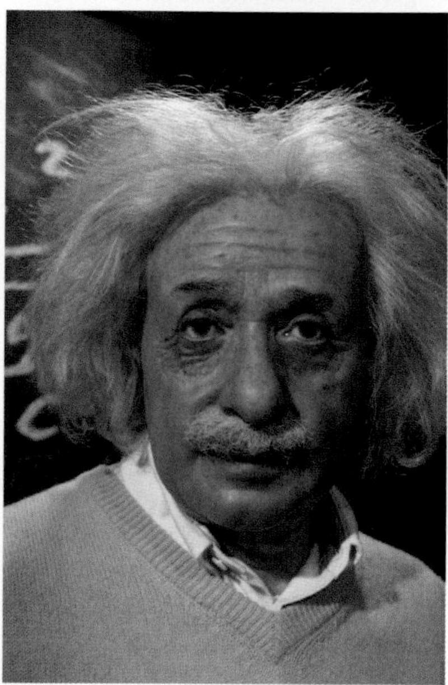

▲ **Figure 1** *Albert Einstein 1879–1955*

### Study tip

Remember that the work function is characteristic of the metal and the threshold frequency relates the work function to the incident radiation.

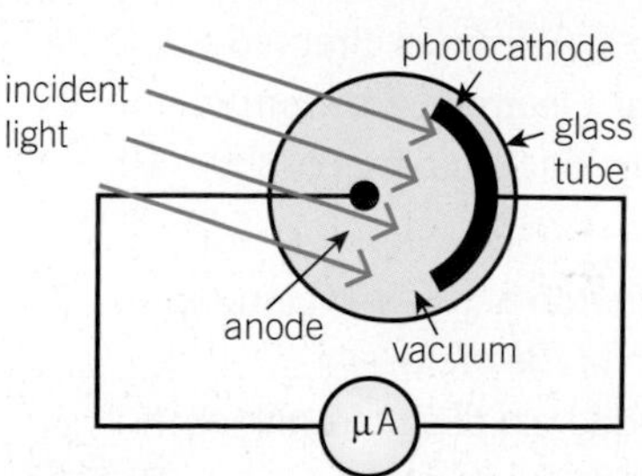


▲ **Figure 2** *Using a vacuum photocell*

## Into the quantum world

At the end of the 19th century, a physicist named Max Plank suggested that the energy of each vibrating atom is **quantised** – only certain levels of energy are allowed. He said the energy could only be in multiples of a basic amount, or quantum, $hf$, where $f$ is the frequency of vibration of the atom and $h$ is a constant, which became known as the Planck constant. He imagined the energy levels to be like the rungs of a ladder, with each atom absorbing or emitting radiation when it moved up or down a level.

There was a problem that the physicists of the time couldn't solve – the photoelectric effect. They couldn't fully explain it using the wave theory of radiation. Einstein did so by inventing a new theory of radiation – the photon model. His key idea was that electromagnetic radiation consists of photons, or wavepackets, of energy $hf$, where $f$ is the frequency of the radiation and $h$ is the Planck constant. Einstein's photon model and Planck's theory of vibrating atoms showed that energy is quantised – a completely new way of thinking about energy.

## More about conduction electrons

The conduction electrons in a metal move about at random, like the molecules of a gas. The average kinetic energy of a conduction electron depends on the temperature of the metal.

The work function of a metal is the *minimum* energy needed by a conduction electron to escape from the metal surface when the metal is at zero potential. The work function of a metal is of the order of $10^{-19}$ J, which is about 20 times greater than the average kinetic energy of a conduction electron in a metal at 300 K.

When a conduction electron absorbs a photon, its kinetic energy increases by an amount equal to the energy of the photon. If the energy of the photon exceeds the work function of the metal, the conduction electron can leave the metal. If the electron does not leave the metal, it collides repeatedly with other electrons and positive ions, and it quickly loses its extra kinetic energy.

## The vacuum photocell

A vacuum photocell is a glass tube that contains a metal plate, referred to as the photocathode, and a smaller metal electrode referred to as the anode. Figure 2 shows a vacuum photocell in a circuit. When light of a frequency greater than the threshold frequency for the metal is directed at the photocathode, electrons are emitted from the cathode and are attracted to the anode. The microammeter in the circuit can be used to measure the photoelectric current. This is proportional to the number of electrons per second that transfer from the cathode to the anode.

- For a photoelectric current $I$, the number of photoelectrons per second that transfer from the cathode to the anode = $I/e$, where $e$ is the charge of the electron.
- The photoelectric current is proportional to the intensity of the light incident on the cathode. The light intensity is a measure of the energy per second carried by the incident light, which is proportional to the number of photons per second incident on the cathode. Because each photoelectron must have absorbed one photon to escape from the metal surface, the number of photoelectrons emitted per second (i.e., the photoelectric current) is therefore proportional to the intensity of the incident light.
- The intensity of the incident light does *not* affect the maximum kinetic energy of a photoelectron. No matter how intense the incident light is, the energy gained by a photoelectron is due to the absorption of one photon only. Therefore, the maximum kinetic energy of a photoelectron is still given by $E_{Kmax} = hf - \phi$.
- The maximum kinetic energy of the photoelectrons emitted for a given frequency of light can be measured using a photocell.
- If the measurements for different frequencies are plotted as a graph of $E_{Kmax}$ against $f$, a straight line of the form $y = mx + c$ is obtained. This is in accordance with the above equation as $y = E_K$ and $x = f$. Note that the gradient of the line $m = h$ and the $y$-intercept, $c = -\phi$. The $x$-intercept is equal to the threshold frequency.

## Synoptic link

The equation charge

$Q$ (C) = current $I$ (A) × time $t$ (s)

will be covered in Topic 12.1, Current and charge.

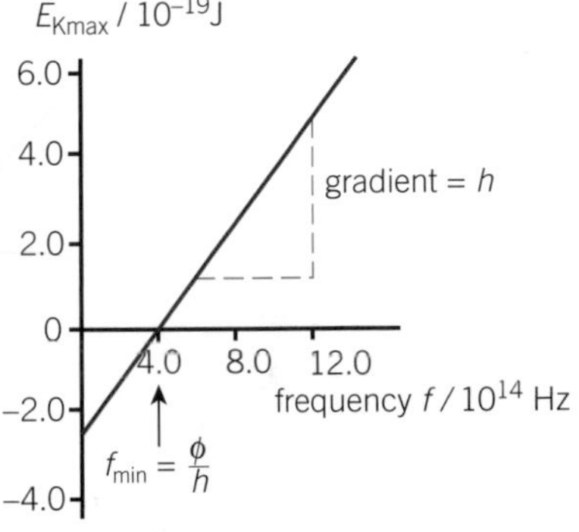


▲ **Figure 3** *A graph of $E_{Kmax}$ against frequency*

## Summary questions

$h = 6.63 \times 10^{-34}$ J s, $c = 3.00 \times 10^{8}$ m s$^{-1}$, $e = 1.6 \times 10^{-19}$ C

**1** A vacuum photocell is connected to a microammeter. Explain the following observations.

**a** When the cathode was illuminated with blue light of low intensity, the microammeter showed a non-zero reading.

**b** When the cathode was illuminated with an intense red light, the microammeter reading was zero.

**2** A vacuum photocell is connected to a microammeter. When light is directed at the photocell, the microammeter reads 0.25 µA.

**a** Calculate the number of photoelectrons emitted per second by the photocathode of the photocell.

**b** Explain why the microammeter reading is doubled if the intensity of the incident light is doubled.

**3** A narrow beam of light of wavelength 590 nm and of power 0.5 mW is directed at the photocathode of a vacuum photocell, which is connected to a microammeter which reads 0.4 µA. Calculate:

**a** the energy of a single light photon of this wavelength

**b** the number of photons incident on the photocathode per second

**c** the number of electrons emitted per second from the photocathode.

**4 a** Use Figure 3 to estimate **i** the threshold frequency, **ii** the work function of the metal.

**b** A metal surface has a work function of $1.9 \times 10^{-19}$ J. Light of wavelength 435 nm is directed at the metal surface. Calculate the maximum kinetic energy of the photoelectrons emitted from this metal surface.

# 3.3 Collisions of electrons with atoms

**Learning objectives:**

→ Explain what is meant by ionisation of an atom.

→ Explain what is meant by excitation of an atom.

→ Explain what happens inside an atom when it becomes excited.

*Specification reference: 3.2.2.2*

## Ionisation

An **ion** is a charged atom. The number of electrons in an ion is not equal to the number of protons. An ion is formed from an uncharged atom by adding or removing electrons from the atom. Adding electrons makes the atom into a negative ion. Removing electrons makes the atom into a positive ion.

Any process of creating ions is called **ionisation**. For example:

- Alpha, beta, and gamma radiation create ions when they pass through substances and collide with the atoms of the substance.
- Electrons passing through a fluorescent tube create ions when they collide with the atoms of the gas or vapour in the tube.

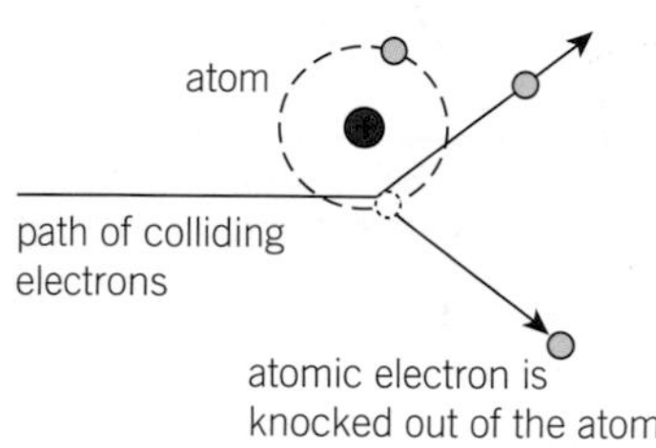


▲ **Figure 1** *Ionisation by collision*

### Measuring ionisation energy 

We can measure the energy needed to ionise a gas atom by making electrons collide at increasing speed with the gas atoms in a sealed tube. The electrons are emitted from a heated filament in the tube and are attracted to a positive metal plate, the anode, at the other end of the tube. The gas needs to be at sufficiently low pressure otherwise there are too many atoms in the tube and the electrons cannot reach the anode.

The potential difference (pd) between the anode and the filament is increased so as to increase the speed of the electrons. The circuit is shown in Figure 2. The ammeter records a very small current due to electrons from the filament reaching the anode. No ionisation occurs until the electrons from the filament reach a certain speed. At this speed, each electron arrives near the anode with just enough kinetic energy to ionise a gas atom, by knocking an electron out of the atom. Ionisation near the anode causes a much greater current to pass through the ammeter.

By measuring the pd, $V$, between the filament and the anode, when the current starts to increase, we can calculate the ionisation energy of a gas atom, as this is equal to the work done $W$ on each electron from the filament (and the work done is transformed to kinetic energy). The work done on each electron from the filament is given by its charge $e$ × the tube potential difference $V$. Therefore, the **ionisation energy of a gas atom =** $eV$.

**Q:** Why does the anode need to be positive relative to the filament?

**Answer:** To attract electrons from the filament.

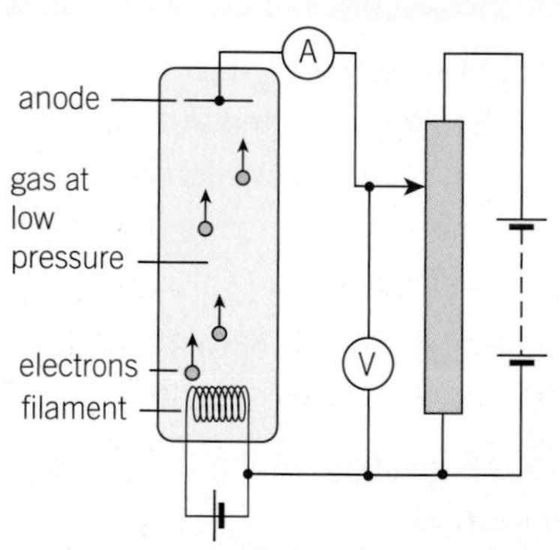


▲ **Figure 2** *Measuring ionisation energy*

**Synoptic link**

You met the notation eV briefly in Topic 2.1, The particle zoo.

## The electron volt

The **electron volt** is a unit of energy equal to the work done when an electron is moved through a pd of 1 V. For a charge $q$ moved through a pd $V$, the work done = $qV$. Therefore, the work done when an electron moves through a potential difference of 1 V is equal to $1.6 \times 10^{-19}$ J ($= 1.6 \times 10^{-19}$ C × 1 V). This amount of energy is defined as 1 electron volt (eV).

For example, the work done on

- an electron when it moves through a potential difference of 1000 V is 1000 eV
- an ion of charge $+2e$ when it moves through a potential difference of 10 V is 20 eV.

## Synoptic link

Potential difference will be studied in detail in Topic 12.2, Potential difference and power.

## Excitation by collision

Using gas-filled tubes with a metal grid between the filament and the anode, we can show that gas atoms can absorb energy from colliding electrons without being ionised. This process, known as **excitation**, happens at certain energies, which are characteristic of the atoms of the gas. If a colliding electron loses all its kinetic energy when it causes excitation, the current due to the flow of electrons through the gas is reduced. If the colliding electron does not have enough kinetic energy to cause excitation, it is deflected by the atom, with no overall loss of kinetic energy.

The energy values at which an atom absorbs energy are known as its **excitation energies**. We can determine the excitation energies of the atoms in the gas-filled tube by increasing the potential difference between the filament and the anode and measuring the pd when the anode current falls. For example, two prominent excitation energies of a mercury atom are 4.9 eV and 5.7 eV. This would mean that the current would fall at 4.9 V and 5.7 V.

When excitation occurs, the colliding electron makes an electron inside the atom move from an inner shell to an outer shell. Energy is needed for this process, because the atomic electron moves away from the nucleus of the atom. The excitation energy is always less than the ionisation energy of the atom, because the atomic electron is not removed completely from the atom when excitation occurs.

## Study tip

Don't confuse atomic electrons with electrons that hit the atom.

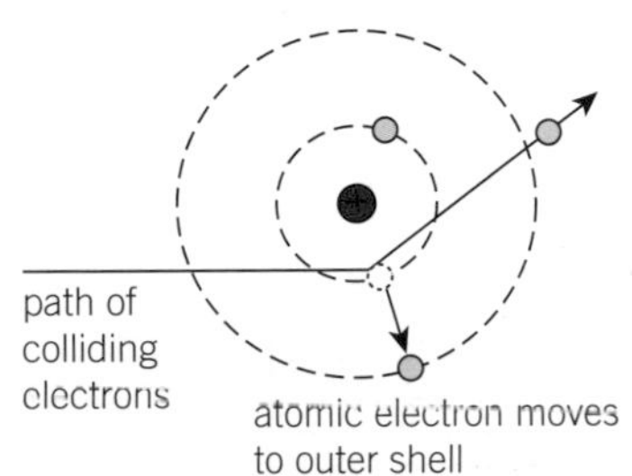


▲ **Figure 3** *A simple model of excitation by collision*

## Summary questions

$e = 1.6 \times 10^{-19}$ C

1 State one difference and one similarity between ionisation and excitation.

2 a The mercury atom has an ionisation energy of 10.4 eV. Calculate this ionisation energy in joules.

b An electron with 12.0 eV of kinetic energy collides with a mercury atom and ionises it. Calculate the kinetic energy, in eV, of the electron after the collision.

3 Complete the sentences below.

a The ionisation energy of neon is much greater than that of sodium. Therefore, ______ energy is needed to remove an electron from a sodium atom than from a neon atom.

b An electron that causes excitation of an atom by colliding with it has ________ kinetic energy before the collision than after the collision. The atom's internal energy _____ and the electron's kinetic energy ______.

4 a Describe what happens to a gas atom when an electron collides with it and causes it to absorb energy from the electron without being ionised.

b Explain why a gas atom cannot absorb energy from a slow-moving electron that collides with it.

# 3.4 Energy levels in atoms

## Learning objectives:

→ Explain what energy levels are.

→ Explain what happens when excited atoms de-excite.

→ Explain how a fluorescent tube works.

*Specification reference: 3.2.2.3*

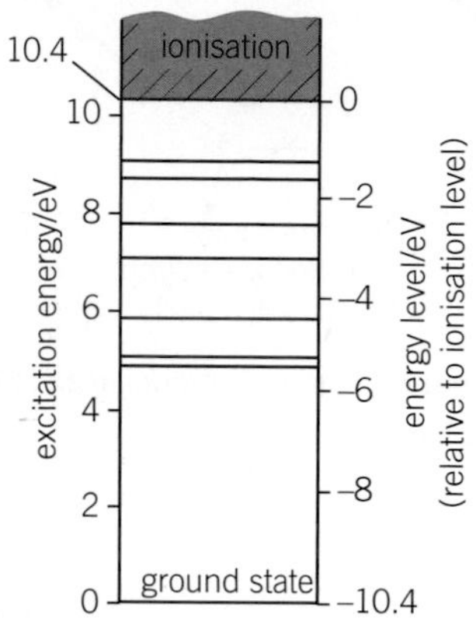


▲ **Figure 1** *The energy levels of the mercury atom*

## Electrons in atoms

The electrons in an atom are trapped by the electrostatic force of attraction of the nucleus. They move about the nucleus in allowed orbits, or shells, surrounding the nucleus. The energy of an electron in a shell is constant. An electron in a shell near the nucleus has less energy than an electron in a shell further away from the nucleus. Each shell can only hold a certain number of electrons. For example, the innermost shell (i.e., the shell nearest to the nucleus) can only hold two electrons and the next nearest shell can only hold eight electrons.

Each type of atom has a certain number of electrons. For example, a helium atom has two electrons. Thus, in its lowest energy state, a helium atom has both electrons in the innermost shell.

The lowest energy state of an atom is called its **ground state**. When an atom in the ground state absorbs energy, one of its electrons moves to a shell at higher energy, so the atom is now in an **excited state**. We can use the excitation energy measurements to construct an **energy level** diagram for the atom, as shown in Figure 1. This shows the allowed energy values of the atom. Each allowed energy corresponds to a certain electron configuration in the atom. Note that the ionisation level may be considered as the zero reference level for energy, instead of the ground state level. The energy levels below the ionisation level would then need to be shown as negative values, as shown on the right-hand side of Figure 1.

## De-excitation

Did you know that gases at low pressure emit light when they are made to conduct electricity? For example, a neon tube emits red-orange light when it conducts. The gas-filled tube used to measure excitation energies in Topic 3.3 emits light when excitation occurs. This happens because the atoms absorb energy as a result of excitation by collision, but they do not retain the absorbed energy permanently.

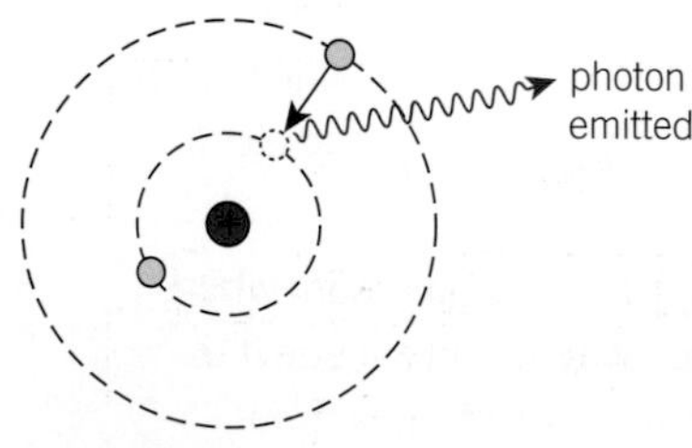


▲ **Figure 2** *De-excitation by photon emission*

The electron configuration in an excited atom is unstable because an electron that moves to an outer shell leaves a vacancy in the shell it moves from. Sooner or later, the vacancy is filled by an electron from an outer shell transferring to it. When this happens, the electron emits a photon. The atom therefore moves to a lower energy level (the process of **de-excitation**).

The energy of the photon is equal to the energy lost by the electron and therefore by the atom. For example, when a mercury atom at an excitation energy level of 4.9 eV de-excites to the ground state, it emits a photon of energy 4.9 eV. The mercury atom could also de-excite indirectly via several energy levels if intermediate energy levels are present. However, since there are no intermediate levels between 4.9 eV and the ground state, the only way that the excited electron in this example can de-excite is to release a photon of energy 4.9 eV.

In general, when an electron moves from energy level $E_1$ to a lower energy level $E_2$,

**the energy of the emitted photon $hf = E_1 - E_2$.**

## Excitation using photons

An electron in an atom can absorb a photon and move to an outer shell where a vacancy exists – but only if the energy of the photon is exactly equal to the gain in the electron's energy (Figure 3). In other words, the photon energy must be exactly equal to the difference between the final and initial energy levels of the atom. If the photon's energy is smaller or larger than the difference between the two energy levels, it will not be absorbed by the electron.

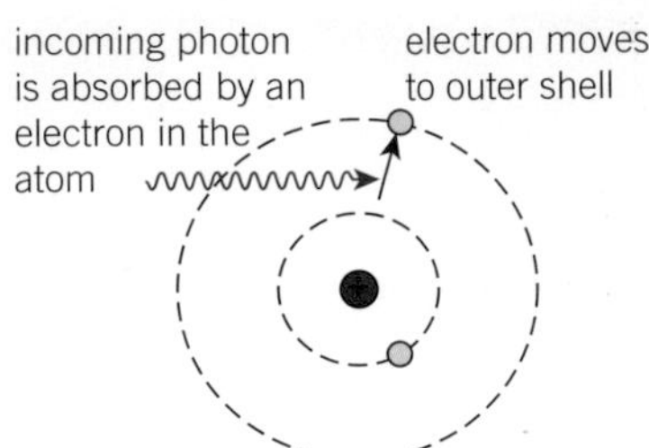


▲ **Figure 3** *Excitation by photon absorption*

> **Hint**
>
> A photon can be absorbed and cause ionisation if its energy is greater than or equal to the difference between the ionisation level and the ground state. However, excitation requires photons of specific energies.

## Fluorescence

An atom in an excited state can de-excite directly or indirectly to the ground state, regardless of how the excitation took place. An atom can absorb photons of certain energies and then emit photons of the same or lesser energies. For example, a mercury atom in the ground state could

- be excited to its 5.7 eV energy level by absorbing a photon of energy 5.7 eV, then
- de-excite to its 4.9 eV energy level by emitting a photon of energy 0.8 eV, then
- de-excite to the ground state by emitting a photon of energy 4.9 eV.

Figure 4 represents these changes on an energy level diagram.

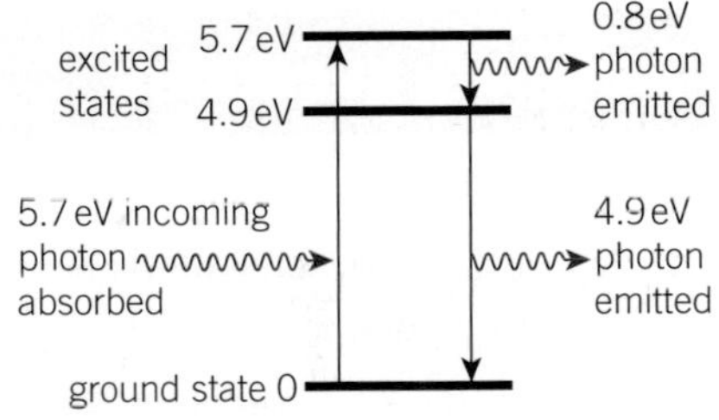


▲ **Figure 4** *Fluorescence*

This overall process explains why certain substances **fluoresce** or glow with visible light when they absorb ultraviolet radiation. Atoms in the substance absorb ultraviolet photons and become excited. When the atoms de-excite, they emit visible photons. When the source of ultraviolet radiation is removed, the substance stops glowing.

The **fluorescent tube** is a glass tube with a fluorescent coating on its inner surface. The tube contains mercury vapour at low pressure. When the tube is on, it emits visible light because:

- ionisation and excitation of the mercury atoms occur as they collide with each other and with electrons in the tube
- the mercury atoms emit ultraviolet photons, as well as visible photons and photons of much less energy, when they de-excite
- the ultraviolet photons are absorbed by the atoms of the fluorescent coating, causing excitation of the atoms
- the coating atoms de-excite in steps and emit visible photons.

Figure 5 shows the circuit for a fluorescent tube. The tube is much more efficient than a filament lamp. A typical 100 W filament lamp releases about 10–15 W of light energy. The rest of the energy supplied to it is wasted as heat. In contrast, a fluorescent tube can produce the same light output with no more than a few watts of power wasted as heat.

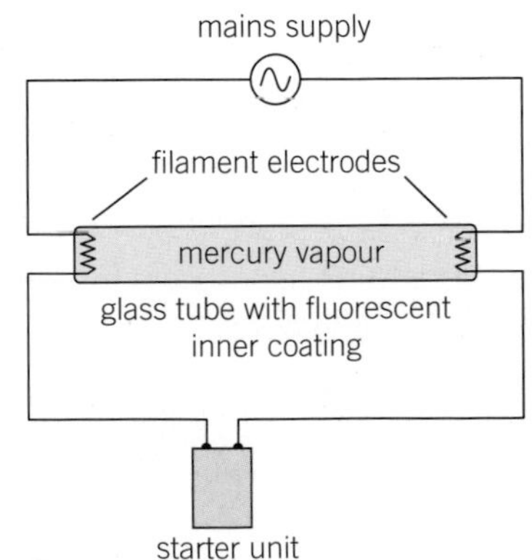


▲ **Figure 5** *The fluorescent tube*

▲ **Figure 6** *A low-energy light bulb and a filament light bulb*

## Starting a fluorescent tube

A fluorescent tube has a filament electrode at each end. A starter unit is necessary because the mains voltage is too small to ionise the vapour in the tube when the electrodes are cold. When the tube is first switched on, the gas (argon) in the starter switch unit conducts and heats a bimetallic strip, making it bend, so the switch closes. The current through the starter unit increases enough to heat the filament electrodes. When the bimetallic switch closes, the gas in the starter unit stops conducting, and so the bimetallic strip cools and the switch opens. The mains voltage now acts between the two electrodes, which are now hot enough for ionisation of the gas to occur.

## Low-energy light bulbs

Did you know that the UK government has stopped most sales of filament light bulbs? We now need to use low-energy light bulbs instead. Such a light bulb uses much less power than a filament light bulb which has the same light output. This is because the light is produced by a folded-up fluorescent tube instead of a glowing filament so less energy is wasted as heat. A 100 W filament bulb emits about 15 W of light and wastes the rest. In contrast, a low-energy light bulb with the same light output of 15 W wastes only about 5 W. Prove for yourself that the low-energy light bulb is five times more efficient than the filament light bulb. Using them at home would cut your electricity bill considerably and help to cut carbon emissions.

### Worked example

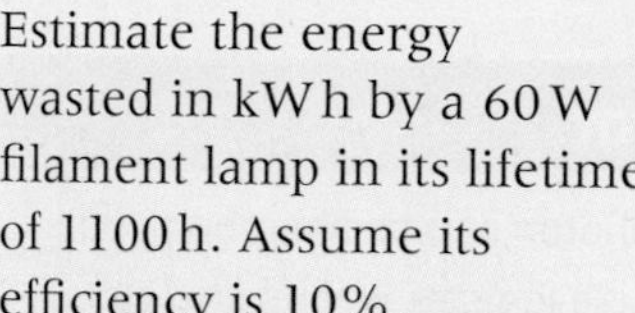

Estimate the energy wasted in kWh by a 60 W filament lamp in its lifetime of 1100 h. Assume its efficiency is 10%.

*Solution*

Energy supplied = power × time = 0.06 kW × 1100 h = 66 kWh

Energy wasted = 90% of 66 kWh ≈ 60 kWh

## Summary questions

$e = 1.6 \times 10^{-19}$ C

1 Figure 1 shows some of the energy levels of the mercury atom.

a Estimate the energy needed to excite the atom from the ground state to the highest excitation level shown in the diagram.

b Mercury atoms in an excited state at 5.7 eV can de-excite directly or indirectly to the ground state. Show that the photons released could have six different energies.

2 a In terms of electrons, state two differences between excitation and de-excitation.

b A certain type of atom has excitation energies of 1.8 eV and 4.6 eV.

i Sketch an energy level diagram for the atom using these energy values.

ii Calculate the possible photon energies from the atom when it de-excites from the 4.6 eV level. Use a downward arrow to indicate on your diagram the energy change responsible for each photon energy.

3 An atom absorbs a photon of energy 3.8 eV and subsequently emits photons of energy 0.6 eV and 3.2 eV.

a Sketch an energy level diagram representing these changes.

b In terms of electrons in the atom, describe how the above changes take place.

4 Explain why the atoms in a fluorescent tube stop emitting light when the electricity supply to it is switched off.

# 3.5 Energy levels and spectra

## A colourful spectrum

A rainbow is a natural display of the colours of the spectrum of sunlight. Raindrops split sunlight into a continuous spectrum of colours. Figure 1 shows how we can use a prism to split a beam of white light from a filament lamp into a continuous spectrum. The wavelength of the light photons that produce the spectrum increases across the spectrum from deep violet at less than 400 nm to deep red at about 650 nm.

If we use a tube of glowing gas as the light source instead of a filament lamp, we see a spectrum of discrete lines of different colours, as shown in Figure 2, instead of a continuous spectrum.

The wavelengths of the lines of a line spectrum of an element are characteristic of the atoms of that element. By measuring the wavelengths of a line spectrum, we can therefore identify the element that produced the light. No other element produces the same pattern of light wavelengths. This is because the energy levels of each type of atom are unique to that atom. So the photons emitted are characteristic of the atom.

- Each line in a line spectrum is due to light of a certain colour and therefore a certain wavelength.
- The photons that produce each line all have the same energy, which is different from the energy of the photons that produce any other line.
- Each photon is emitted when an atom de-excites due to one of its electrons moving to an inner shell.
- If the electron moves from energy level $E_1$ to a lower energy level $E_2$

**the energy of the emitted photon $hf = E_1 - E_2$**

For each wavelength $\lambda$, we can calculate the energy of a photon of that wavelength as its frequency $f = c/\lambda$, where $c$ is the speed of light. Given the energy level diagram for the atom, we can therefore identify on the diagram the transition that causes a photon of that wavelength to be emitted.

### Worked example:

**$c = 3.0 \times 10^8\,\text{m s}^{-1}$, $e = 1.6 \times 10^{-19}\,\text{C}$, $h = 6.63 \times 10^{-34}\,\text{J s}$**

A mercury atom de-excites from its 4.9 eV energy level to the ground state. Calculate the wavelength of the photon released.

### *Solution*

$E_1 - E_2 = 4.9 - 0 = 4.9\,\text{eV} = 4.9 \times 1.6 \times 10^{-19}\,\text{J} = 7.84 \times 10^{-19}\,\text{J}$

Therefore, $f = \dfrac{E_1 - E_2}{h} = \dfrac{7.84 \times 10^{-19}}{6.63 \times 10^{-34}} = 1.18 \times 10^{15}\,\text{Hz}$

$\lambda = \dfrac{c}{f} = \dfrac{3.0 \times 10^8}{1.18 \times 10^{15}} = 2.54 \times 10^{-7}\,\text{m} = 254\,\text{nm}$

### Learning objectives:

→ Define a line spectrum.

→ Explain why atoms emit characteristic line spectra.

→ Calculate the wavelength of light for a given electron transition.

*Specification reference: 3.2.2.3*

▲ **Figure 1** *Observing a continuous spectrum*

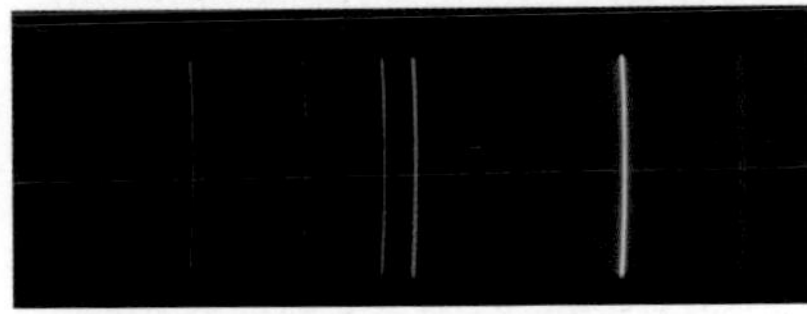

▲ **Figure 2** *A line spectrum*

### Synoptic link

You have met energy levels in more detail in Topic 3.4, Energy levels in atoms.

### Synoptic link

You will see how to measure the wavelengths in a line spectrum in Topic 5.7, The diffraction grating.

## Study tip

Don't forget that lines in a spectrum are always due to energy levels.

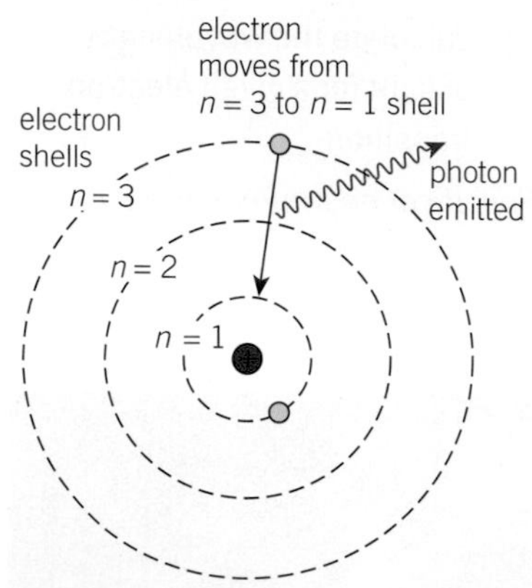


▲ **Figure 3** *An electron transition in the hydrogen atom*

## Summary questions

$h = 6.63 \times 10^{-34}$ J s

$c = 3.00 \times 10^{8}$ m s$^{-1}$

$e = 1.6 \times 10^{-19}$ C

1 State two differences between a continuous spectrum and a line spectrum.

2  A mercury atom de-excites from 5.7 eV to 4.9 eV. For the photon emitted, calculate

 a its energy in J

 b its wavelength.

3 A line spectrum has a line at a wavelength of 620 nm. Calculate

 a the energy, in J, of a photon of this wavelength

 b the energy loss, in eV, of an atom that emits a photon of this wavelength.

4 Explain why the line spectrum of an element is unique to that element and can be used to identify it.

## The Bohr atom

The hydrogen atom is the simplest type of atom – just one proton as its nucleus and one electron. The energy levels of the hydrogen atom, relative to the ionisation level, are given by the general formula:

$$E = -\frac{13.6\text{ eV}}{n^2}$$

where $n = 1$ for the ground state, $n = 2$ for the next excited state, etc.

Therefore, when a hydrogen atom de-excites from energy level $n_1$ to a lower energy level $n_2$, the energy of the emitted photon is given by

$$E = \left(\frac{1}{n_2^{\,2}} - \frac{1}{n_1^{\,2}}\right) \times 13.6\text{ eV}$$

Each energy level corresponds to the electron in a particular shell. Thus the above formula gives the energy of a photon released when an electron in the hydrogen atom moves from one shell to a shell at lower energy.

Figure 3 shows an example of a transition that can take place in an excited hydrogen atom. The energy level formula for hydrogen was first deduced from the measurements of the wavelengths of the lines. Later, the Danish physicist Niels Bohr applied the quantum theory to the motion of the electron in the hydrogen atom, and so produced the first theoretical explanation of the energy level formula for hydrogen.

**Q:** Which transition between adjacent levels in an excited hydrogen atom gives the highest energy of the released photon?

**Answer:** From $n = 2$ to $n = 1$.

## The discovery of helium

Measurements of the wavelength of light are important in branches of science such as astronomy and forensic science, as they enable us to identify the chemical elements in the light source. Helium was discovered from the spectrum of sunlight. A pattern of lines in the spectrum was observed at wavelengths that had never been observed from any known gas, and were therefore due to the presence of a previously unknown element in the Sun. Helium is produced as a result of the nuclear fusion of hydrogen nuclei in the Sun, and was given the name helium from *helios*, the Greek word for sun. Helium is also present in the Earth, produced as alpha particles from the radioactive decay of elements such as uranium. It can be collected at oil wells and stored for use in fusion reactors, helium–neon lasers, and for very low temperature devices. In the liquid state, below a temperature of 2.17 K, it becomes a superfluid. This is a fluid with no resistance to flow, which escapes from an open container by creeping – as a thin film – up and over the sides of the container.

# 3.6 Wave–particle duality

## The dual nature of light

Light is part of the electromagnetic spectrum of waves. The theory of electromagnetic waves predicted the existence of electromagnetic waves beyond the visible spectrum. The subsequent discovery of X-rays and radio waves confirmed these predictions and seemed to show that the nature of light had been settled. Many scientists in the late 19th century reckoned that all aspects of physics could be explained using Newton's laws of motion and the theory of electromagnetic waves. They thought that the few minor problem areas, such as the photoelectric effect, would be explained sooner or later using Newton's laws of motion and Maxwell's theory of electromagnetic waves. However, as explained in Topic 3.1, the photoelectric effect was not explained until Einstein put forward the radical theory that light consists of photons, which are particle-like packets of electromagnetic waves. Light has a dual nature, in that it can behave as a wave or as a particle, according to circumstances.

- The **wave-like nature** is observed when **diffraction** of light takes place. This happens, for example, when light passes through a narrow slit. The light emerging from the slit spreads out in the same way as water waves spread out after passing through a gap. The narrower the gap or the longer the wavelength, the greater the amount of diffraction.

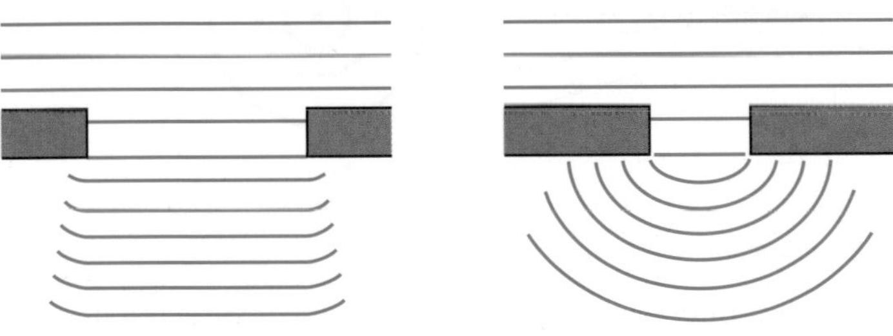

▲ **Figure 1** *Diffraction*

- The **particle-like nature** is observed, for example, in the photoelectric effect. When light is directed at a metal surface and an electron at the surface absorbs a photon of frequency $f$, the kinetic energy of the electron is increased from a negligible value by $hf$. The electron can escape if the energy it gains from a photon exceeds the work function of the metal.

## Matter waves

If light has a dual wave–particle nature, perhaps particles of matter also have a dual wave–particle nature. Electrons in a beam can be deflected by a magnetic field. This is evidence that electrons have a particle-like nature. The idea that matter particles also have a wave-like nature was first considered by Louis de Broglie in 1923 (**de Broglie hypothesis**).

### Learning objectives:

- → Explain why we say photons have a dual nature.
- → Describe how we know that matter particles have a dual nature.
- → Discuss why we can change the wavelength of a matter particle but not that of a photon.

*Specification reference: 3.2.2.4*

## Study tip

Don't mix up matter waves and electromagnetic waves and don't confuse their equations.

By extending the ideas of duality from photons to matter particles, de Broglie put forward the hypothesis that:

- matter particles have a dual wave–particle nature
- the wave-like behaviour of a matter particle is characterised by a wavelength, its **de Broglie wavelength**, $\lambda$, which is related to the momentum, $p$, of the particle by means of the equation.

$$\lambda = \frac{h}{p}$$

Since the momentum of a particle is defined as its mass × its velocity, according to de Broglie's hypothesis, a particle of mass $m$ moving at velocity $v$ has a de Broglie wavelength given by

$$\lambda = \frac{h}{mv}$$

***Note:***
The de Broglie wavelength of a particle can be altered by changing the velocity of the particle.

## Evidence for de Broglie's hypothesis

The wave-like nature of electrons was discovered when, three years after de Broglie put forward his hypothesis, it was demonstrated that a beam of electrons can be diffracted. Figure 2 shows in outline how this is done. After this discovery, further experimental evidence, using other types of particles, confirmed the correctness of de Broglie's theory.

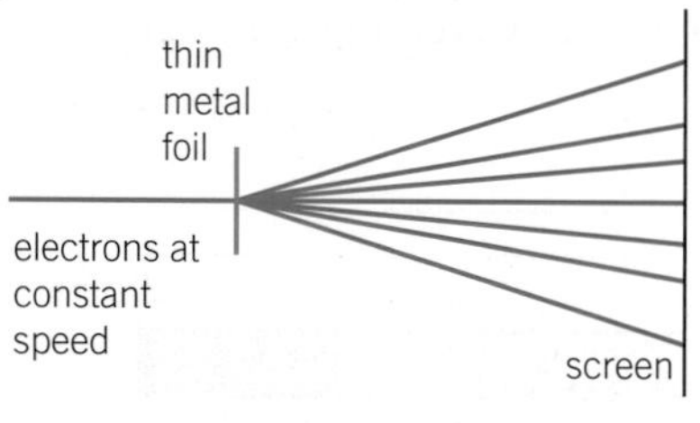


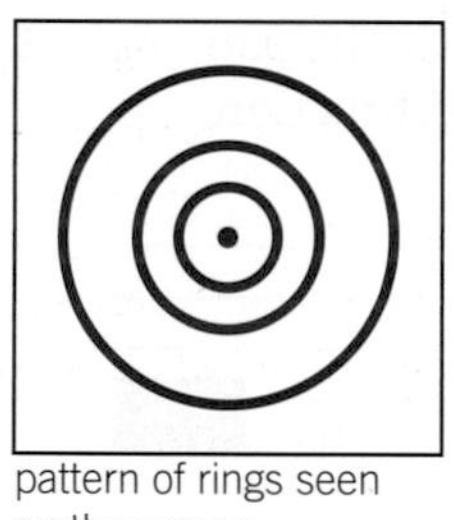


▲ **Figure 2** *Diffraction of electrons*

- A narrow beam of electrons in a vacuum tube is directed at a thin metal foil. A metal is composed of many tiny crystalline regions. Each region, or grain, consists of positive ions arranged in fixed positions in rows in a regular pattern. The rows of atoms cause the electrons in the beam to be diffracted, just as a beam of light is diffracted when it passes through a slit.
- The electrons in the beam pass through the metal foil and are diffracted in certain directions only, as shown in Figure 2. They form a pattern of rings on a fluorescent screen at the end of the tube. Each ring is due to electrons diffracted by the same amount from grains of different orientations, at the same angle to the incident beam.
- The beam of electrons is produced by attracting electrons from a heated filament wire to a positively charged metal plate, which has a small hole at its centre. Electrons that pass through the hole form the beam. The speed of these electrons can be increased by increasing the potential difference between the filament and the metal plate. This makes the diffraction rings smaller, because the increase of speed makes the de Broglie wavelength smaller. So less diffraction occurs and the rings become smaller.

## Synoptic link

You have met the PET scanner in Topic 1.4, Particles and antiparticles.

## Energy levels explained

An electron in an atom has a fixed amount of energy that depends on the shell it occupies. Its de Broglie wavelength has to fit the shape and size of the shell. This is why its energy depends on the shell it occupies.

For example, an electron in a spherical shell moves round the nucleus in a circular orbit. The circumference of its orbit must be equal to a whole number of de Broglie wavelengths (circumference = $n\lambda$, where $n = 1$ or 2 or 3, etc.). This condition can be used to derive the energy level formula for the hydrogen atom – and it gives you a deeper insight into quantum physics.

**Q:** Does the de Broglie wavelength of an electron increase or decrease when it moves to an orbit where it travels faster?

**Answer:** The wavelength becomes smaller because its momentum becomes greater.

## Quantum technology

The PET scanner is an example of quantum physics in use. Some further applications of quantum technology include

- The STM (scanning tunneling microscope) is used to map atoms on solid surfaces. The wave nature of electrons allows them to tunnel between the surface and a metal tip a few nanometres above the surface as the tip scans across the surface.
- The TEM (transmission electron microscope) is used to obtain very detailed images of objects and surface features too small to see with optical microscopes. Electrons are accelerated in a TEM to high speed so their de Broglie wavelength is so small that they can give very detailed images.
- The MR (magnetic resonance) body scanner used in hospitals detects radio waves emitted when hydrogen atoms in a patient in a strong magnetic field flip between energy levels.
- SQUIDs (superconducting quantum interference devices) are used to detect very very weak magnetic fields, for example SQUIDs are used to detect magnetic fields produced by electrical activity in the brain.

▲ **Figure 3** *An example of a SQUID*

## Summary questions

**$h = 6.6 \times 10^{-34}$ J s**
**the mass of an electron = $9.1 \times 10^{-31}$ kg**
**the mass of a proton = $1.7 \times 10^{-27}$ kg**

**1** With the aid of an example in each case, explain what is meant by the dual wave–particle nature of
- **a** light
- **b** matter particles, for example, electrons.

**2** State whether each of the following experiments demonstrates the wave nature or the particle nature of matter or of light:
- **a** the photoelectric effect
- **b** electron diffraction.

**3** Calculate the de Broglie wavelength of
- **a** an electron moving at a speed of $2.0 \times 10^7$ m s$^{-1}$
- **b** a proton moving at the same speed.

**4** Calculate the momentum and speed of
- **a** an electron that has a de Broglie wavelength of 500 nm
- **b** a proton that has the same de Broglie wavelength.

# Practice questions: Chapter 3

1 When light at sufficiently high frequency, $f$, is incident on a metal surface, the maximum kinetic energy, $E_{Kmax}$, of a photoelectron is given by

$E_{Kmax} = hf - \phi$, where $\phi$ is the work function of the metal.

**(a)** State what is meant by the work function $\phi$.

**(b)** The following results were obtained in an experiment to measure $E_{Kmax}$ for different frequencies, $f$: *(1 mark)*

| $f$ / $10^{14}$ Hz | 5.6 | 6.2 | 6.8 | 7.3 | 8.3 | 8.9 |
|---|---|---|---|---|---|---|
| $E_{Kmax}$ / $10^{-19}$ J | 0.8 | 1.2 | 1.7 | 2.1 | 2.9 | 3.4 |

(i) Use these results to plot a graph of $E_{Kmax}$ against $f$.

(ii) Explain why your graph confirms the equation above.

(iii) Use your graph to determine the value of $h$ and to calculate the work function of the metal. *(13 marks)*

**(c)** Measurements like those above were first made to test the correctness of Einstein's explanation of the photoelectric effect. The equation above was a prediction by Einstein using the photon theory.
Why is it important to test any new theory by testing its predictions experimentally? *(2 marks)*

2 **(a)** One quantity in the photoelectric equation is a characteristic property of the metal that emits photoelectrons. Name and define this quantity. *(2 marks)*

**(b)** A metal is illuminated with monochromatic light. Explain why the kinetic energy of the photoelectrons emitted has a range of values up to a certain maximum. *(3 marks)*

**(c)** A gold surface is illuminated with monochromatic ultraviolet light of frequency $1.8 \times 10^{15}$ Hz. The maximum kinetic energy of the emitted photoelectrons is $4.2 \times 10^{-19}$ J. Calculate, for gold:

(i) the work function, in J

(ii) the threshold frequency. *(5 marks)*

AQA, 2006

3 **Figure 1** shows how the maximum kinetic energy of electrons emitted from the cathode of a photoelectric cell varies with the frequency of the incident radiation.

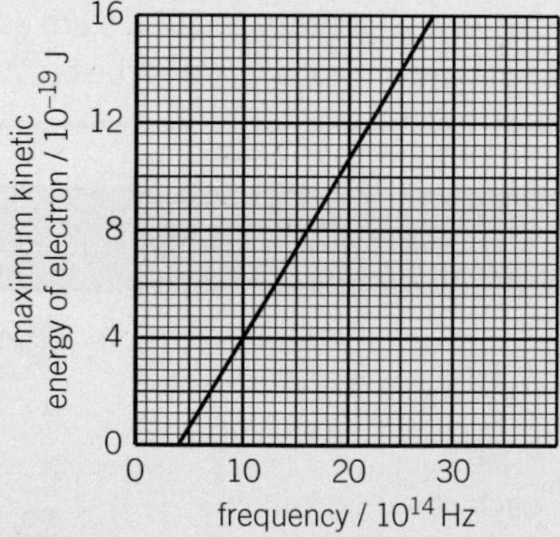


▲ **Figure 1**

**(a)** Calculate the maximum wavelength of electromagnetic radiation that can release photoelectrons from the cathode surface.

speed of electromagnetic radiation in a vacuum = $3.0 \times 10^8\,\mathrm{m\,s^{-1}}$ *(3 marks)*

**(b)** Another photoelectric cell uses a different metal for the photocathode. This metal requires twice the minimum energy for electron release compared to the metal in the first cell.

(i) If drawn on the same axes, how would the graph line obtained for this second cell compare with the one for the first cell?

(ii) Explain your answer with reference to the Einstein photoelectric equation. *(3 marks)*

AQA, 2003

**4** A fluorescent light tube contains mercury vapour at low pressure. The tube is coated on the inside, and contains two electrodes.

**(a)** Explain why the mercury vapour is at a low pressure. *(1 mark)*

**(b)** Explain the purpose of the coating on the inside of the tube. *(3 marks)*

AQA, 2003

**5** The lowest energy levels of a mercury atom are shown below. The diagram is *not* to scale.

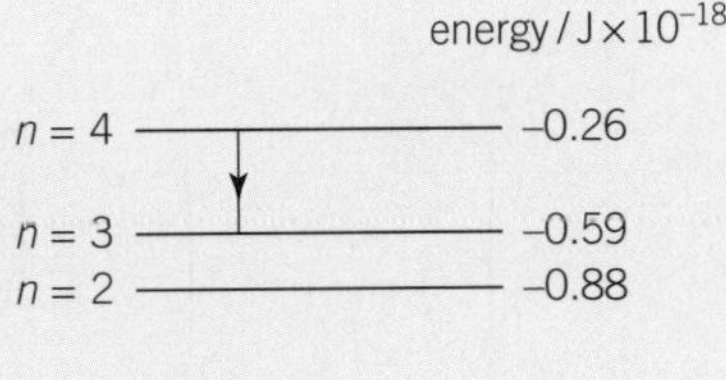


**(a)** Calculate the frequency of an emitted photon due to a transition, shown by an arrow, from level $n = 4$ to level $n = 3$. *(2 marks)*

**(b)** Which transition would cause the emission of a photon of a longer wavelength than that emitted in the transition from level $n = 4$ to level $n = 3$? *(1 mark)*

AQA, 2002

**6** **(a)** State what is meant by the duality of electrons. Give *one* example of each type of behaviour. *(3 marks)*

**(b)** (i) Calculate the speed of an electron which has a de Broglie wavelength of $1.3 \times 10^{-10}$ m.

(ii) A particle when travelling at the speed calculated in (b)(i) has a de Broglie wavelength of $8.6 \times 10^{-14}$ m.
Calculate the mass of the particle. *(4 marks)*

AQA, 2007

**7** Electrons travelling at a speed of $5.00 \times 10^5\,\mathrm{m\,s^{-1}}$ exhibit wave properties.

**(a)** What phenomenon can be used to demonstrate the wave properties of electrons? Details of any apparatus used are not required. *(1 mark)*

**(b)** Calculate the wavelength of these electrons. *(2 marks)*

**(c)** Calculate the speed of muons with the same wavelength as these electrons.
mass of muon = 207 × mass of electron *(3 marks)*

**(d)** Both electrons and muons were accelerated from rest by the same potential difference. Explain why they have different wavelengths. *(2 marks)*

AQA, 2003

# Section 1 Summary

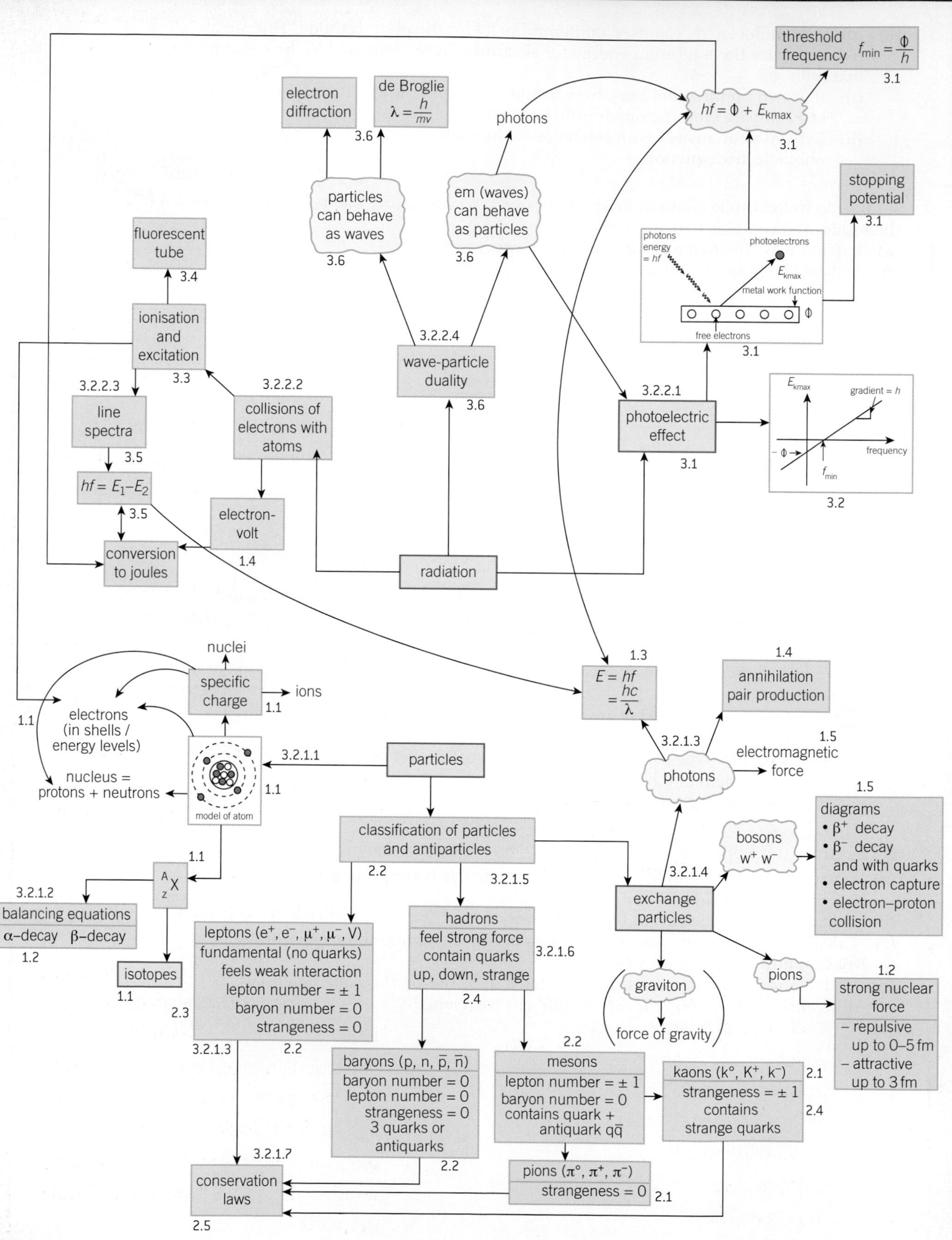

## Practical skills

In this section you have met the following skills:

- use of a cloud chamber or spark counter or Geiger counter to detect alpha particles to observe their range
- use of a microammeter to measure photoelectric current
- record the precision of a microammeter.

## Maths skills

In this section you have met the following skills:

- use of standard form and conversion to standard form from units with prefixes (e.g. wavelength in nm)
- use calculators to solve calculations involving powers of ten (e.g. photoelectric effect/line spectra)
- use of an appropriate number of significant figures in all answers to calculations
- use of appropriate units in all answers to calculations
- converting an answer where required from one unit to another (e.g. eV to Joules)
- change the subject of an equation in order to solve calculations (finding $\varphi$ from photoelectric equation)
- plotting a graph from data provided or found experimentally (e.g. $E_{kmax}$ against frequency)
- relating $y = mx + c$ to a linear graph with physics variables to find the gradient and intercept and understanding what the gradient and intercept represent in physics (e.g. $E_{kmax}$ against frequency graph).

## Extension task

Research one or more of the following topics using text books and the internet and produce a presentation suitable to show your class.

1 Particles are used in both PET and MRI scans in Medical Physics.
Include in your presentation
- the principles of physics involved in each type of scan.
- which medical conditions are suited to each type of scan? Give examples of the images produced and the resultant diagnoses.
- costs of purchasing and using each of the scanners.

2 Many particles were first discovered during research at CERN in Geneva and at other particle accelerators in America including SLAC.
Include in your presentation
- the principles of physics which enable the linear accelerator, cyclotron and synchrotron to accelerate particles and where the most famous accelerators are located around the world.
- limitations of each type of accelerator and the reasons for the use of each type in a particular situation.
- why was the original LEP accelerator at CERN dismantled and the LHC built instead?

# Practice questions: Particles and radiation

**1** **(a)** An ion of plutonium $^{239}_{94}$Pu has an overall charge of $+1.6 \times 10^{-19}$ C. For this ion state the number of
(i) protons
(ii) neutrons
(iii) electrons. *(3 marks)*

**(b)** Plutonium has several *isotopes*. Explain the meaning of the word isotopes. *(2 marks)*

AQA, 2006

**2** **(a)** The equation below represents the decay of the radioactive copper isotope $^{64}_{29}$Cu by the emission of a $\beta^+$ particle:

$$^{64}_{29}\text{Cu} \rightarrow \ ^{\dots\dots}_{\dots\dots}\text{Ni} + \ ^{0}_{\dots\dots}\beta + \text{X}$$

(i) Identify particle X. Copy and complete the equation. *(3 marks)*
(ii) State and explain which nucleus – the copper nucleus or the nickel nucleus – has the greater specific charge. *(2 marks)*
(iii) Calculate the specific charge of the $^{64}_{29}$Cu nucleus. *(3 marks)*

**(b)** (i) State the quark composition of a neutron. *(1 mark)*
(ii) With the aid of a labelled diagram, explain the change in terms of quarks and other fundamental particles that takes place when the copper nucleus $^{64}_{29}$Cu emits a $\beta^+$ particle. *(3 marks)*

**3** A physicist, who is attempting to analyse a nuclear event, suggests that a $\pi^-$ particle and a proton collided and were annihilated with the creation of a neutron, a $\pi^+$ particle, and a $K^-$ particle.
$\pi$ and K particles are mesons. The baryon and lepton numbers of both these mesons are zero.
**(a)** Write down the equation that represents this interaction. *(1 mark)*
**(b)** Show in terms of the conservation of charge, baryon number, and lepton number that this transformation is permitted. *(4 marks)*

AQA, 2002

**4** **(a)** A negative muon, $\mu^-$, is 207 times more massive than an electron. Calculate the de Broglie wavelength of a negative muon travelling at $3.0 \times 10^6$ m s$^{-1}$.

**(b)** Using values from the data sheet (page 262–265) calculate the ratio $\dfrac{\text{rest mass of } \pi^0}{\text{rest mass of } \mu^-}$ where $\pi^0$ is a neutral pion.

**(c)** Calculate the speed necessary for a $\pi^0$ to have the same de Broglie wavelength as that of the $\mu^-$ in **(a)**. *(6 marks)*

AQA, 2006

**5** **(a)** State what is meant by the *photoelectric effect* and explain how *two* observations made in photoelectric experiments suggest that electromagnetic radiation behaves like a stream of particles rather than a wave. *(6 marks)*

**(b)** Photons with energy $1.1 \times 10^{-18}$ J are incident on a metal surface. The maximum energy of electrons emitted from the surface is $4.8 \times 10^{-19}$ J.
(i) Calculate the work function of the metal.
(ii) Calculate the wavelength of the de Broglie wave associated with the emitted electrons of maximum energy. *(4 marks)*

AQA, 2006

6 The quantum theory suggests that the electron in a hydrogen atom can only exist in certain well-defined energy states. Some of these are shown in **Figure 1**.

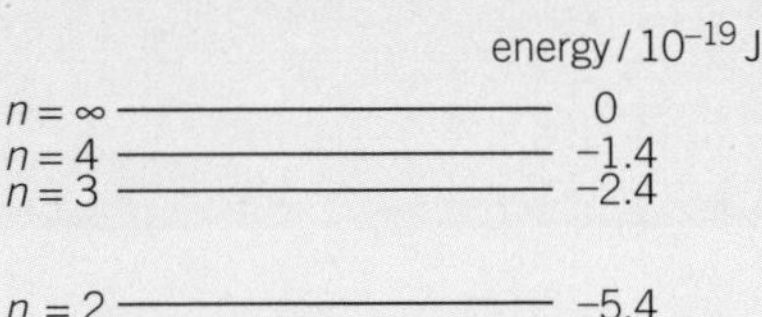

▲ **Figure 1**

An electron E of energy $2.5 \times 10^{-18}$ J collides with a hydrogen atom that is in its ground state and excites the electron in the hydrogen atom to the $n = 3$ level.
Calculate

(i) the energy that is needed to excite an electron in the hydrogen atom from the ground state to the $n = 3$ level,

(ii) the kinetic energy of the incident electron E after the collision,

(iii) the wavelength of the lowest energy photon that could be emitted as the excited electron returns to the ground state. *(5 marks)*

AQA, 2006

7 **Figure 2** shows data for the variation of the power output of a photovoltaic cell with load resistance. The data was obtained by placing the cell in sunlight. The intensity of the energy from the Sun incident on the surface of the cell was constant.

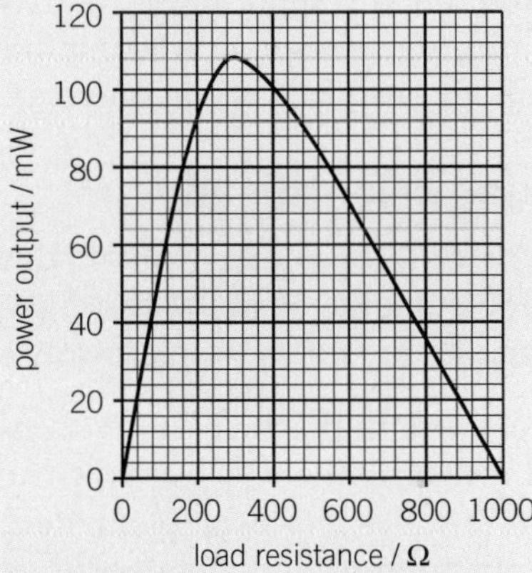


▲ **Figure 2**

**(a)** Use data from **Figure 2** to calculate the current in the load at the peak power. *(3 marks)*

**(b)** The intensity of the Sun's radiation incident on the cell is $730\,W\,m^{-2}$.
The active area of the cell has dimensions of 60 mm × 60 mm.

Calculate, at the peak power, the ratio $\dfrac{\text{electrical energy delivered by the cell}}{\text{energy arriving at the cell}}$. *(3 marks)*

**(c)** The average wavelength of the light incident on the cell is 500 nm. Estimate the number of photons incident on the active area of the cell every second. *(2 marks)*

**(d)** The measurements of the data in **Figure 2** were carried out when the rays from the Sun were incident at 90° to the surface of the panel. A householder wants to generate electrical energy using a number of solar panels to produce a particular power output. Discuss two pieces of information scientists could provide to inform the production of a suitable system. *(2 marks)*

AQA Specimen paper 2, 2015

# Section 2

## Waves and optics

**Chapters in this section:**


**4** Waves

**5** Optics


### Introduction

In this section, you will look at wave measurements, longitudinal waves and transverse waves, and discuss why polarisation is a property of transverse waves but not of longitudinal waves. You will also consider general wave properties – properties that apply to all waves – including **refraction**, **diffraction**, and **interference**. You will also find out how to create **stationary waves**. You will look at some well-known applications that use the properties of waves, including satellite TV dishes and the sounds created by the vibrating strings of musical instruments. In Chapter 5, you will examine the refraction of light, including total internal reflection, and applications such as the endoscope and fibre optic communications. Experiments that demonstrate the wave nature of light will be discussed to show how you can accurately measure the wavelength of different colours of light. By measuring the wavelength of light from stars and galaxies, physicists discovered that the distant galaxies are all moving away from each other. This led them to conclude that the universe is expanding. The study of waves and optics therefore helps us to understand many applications that we all use every day, as well as helping us to understand fundamental discoveries that have been made about our place in the universe.

### Working scientifically

In this part of the course, you will develop your knowledge and understanding of waves through practical work and related mathematics including problem-solving and calculations.

Practical work in this section involves making observations and careful measurements by using instruments such as protractors and millimetre scales. For example, when you observe light in polarisation or diffraction tests, you will need to consider how the intensity and/or the colour of the light changes when some part of the arrangement, such as the width of a slit, is changed. In some experiments, you will use instruments to make careful measurements, and you will be expected to assess the accuracy of your measurements and the results of your calculations that use your measured data. In some experiments, such as measuring the refractive index, you will be able to test the accuracy of your results by comparing them with accepted values. By estimating how accurate a result is (i.e., its uncertainty) you will know how close your result is to the accepted value. Extra care is needed where angles are to be measured, because a measurement error of 0.5° can be very significant if we calculate the sines, cosines, and tangents of angles. In these kinds of experiments, as well as in many other practical experiments, it will help you to be aware of the approximate values of physical quantities. For example, the refractive index of ordinary glass is about 1.5. So, if in your experiment you get a value of about 15 or 16, you will know straight away that you have made an error, either in your measurements or in your calculations.

You will develop your maths skills in this section by rearranging equations and carrying out calculations like trigonometry. In some experiments, such as investigating the vibrations of a stretched string, you will be expected to plot a quantity from your measurements to create a straight line graph as predicted by theory. You will then determine the graph gradient (and/or intercept) and relate the values you get to a physical quantity. The notes and exercises in Chapter 16, including the section on straight line graphs, will help you with these maths skills.

### What you already know

From your GCSE studies on waves, you should know that:

- ☐ sound is a longitudinal wave, and all electromagnetic waves are transverse waves
- ☐ a complete wave is a wave from one peak to the next peak
- ☐ wavelength is the distance between adjacent wave peaks
- ☐ frequency is the number of complete waves passing a point per second
- ☐ the speed of a wave = its wavelength × its frequency
- ☐ all waves undergo reflection, refraction, and diffraction
- ☐ the spectrum of electromagnetic waves, in order of increasing frequency, is:
    - radio waves
    - microwaves
    - infrared radiation
    - visible light
    - ultraviolet radiation
    - X-rays
    - gamma radiation.

# 4 Waves

## 4.1 Waves and vibrations

**Learning objectives:**

- → Explain the differences between transverse and longitudinal waves.
- → Define a plane-polarised wave.
- → Describe a physics test that can distinguish transverse waves from longitudinal waves.

*Specification reference: 3.3.1.1; 3.3.1.2*

### Types of waves

Waves that pass through a substance are vibrations which pass through that substance. For example, sound waves in air are created by making a surface vibrate so it sends compression waves through the surrounding air. Sound waves, seismic waves, and waves on strings are examples of waves that pass through a substance. These types of waves are often referred to as **mechanical waves**. When waves progress through a substance, the particles of the substance vibrate in a certain way which makes nearby particles vibrate in the same way, and so on.

**Electromagnetic waves** are oscillating electric and magnetic fields that progress through space without the need for a substance. The vibrating electric field generates a vibrating magnetic field, which generates a vibrating electric field further away, and so on. Electromagnetic waves include radio waves, microwaves, infrared radiation, light, ultraviolet radiation, X-rays, and gamma radiation.

**Synoptic link**

You have met the full spectrum of electromagnetic waves in more detail in Topic 1.3, Photons.

### Longitudinal and transverse waves

**Longitudinal waves** are waves in which the direction of vibration of the particles is **parallel** to (along) the direction in which the wave travels. Sound waves, primary seismic waves and compression waves on a slinky toy are all longitudinal waves. Figure 2 shows how to send longitudinal waves along a slinky. When one end of the slinky is moved to and fro repeatedly, each 'forward' movement causes a compression wave to pass along the slinky as the coils push into each other. Each 'reverse' movement causes the coils to move apart so rarefaction (expansion) wave passes along the slinky.

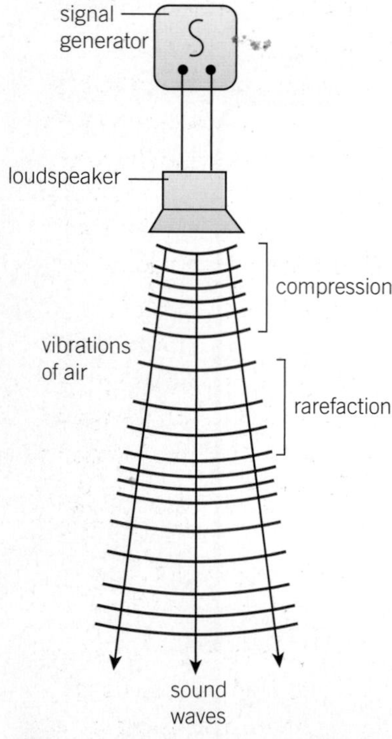


▲ **Figure 1** *Creating sound waves in air*

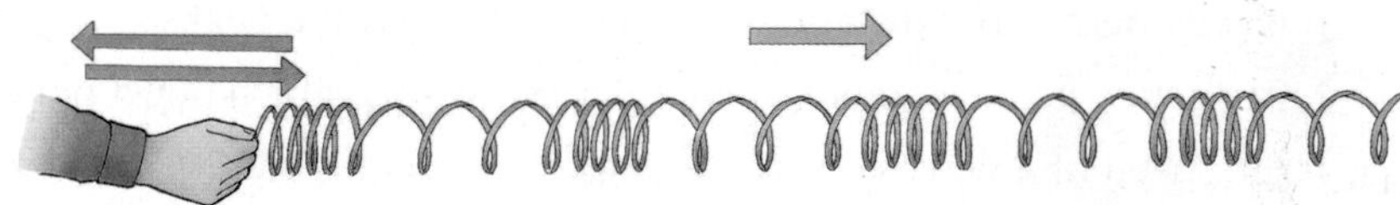

▲ **Figure 2** *Longitudinal waves on a slinky*

**Transverse waves** are waves in which the direction of vibration is **perpendicular** to the direction in which the wave travels. Electromagnetic waves, secondary seismic waves, and waves on a string or a wire are all transverse waves.

Figure 3 shows transverse waves travelling along a rope. When one end of the rope is moved from side to side repeatedly, these sideways movements travel along the rope, as each unaffected part of the rope is pulled sideways when the part next to it moves sideways.

**Study tip**

Vibrations of the particles in a longitudinal wave are in the same direction as that along which the wave travels.

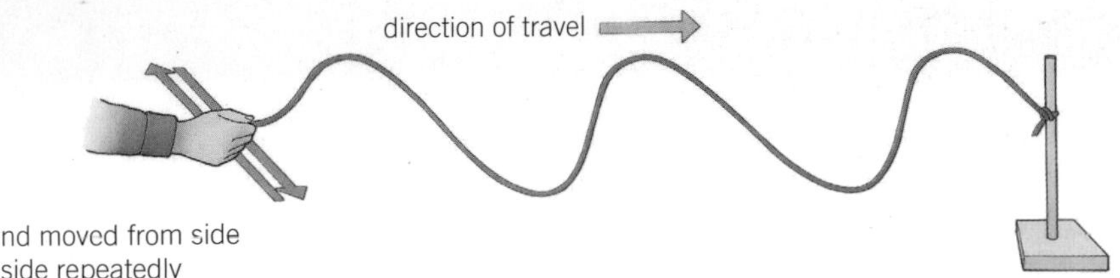


▲ **Figure 3** *Making rope waves*

## Polarisation

Transverse waves are **plane-polarised** if the vibrations stay in one plane only. If the vibrations change from one plane to another, the waves are **unpolarised**. Longitudinal waves cannot be polarised.

Figure 4 shows unpolarised waves travelling on a rope. When they pass through a slit in a board, as in Figure 4, they become polarised because only the vibrations parallel to the slit pass through it.

Light from a filament lamp or a candle is unpolarised. If unpolarised light is passed through a polaroid filter, the transmitted light is polarised as the filter only allows through light which vibrates in a certain direction, according to the alignment of its molecules.

If unpolarised light is passed through two polaroid filters, the transmitted light intensity changes if one polaroid is turned relative to the other one. The filters are said to be crossed when the transmitted intensity is a minimum. At this position, the polarised light from the first filter cannot pass through the second filter, as the alignment of molecules in the second filter is at 90° to the alignment in the first filter. This is like passing rope waves through two letter boxes at right angles to each other, as shown in Figure 4.

Light is part of the spectrum of electromagnetic waves. The plane of polarisation of an electromagnetic wave is defined as the plane in which the electric field oscillates.

Polaroid sunglasses reduce the glare of light reflected by water or glass. The reflected light is polarised and the intensity is reduced when it passes through the polaroid sunglasses.

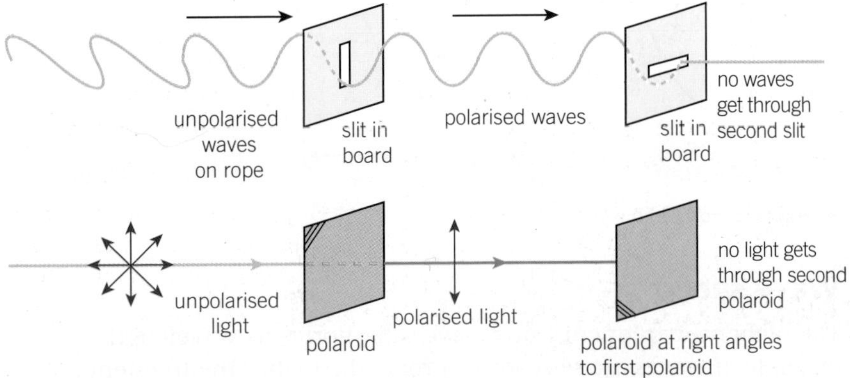


▲ **Figure 4** *Rope model and diagram to explain polarisation*

### Good reception

Radio waves from a transmitter are polarised. The aerial of a radio receiver needs to be aligned in the same plane as the radio waves to obtain the best reception.

## Summary questions

1. Classify the following types of waves as either longitudinal or transverse: **a** radio waves, **b** microwaves, **c** sound waves, **d** secondary seismic waves.
2. Sketch a snapshot of a longitudinal wave travelling on a slinky coil, indicating the direction in which the wave is travelling and areas of high density (compression) and low density (rarefaction).
3. Sketch a snapshot of a transverse wave travelling along a rope, indicating the direction in which the wave is travelling and the direction of motion of the particles at the points of zero displacement.
4. **a** What is meant by a polarised wave?

   **b** A light source is observed through two pieces of polaroid which are initially aligned parallel to each other. Describe and explain what you would expect to observe when one of the polaroids is rotated through 360°.

# 4.2 Measuring waves

## Learning objectives:

- → Explain what is meant by the amplitude of a wave.
- → Explain what is meant by the wavelength of a wave.
- → Calculate the frequency of a wave from its period.

*Specification reference: 3.3.1.1*

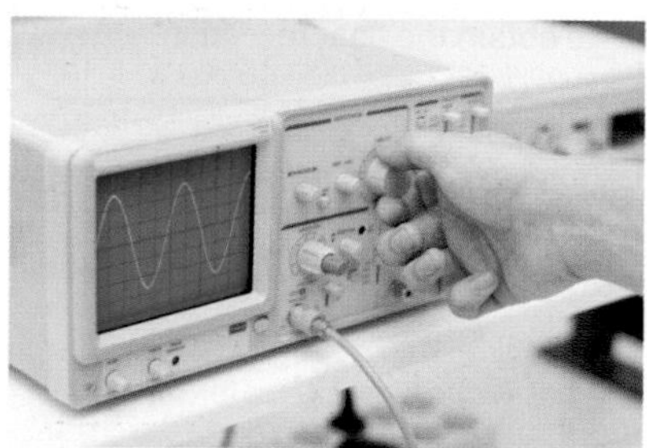

▲ **Figure 1** *Using a cathode ray oscilloscope (CRO) to give a voltage–time graph of a wave*

When we make a phone call, sound waves are converted to electrical waves. In an intercontinental phone call, these waves are carried by electromagnetic waves from ground transmitters to satellites in space and back to ground receivers, where they are converted back to electrical waves then back to sound waves. The electronic circuits ensure that these sound waves are very similar to the original sound waves. The engineers who design and maintain communications systems need to measure the different types of waves at different stages, to make sure the waves are not distorted.

## Key terms

The following terms, some of which are illustrated in Figure 2, are used to describe waves.

- The **displacement** of a vibrating particle is its distance and direction from its equilibrium position.
- The **amplitude** of a wave is the maximum displacement of a vibrating particle. For a transverse wave, this is the height of a wave crest or the depth of a wave trough from its equilibrium position.
- The **wavelength** of a wave is the least distance between two adjacent vibrating particles with the same displacement and velocity at the same time (e.g., distance between adjacent crests).
- One complete **cycle** of a wave is from maximum displacement to the next maximum displacement (e.g., from one wave peak to the next).
- The **period** of a wave is the time for one complete wave to pass a fixed point.
- The **frequency** of a wave is the number of cycles of vibration of a particle per second, or the number of complete waves passing a point per second. The unit of frequency is the hertz (Hz).

**For waves of frequency $f$, the period of the wave = $\frac{1}{f}$**

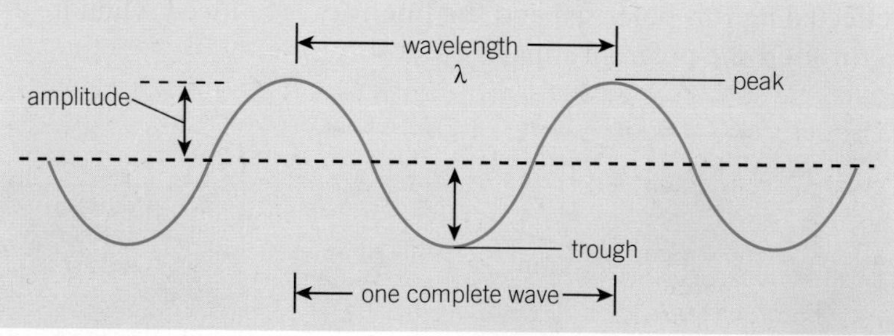


▲ **Figure 2** *Parts of a wave*

## Study tip

Amplitude is measured from the equilibrium position to maximum positive or maximum negative, not from maximum positive to maximum negative. Phase difference is measured as an angle (in radians or degrees), not in terms of wavelength.

## Wave speed

The higher the frequency of a wave, the shorter its wavelength. For example, if waves are sent along a rope, the higher the frequency at which they are produced, the closer together the wave peaks are. The same effect can be seen in a ripple tank when straight waves are produced at a constant frequency. If the frequency is raised to a higher value, the waves are closer together.

Figure 3 represents the crests of straight waves travelling at a constant speed in a ripple tank.

- Each wave crest travels a distance equal to one wavelength ($\lambda$) in the time taken for one cycle.
- The time taken for one cycle = $1/f$, where $f$ is the frequency of the waves.

Therefore, the speed of the waves, $c = \dfrac{\text{distance travelled in one cycle}}{\text{time taken for one cycle}}$

$$= \frac{\lambda}{1/f} = f\lambda$$

For waves of frequency $f$ and wavelength $\lambda$

**wave speed $c = f\lambda$**

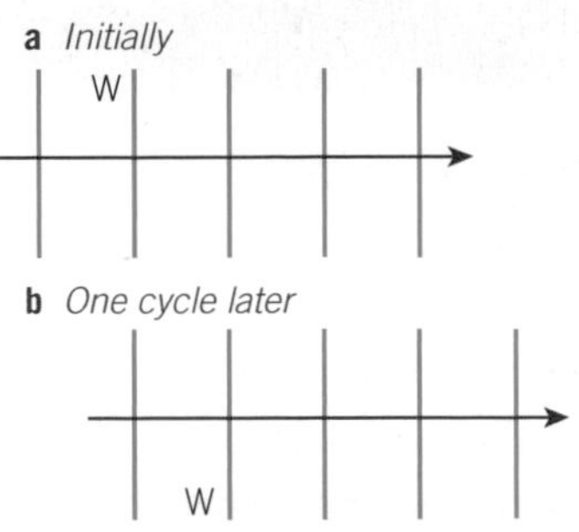


▲ **Figure 3** *Wave speed*

## Study tip

The symbol $c$ is used as a general symbol for the speed of waves. In the data booklet, $c$ refers to the speed of electromagnetic waves in free space.

## Phase difference

The **phase** of a vibrating particle at a certain time is the fraction of a cycle it has completed since the start of the cycle. The **phase difference** between two particles vibrating at the same frequency is the fraction of a cycle between the vibrations of the two particles, measured either in degrees or radians, where 1 cycle = 360° = 2π **radians**. For two points at distance $d$ apart along a wave of wavelength $\lambda$

$$\textbf{the phase difference in radians} = \frac{2\pi d}{\lambda}$$

Figure 4 shows three successive snapshots of the particles of a transverse wave progressing from left to right across the diagram. Particles O, P, Q, R, and S are spaced approximately $\frac{1}{4}$ of a wavelength apart. Table 1 shows the phase difference between O and each of the other particles.

▼ **Table 1** *Phase differences*

| | P | Q | R | S |
|---|---|---|---|---|
| distance from O in terms of wavelength, $\lambda$ | $\frac{1}{4}\lambda$ | $\frac{1}{2}\lambda$ | $\frac{3}{4}\lambda$ | $\lambda$ |
| phase difference relative to O / radians | $\frac{1}{2}\pi$ | $\pi$ | $\frac{3}{2}\pi$ | $2\pi$ |

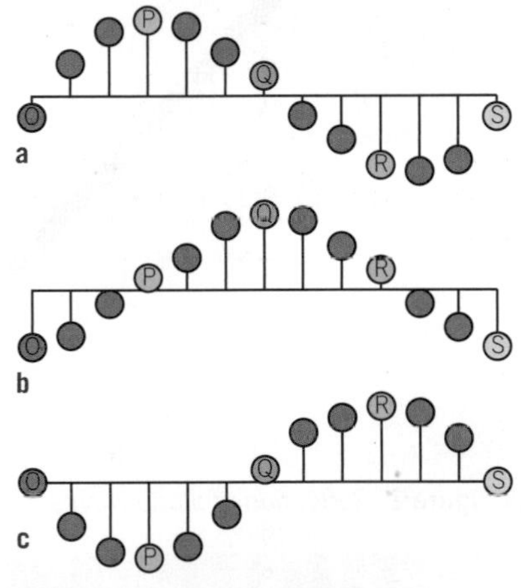


▲ **Figure 4** *Progressive waves*

## Summary questions

1. Sound waves in air travel at a speed of 340 m s$^{-1}$ at 20 °C. Calculate the wavelength of sound waves in air which have a frequency of **a** 3400 Hz **b** 18 000 Hz.
2. Electromagnetic waves in air travel at a speed of $3.0 \times 10^8$ m s$^{-1}$. Calculate the frequency of light waves of wavelength **a** 0.030 m **b** 600 nm.
3. Figure 5 shows a waveform on an oscilloscope screen when the $y$-sensitivity of the oscilloscope was 0.50 V cm$^{-1}$ and the time base was set at 0.5 ms cm$^{-1}$. Determine the amplitude and the frequency of this waveform.
4. **a** For the waves in Figure 4, measure
   **i** the amplitude and the wavelength
   **ii** the phase difference between P and R
   **iii** the phase difference between P and S.
   **b** What would be the displacement and direction of motion of Q three-quarters of a period after the last snapshot?

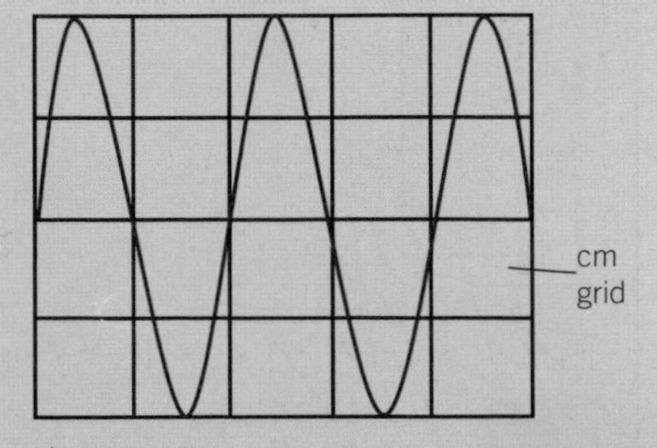


▲ **Figure 5**

# 4.3 Wave properties 1

## Learning objectives:

- → Explain what causes waves to refract when they pass across a boundary.
- → Demonstrate the direction light waves bend when they travel out of glass and into air.
- → Explain what we mean by diffraction.

*Specification reference: 3.3.1.1; 3.3.1.2*

Wave properties such as reflection, refraction, and diffraction occur with many different types of waves. A **ripple tank** may be used to study these wave properties. The tank is a shallow transparent tray of water with sloping sides. The slopes prevent waves reflecting off the sides of tank. If they did reflect, it would be difficult to see the waves.

- The waves observed in a ripple tank are referred to as **wavefronts**, which are lines of constant phase (e.g., crests).
- The direction in which a wave travels is at right angles to the wavefront.

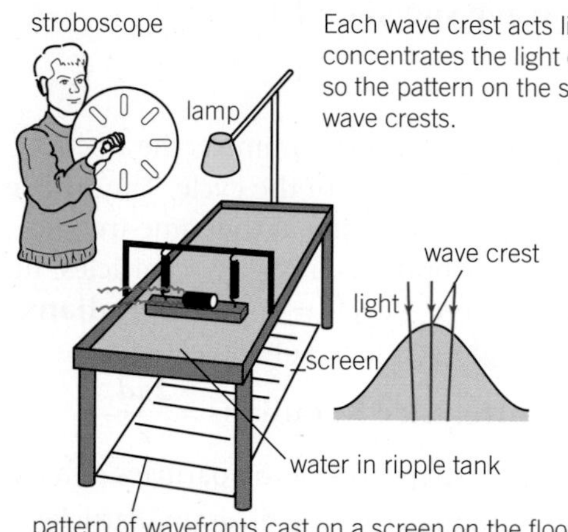


▲ **Figure 1** *The ripple tank*

▲ **Figure 2** *Reflection of plane waves*

## Reflection

Straight waves directed at a certain angle to a hard flat surface (the reflector) reflect off at the same angle, as shown in Figure 2. The angle between the reflected wavefront and the surface is the same as the angle between the incident wavefront and the surface. Therefore the direction of the reflected wave is at the same angle to the reflector as the direction of the incident wave. This effect is observed when a light ray is directed at a plane mirror. The angle between the incident ray and the mirror is equal to the angle between the reflected ray and the mirror.

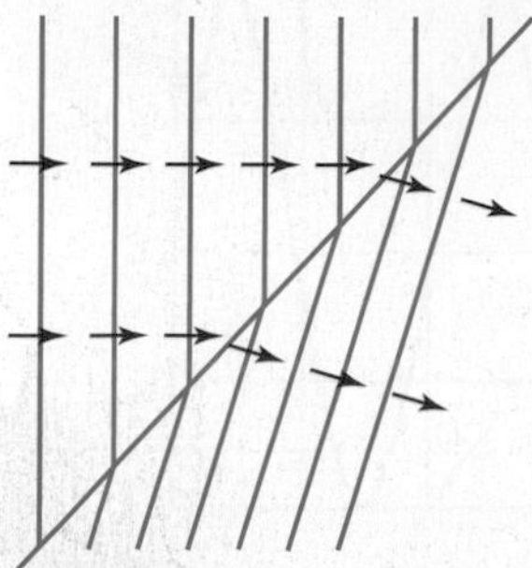

▲ **Figure 3** *Refraction*

## Refraction

When waves pass across a boundary at which the wave speed changes, the wavelength also changes. If the wavefronts approach at an angle to the boundary, they change direction as well as changing speed. This effect is known as **refraction**.

Figure 3 shows the refraction of water waves in a ripple tank when they pass across a boundary from deep to shallow water at an angle to the boundary. Because they move more slowly in the shallow water, the wavelength is smaller in the shallow water and therefore they change direction.

Refraction of light is observed when a light ray is directed into a glass block at an angle (i.e., not along the normal). The light ray changes direction when it crosses the glass boundary. This happens because light waves travel more slowly in glass than in air.

## Diffraction

Diffraction occurs when waves spread out after passing through a gap or round an obstacle. The effect can be seen in a ripple tank when straight waves are directed at a gap, as shown in Figure 4.

- The narrower the gap, the more the waves spread out.
- The longer the wavelength, the more the waves spread out.

To explain why the waves are diffracted on passing through the gap, consider each point on a wavefront as a secondary emitter of wavelets. The wavelets from the points along a wavefront travel only in the direction in which the wave is travelling, not in the reverse direction, and they combine to form a new wavefront spreading beyond the gap.

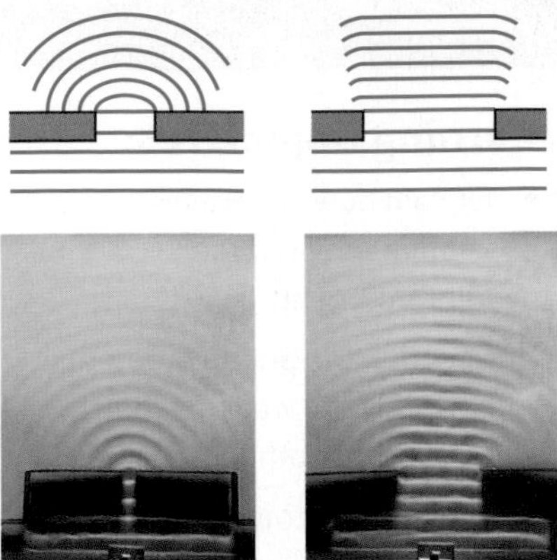

▲ **Figure 4** *The effect of the gap width*

### Dish design

Satellite TV dishes in Europe need to point south, because the satellites orbit the Earth directly above the equator. The bigger the dish, the stronger the signal it can receive, because more radio waves are reflected by the dish onto the aerial. But a bigger dish reflects the radio waves to a smaller focus, because it diffracts the waves less. The dish therefore needs to be aligned more carefully than a smaller dish, otherwise it will not focus the radio waves onto the aerial.

**Synoptic link**

You will meet refraction and reflection of light in more detail in Topic 5.1, refraction of light, and Topic 5.2, More about refraction.

**Study tip**

Remember from your GCSE that the normal is an imaginary line perpendicular to a boundary between two materials or a surface.

**Synoptic link**

You will meet diffraction in more detail in Topic 5.6, Diffraction.

## Summary questions

1. Copy and complete the diagram in Figure 6 by showing the wavefront after it has reflected from the straight reflector. Also show the direction of the reflected wavefront.

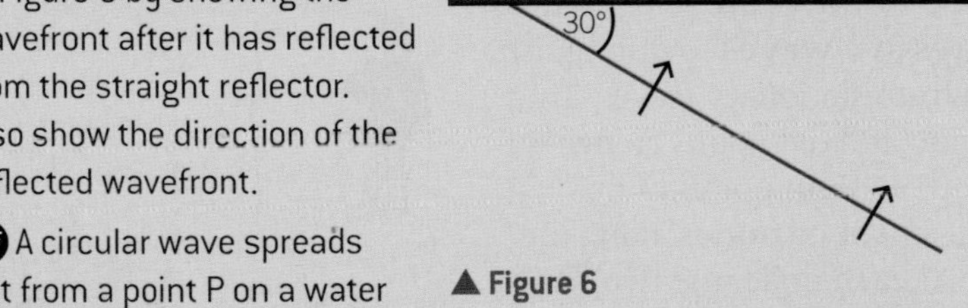


▲ **Figure 6**

2. A circular wave spreads out from a point P on a water surface, which is 0.50 m from a flat reflecting wall. The wave travels at a speed of $0.20\ \text{m s}^{-1}$. Sketch the arrangement and show the position of the wavefront **a** 2.5 s, **b** 4.0 s after the wavefront was produced at P.

3. Copy and complete the diagram in Figure 7 by showing the wavefronts after they passed across the boundary and have been refracted. Also show the direction of the refracted waves.

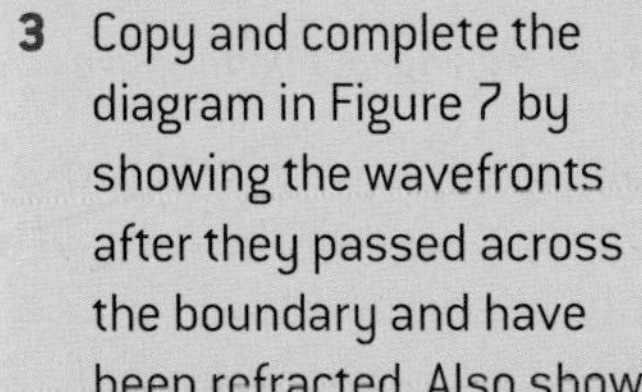

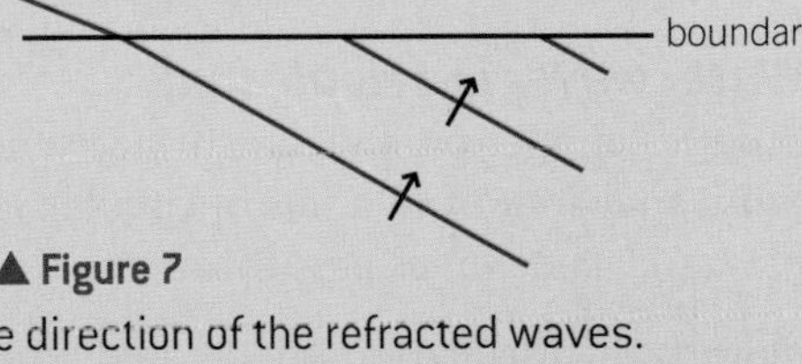


▲ **Figure 7**

4. Water waves are diffracted on passing through a gap. How is the amount of diffraction changed as a result of:
   - **a** widening the gap without changing the wavelength
   - **b** increasing the wavelength of the water waves without changing the gap width
   - **c** increasing the wavelength of the water waves and reducing the gap width
   - **d** widening the gap and increasing the wavelength of the waves?

▲ **Figure 5** *A satellite TV dish*

# 4.4 Wave properties 2

## Learning objectives:

- → Explain how two waves combine to produce reinforcement.
- → Describe the phase difference between two waves if they cancel each other out.
- → Explain why total cancellation is rarely achieved in practice.

*Specification reference: 3.3.1.1; 3.3.1.2*

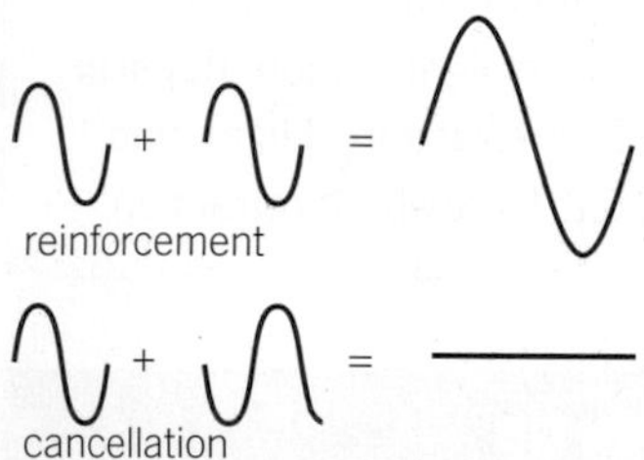


▲ **Figure 1** *Superposition*

## The principle of superposition

When waves meet, they pass through each other. At the point where they meet, they combine for an instant before they move apart. This combining effect is known as **superposition**. Imagine a boat hit by two wave crests at the same time from different directions. Anyone on the boat would know it had been hit by a supercrest, the combined effect of two wave crests.

**The principle of superposition states that when two waves meet, the total displacement at a point is equal to the sum of the individual displacements at that point.**

- Where a crest meets a crest, a **supercrest** is created – the two waves reinforce each other.
- Where a trough meets a trough, a **supertrough** is created – the two waves reinforce each other.
- Where a crest meets a trough of the same amplitude, the resultant displacement is **zero**; the two waves cancel each other out. If they are not the same amplitude, the resultant is called a minimum.

### Further examples of superposition

#### *1 Stationary waves on a rope*

**Stationary waves** are formed on a rope if two people send waves continuously along a rope from either end, as shown in Figure 2. The two sets of waves are referred to as **progressive waves** to distinguish them from stationary waves. They combine at fixed points along the rope to form points of no displacement or **nodes** along the rope. At each node, the two sets of waves are always 180° out of phase, so they cancel each other out. Stationary waves are described in more detail in Topics 4.5 and 4.6.

▲ **Figure 2** *Making stationary waves*

#### *2 Water waves in a ripple tank*

A vibrating dipper on a water surface sends out circular waves. Figure 3 shows a snapshot of two sets of circular waves produced in this way in a ripple tank. The waves pass through each other continuously.

- Points of cancellation are created where a crest from one dipper meets a trough from the other dipper. These points of cancellation are seen as gaps in the wavefronts.
- Points of reinforcement are created where a crest from one dipper meets a crest from the other dipper, or where a trough from one dipper meets a trough from the other dipper.

As the waves are continuously passing through each other at constant frequency and at a constant phase difference, cancellation and reinforcement occurs at fixed positions. This effect is known as **interference**. **Coherent** sources of waves produce an interference pattern where they overlap, because they vibrate at the same frequency

### Study tip

Be careful when using the term 'minimum'. When two troughs meet at the same point, they superpose to produce maximum negative displacement, not a minimum.

### Synoptic link

You will meet the interference of light in more detail in Topic 5.4, Double slit interference.

with a constant phase difference. If the phase difference changed at random, the points of cancellation and reinforcement would move about at random, and no interference pattern would be seen.

▲ **Figure 3** *Interference of water waves*

## Tests using microwaves

A microwave transmitter and receiver can be used to demonstrate reflection, refraction, diffraction, interference, and polarisation of microwaves. The transmitter produces microwaves of wavelength 3.0 cm. The receiver can be connected to a suitable meter, which gives a measure of the intensity of the microwaves at the receiver.

1 Place the receiver in the path of the microwave beam from the transmitter. Move the receiver gradually away from the transmitter and note that the receiver signal decreases with distance from the transmitter. This shows that the microwaves become weaker as they travel away from the transmitter.
2 Place a metal plate between the transmitter and the receiver to show that microwaves cannot pass through metal.
3 Use two metal plates to make a narrow slit and show that the receiver detects microwaves that have been diffracted as they pass through the slit. Show that if the slit is made wider, less diffraction occurs.
4 Use a narrow metal plate with the two plates from step 3 above to make a pair of slits as in Figure 4. Direct the transmitter at the slits and use the receiver to find points of cancellation and reinforcement, where the microwaves from the two slits overlap.

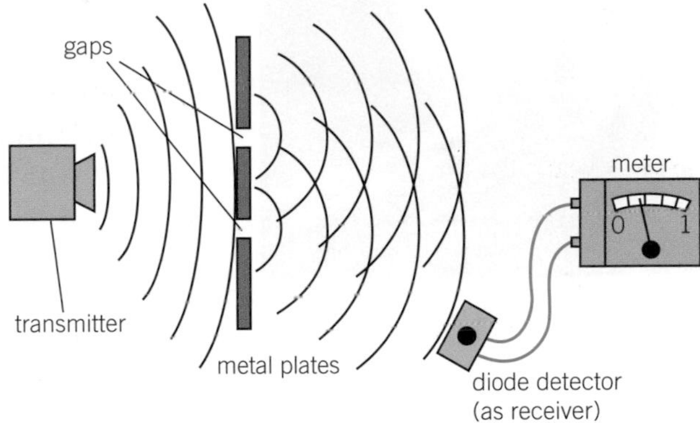


▲ **Figure 4** *Interference of microwaves*

## Summary questions

1 Figure 5 shows two wave pulses on a rope travelling towards each other. Sketch a snapshot of the rope

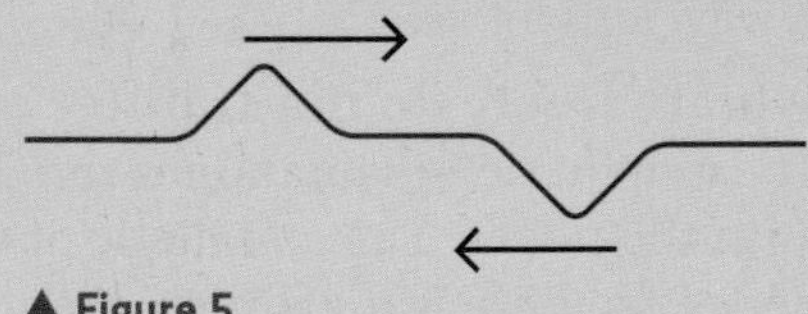

▲ **Figure 5**

**a** when the two waves are passing through each other

**b** when the two waves have passed through each other.

2 How would you expect the interference pattern in Figure 3 to change if

**a** the two dippers are moved further apart

**b** the frequency of the waves produced by the dippers is reduced?

3 Microwaves from a transmitter are directed at a narrow slit between two metal plates. A receiver is placed in the path of the diffracted microwaves, as shown in Figure 6.

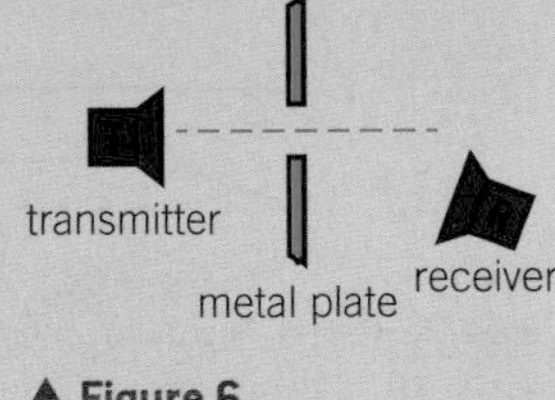


▲ **Figure 6**

How would you expect the receiver signal to change if **a** the receiver is moved in a line parallel to the metal plates, **b** the slit is then made narrower?

4 Microwaves from a transmitter are directed at two parallel slits in a metal plate (Figure 4). A receiver is placed on the other side of the metal plate on a line parallel to the plate. When the receiver is moved a short distance, the receiver signal decreases then increases again.

**a** Explain why the signal decreased when it was first moved.

**b** Explain why the signal increased as it continued to move.

# 4.5 Stationary and progressive waves

**Learning objectives:**

- → Describe the necessary condition for the formation of a stationary wave.
- → Deduce whether a stationary wave is formed by superposition.
- → Explain why nodes are formed in fixed positions.

*Specification reference: 3.3.1.3*

▲ **Figure 1** *Vibrations of a guitar string*

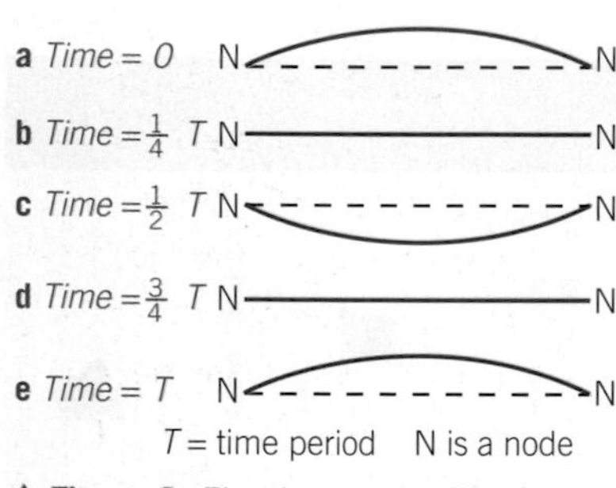


▲ **Figure 2** *First harmonic vibrations*

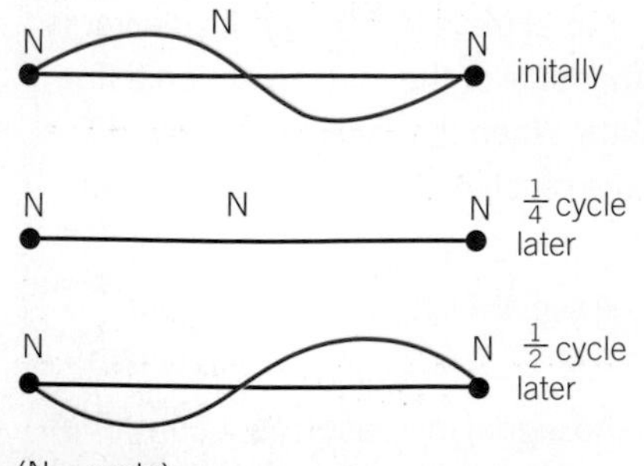


▲ **Figure 3** *A stationary wave of two loops*

## Formation of stationary waves

When a guitar string is plucked, the sound produced depends on the way in which the string vibrates. If the string is plucked gently at its centre, a stationary wave of constant frequency is set up on the string. The sound produced therefore has a constant frequency. If the guitar string is plucked harshly, the string vibrates in a more complicated way and the note produced contains other frequencies, as well as the frequency produced when it is plucked gently.

As explained in Topic 4.4, a stationary wave is formed when two progressive waves pass through each other. This can be achieved on a string in tension by fixing both ends and making the middle part vibrate, so progressive waves travel towards each end, reflect at the ends, and then pass through each other.

The simplest stationary wave pattern on a string is shown in Figure 2. This is called the first harmonic of the string (sometimes referred to as its **fundamental mode of vibration**). It consists of a single loop that has a **node** (a point of no displacement) at either end. The string vibrates with maximum amplitude midway between the nodes. This position is referred to as an **antinode**. In effect, the string vibrates from side to side repeatedly. For this pattern to occur, the distance between the nodes at either end (i.e., the length of the string) must be equal to one half-wavelength of the waves on the string.

$$\textbf{Distance between adjacent nodes} = \frac{1}{2}\lambda$$

If the frequency of the waves sent along the rope from either end is raised steadily, the pattern in Figure 2 disappears and a new pattern is observed with two equal loops along the rope. This pattern, shown in Figure 3, has a node at the centre as well as at either end. It is formed when the frequency is twice as high as in Figure 2, corresponding to half the previous wavelength. Because the distance from one node to the next is equal to half a wavelength, the length of the rope is therefore equal to one full wavelength.

**Stationary waves that vibrate freely do not transfer energy to their surroundings.** The amplitude of vibration is zero at the nodes so there is no energy at the nodes. The amplitude of vibration is a maximum at the antinodes, so there is maximum energy at the antinodes. Because the nodes and antinodes are at fixed positions, no energy is transferred in a freely vibrating stationary wave pattern.

## Explanation of stationary waves

Consider a snapshot of two progressive waves passing through each other.

- When they are in phase, they reinforce each other to produce a large wave, as shown in Figure 4a.
- A quarter of a cycle later, the two waves have each moved one-quarter of a wavelength in opposite directions. As shown in Figure 4b, they are now in antiphase so they cancel each other.
- After a further quarter cycle, the two waves are back in phase. The resultant is again a large wave as in Figure 4a, except reversed.

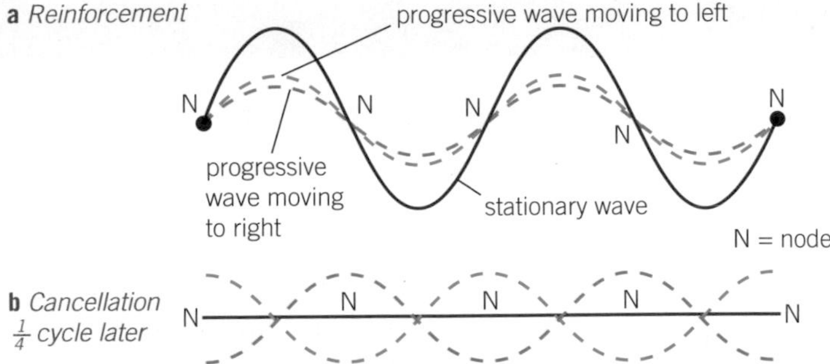


▲ **Figure 4** *Explaining stationary waves*

The points where there is no displacement (i.e., the nodes) are fixed in position throughout. Between these points, the stationary wave oscillates between the nodes. In general, in any stationary wave pattern:

1 The amplitude of a vibrating particle in a stationary wave pattern varies with position from zero at a node to maximum amplitude at an antinode.
2 The phase difference between two vibrating particles is
   - zero if the two particles are between adjacent nodes or separated by an even number of nodes
   - 180° (= π radians) if the two particles are separated by an odd number of nodes.

| | Stationary waves | Progressive waves |
|---|---|---|
| frequency | all particles except those at the nodes vibrate at the same frequency | all particles vibrate at the same frequency |
| amplitude | the amplitude varies from zero at the nodes to a maximum at the antinodes | the amplitude is the same for all particles |
| phase difference between two particles | equal to $m\pi$, where $m$ is the number of nodes between the two particles | equal to $2\pi d/\lambda$, where $d$ = distance apart and $\lambda$ is the wavelength |

## More examples of stationary waves

### 1 Sound in a pipe

Sound resonates at certain frequencies in an air-filled tube or pipe. In a pipe closed at one end, these resonant frequencies occur when there is an antinode at the open end and a node at the other end.

### 2 Using microwaves

Microwaves from a transmitter are directed normally at a metal plate, which reflects the microwaves back towards the transmitter. When a detector is moved along the line between the transmitter and the metal plate, the detector signal is found to be zero (or at a minimum) at equally spaced positions along the line. The reflected waves and the waves from the transmitter form a stationary wave pattern. The positions where no signal (or a minimum) is detected are where nodes occur. They are spaced at intervals of half a wavelength.

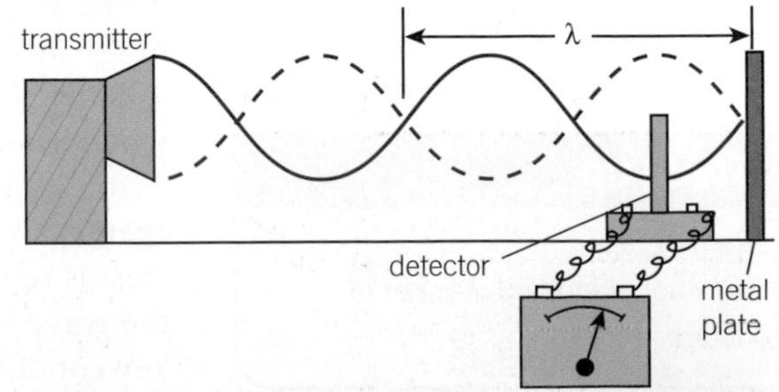


▲ **Figure 6** *Using microwaves*

## Summary questions

**1 a** Sketch the stationary wave pattern seen on a rope when there is a node at either end and an antinode at its centre.

**b** If the rope in **a** is 4.0 m in length, calculate the wavelength of the waves on the rope.

**2** The stationary wave pattern shown in Figure 5 is set up on a rope of length 3.0 m.

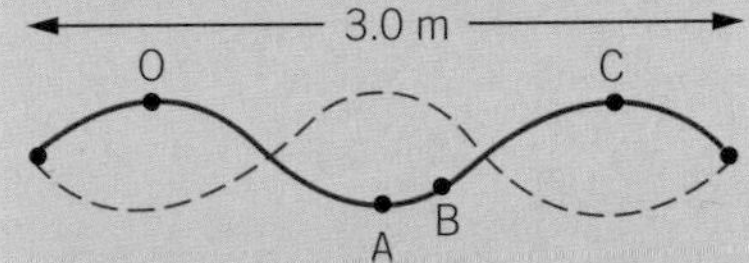


▲ **Figure 5**

**a** Calculate the wavelength of these waves.

**b** State the phase difference between the particle vibrating at O and the particle vibrating at
**i** A **ii** B **iii** C.

**3** State two differences between a stationary wave and a progressive wave in terms of the vibrations of the particles.

**4** The detector in Figure 6 is moved along the line between the transmitter and the metal plate. The detector signal is zero at positions 15 mm apart.

**a** Explain why the signal is at a minimum at certain positions.

**b** Calculate the wavelength of the microwaves.

# 4.6 More about stationary waves on strings

## Learning objectives:

- → Explain what condition must be satisfied at both ends of the string.
- → Describe the simplest possible stationary wave pattern that can be formed.
- → Compare the frequencies of higher harmonics with the first harmonic frequency.

*Specification reference: 3.3.1.3*

## Stationary waves on a vibrating string

A controlled arrangement for producing stationary waves is shown in Figure 1. A string or wire is tied at one end to a mechanical vibrator connected to a frequency generator. The other end of the string passes over a pulley and supports a weight, which keeps the tension in the string constant. As the frequency of the generator is increased from a very low value, different stationary wave patterns are seen on the string. In every case, the length of string between the pulley and the vibrator has a node at either end.

- The **first harmonic pattern of vibration** is seen at the lowest possible frequency that gives a pattern. This has an antinode at the middle as well as a node at either end. Because the length $L$ of the vibrating section of the string is between adjacent nodes and the distance between adjacent nodes is $\frac{1}{2}\lambda_1$

  the wavelength of the waves that form this pattern, the first harmonic wavelength, $\boldsymbol{\lambda_1 = 2L}$.

  Therefore, the first harmonic frequency $f = \frac{c}{\lambda_1} = \frac{c}{2L}$, where $c$ is the speed of the progressive waves on the wire.

- The next stationary wave pattern, the **second harmonic**, is where there is a node at the middle, so the string is in two loops. The wavelength of the waves that form this pattern $\lambda_2 = L$ because each loop has a length of half a wavelength.

  Therefore, the frequency of the second harmonic vibrations

$$f_2 = \frac{c}{\lambda_2} = \frac{c}{L} = 2f_1.$$

- The next stationary wave pattern, the **third harmonic**, is where there are nodes at a distance of $\frac{1}{3}L$ from either end and an antinode at the middle. The wavelength of the waves that form this pattern $\lambda_3 = \frac{2}{3}L$ because each loop has a length of half a wavelength.

  Therefore, the frequency of the third harmonic vibrations

$$f_3 = \frac{c}{\lambda_3} = \frac{3c}{2L} = 3f_1.$$

In general, stationary wave patterns occur at frequencies $f_1$, $2f_1$, $3f_1$, $4f_1$, and so on, where $f$ is the first harmonic frequency of the fundamental vibrations. This is the case in any vibrating linear system that has a node at either end.

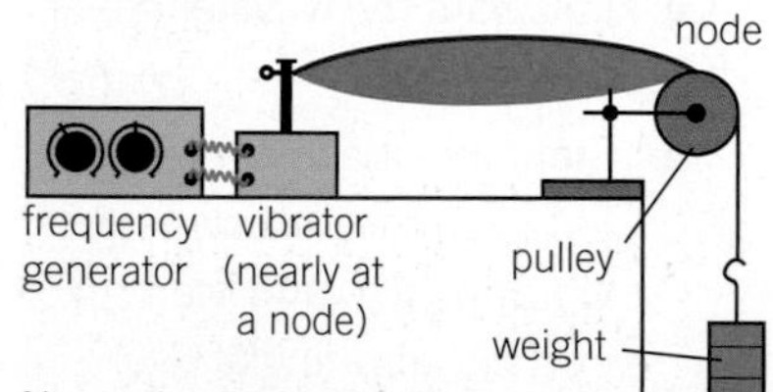


N = node A = antinode
(dotted line shows string half a cycle earlier)

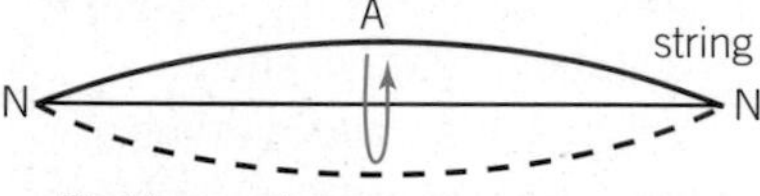


**a** *First harmonic*

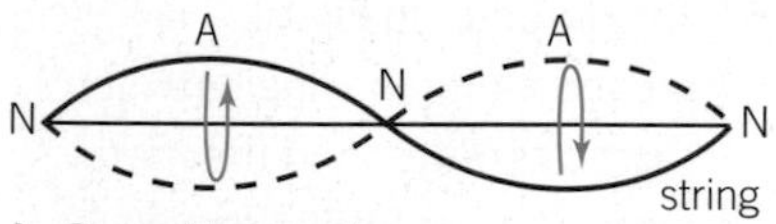


**b** *Second harmonic*

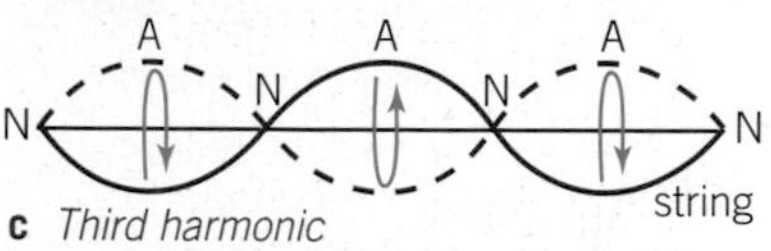


**c** *Third harmonic*

▲ **Figure 1** *Stationary waves on a string*

## Explanation of the stationary wave patterns on a vibrating string

In the arrangement shown in Figure 1, consider what happens to a progressive wave sent out by the vibrator. The crest reverses its phase when it reflects at the fixed end and travels back along the string as a trough. When it reaches the vibrator, it reflects and reverses phase again, travelling away from the vibrator once more as a crest. If this crest is reinforced by a crest created by the vibrator, the amplitude of the wave is increased. This is how a stationary wave is formed. The key condition is that the time taken for a wave to travel along the

### Study tip

Remember
Number of loops = number of nodes – 1.

string and back should be equal to the time taken for a whole number of cycles of the vibrator.

- The time taken for a wave to travel along the string and back is $t = 2L/c$, where $c$ is the speed of the waves on the string.
- The time taken for the vibrator to pass through a whole number of cycles $= m/f$, where $f$ is the frequency of the vibrator and $m$ is a whole number.

Therefore the key condition may be expressed as $\frac{2L}{c} = \frac{m}{f}$.

Rearranging this equation gives $f = \frac{mc}{2L} = mf$ and $\lambda = \frac{c}{f} = \frac{2L}{m}$.

In other words,

- stationary waves are formed at frequencies $f$, $2f$, $3f$, etc.
- the length of the vibrating section of the string $L = \frac{m\lambda}{2}$ = a whole number of half wavelengths.

## Making music

A guitar produces sound when its strings vibrate as a result of being plucked. When a stretched string or wire vibrates, its pattern of vibration is a mix of its first and higher harmonics. The sound produced is the same mix of frequencies which change with time as the pattern of vibration changes. A **spectrum analyser** can be used to show how the intensity of a sound varies with frequency and with time. Combined with a **sound synthesiser**, the original sound can be altered by amplifying or suppressing different frequency ranges.

The **pitch** of a note corresponds to frequency. This means that the pitch of the note from a stretched string can be altered by changing the tension of the string or by altering its length.

- Raising the tension or shortening the length increases the pitch.
- Lowering the tension or increasing the length lowers the pitch.

By changing the length or altering the tension, a vibrating string or wire can be tuned to the same pitch as a tuning fork. However, the sound from a vibrating string includes all the harmonic frequencies, whereas a tuning fork vibrates only at a single frequency. The wire is tuned when its first harmonic frequency is the same as the tuning fork frequency. It can be shown that the first harmonic frequency $f$ depends on the tension $T$ in the wire and its mass per unit length $\mu$ according to the equation $f = \frac{1}{2l}\sqrt{\frac{T}{\mu}}$

### Note

A simple visual check when using a tuning fork to tune a wire is to balance a small piece of paper on the wire at its centre. Placing the base of the vibrating tuning fork on one end of the wire will cause the paper to fall off if the wire is tuned correctly.

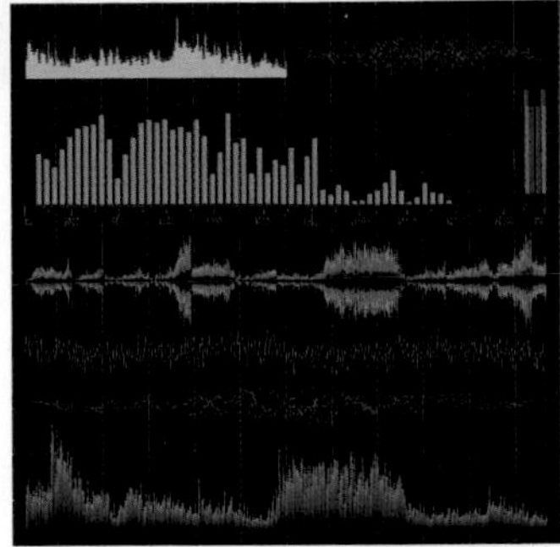

▲ **Figure 2** *A spectrum analyser*

## Summary questions

1. A stretched wire of length 0.80 m vibrates at its first harmonic with a frequency of 256 Hz. Calculate **a** the wavelength of the progressive waves on the wire, **b** the speed of the progressive waves on the wire.
2. The first harmonic frequency of vibration of a stretched wire is inversely proportional to the length of the wire. For the wire in **1** at the same tension, calculate the length of the wire to produce a frequency of **a** 512 Hz, **b** 384 Hz.
3. The tension in the wire in Q1 was 40 N. Calculate **a** the mass per unit length of the wire, **b** the diameter of the wire if its density (mass per unit volume) was 7800 kg m$^{-3}$.
4. The speed, $c$, of the progressive waves on a stretched wire varies with the tension $T$ in the wire, in accordance with the equation $c = (T/\mu)^{1/2}$, where $\mu$ is the mass per unit length of the wire. Use this formula to explain why a nylon wire and a steel wire of the same length, diameter, and tension produce notes of different pitch. State, with a reason, which wire would produce the higher pitch.

# 4.7 Using an oscilloscope

## Learning objectives:

- → Describe how an oscilloscope can be used.
- → Interpret waveforms on an oscilloscope to give peak voltage and wavelength.

An oscilloscope consists of a specially made electron tube and associated control circuits. An electron gun at one end of the glass tube emits electrons in a beam towards a fluorescent screen at the other end of the tube, as shown in Figure 1. Light is emitted from the spot on the screen where the beam hits the screen.

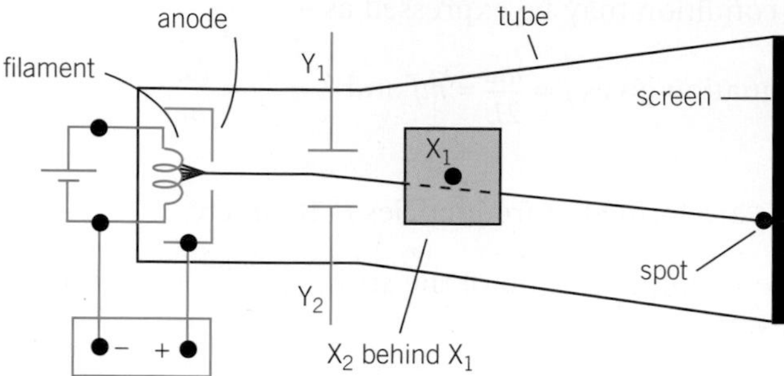


▲ **Figure 1** *An oscilloscope tube*

## How to use an oscilloscope

The position of the spot of light on the screen is affected by the pd across either pair of deflecting plates. With no pd across either set of deflecting plates, the spot on the screen stays in the same position. If a pd is applied across the X-plates, the spot deflects horizontally. A pd across the Y-plates makes it deflect vertically. In both cases, **the displacement of the spot is proportional to the applied pd.**

To display a waveform:

- the X-plates are connected to the oscilloscope's **time base circuit** which makes the spot move at constant speed left to right across the screen, then back again much faster. Because the spot moves at constant speed across the screen, the *x*-scale can be calibrated, usually in milliseconds or microseconds per centimetre.
- the pd to be displayed is connected to the Y-plates via the Y-input so the spot moves up and down as it moves left to right across the screen. As it does so, it traces out the waveform on the screen. Because the vertical displacement of the spot is proportional to the pd applied to the Y-plates, the Y-input is calibrated in volts per centimetre (or per division if the grid on the oscilloscope screen is not a centimetre grid). The calibration value is usually referred to as the Y-sensitivity or Y-gain of the oscilloscope.

Figure 2 shows the trace produced when an alternating pd is applied to the Y-input. The screen is marked with a centimetre grid.

1. To measure the peak pd, observe that the waveform height from the bottom to the top of the wave is 3.2 cm. The amplitude (i.e., peak height) of the wave is therefore 1.6 cm. As the Y-gain is set at $5.0\ \text{V cm}^{-1}$, the peak pd is therefore 8.0 V ($= 5.0\ \text{V cm}^{-1} \times 1.6\ \text{cm}$).
2. To measure the frequency of the alternating pd, we need to measure the time period $T$ (the time for one full cycle) of the waveform. Then we can calculate the frequency $f$ as $f = 1/T$.

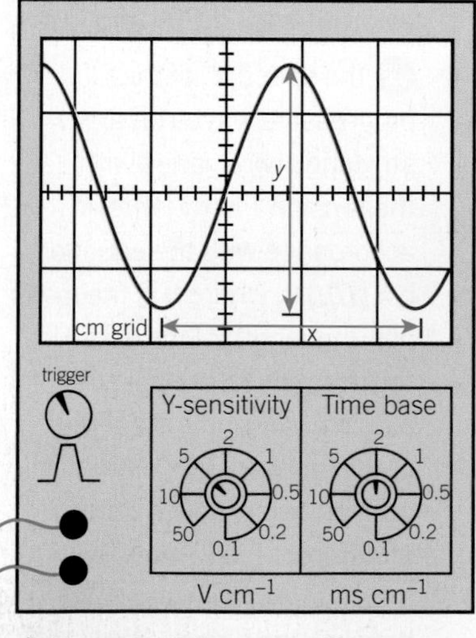


▲ **Figure 2** *Using an oscilloscope*

3 We can see from the waveform that one full cycle corresponds to a distance of 3.8 cm across the screen horizontally. As the time base control is set at $2\,\text{ms cm}^{-1}$, the time period $T$ is therefore 7.6 ms ($= 2\,\text{ms cm}^{-1} \times 3.8\,\text{cm}$). Therefore, the frequency of the alternating pd is 132 Hz.

## Measuring the speed of ultrasound

The time base circuit of an oscilloscope can be used to trigger an ultrasonic transmitter so it sends out a short pulse of ultrasonic waves. An ultrasonic receiver can be used to detect the transmitted pulse. If the receiver signal is applied to the Y-input of the oscilloscope, the waveform of the received pulse can be seen on the oscilloscope screen, as shown in Figure 3.

Because the pulse takes time to travel from the transmitter to the receiver, it is displayed on the screen at the point reached by the spot as it sweeps across from left to right. By measuring the horizontal distance on the screen from the leading edge of the pulse to the start of the spot's sweep, the travel time of the pulse from the transmitter to the receiver can be determined. For example, if the pulse is 3.5 cm from the start of the spot's sweep and the time base control is set at $0.2\,\text{ms cm}^{-1}$, the travel time of the pulse must be 0.7 ms ($= 0.2\,\text{ms cm}^{-1} \times 3.5\,\text{cm}$). If the distance from the transmitter to the receiver is known, the speed of ultrasound can be calculated (from distance / travel time).

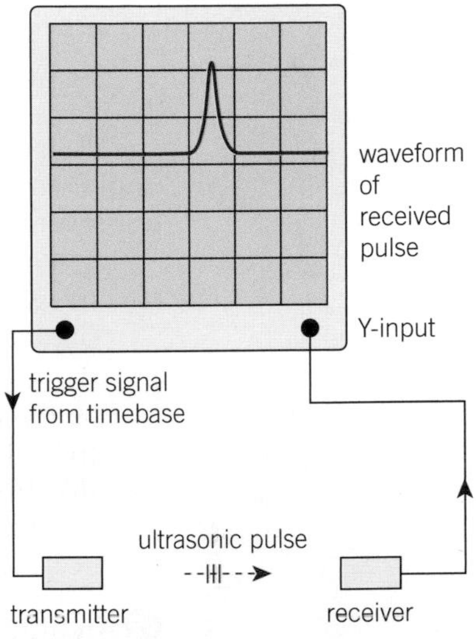


▲ **Figure 3** *Measuring the speed of ultrasound*

## Summary questions

1 The trace of an oscilloscope is displaced vertically by 0.9 cm when a pd of 4.5 V is applied to the Y-input.

a Calculate **i** the Y-gain of the oscilloscope, **ii** the displacement of the spot when a pd of 12 V is applied to the Y-input.

b An alternating pd is applied to the Y-input instead. The height of the waveform from the bottom to the top is 6.5 cm when the Y-gain is $0.5\,\text{V cm}^{-1}$. Calculate the peak value of the alternating pd.

2 The time base control of an oscilloscope is set at $10\,\text{ms cm}^{-1}$ and an alternating pd is applied to the Y-input. The horizontal distance across two complete cycles is observed to be 4.4 cm. Calculate:

a the time period of the alternating pd

b its frequency.

3 Figure 4 shows the waveform on an oscilloscope screen when an alternating pd was applied to the Y-plates.

a The Y-gain of the oscilloscope is $5.0\,\text{V cm}^{-1}$. Calculate the peak value of the alternating pd.

b The time base setting of the oscilloscope was $5\,\text{ms cm}^{-1}$. Calculate the time period and the frequency of the alternating pd. (Assume each square represents $1\,\text{cm}^2$.)

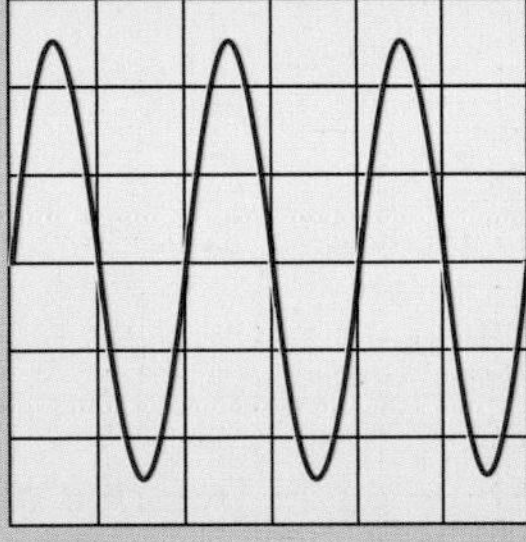

▲ **Figure 4**

4 Copy the grid of Figure 4 and sketch the trace you would observe if a constant pd of 10.0 V was applied to the Y-input.

# Practice questions: Chapter 4

1 A student adjusted the tension of a stretched metal wire of length 820 mm so that when it vibrated it emitted sound at the same frequency as a 256 Hz tuning fork.

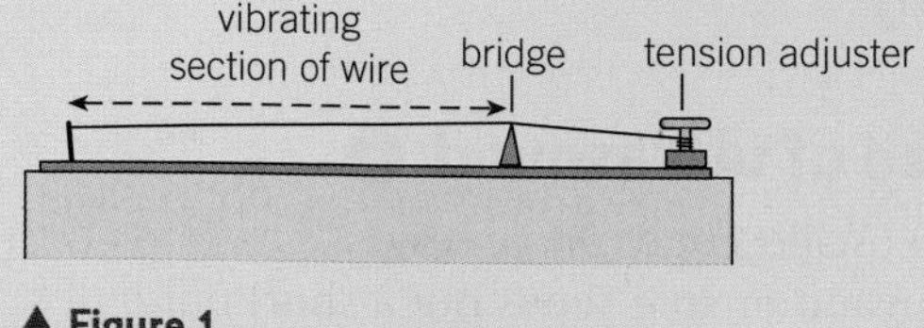


▲ **Figure 1**

She then altered the length $L$ of the vibrating section of the wire as shown in **Figure 1**, until it emitted sound at the same frequency as a 512 Hz tuning fork. She repeated the test several times and obtained the following measurements for the length $L$:

425 mm 407 mm 396 mm 415 mm 402 mm

**(a)** (i) Calculate the mean length $L$ at 512 Hz.
(ii) Estimate the uncertainty in this length measurement. *(2 marks)*

**(b)** The student thought the measurements showed that the frequency of sound $f$ emitted by the wire is inversely proportional to the length $L$ of the vibrating section.
(i) Discuss whether or not the measurements support this hypothesis.
(ii) In order to test the hypothesis further, state what further measurements the student could make and show how these measurements should be used. Assume further calibrated tuning forks are available. *(9 marks)*

2 An ultrasonic signal from a ship travels vertically downwards through the water. The wavelength of the waves is $5.3 \times 10^{-2}$ m and the frequency of the waves is 29 kHz.

**(a)** Calculate the speed of the sound through the water. *(3 marks)*

**(b)** The sound is reflected from the sea bed and is received back at the ship 0.23 s after it is transmitted. Calculate the depth of the water. *(2 marks)*

AQA, 2007

3 **(a)** State the characteristic features of
(i) longitudinal waves
(ii) transverse waves. *(3 marks)*

**(b)** Daylight passes horizontally through a fixed polarising filter **P**. An observer views the light emerging through a second polarising filter **Q**, which may be rotated in a vertical plane about point **X** as shown in **Figure 2**.

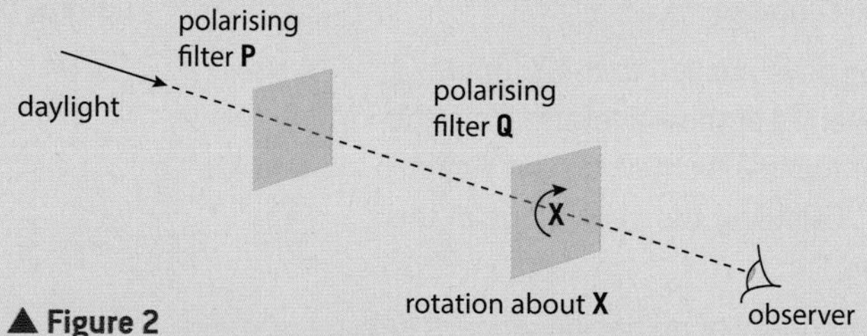


▲ **Figure 2**

Describe what the observer would see as **Q** is rotated slowly through 360°. *(2 marks)*

AQA, 2005

4 Polarisation is a property of one type of wave.

**(a)** There are two general classes of wave, longitudinal and transverse. Which class of wave can be polarised? *(1 mark)*

**(b)** Give *one* example of the type of wave that can be polarised. *(1 mark)*

**(c)** Explain why some waves can be polarised but others cannot. *(3 marks)*

AQA, 2002

5 **Figure 3** shows three particles in a medium that is transmitting a sound wave. Particles **A** and **C** are separated by one wavelength and particle **B** is halfway between them when no sound is being transmitted.

(a) Name the type of wave that is involved in the transmission of this sound. *(1 mark)*

(b) At one instant particle **A** is displaced to the point **A′** indicated by the tip of the arrow in **Figure 3**. Show on **Figure 3** the displacements of particles **B** and **C** at the same instant. Label the positions **B′** and **C′**, respectively. *(2 marks)*

AQA, 2005

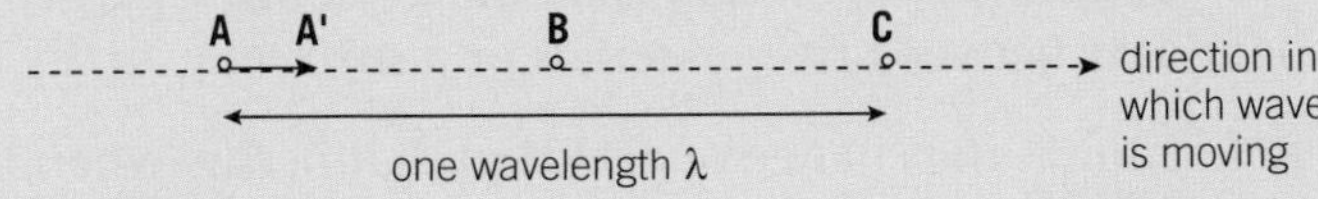


▲ **Figure 3**

6 **Figure 4** represents a stationary wave on a stretched string. The continuous line shows the position of the string at a particular instant when the displacement is a maximum. P and S are the fixed ends of the string. Q and R are the positions of the nodes. The speed of the waves on the string is $200\,\mathrm{m\,s^{-1}}$.

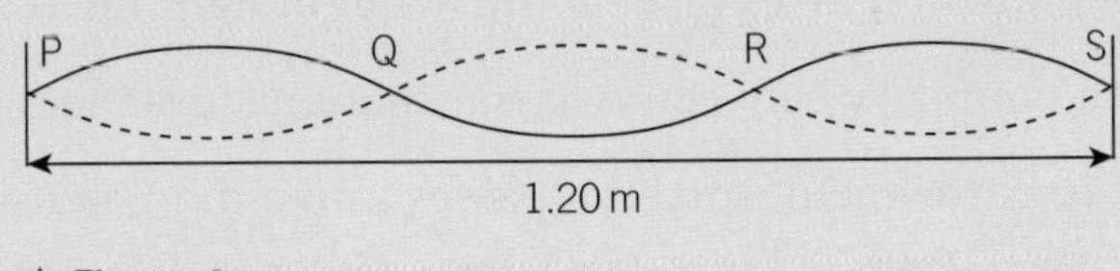


▲ **Figure 4**

(a) State the wavelength of the waves on the string.

(b) Calculate the frequency of vibration.

(c) Draw on a copy of the diagram the position of the string 3.0 ms later than the position shown.
Explain how you arrive at your answer. *(5 marks)*

AQA, 2004

7 Short pulses of sound are reflected from the wall of a building 18 m away from the sound source. The reflected pulses return to the source after 0.11 s.

(a) Calculate the speed of sound. *(3 marks)*

(b) The sound source now emits a continuous tone at a constant frequency. An observer, walking at a constant speed from the source to the wall, hears a regular rise and fall in the intensity of the sound. Explain how the *minima* of intensity occur. *(3 marks)*

AQA, 2002

8 A microwave transmitter directs waves towards a metal plate, as shown in **Figure 5**. When a microwave detector is moved along a line normal to the transmitter and the plate, it passes through a sequence of equally spaced maxima and minima of intensity.

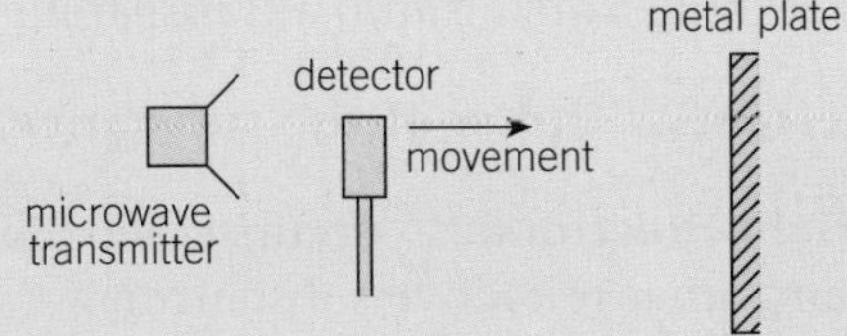


▶ **Figure 5**

(a) Explain how these maxima and minima are formed. *(4 marks)*

(b) The detector is placed at a position where the intensity is a minimum. When it is moved a distance of 144 mm it passes through nine maxima and reaches the ninth minimum from the starting point. Calculate:

(i) the wavelength of the microwaves

(ii) the frequency of the microwave transmitter. *(3 marks)*

AQA, 2003

# 5 Optics

## 5.1 Refraction of light

**Learning objectives:**

- → Explain what we mean by rays.
- → State Snell's law.
- → Comment on whether refraction is different for a light ray travelling from a transparent substance into air.

*Specification reference: 3.3.2.3*

The wave theory of light can be used to explain reflection and refraction of light. However, when we consider the effect of lenses or mirrors on the path of light, we usually prefer to draw diagrams using light rays and normals. Light rays represent the direction of travel of wavefronts. The **normal** is an imaginary line perpendicular to a boundary between two materials or a surface.

Refraction is the change of direction that occurs when light passes at an angle across a boundary between two transparent substances. Figure 2 shows the change of direction of a light ray when it enters and when it leaves a rectangular glass block in air. The light ray bends:

- towards the normal when it passes from air into glass
- away from the normal when it passes from glass into air.

No refraction takes place if the incident light ray is along the normal.

At a boundary between two transparent substances, the light ray bends:

- towards the normal if it passes into a more dense substance
- away from the normal if it passes into a less dense substance.

### Study tip

Remember from your GCSE that the normal is an imaginary line perpendicular to a boundary between two materials or a surface.

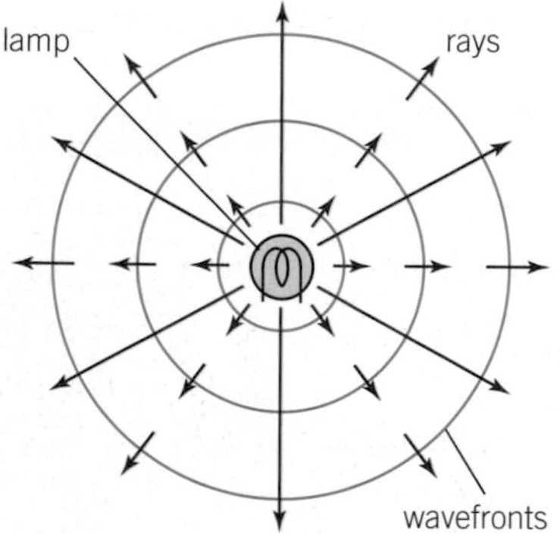


▲ **Figure 1** *Rays and waves*

### Investigating the refraction of light by glass

Use a ray box to direct a light ray into a rectangular glass block at different angles of incidence at the midpoint P of one of the longer sides, as shown in Figure 2. Note that the angle of incidence is the angle between the incident light ray and the normal at the point of incidence.

For each angle of incidence at P, mark the point Q where the light ray leaves the block. The angle of refraction is the angle between the normal at P and the line PQ. Measurements of the angles of incidence and refraction for different incident rays show that:

- the angle of refraction, $r$, at P is always less than the angle of incidence, $i$
- the ratio of $\frac{\sin i}{\sin r}$ is the same for each light ray. This is known as Snell's law. The ratio is referred to as the **refractive index**, $n$, of glass.

For a light ray travelling from air into a transparent substance,

$$\textbf{the refractive index of the substance, } n = \frac{\sin i}{\sin r}$$

Notice that **partial reflection** also occurs when a light ray in air enters glass (or any other refractive substance).

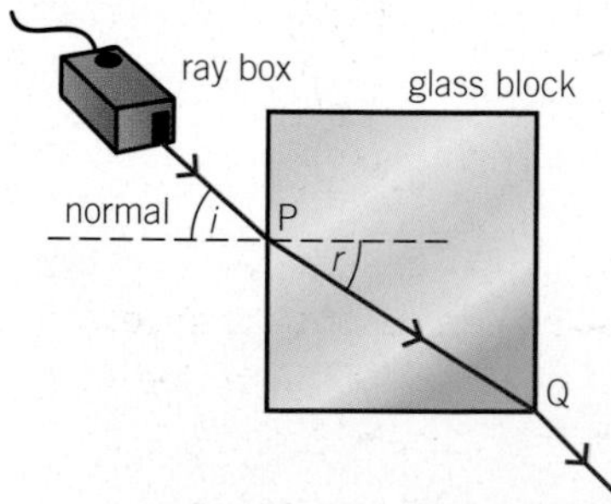


▲ **Figure 2** *Investigating refraction*

### Study tip

Set your calculator mode to degrees when using Snell's law.

### Worked example

A light ray is directed into a glass block of refractive index 1.5 at an angle of incidence of 40°. Calculate the angle of refraction of this light ray.

### Solution

$i = 40°$, $n = 1.5$

Rearranging: $\frac{\sin i}{\sin r} = n$ gives $\sin r = \frac{\sin i}{n} = \frac{\sin 40}{1.5} = \frac{0.643}{1.5} = 0.429$

Therefore $r = 25°$.

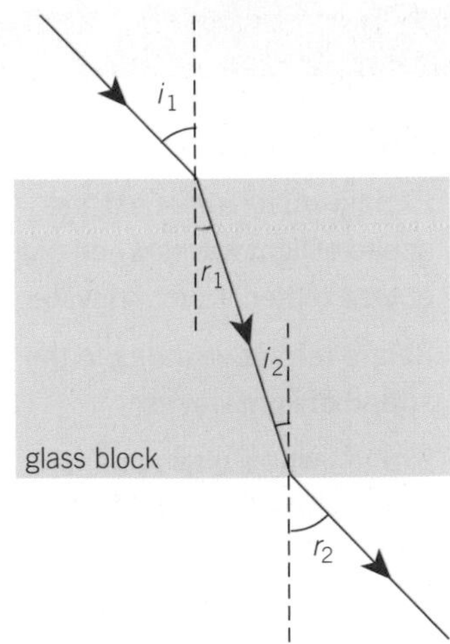


▲ **Figure 3** *Comparing glass to air refraction with air to glass refraction*

## Comparing glass to air refraction with air to glass refraction

In Figure 3, notice that the angle of refraction of the light ray emerging from the block is the same as the angle of incidence of the light ray entering the block. This is because the two sides of the block at which refraction occurs are parallel to each other.

- If $i_1$ and $r_1$ are the angles of incidence and refraction at the point where the light ray enters the block, then the refractive index of the glass $n = \frac{\sin i_1}{\sin r_1}$.
- At the point where the light ray leaves the block, $i_2 = r_1$ and $r_2 = i_1$, so $\frac{\sin i_2}{\sin r_2} = \frac{1}{n}$.

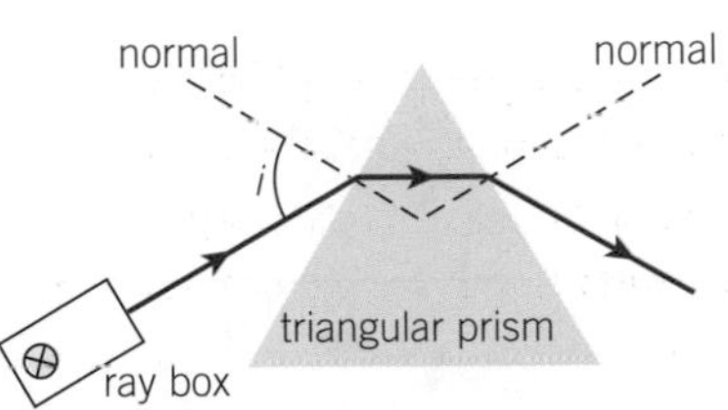


▲ **Figure 4** *Refraction by a glass prism*

## Refraction of a light ray by a triangular prism

Figure 4 shows the path of a monochromatic light ray through a triangular prism. The light ray refracts towards the normal where it enters the glass prism then refracts away from the normal where it leaves the prism.

## Summary questions

**1** **a** A light ray in air is directed into water from air. The refractive index of water is 1.33. Calculate the angle of refraction of the light ray in the water for an angle of incidence of

**i** 20° **ii** 40° **iii** 60°.

**b** A light ray in water is directed at the water surface at an angle of incidence of 40°.

**i** Calculate the angle of refraction of the light ray at this surface.

**ii** Sketch the path of the light ray showing the normal and the angles of incidence and refraction.

**2** A light ray is directed from air into a glass block of refractive index 1.5. Calculate:

**a** the angle of refraction at the point of incidence if the angle of incidence is **i** 30° **ii** 60°

**b** the angle of incidence if the angle of refraction at the point of incidence is **i** 35° **ii** 40°.

**3** A light ray in air was directed at the flat side of a semicircular glass block at an angle of incidence of 50°, as shown in Figure 5. The angle of refraction at the point of incidence was 30°.

**a** Calculate the refractive index of the glass.

**b** Calculate the angle of refraction if the angle of incidence was changed to 60°.

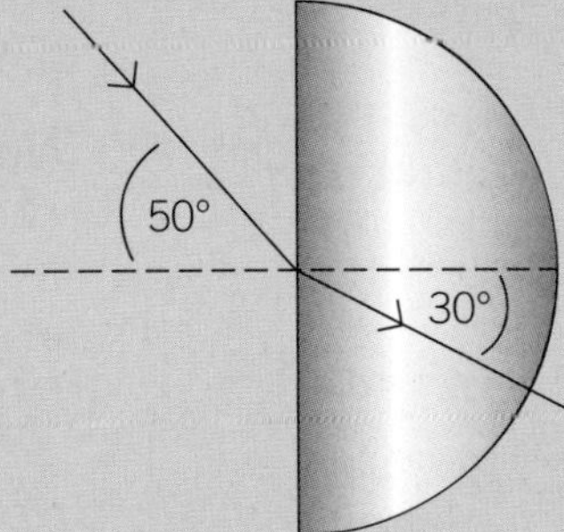


▲ **Figure 5**

**4** A light ray enters an equilateral glass prism of refractive index 1.55 at the midpoint of one side of the prism at an angle of incidence of 35°.

**a** Sketch this arrangement and show that the angle of refraction of the light ray in the glass is 22°.

**b** **i** Show that the angle of incidence where the light ray leaves the glass prism is 38°.

**ii** Calculate the angle of refraction of the light ray where it leaves the prism.

# 5.2 More about refraction

## Learning objectives:

- → Explain what happens to the speed of light waves when they enter a material such as water.
- → Relate refractive index to the speed of light waves.
- → Explain why a glass prism splits white light into the colours of a spectrum.

*Specification reference: 3.3.2.3*

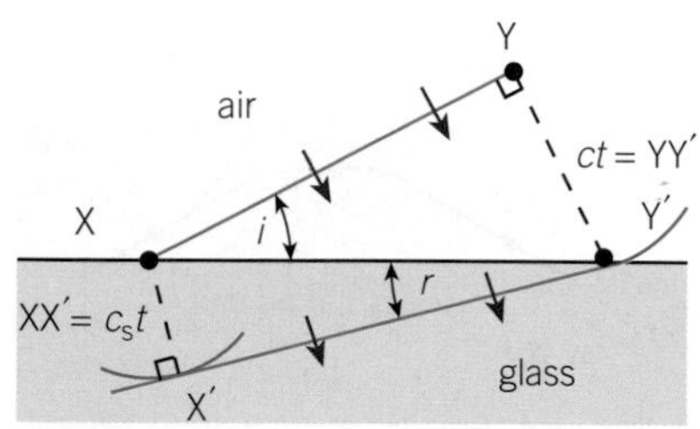


▲ **Figure 1** *Explaining refraction*

## Explaining refraction

Refraction occurs because the speed of the light waves is different in each substance. The amount of refraction that takes place depends on the speed of the waves in each substance.

Consider a wavefront of a light wave when it passes across a straight boundary from a vacuum (or air) into a transparent substance, as shown in Figure 1. Suppose the wavefront moves from XY to X′Y′ in time *t*, crossing the boundary between X and Y′. In this time, the wavefront moves

- a distance $ct$ at speed $c$ in a vacuum from Y to Y′
- a distance $c_s t$ at speed $c_s$ in the substance from X to X′.

Considering triangle XYY′, since YY′ is the direction of the wavefront in the vacuum and is therefore perpendicular to XY, then YY′ = XY′ sin *i*, where *i* = angle YXY′.

$$ct = XY' \sin i$$

Considering triangle XX′Y′, since XX′ is the direction of the wavefront in the substance and is therefore perpendicular to X′Y′, then XX′ = XY′ sin *r*, where *r* = angle XY′X′.

$$c_s t = XY' \sin r$$

Combining these two equations therefore gives

$$\frac{\sin i}{\sin r} = \frac{c}{c_s}$$

Therefore the refractive index of the substance, $n_s = \dfrac{c}{c_s}$

This equation shows that the smaller the speed of light is in a substance, the greater is the refractive index of the substance.

### *Note*

The frequency $f$ of the waves does not change when refraction occurs. As $c = \lambda f$ and $c_s = \lambda_s f$, where $\lambda$ and $\lambda_s$ are the wavelengths of the waves in a vacuum and in the substance, respectively, then it follows that the refractive index of the substance,

$$n_s = \frac{c}{c_s} = \frac{\lambda}{\lambda_s}$$

### Worked example

The speed of light is $3.00 \times 10^8\,\mathrm{m\,s^{-1}}$ in a vacuum. A certain type of glass has a refractive index of 1.62. Calculate the speed of light in the glass.

### *Solution*

Rearranging $n_s = \dfrac{c}{c_s}$ gives $c_s = \dfrac{c}{n_s} = \dfrac{3.0 \times 10^8}{1.62} = 1.85 \times 10^8\,\mathrm{m\,s^{-1}}$

## Refraction at a boundary between two transparent substances

Consider a light ray crossing a boundary from a substance in which the speed of light is $c_1$ to a substance in which the speed of light is $c_2$. Using the same theory as on the previous page gives

$$\frac{\sin i}{\sin r} = \frac{c_1}{c_2}$$

where $i$ = the angle between the incident ray and the normal and $r$ = the angle between the refracted ray and the normal.

This equation may be rearranged as $\frac{1}{c_1}\sin i = \frac{1}{c_2}\sin r$.

Multiplying both sides of this equation by $c$, the speed of light in a vacuum, gives

$$\frac{c}{c_1}\sin i = \frac{c}{c_2}\sin r$$

Substituting $n_1$ for $\frac{c}{c_1}$, where $n_1$ is the refractive index of substance 1, and $n_2$ for $\frac{c}{c_2}$, where $n_2$ is the refractive index of substance, gives Snell's law,

$$n_1 \sin\theta_1 = n_2 \sin\theta_2$$

where $\theta_1 = i$ and $\theta_2 = r$.

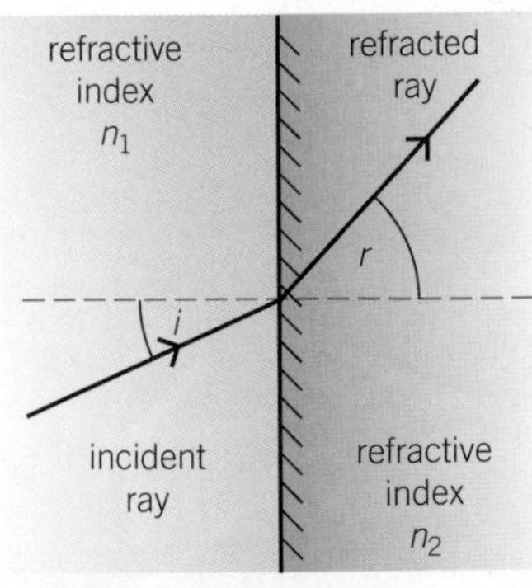


▲ **Figure 2** *The n sin i rule*

**Study tip**

Always use the equation $n_1 \sin\theta_1 = n_2 \sin\theta_2$ in calculations if you know three of its four quantities.

### Worked example

A light ray crosses the boundary between water of refractive index 1.33 and glass of refractive index 1.50 at an angle of incidence of 40°. Calculate the angle of refraction of this light ray.

#### *Solution*

$n_1 = 1.33$, $n_2 = 1.50$, $\theta_1 = 40°$

Using $n_1 \sin\theta_1 = n_2 \sin\theta_2$ gives $1.33 \sin 40 = 1.50 \sin\theta_2$

$$\therefore \sin\theta_2 = \frac{1.33 \sin 40}{1.5} = 0.57$$

$$\theta_2 = 35°$$

**Study tip**

During refraction, the speed and wavelength both change, but frequency stays constant.

***Note:***

When a light ray passes from a vacuum into a transparent substance of refractive index $n$,

$$\frac{\sin\theta_1}{\sin\theta_2} = n$$

where $\theta_1$ = the angle between the incident ray and the normal,

$\theta_2$ = the angle between the refracted ray and the normal.

$$n = \frac{\textbf{the speed of light in a vacuum}}{\textbf{the speed of light in the transparent substance}}$$

**Study tip**

Remember that the refractive index of air is approximately 1. This is very similar to the refractive index of a vacuum.

▲ **Figure 3** *Dispersion of light*

The speed of light in air at atmospheric pressure is 99.97% of the speed of light in a vacuum.

Therefore, the refractive index of air is 1.0003. For most purposes, the refractive index of air may be assumed to be 1.

## The white light spectrum

We can use a prism to split a beam of white light from a filament lamp (or sunlight) into the colours of the spectrum by a glass prism, as shown in Figure 3. This happens because white light is composed of light with a continuous range of wavelengths, from red at about 650 nm to violet at about 350 nm. The glass prism refracts light by different amounts, depending on its wavelength. The shorter the wavelength in air, the greater the amount of refraction. So each colour in the white light beam is refracted by a different amount. This dispersive effect occurs because the speed of light in glass depends on wavelength. Violet light travels more slowly than red light in glass so the refractive index of violet light is greater than the refractive index of red light.

## Summary questions

**1** Water waves of frequency 4.0 Hz travelling at a speed of $0.16\,\mathrm{m\,s^{-1}}$ travel across a boundary from deep to shallow water where the speed is $0.12\,\mathrm{m\,s^{-1}}$.

- **a** Calculate the wavelength of these waves
  - **i** in the deep water **ii** in the shallow water.
- **b** The incident wavefronts cross the boundary at an angle of 25° to the boundary. Calculate the angle of the refracted wavefronts to the boundary.

**2** **a** The speed of light in a vacuum is $3.00 \times 10^8\,\mathrm{m\,s^{-1}}$. Calculate the speed of light in
  - **i** glass of refractive index 1.52 **ii** water.

  The refractive index of water = 1.33.

- **b** A light ray passes across a plane boundary from water into glass at an angle of incidence of 55°. Use the refractive index values from **a** to show that the angle of refraction of this light ray is 46°.

**3** Calculate the angle of refraction for a light ray entering glass of refractive index 1.50 at an angle of incidence of 40° in:

- **a** air **b** water of refractive index 1.33.

**4** A white light ray is directed through the curved side of a semicircular glass block at the midpoint of the flat side, as shown in Figure 4. The angle of incidence of the light ray at the flat side is 30°.

The refractive index of the glass for red light is 1.52 and for blue light is 1.55.

- **a** Calculate the angle of refraction at the midpoint of
  - **i** the red component, and
  - **ii** the blue component of the light ray.
- **b** Hence show that the angle between the red and blue components of the refracted light ray is 1.3°.

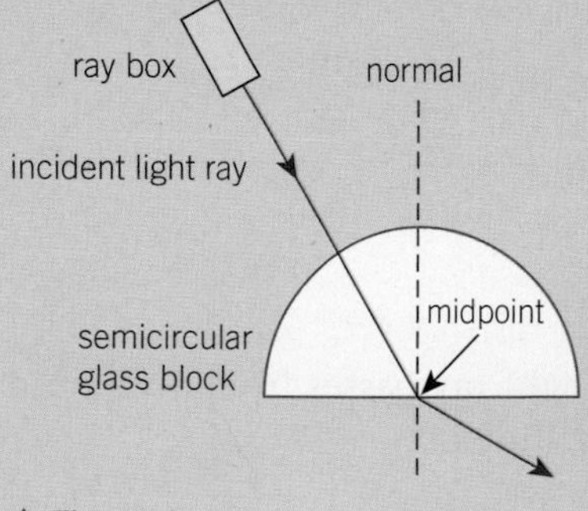


▲ **Figure 4**

# 5.3 Total internal reflection

## Investigating total internal reflection

When a light ray travels from glass into air, it refracts away from the normal. If the angle of incidence is increased to a certain value known as the **critical angle**, the light ray refracts along the boundary. Figure 1 shows this effect. If the angle of incidence is increased further, the light ray undergoes **total internal reflection** at the boundary, the same as if the boundary were replaced by a plane mirror.

In general, total internal reflection can only take place if:

1 the incident substance has a **larger refractive index** than the other substance

2 the angle of incidence **exceeds the critical angle**.

At the critical angle $i_c$, the angle of refraction is 90° because the light ray emerges along the boundary. Therefore, $n_1 \sin i_c = n_2 \sin 90$ where $n_1$ is the refractive index of the incident substance and $n_2$ is the refractive index of the other substance. Since $\sin 90 = 1$, then

$$\sin \theta_c = \frac{n_2}{n_1}$$

Prove for yourself that the critical angle for the boundary between glass of refractive index 1.5 and air (refractive index = 1) is 42°. This means that if the angle of incidence of a light ray in the glass is greater than 42°, the light ray undergoes total internal reflection back into the glass at the boundary.

### Why do diamonds sparkle when white light is directed at them?

When white light enters a diamond, it is split into the colours of the spectrum. Diamond has a very high refractive index of 2.417 so it separates the colours more than any other substance does. In addition, the high refractive index gives diamond a critical angle of 24.4°. So a light ray in a diamond may be totally internally reflected many times before it emerges, which means its colours spread out more and more. So the diamond sparkles with different colours.

## Optical fibres

**Optical fibres** are used in medical **endoscopes** to see inside the body, and in communications to carry light signals. Figure 2 (overleaf) shows the path of a light ray along an optical fibre. The light ray is totally internally reflected each time it reaches the fibre boundary, even where the fibre bends, unless the radius of the bend is too small. At each point where the light ray reaches the boundary, the angle of incidence exceeds the critical angle of the fibre.

A communications optical fibre allows pulses of light that enter at one end, from a transmitter, to reach a receiver at the other end. Such fibres need to be highly transparent to minimise absorption of light, which would otherwise reduce the amplitude of the pulses progressively the further they travel in the fibre. Each fibre consists of a core surrounded by a layer of cladding of lower refractive index to reduce light loss from the core. Light loss would also reduce the amplitude of the pulses.

## Learning objectives:

- → State the conditions for total internal reflection.
- → Relate the critical angle to refractive index.
- → Explain why diamonds sparkle.

*Specification reference: 3.3.2.3*

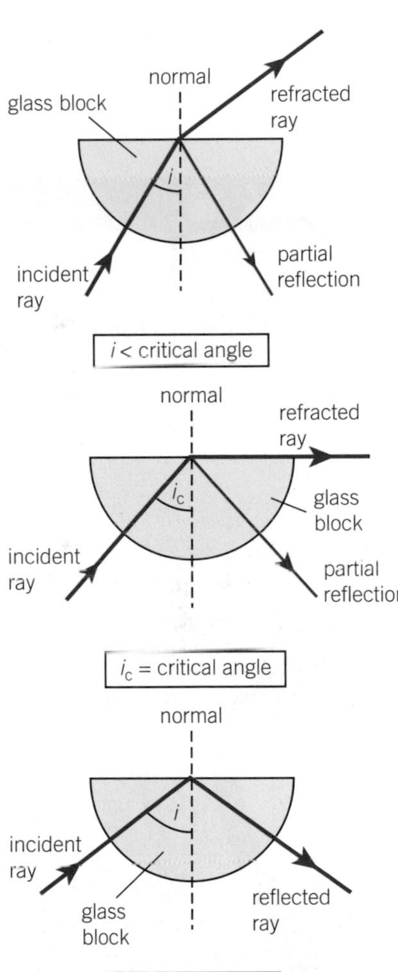


▲ **Figure 1** *Total internal reflection*

### Study tip

Partial internal reflection always occurs at a boundary when the angle of incidence is less than or equal to the critical angle.

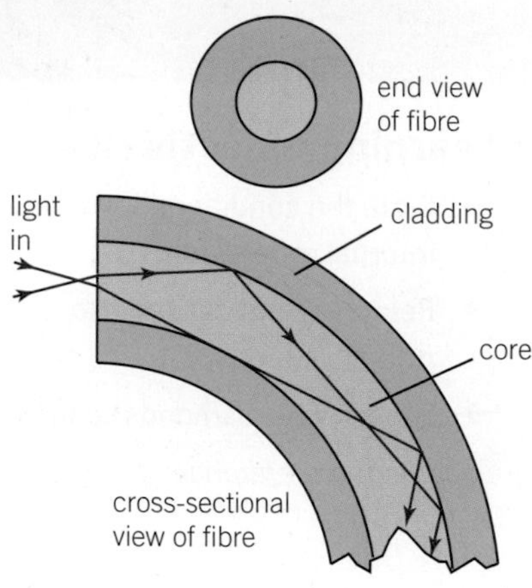


▲ **Figure 2** *Fibre optics*

- Total internal reflection takes place at the core–cladding boundary. At any point where two fibres are in direct contact, light would cross from one fibre to the other if there were no cladding. Such crossover would mean that the signals would not be secure, as they would reach the wrong destination.
- The core must be very narrow to prevent **modal** (i.e., multipath) **dispersion**. This occurs in a wide core because light travelling along the axis of the core travels a shorter distance per metre of fibre than light that repeatedly undergoes total internal reflection. A pulse of light sent along a wide core would become longer than it ought to be. If it was too long, it would merge with the next pulse.

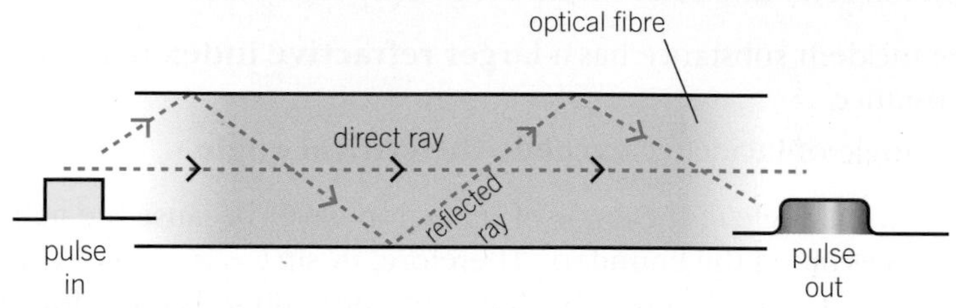


▲ **Figure 3** *Modal dispersion*

## Maths link

**Q:** Make an order-of-magnitude estimate of the number of fibres of diameter 10 μm in a bundle of fibres of diameter 10 mm.

**A:** $10^6$

### *Note:*

Pulse dispersion also occurs if white light is used instead of monochromatic light (light of a single wavelength). This **material dispersion** (sometimes referred to as spectral dispersion) is because the speed of light in the glass of the optical fibre depends on the wavelength of light travelling through it. Violet light travels more slowly than red light in glass. The difference in speed would cause white light pulses in optical fibres to become longer, as the violet component falls behind the faster red component of each pulse. So the light (or infrared radiation) used must be monochromatic to prevent pulse merging.

The **medical endoscope** contains two bundles of fibres. The endoscope is inserted into a body cavity, which is then illuminated using light sent through one of the fibre bundles. A lens over the end of the other fibre bundle is used to form an image of the body cavity on the end of the fibre bundle. The light that forms this image travels along the fibres to the other end of the fibre bundle where the image can be observed. This fibre bundle needs to be a **coherent bundle**, which means that the fibre ends at each end are in the same relative positions.

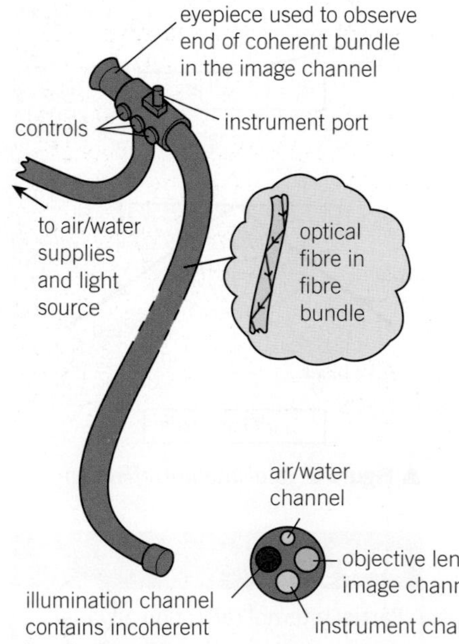


**a** *Endoscope design*

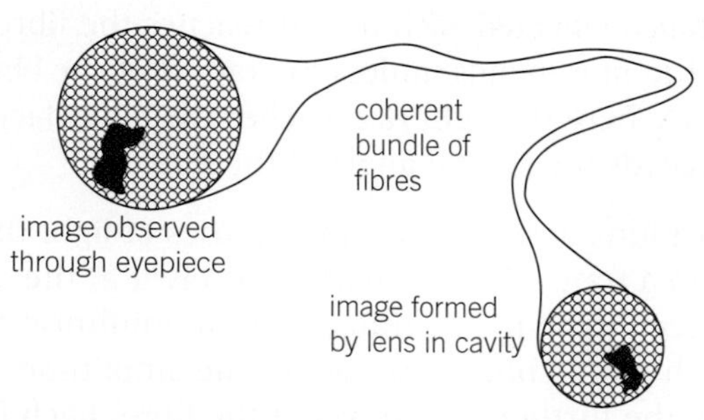


**b** *A coherent bundle*

▲ **Figure 4** *The endoscope*

## Summary questions

**1 a** State two conditions for a light ray to undergo total internal reflection at a boundary between two transparent substances.

**b** Calculate the critical angle for

**i** glass of refractive index 1.52 and air

**ii** water (refractive index = 1.33) and air.

**2 a** Show that the critical angle at a boundary between glass of refractive index 1.52 and water (refractive index = 1.33) is 61°.

**b** Figure 5 shows the path of a light ray in water of refractive index 1.33 directed at an angle of incidence of 40° at a thick glass plate of refractive index 1.52.

**i** Calculate the angle of refraction of the light ray at P.

**ii** State the angle of incidence of the light ray at Q.

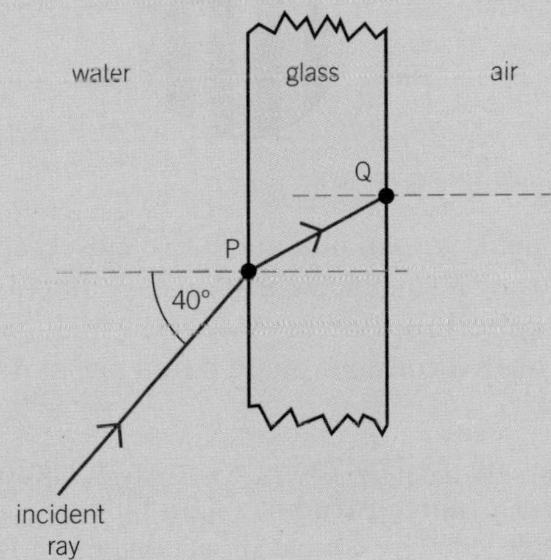


▲ **Figure 5**

**c** Sketch the path of the light ray beyond Q.

**3** A window pane made of glass of refractive index 1.55 is covered on one side only with a transparent film of refractive index 1.40.

**a** Calculate the critical angle of the film–glass boundary,

**b** A light ray in air is directed at the film at an angle of incidence of 45° as shown in Figure 6. Calculate

**i** the angle of refraction in the film

**ii** the angle of refraction of the ray where it leaves the pane.

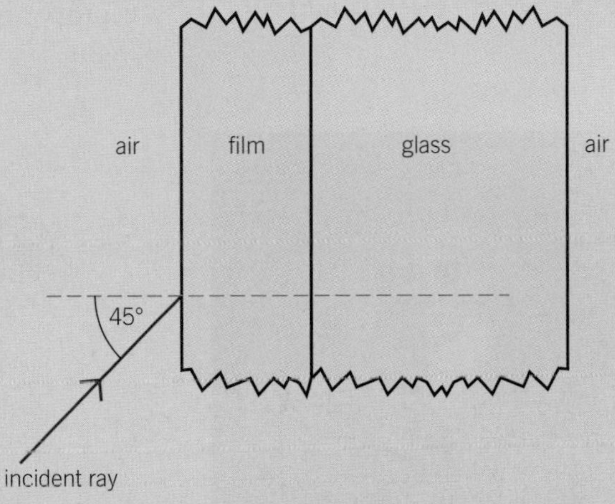


▲ **Figure 6**

**4 a** In a medical endoscope, the fibre bundle used to view the image is coherent.

**i** What is meant by a coherent fibre bundle?

**ii** Explain why this fibre bundle needs to be coherent.

**b i** Why is an optical fibre used in communication composed of a core surrounded by a layer of cladding of lower refractive index?

**ii** Why is it necessary for the core of an optical communications fibre to be narrow?

# 5.4 Double slit interference

## Learning objectives:

- → State the general condition for the formation of a bright fringe.
- → Describe the Young's double slit experiment.
- → Describe what factors could be (i) increased or (ii) decreased, to increase the fringe spacing.

*Specification reference: 3.3.2.1*

## Young's double slit experiment

The wave nature of light was first suggested by Christiaan Huygens in the 17th century but it was rejected at the time in favour of Sir Isaac Newton's corpuscular theory of light. Newton considered that light was composed of tiny particles, which he referred to as corpuscles, and he was able to explain reflection and refraction using his theory. Huygens was also able to explain reflection and refraction using his wave theory. However, because of Newton's much stronger scientific reputation, Newton's theory of light remained unchallenged for over a century until 1803, when Thomas Young demonstrated interference of light.

▲ **Figure 1** *Thomas Young 1773–1829*

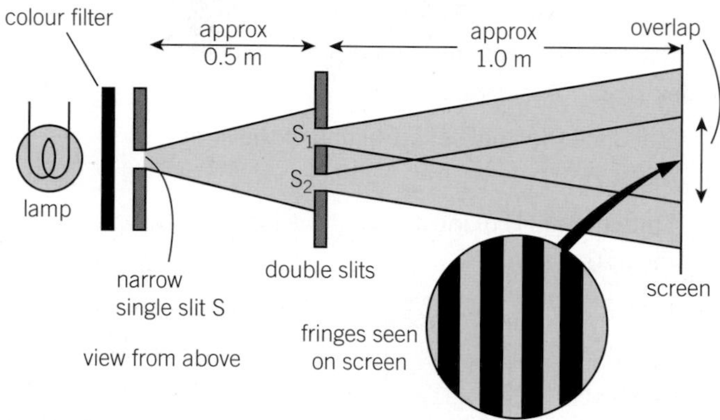


▲ **Figure 2** *Young's double slit experiment*

To observe interference of light, we can illuminate two closely spaced parallel slits (double slits) using a suitable light source, as described below. The two slits act as **coherent** sources of waves, which means that they emit light waves with a constant phase difference and the same frequency.

1 An arrangement like the one used by Young is shown in Figure 2. Young would have used a candle instead of a light bulb to illuminate a narrow single slit. The double slit arrangement is illuminated by light from the narrow single slit. Alternate bright and dark fringes, referred to as **Young's fringes**, can be seen on a white screen placed where the diffracted light from the double slits overlaps. The fringes are evenly spaced and parallel to the double slits.

### *Note:*

If the single slit is too wide, each part of it produces a fringe pattern, which is displaced slightly from the pattern due to adjacent parts of the single slit. As a result, the dark fringes of the double slit pattern become narrower than the bright fringes, and contrast is lost between the dark and the bright fringes.

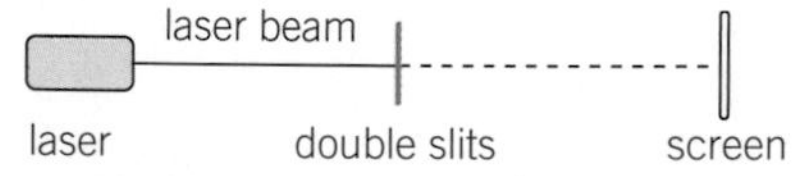


▲ **Figure 3** *Using a laser to demonstrate interference*

2 A laser beam from a low power laser could be used instead of the light bulb and the single slit. Figure 3 shows the arrangement. The fringes **must** be displayed on a screen, as a beam of laser light will damage the retina if it enters the eye. See Topic 5.5 for more information about laser light.

### Practical link

**Warning:** Never look along a laser beam, even after reflection.

The fringes are formed due to **interference of light** from the two slits:

- **Where a bright fringe is formed**, the light from one slit reinforces the light from the other slit. In other words, the light waves from each slit arrive in phase with each other.
- **Where a dark fringe is formed**, the light from one slit cancels the light from the other slit. In other words, the light waves from the two slits arrive **180° out of phase**.

**Synoptic link**

You have met the principle of superposition in more detail in Topic 4.4, Wave properties 2.

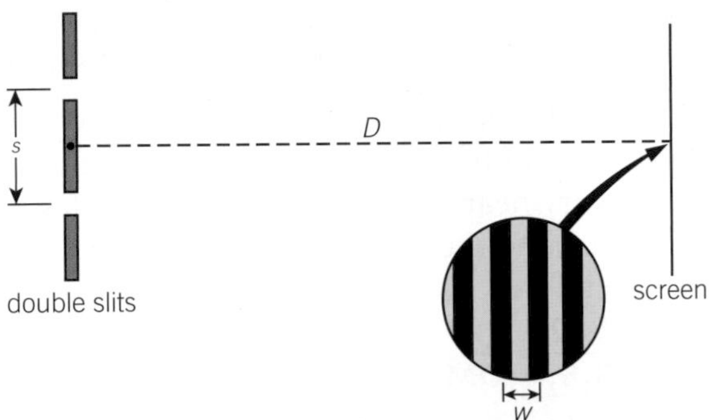


▲ **Figure 4** *Diagram to show w, D, and s for a Young's double slit experiment*

The distance from the centre of a bright fringe to the centre of the next bright fringe is called the fringe separation. This depends on the slit spacing $s$ and the distance $D$ from the slits to the screen, in accordance with the equation

**fringe separation,** $w = \frac{\lambda D}{s}$**, where** $\lambda$ **is the wavelength of light**

The equation shows that the fringes become more widely spaced if:

- the distance $D$ from the slits to the screen is increased
- the wavelength $\lambda$ of the light used is increased
- the slit spacing, $s$, is reduced. Note that the slit spacing is the distance between the centres of the slits.

**Study tip**

Convert all distances into metres when substituting into the fringe spacing equation. The wavelength of visible light is usually quoted in nanometres ($1\text{ nm} = 10^{-9}\text{ m}$).

## The theory of the double slit equation

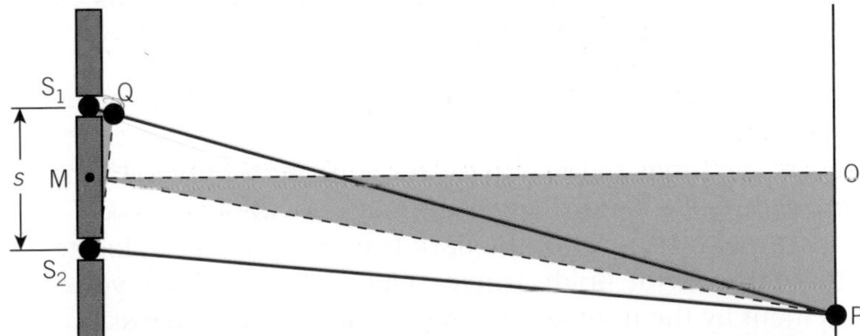


▲ **Figure 5** *The theory of the double slit experiment*

Consider the two slits $S_1$ and $S_2$ shown in Figure 5. At a point P on the screen where the fringes are observed, light emitted from $S_1$ arrives later than light from $S_2$ emitted at the same time. This is because the distance $S_1P$ is greater than the distance $S_2P$. The difference between distances $S_1P$ and $S_2P$ is referred to as the path difference.

- **For reinforcement at P, the path difference** $S_1P - S_2P = m\lambda$**,** where $m = 0, 1, 2$, etc.

## Study tip

Wavelengths of light are usually expressed in nanometers (nm), where
$1\,\text{nm} = 10^{-9}\,\text{m}$.

Light emitted simultaneously from $S_1$ and $S_2$ arrives in phase at P so reinforcement occurs at this point.

- **For cancellation at P, the path difference $S_1P - S_2P = (m + \frac{1}{2})\lambda$,** where $m = 0, 1, 2$, etc.

Light emitted simultaneously from $S_1$ and $S_2$ arrives at P out of phase by 180°, so cancellation occurs at P.

In Figure 4, a point Q along line $S_1P$ has been marked such that $QP = S_2P$.

Therefore, the path difference $S_1P - S_2P$ is represented by the distance $S_1Q$.

Consider triangles $S_1S_2Q$ and MOP, where M is the midpoint between the two slits and O is the midpoint of the central bright fringe of the pattern. The two triangles are very nearly similar in shape, as angles $S_1\hat{S}_2Q$ and $P\hat{M}O$ are almost equal and the long sides of each triangle are of almost equal length. Therefore

$$\frac{S_1Q}{S_1S_2} = \frac{OP}{OM}$$

If P is the $m$th bright fringe from the centre (where $m = 0, 1, 2$, etc.), then $S_1Q = m\lambda$ and $OP = mw$, where $w$ is the distance between the centres of adjacent bright fringes. Also, OM = distance $D$ and $S_1S_2$ = slit spacing $s$.

Therefore,

$$\frac{m\lambda}{s} = \frac{mw}{D}$$

Rearranging this equation gives

$$\lambda = \frac{sw}{D}$$

or

$$w = \frac{\lambda D}{s}$$

By measuring the slit spacing, $s$, the fringe separation, $w$, and the slit–screen distance $D$, the wavelength $\lambda$ of the light used can be calculated. The formula is valid only if the fringe separation, $w$, is much less than the distance $D$ from the slits to the screen. This condition is to ensure that the triangles $S_1S_2Q$ and MOP are very nearly similar in shape.

### Notes

1 To measure the fringe separation, $w$, measure across several fringes from the centre of a dark fringe to the centre of another dark fringe, because the centres of the dark fringes are easier to locate than the centres of the bright fringes. Obtain $w$ by dividing your measurement by the number of fringes you measured across. This means that the derivation above can also be done using the distance between the centres of adjacent dark fringes.

2 Two loudspeakers connected to the *same* signal generator can be used to demonstrate interference as they are coherent sources of sound waves. You can detect points of cancellation and reinforcement by ear as you move across in front of the speakers.

The Young's slit equation can then be used to estimate the wavelength of the sound waves (provided the wavelength is small compared with the distance between the speakers).

## Summary questions

1 In a double slit experiment using red light, a fringe pattern is observed on a screen at a fixed distance from the double slits. How would the fringe pattern change if:

  a the screen is moved closer to the slits

  b one of the double slits is blocked completely?

2 The following measurements were made in a double slit experiment:

  slit spacing $s = 0.4$ mm, fringe separation $w = 1.1$ mm, slit–screen distance $D = 0.80$ m.

  Calculate the wavelength of light used.

3 In 2, the double slit arrangement was replaced by a pair of slits with a slit spacing of 0.5 mm. Calculate the fringe separation for the same slit–screen distance and wavelength.

4 The following measurements were made in a double slit experiment:

  slit spacing $s = 0.4$ mm, fringe separation $w = 1.1$ mm, wavelength of light used = 590 nm. Calculate the distance from the slits to the screen.

# 5.5 More about interference

## Learning objectives:

- → Identify coherent sources.
- → Explain why slits, rather than two separate light sources, are used in Young's double slit experiment.
- → Describe the roles of diffraction and interference when producing Young's fringes.

*Specification reference: 3.3.2.1*

## Coherence

The double slits are described as **coherent sources** because they emit light waves of the same frequency with a constant phase difference, provided we illuminate the double slits with laser light, or with light from a narrow single slit if we are using non-laser light. Each wave crest or wave trough from the single slit always passes through one of the double slits a fixed time after it passes through the other slit. The double slits therefore emit wavefronts with a constant phase difference.

The double slit arrangement is like the ripple tank demonstration in Figure 1. Straight waves from the beam vibrating on the water surface diffract after passing through the two gaps in the barrier, and produce an interference pattern where the diffracted waves overlap. If one gap is closer to the beam than the other, each wavefront from the beam passes through the nearer gap first. However, the time interval between the same wavefront passing through the two gaps is always the same so the waves emerge from the gaps with a constant phase difference.

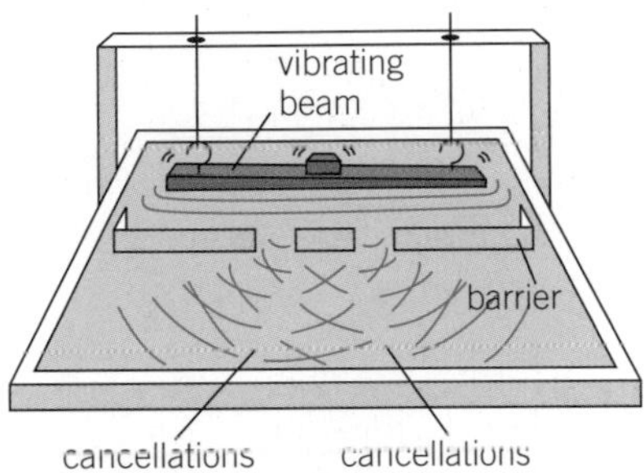


▲ **Figure 1** *Interference of water waves*

Light from two nearby lamp bulbs could not form an interference pattern because the two light sources emit light waves at random. The points of cancellation and reinforcement would change at random, so no interference pattern is possible.

## Wavelength and colour

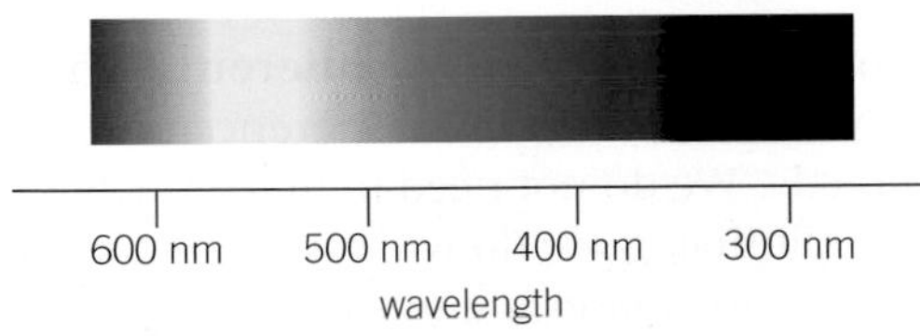


▲ **Figure 2** *Wavelength and colour*

### Study tip

It isn't necessary for two sources to be in phase for them to be coherent – a constant phase difference between them is sufficient. However, the waves must be of the same frequency.

In the double slit experiment, the fringe separation depends on the colour of light used. White light is composed of a continuous spectrum of colours, corresponding to a continuous range of wavelengths from about 350 nm for violet light to about 650 nm for red light. Each colour of light has its own wavelength, as shown in Figure 2.

The fringe pattern photographs in Figure 3 show that the fringe separation is greater for red light than for blue light. This is because red light has a longer wavelength than blue light. The fringe spacing, $w$, depends on the wavelength $\lambda$ of the light, according to the formula

$$w = \frac{\lambda D}{s}$$

as explained in Topic 5.4, so the longer the wavelength of the light used, the greater the fringe separation.

▲ **Figure 3** *The double slit fringe pattern*

## Light sources

**Vapour lamps and discharge tubes** produce light with a dominant colour. For example, the sodium vapour lamp produces a yellow/orange glow, which is due to light of wavelength 590 nm. Other wavelengths of light are also emitted from a sodium vapour lamp but the colour due to light of wavelength 590 nm is much more intense than any other colour. A sodium vapour lamp is in effect a monochromatic light source, because its spectrum is dominated by light of a certain colour.

**Light from a filament lamp or from the Sun** is composed of the colours of the spectrum and therefore covers a continuous range of wavelengths from about 350 nm to about 650 nm. If a beam of white light is directed at a colour filter, the light from the filter is a particular colour because it contains a much narrower range of wavelengths than white light does.

**Light from a laser** differs from non-laser light in two main ways:

- Laser light is highly **monochromatic**, which means we can specify its wavelength to within a nanometre. The wavelength depends on the type of laser that produces it. For example, a helium-neon laser produces red light of wavelength 635 nm. Because a laser beam is almost perfectly parallel and monochromatic, a convex lens can focus it to a very fine spot. The beam power is then concentrated in a very small area. This is why a laser beam is very dangerous if it enters the eye. The eye lens would focus the beam on a tiny spot on the retina and the intense concentration of light at that spot would destroy the retina.

  **Never look along a laser beam, even after reflection.**

- **A laser is a convenient source of coherent light.** When we use a laser to demonstrate double slit interference, we can illuminate the double slits directly. We do not need to make the light pass through a narrow single slit first as we do with light from a non-laser light source. This is because a laser is a source of coherent light. A light source emits light as a result of electrons inside its atoms moving to lower energy levels inside the atoms. Each such electron emits a photon, which is a packet of electromagnetic waves of constant frequency. Inside a laser, each emitted photon causes more photons to be emitted as it passes through the light-emitting substance. These stimulated photons are in phase with the photon that caused them. As a result, the photons in a laser beam are in phase with each other. So the laser is a coherent source of light. In comparison, in a non-laser light source, the atoms emit photons at random so the photons in such a beam have random phase differences.

### Synoptic link

You have met photons in more detail in Topic 1.3, Photons.

### DVDs

Laser light is used to read DVDs in computers and DVD players. The track of a DVD is very narrow and is encoded with pits representing digital information. Laser light is reflected from the track. The pits cause the laser light to change in intensity. This change is then converted by a detector into an electronic signal to produce the output.

### Study tip

You need to know that a laser is a coherent source of light and a non-laser source is not a coherent source.

## White light fringes

Figure 3 shows the fringe patterns observed with blue light and with red light. As explained above, the blue light fringes are closer together than the red light fringes. However, the central fringe of each pattern is in the same position on the screen. The fringe pattern produced by white light is shown in Figure 4. Each component colour of white light produces its own fringe pattern, and each pattern is centred on the screen at the same position. As a result:

The central fringe is white because every colour contributes at the centre of the pattern.

The inner fringes are tinged with blue on the inner side and red on the outer side. This is because the red fringes are more spaced out than the blue fringes and the two fringe patterns do not overlap exactly.

The outer fringes merge into an indistinct background of white light becoming fainter with increasing distance from the centre. This is because, where the fringes merge, different colours reinforce and therefore overlap.

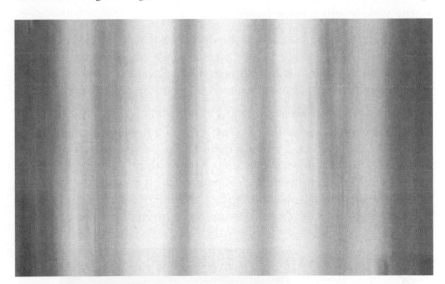

▲ **Figure 4** *White light fringes*

### Summary questions

1 **a** Sketch an arrangement that may be used to observe the fringe pattern when light from a narrow slit illuminated by a sodium vapour lamp is passed through a double slit.

**b** Describe the fringe pattern you would expect to observe in **a**.

2 Describe how the fringe pattern would change in **1a** if the narrow single slit is replaced by a wider slit.

3 Double slit interference fringes are observed using light of wavelength 590 nm and a double slit of slit spacing 0.50 mm. The fringes are observed on a screen at a distance of 0.90 m from the double slits. Calculate the fringe separation of these fringes.

4 Describe and explain the fringe pattern that would be observed in **3** if the light source were replaced by a white light source.

# 5.6 Diffraction

## Learning objectives:

- → Explain why diffraction of light is important in the design of optical instruments.
- → Compare the single slit diffraction pattern with the pattern of Young's fringes.
- → Describe the effect of the single slit pattern on the brightness of Young's fringes.

*Specification reference: 3.3.2.2*

### Synoptic link

You have met the study of water waves using a ripple tank in more detail in Topic 4.3, Wave properties 1.

### Hint

The central bright fringe of the single slit pattern is twice as wide as the others, and much brighter.

## Observing diffraction

Diffraction is the spreading of waves when they pass through a gap or by an edge. This general property of all waves is very important in the design of optical instruments, such as cameras, microscopes, and telescopes. For example, when a telescope is used to observe planets, we can often see features that are not evident when observed directly. This is partly because less diffraction occurs when waves pass through a wide gap than through a narrow gap. Therefore, because a telescope is much wider than the eye pupil, much less diffraction occurs when using a telescope than when observing with the unaided eye.

**Diffraction of water waves through a gap** can be observed using a ripple tank. This arrangement shows that the diffracted waves spread out more if:

- the gap is made narrower, or
- the wavelength is made larger.

In addition, close examination of the diffracted waves should reveal that each diffracted wavefront has breaks either side of the centre. These breaks are due to waves diffracted by adjacent sections on the gap being out of phase and cancelling each other out in certain directions.

**Diffraction of light by a single slit** can be demonstrated by directing a parallel beam of light at the slit. The diffracted light forms a pattern that can be observed on a white screen. The pattern shows a central fringe with further fringes either side of the central fringe, as shown in Figure 2. The intensity of the fringes is greatest at the centre of the central fringe. Figure 2 also shows the variation of the intensity of the diffracted light with the distance from the centre of the fringe pattern.

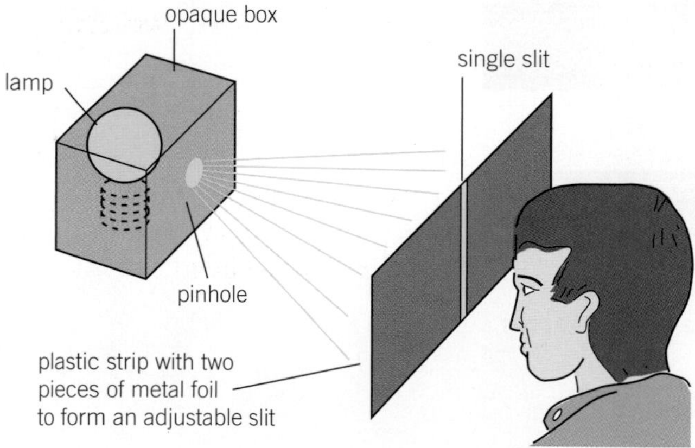


▲ **Figure 1**

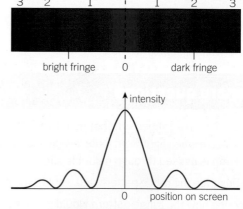

▲ **Figure 2** *Single slit diffraction*

### *Note*

- The central fringe is twice as wide as each of the outer fringes (measured from minimum to minimum intensity).
- The peak intensity of each fringe decreases with distance from the centre.
- Each of the outer fringes is the same width.
- The outer fringes are much less intense than the central fringe.

## More about single slit diffraction

If the single slit pattern is observed:

- Using different sources of monochromatic light in turn, the observations show that the greater the wavelength, the wider the fringes.
- Using an adjustable slit, the observations show that making the slit narrower makes the fringes wider.

It can be shown theoretically that the width $W$ of the central fringe observed on a screen at distance $D$ from the slit is given by

$$W = \frac{\textbf{the wavelength of the light } (\lambda)}{\textbf{the width of the single slit } (a)} \times 2D$$

Therefore, the width of each fringe is proportional to $\lambda / a$. For this reason, the fringes are narrower using blue light than if red light had been used.

## Single slit diffraction and Young's fringes

In the double slit experiment in Topic 5.4, light passes through the two slits of the double slit arrangement and produces an interference pattern. However, if the slits are too wide and too far apart, no interference pattern is observed. This is because interference can only occur if the light from the two slits overlaps. For this to be the case:

- each slit must be narrow enough to make the light passing through it diffract sufficiently
- the two slits must be close enough so the diffracted waves overlap on the screen.

In general, for monochromatic light of wavelength $\lambda$, incident on two slits of aperture width $a$ at slit separation $s$ (from centre to centre),

- the fringe spacing of the interference fringes, $w = \frac{\lambda D}{s}$
- the width of the central diffraction fringe, $W = \frac{2\lambda D}{a}$, where $D$ is the slit–screen distance.

Figure 4 shows the intensity variation with distance across the screen in terms of distance from the centre of the pattern. Using the expressions above, you should be able to see that only a few interference fringes will be observed in the central diffraction fringe if the slit separation $s$ is small compared to the slit width $a$.

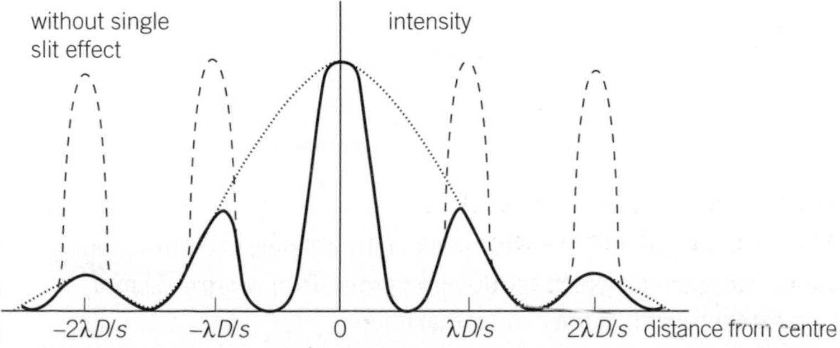


▲ **Figure 4** *Intensity distribution for Young's fringes*

### Microscopes and diffraction

Microscopes are often fitted with a blue filter as greater detail (better resolution) can be seen in microscope images observed with blue light than with white light. In an electron microscope, the resolution is increased when a higher voltage setting is used. This is because the higher the voltage, the greater the speed of the electrons in the microscope beam and therefore the smaller the de Broglie wavelength of the electrons. Hence the electrons are diffracted less as they pass through the microscope's magnetic lenses.

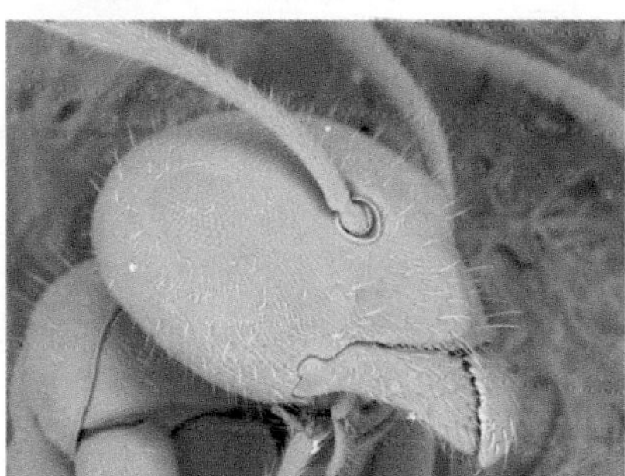

▲ **Figure 3** *An electron microscope image*

### Synoptic link

You have met the de Broglie wavelength in Topic 3.6, Wave–particle duality.

## Summary questions

1 Red light from a laser is directed normally at a slit that can be adjusted in width. The diffracted light from the slit forms a pattern of diffraction fringes on a white screen. Describe how the appearance of the fringes changes if the slit is made wider.

2 A narrow beam of white light is directed normally at a single slit. The diffracted light forms a pattern on a white screen.

   a A blue filter is placed in the path of the beam before it reaches the slit. The distance across five fringes including the central fringe is 18 mm. Calculate the width of the central fringe.

   b The blue filter is replaced by a red filter. Compare the red fringe pattern with the blue fringe pattern.

3 Figure 3 on page 77 shows the interference fringes observed in a Young's fringes experiment.

   a Explain why the fringes at the outer edges are dimmer than the fringes nearer the centre.

   b Describe how the appearance of the fringes would change if the slits were made wider without changing their centre-to-centre separation.

   c Sketch a graph to show how the intensity of the fringes varies with distance from the centre of the central fringe.

## Explaining single slit diffraction

In Figure 2, the intensity minima occur at evenly spaced positions on the screen. To explain this, consider wavelets emitted simultaneously by equally spaced point sources across the slit each time a wavefront arrives at the slit. At a certain position, P, on the screen, the wavelet from each point source arrives a short time after the wavelet from the adjacent point source nearer P. Therefore, at any given time, the phase difference $\phi$ between the contributions from any two adjacent point sources is constant.

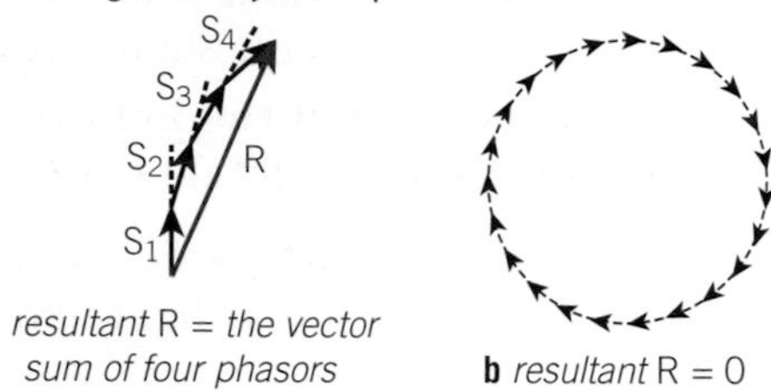


**a** *resultant* R = *the vector sum of four phasors*

**b** *resultant* R = 0

▲ **Figure 5**

The contributions at P from each point source can be represented by phased vectors or **phasors** as shown in Figure 5. Each phasor is at angle $\phi$ to each adjacent phasor. In Figure 5a, the resultant is the vector sum of four phasors, which is non-zero. In Figure 5b, the resultant is zero because $\phi \times$ the number of phasors is equal to $2\pi$.

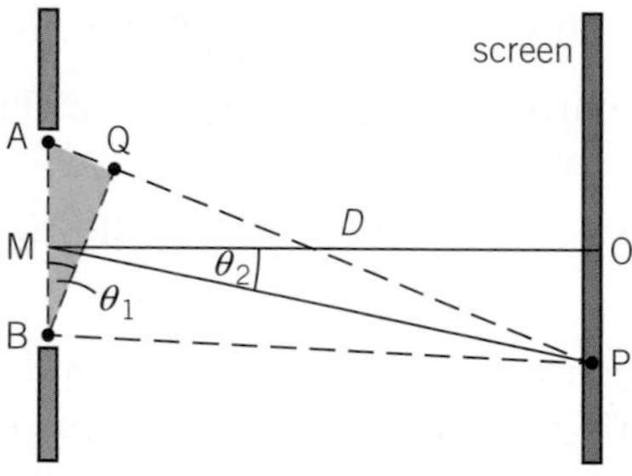


▲ **Figure 6**

More generally, intensity minima occur at positions P so that the phase difference between contributions from two point sources A and B at opposite sides of the slit is equal to $2\pi m$, where $m$ is a whole number (see Figure 6), so the path difference AP – BP = $m\lambda$.

In Figure 6, point Q on the line AP is such that QP = BP.
Therefore AQ = AP – QP = AP – BP = $m\lambda$.

Because $A\widehat{Q}B$ is almost 90°, $\sin\theta_1$ = AQ/AB = $m\lambda/a$, where the aperture width $a$ = AB, and angle $\theta_1 = A\widehat{B}Q$.

Consider triangles AQB and MOP, where M is the midpoint of the slit, and O is the point on the screen where the intensity is greatest. The two triangles are almost identical. This means that OP/OM = AQ/AB = $m\lambda/a$.

Intensity minima occurs at evenly spaced positions P such that

**OP = $m\lambda D/a$   where D = distance OM.**

**Question**

1 Light of wavelength 620 mm is used to form a single slit diffraction pattern on a screen which is 800 mm from a slit of width 0.24 mm. Calculate the width of the central maximum.

# 5.7 The diffraction grating

## Testing a diffraction grating

A **diffraction grating** consists of a plate with many closely spaced parallel slits ruled on it. When a parallel beam of monochromatic light is directed normally at a diffraction grating, light is transmitted by the grating in certain directions only. This is because:

- the light passing through each slit is diffracted
- the diffracted light waves from adjacent slits reinforce each other in certain directions only, including the incident light direction, and cancel out in all other directions.

Figure 1 shows one way to observe the effect of a diffraction grating on the incident light beam.

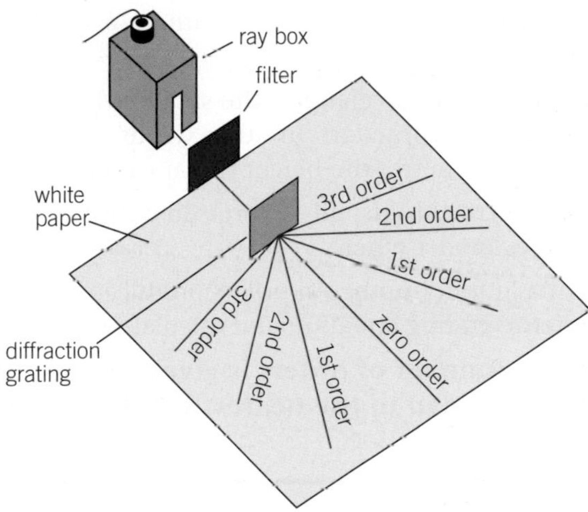


▲ **Figure 1** *The diffraction grating*

The central beam, referred to as the zero order beam, is in the same direction as the incident beam. The other transmitted beams are numbered outwards from the zero order beam. The angle of diffraction between each transmitted beam and the central beam increases if:

- light of a longer wavelength is used (e.g., by replacing a blue filter with a red filter)
- a grating with closer slits is used.

## The diffraction grating equation

Consider a magnified view of part of a diffraction grating and a snapshot of the diffracted waves, as shown in Figure 2. Each slit diffracts the light waves that pass through it. As each diffracted wavefront emerges from a slit, it reinforces a wavefront from a slit adjacent to it. In this example, the wavefront emerging at P reinforces the wavefront emitted from Q one cycle earlier, which reinforces the wavefront emitted from R one cycle earlier, and so on. The effect is to form a new wavefront PYZ.

### Learning objectives:

- → Explain why a diffraction grating diffracts monochromatic light in certain directions only.
- → If a coarser grating is used, explain the effect on the number of diffracted beams produced and on the spread of each diffracted beam.
- → Determine the grating spacing for any given grating, if it is not known.

*Specification reference: 3.3.2.2*

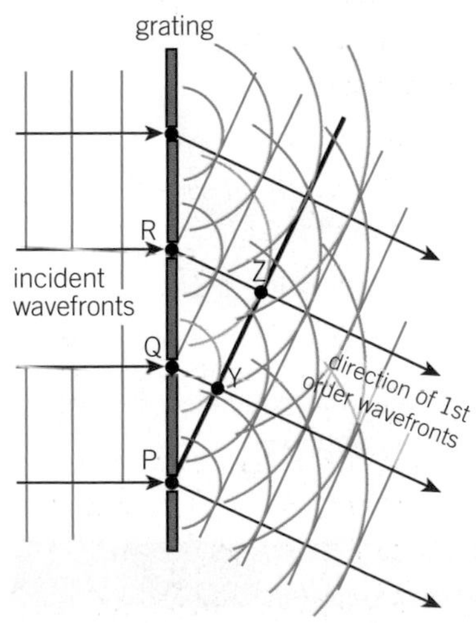


▲ **Figure 2** *Formation of the first order wavefront*

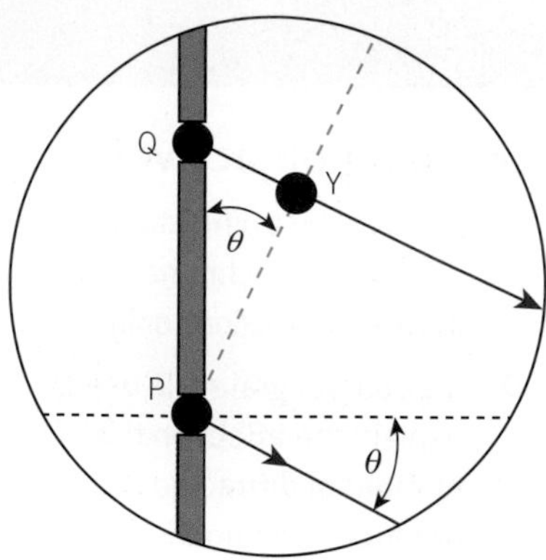


▲ **Figure 3** *The nth order wavefront*

Figure 3 shows the formation of a wavefront of the $n^{th}$ order beam. The wavefront emerging from slit P reinforces a wavefront emitted $n$ cycles earlier by the adjacent slit Q. This earlier wavefront therefore must have travelled a distance of $n$ wavelengths from the slit. Thus the perpendicular distance QY from the slit to the wavefront is equal to $n\lambda$, where $\lambda$ is the wavelength of the light waves.

Since the angle of diffraction of the beam, $\theta$, is equal to the angle between the wavefront and the plane of the slits, it follows that $\sin\theta = QY/QP$, where QP is the grating spacing $d$.

Substituting $d$ for QP and $n\lambda$ for QY into the equation above gives $\sin\theta = n\lambda/d$. Rearranging this equation gives the diffraction grating equation for the angle of diffraction of the $n^{th}$ order beam

$$d\sin\theta = n\lambda$$

### Notes:

1 The number of slits per metre on the grating, $N = 1/d$, where $d$ is the grating spacing.

2 For a given order and wavelength, the smaller the value of $d$, the greater the angle of diffraction. In other words, the larger the number of slits per metre, the bigger the angle of diffraction.

3 Fractions of a degree are usually expressed either as a decimal or in minutes (abbreviated ′) where $1° = 60'$.

4 To find the maximum number of orders produced, substitute $\theta = 90°$ ($\sin\theta = 1$) in the grating equation and calculate $n$ using $n = d/\lambda$.

**The maximum number of orders is given by the value of $d/\lambda$ rounded down to the nearest whole number.**

## Hint

A minimum in the single slit diffraction will suppress a diffracted order from the grating, and there will be a missing order in the pattern as a result.

## Study tip

The number of maxima observed is $2n + 1$, where $n$ is the greatest order.

## Diffraction gratings in action

Diffraction gratings can be made by cutting parallel grooves very close together on a smooth glass plate. Each groove transmits some incident light and reflects or scatters some. So the grooves act as coherent emitters of waves just as if they were slits. However, the effective slit width needs to be much smaller than the grating spacing. This is so that the diffracted waves from each groove spread out widely. If they did not, the higher order beams would be much less intense than the lower order beams.

We can use a diffraction grating in a **spectrometer** to study the spectrum of light from any light source and to measure light wavelengths very accurately. A spectrometer is designed to measure angles to within 1 arc minute, which is a sixtieth of a degree. The angle of diffraction of a diffracted beam can be measured very accurately. The grating spacing of a diffraction grating can therefore be measured very accurately using light of a known wavelength. The grating can then be used to measure light of any wavelength.

Most industrial and research laboratories now use a **spectrum analyser**, which is an electronic spectrometer linked to a computer that gives a visual display of the variation of intensity with wavelength. You do not need to know how spectrometers or spectrum analysers work or are used.

▲ **Figure 4** *A spectrometer fitted with a diffraction grating*

## Synoptic link

You have met line spectra in more detail in Topic 3.5, Energy levels and spectra.

## Types of spectra

### Continuous spectra

The spectrum of light from a filament lamp is a continuous spectrum of colour from deep violet at about 350 nm to deep red at about 650 nm, as shown in Figure 5. The most intense part of the spectrum depends on the temperature of the light source. The hotter the light source, the shorter the wavelength of the brightest part of the spectrum. By measuring the wavelength of the brightest part of a continuous spectrum, we can therefore measure the temperature of the light source.

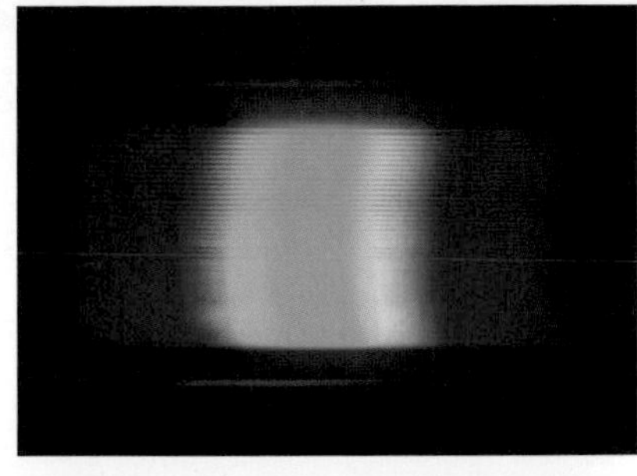

▲ **Figure 5** *A continuous spectrum*

### Line emission spectra

A glowing gas in a vapour lamp or a discharge tube emits light at specific wavelengths so its spectrum consists of narrow vertical lines of different colours as shown in Figure 6. The wavelengths of the lines are characteristic of the chemical element that produced the light. If a glowing gas contains more than one element, the elements in the gas can be identified by observing its line spectrum.

▲ **Figure 6** *A line emission spectrum*

### Line absorption spectra

A line absorption spectrum is a continuous spectrum with narrow dark lines at certain wavelengths. For example, if the spectrum of light from a filament lamp is observed after passing it through a glowing gas, thin dark vertical lines are observed superimposed on the continuous spectrum, as shown in Figure 7. The pattern of the dark lines is due to the elements in the glowing gas. These elements absorb light of the same wavelengths they can emit at so the transmitted light is missing these wavelengths. The atoms of the glowing gas that absorb light then emit the light subsequently but not necessarily in the same direction as the transmitted light.

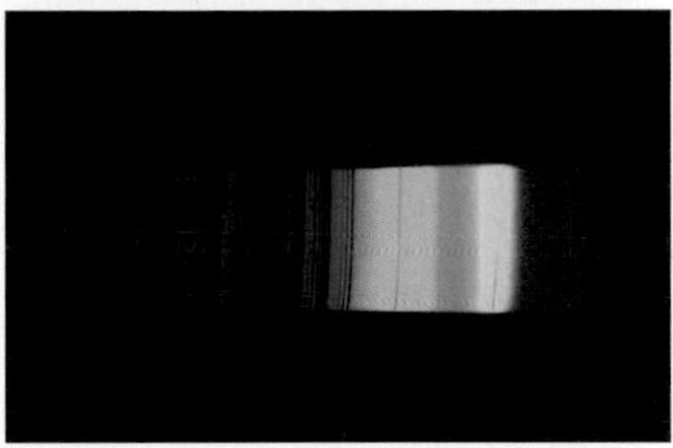

▲ **Figure 7** *A line absorption spectrum*

## Summary questions

**1** $\sqrt{x}$ A laser beam of wavelength 630 nm is directed normally at a diffraction grating with 300 lines per millimetre. Calculate:

- **a** the angle of diffraction of each of the first two orders
- **b** the number of diffracted orders produced.

**2** $\sqrt{x}$ Light directed normally at a diffraction grating contains wavelengths of 580 and 586 nm only. The grating has 600 lines per mm.

- **a** How many diffracted orders are observed in the transmitted light?
- **b** For the highest order, calculate the angle between the two diffracted beams.

**3** $\sqrt{x}$ Light of wavelength 430 nm is directed normally at a diffraction grating. The first order transmitted beams are at 28° to the zero order beam. Calculate:

- **a** the number of slits per millimetre on the grating
- **b** the angle of diffraction for each of the other diffracted orders of the transmitted light.

**4** $\sqrt{x}$ A diffraction grating is designed with a slit width of 0.83 μm. When used in a spectrometer to view light of wavelength 430 nm, diffracted beams are observed at angles of 14° 55′ and 50° 40′ to the zero order beam. (Reminder: 1 degree = 60 minutes.)

- **a** Assuming the low-angle diffracted beam is the first order beam, calculate the number of lines per mm on the grating.
- **b** Explain why there is no diffracted beam between the two observed beams. What is the order number for the beam at 50° 40′?

# Practice questions: Chapter 5

1 In a secure communications system, an optical fibre is used to transmit digital signals in the form of infrared pulses. The fibre has a thin core, which is surrounded by cladding that has a lower refractive index than the core.

**(a)** **Figure 1** shows a cross section of a straight length of the fibre.

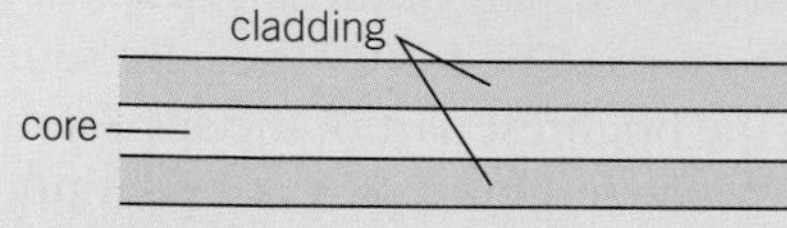


▲ **Figure 1**

(i) The core has a refractive index of 1.52 and the cladding has a refractive index of 1.35. Show that the critical angle at the core–cladding boundary is 62.6°.

(ii) Sketch the path of a light ray in the core that is totally internally reflected when it reaches the core–cladding boundary. *(6 marks)*

**(b)** (i) Explain why the cladding is necessary.

(ii) An optical fibre is used to transmit digital images from a security camera to a computer where the images are stored. A prominent notice near the camera informs people that the camera is in use. Discuss the benefits and drawbacks associated with making and storing images of people in this situation. *(7 marks)*

2 **Figure 2** shows a cube of glass. A ray of light, incident at the centre of a face of the cube, at an angle of incidence $\theta$, goes on to meet another face at an angle of incidence of 50°, as shown in **Figure 2**.
Critical angle at the glass–air boundary = 45°

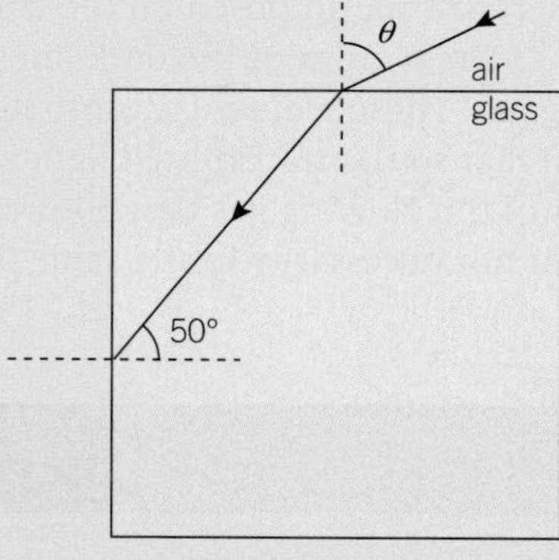


▲ **Figure 2**

**(a)** Draw on the diagram the continuation of the path of the ray, showing it passing through the glass and out into the air. *(3 marks)*

**(b)** Show that the refractive index of the glass is 1.41. *(2 marks)*

**(c)** Calculate the angle of incidence, $\theta$. *(3 marks)*

AQA, 2005

3 **Figure 3** shows a ray of monochromatic light, in the plane of the paper, incident on the end face of an optical fibre.

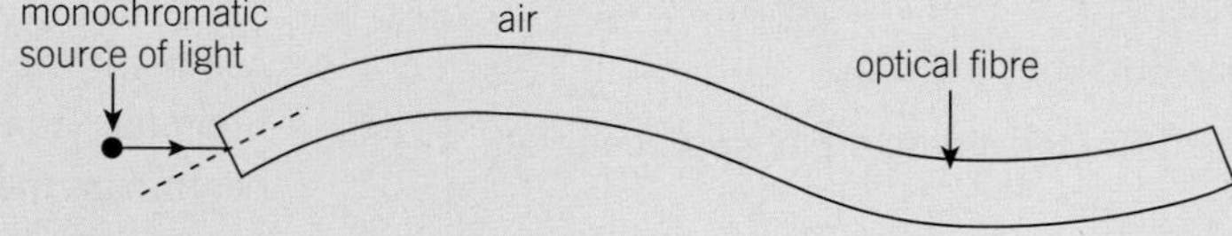


▲ **Figure 3**

**(a)** (i) Draw on a copy of the diagram the complete path followed by the incident ray, showing it entering into the fibre and emerging from the fibre at the far end.

(ii) State any changes that occur in the speed of the ray as it follows this path from the source.
Calculations are not required. *(4 marks)*

**(b)** √x (i) Calculate the critical angle for the optical fibre at the air boundary. refractive index of the optical fibre glass = 1.57.
(ii) The optical fibre is now surrounded by cladding of refractive index 1.47. Calculate the critical angle at the core–cladding boundary.
(iii) State *one* advantage of cladding an optical fibre. *(6 marks)*
AQA, 2004

**4** **(a)** State what is meant by coherent sources of light. *(2 marks)*

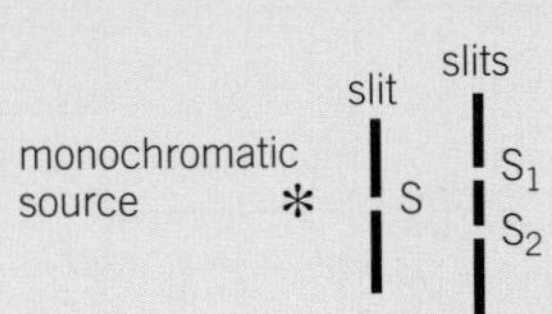


▲ **Figure 4**

**(b)** Young's fringes are produced on the screen from the monochromatic source by the arrangement shown in **Figure 4**.
(i) Explain why slit S should be narrow.
(ii) Why do slits $S_1$ and $S_2$ act as coherent sources? *(4 marks)*

**(c)** The pattern on the screen may be represented as a graph of intensity against position on the screen. The central fringe is shown on the graph in **Figure 5**. Copy and complete this graph to represent the rest of the pattern. *(2 marks)*

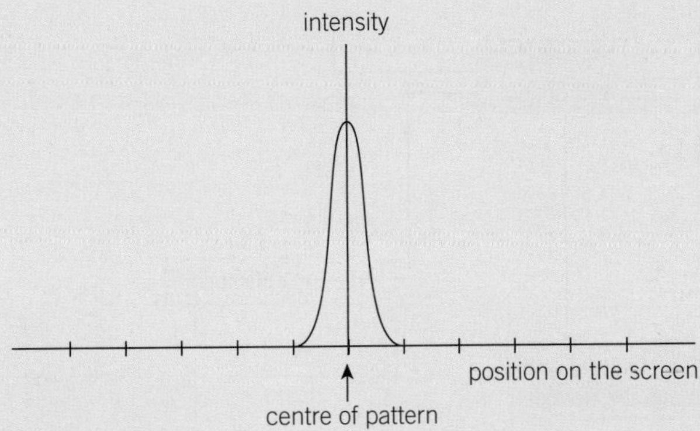


▲ **Figure 5**

AQA, 2005

**5** √x A diffraction grating has 940 lines per mm.
**(a)** Calculate the distance between adjacent lines on the grating. *(1 mark)*
**(b)** Monochromatic light is incident on the grating and a second order spectral line is formed at an angle of 55° from the normal to the grating. Calculate the wavelength of the light. *(3 marks)*
AQA, 2006

# Section 2 Summary

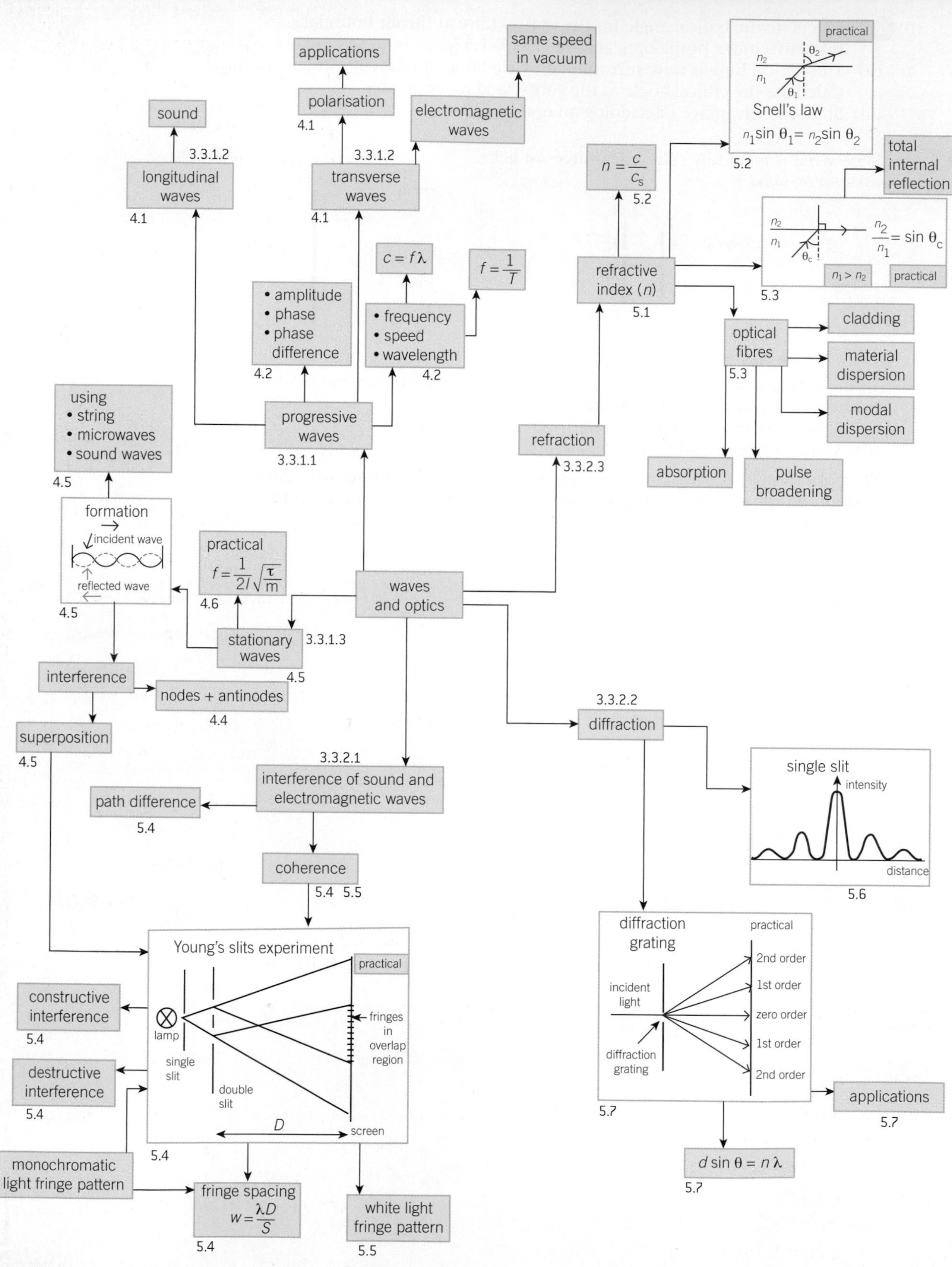

## Practical skills

In this section you have met the following skills:

- use of a tuning fork/microphone/ripple tank/ signal generator/microwave source to generate and observe waves
- use of a light source (including a laser) to investigate refraction, interference and diffraction.
- use of electrical equipment and lasers safely. (e.g. use of a ray box)
- use of a protractor to measure angles.
- use of a metre rule to measure distance (e.g. for stationary waves)
- use of an oscilloscope to observe waveforms (e.g. sound waves from a microphone) (note: oscilloscope measurements are A Level not AS Level)
- use of double slits and a single slit to observe fringe patterns (e.g. Young's fringes)

In all practical experiments:

- experimental data must be displayed in a table with headings and units and significant figures used correctly
- repeating results improves reliability
- the precision of any instrument used must be recorded
- plot a graph from experimental and/or processed data, labelling axes with quantity and units appropriate scales
- consider the accuracy of your results/ conclusion, compared to a known value, and consider reasons for and types of error
- identify uncertainties in measurements and combine them to find the uncertainty in the overall experiment.

## Maths skills

In this section you have met the following skills:

- use of standard form and conversion to standard form from units with prefixes
- use calculators to solve calculations involving powers of ten and sin $x$ (e.g. $d \sin \theta = n\lambda$)
- use of the appropriate number of significant figures in all calculations
- use of appropriate units in all calculations
- use of protractor to measure angles
- quote phase differences in degrees or radians and convert from one to the other
- calculation of a mean from a set of repeat readings to find fringe spacing
- change the subject of an equation in order to solve calculations (e.g. $d \sin \theta = n\lambda$)
- plot a graph from data provided or found experimentally (e.g. sin $i$ against sin $r$)
- relate $y = mx + c$ to a linear graph with gradient $m$ and intercept $c$, and understand what these represent in physics (e.g. sin $i$ against sin $r$).

## Extension task

Research seismic waves using text books and the internet and produce a presentation suitable for showing to your class.

Include in your presentation:

- types of seismic wave and whether they are longitudinal or transverse
- where an earthquake starts and how it progresses
- why some earthquakes are more devastating than others
- how they are measured
- what a tsunami is and why it occurs.

# Practice questions: Waves and optics

1 (a) **Figure 1** shows an arrangement used to investigate the properties of microwaves.

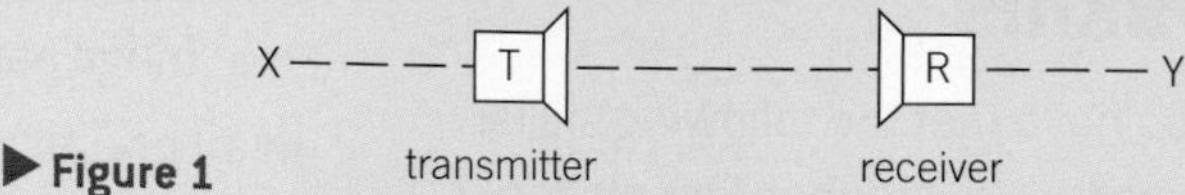


▶ **Figure 1**

When the transmitter T was rotated through 90° about the straight line XY, the receiver signal decreased to zero. Explain why this happened and state the property of microwaves responsible for this effect. *(3 marks)*

(b) A microwave oven produces microwaves of wavelength 0.12 m in air.

(i) Calculate the frequency of these microwaves.

(ii) In a certain oven, explain why food heated in a fixed position in this oven would be uncooked at certain points if stationary waves were allowed to form. *(3 marks)*

AQA, 2004

2 **Figure 2** shows a cross-sectional view of the base of a glass tank containing water. A point monochromatic light source is in contact with the base, and ray, $R_1$, from the source has been drawn up to the point where it emerges along the surface of the water.

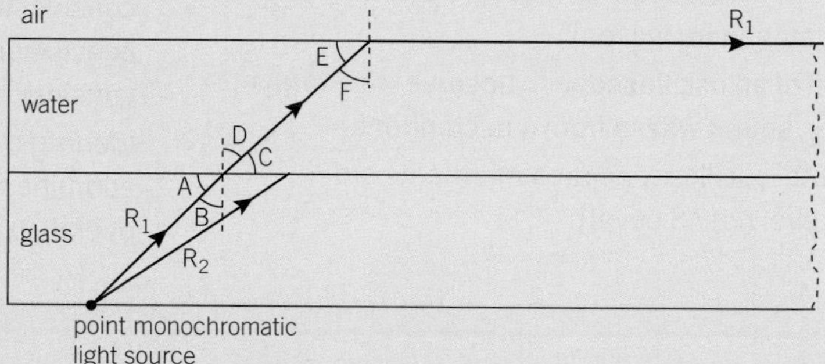


▶ **Figure 2**

(a) (i) Which angle, out of angles A to F, is a critical angle?

(ii) Explain how the path of $R_1$ demonstrates that the refractive index of glass is greater than the refractive index of water. *(2 marks)*

(b) Using the following information:

A = 47.1°  B = 42.9°  C = E = 41.2°  D = F = 48.8°

calculate

(i) the refractive index of water

(ii) the ratio $\dfrac{\text{speed of light in water}}{\text{speed of light in glass}}$. *(5 marks)*

(c) Ray $R_2$ emerges from the source a few degrees away from ray $R_1$ as shown. Draw the continuation of ray $R_2$. Where possible show the ray being refracted. *(2 marks)*

AQA, 2004

3 Red light from a laser is passed through a single narrow slit, as shown in **Figure 3**. A pattern of bright and dark regions can be observed on the screen which is placed several metres beyond the slit.

(a) The pattern on the screen may be represented as a graph of intensity against distance along the screen. The graph has been started in outline in **Figure 4**. The central bright region is already shown. Copy and complete this graph to represent the rest of the pattern. *(4 marks)*

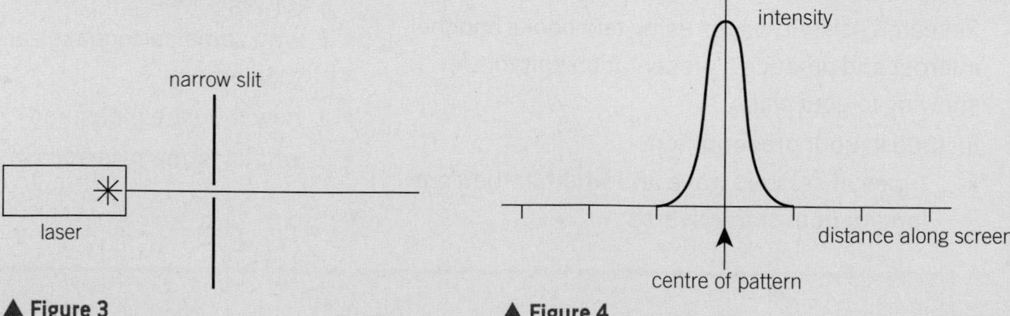


▲ **Figure 3**  ▲ **Figure 4**

**(b)** State the effect on the pattern if each of the following changes is made separately.

(i) The width of the narrow slit is reduced.

(ii) With the original slit width, the intense red source is replaced with an intense source of green light. *(3 marks)*

AQA, 2002

**4** A student tests a fence consisting of evenly spaced wooden strips to find out if it diffracts sound waves. She stands on one side of a fence that consists of evenly spaced wooden strips. The fence behaves as a diffraction grating for sound waves.

**(a)** She uses a loudspeaker to send a sound wave of frequency 2.4 kHz towards the fence.
Calculate the wavelength of this sound wave.
speed of sound in air, $v = 340\,\mathrm{m\,s^{-1}}$.

**(b)** A diffracted first order maximum is observed at an angle of 28°.
Calculate the spacing between adjacent wooden strips.

**(c)** Determine the highest order of maximum that could be heard. *(6 marks)*

AQA, 2006

**5** **Figure 5** and **Figure 6** show a version of Quincke's tube, which is used to demonstrate interference of sound waves.

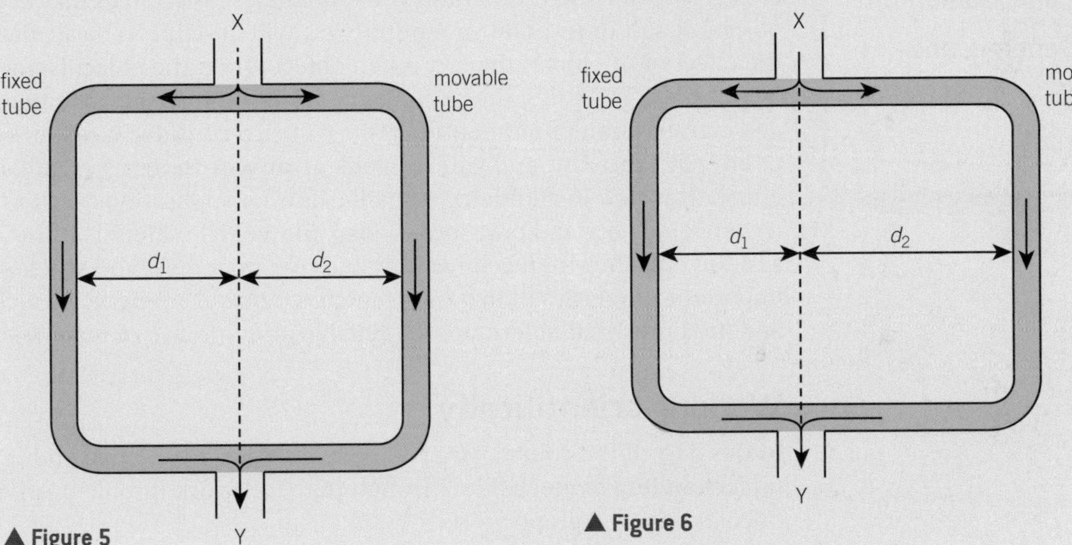


▲ **Figure 5**

▲ **Figure 6**

A loudspeaker at X produces sound waves of one frequency. The sound waves enter the tube and the sound energy is divided equally before travelling along the fixed and movable tubes. The two waves superpose and are detected by a microphone at Y.

**(a)** The movable tube is adjusted so that $d_1 = d_2$ and the waves travel the same distance from X to Y, as shown in **Figure 5**. As the movable tube is slowly pulled out, as shown in **Figure 6**, the sound detected at Y gets quieter and then louder. Explain the variation in the loudness of the sound at Y as the movable tube is slowly pulled out. *(4 marks)*

**(b)** The tube starts in the position shown in **Figure 5**. Calculate the minimum distance moved by the movable tube for the sound detected at Y to be at its quietest.
frequency of sound from loudspeaker = 800 Hz
speed of sound in air = $340\,\mathrm{m\,s^{-1}}$ *(3 marks)*

**(c)** Quincke's tube can be used to determine the speed of sound.
State and explain the measurements you would make to obtain a reliable value for the speed of sound using Quincke's tube and a sound source of known frequency. *(4 marks)*

AQA Specimen paper 1, 2015

# Section 3

## Mechanics and materials

### Chapters in this section:


6 Forces in equilibrium

7 On the move

8 Newton's laws of motion

9 Force and momentum

10 Work, energy, and power

11 Materials


### Introduction

In this section, you will look at the principles and applications of mechanics and materials. These subject areas underpin many work-related areas including engineering, transport, and technology. A lot of new technologies and devices have been developed in these subject areas, including vehicle safety features and nanotechnology, which is about devices that are too small for us to see without a powerful microscope.

This section builds on your GCSE studies on motion, force, and energy. In studying mechanics, you will analyse the forces that keep objects at rest or in uniform motion. You will describe and calculate the effect of the forces that act on an object when the object is not in equilibrium, including the relationship between force and momentum, and conservation of momentum. You will also consider **work** in terms of **energy transfer**, and you will look at **power** in terms of rate of energy transfer. In addition, you will examine applications such as terminal velocity and drag forces, road and vehicle safety features, and the efficiency of machines. By studying materials, you will look at important properties of materials, such as density, strength, elasticity, and the limits that determine the reliable and safe use of materials.

### Working scientifically

In this part of the course, you will develop your knowledge and understanding of mechanics through practical work, problem-solving, calculations, and graph work.

Practical work in this section involves making careful measurements by using instruments such as micrometers, top pan balances, light gates, and stopwatches. You will also be expected to assess the accuracy of your measurements and the results of calculations from your measured data. These practical skills are part of the everyday work of scientists. In some experiments, such as measurements of density, you will test the accuracy of your results by comparing them with accepted values. By estimating how accurate a result is (i.e., its uncertainty), you will know how close your result is to the accepted value. In this way, you will develop and improve your practical skills to a level where you can have confidence in a result if the accepted value is not known. Chapter 14 gives you lots of information and tips about practical skills.

You will develop your maths skills in this section by rearranging equations and in graph work. In some experiments, you will use your measurements to plot a graph that is predicted to be a straight line. Sometimes, a quantity derived from the measurements (e.g., $t^2$, where time $t$ is measured) is plotted to give a straight line graph. In these experiments, you may be asked to determine the graph gradient (and/or intercept) and then relate the values you get to a physical quantity. Make good use of the notes and exercises in Chapter 16, including the section on straight line graphs, to help you with your maths skills.

## What you already know

From your GCSE studies on force and energy, you should know that:

- ▢ speed is distance travelled per unit time, and velocity is speed in a given direction
- ▢ the acceleration of an object is its rate of change of velocity
- ▢ an object acted on by two equal and opposite forces (i.e., when the result fore is zero) is at rest or moves at constant velocity
- ▢ when an object accelerates or decelerates:
  - the greater the resultant force acting on the object, the greater is its acceleration
  - the greater the mass of the object, the smaller is its acceleration.
- ▢ the gravitational field strength $g$ at any point is the force per unit mass on a small object caused by the Earth's gravitational attraction on the object
- ▢ the weight of an object in newtons (N) = its mass in kilograms (kg) × $g$
- ▢ work is done by a force when the force moves its point of application in the direction of the force
- ▢ energy is transferred by a force when it does work
- ▢ energy cannot be created or destroyed – it can only be transferred from one type of store into other types of stores
- ▢ power is the rate of transfer of energy.

# 6 Forces in equilibrium

## 6.1 Vectors and scalars

### Learning objectives:

- → Define a vector quantity.
- → Describe how we represent vectors.
- → Explain how we add and resolve vectors.

*Specification reference: 3.4.1.1*

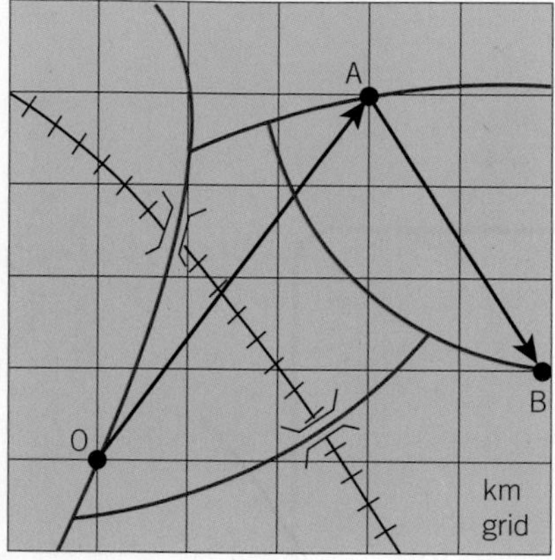


▲ **Figure 1** *Map of locality*

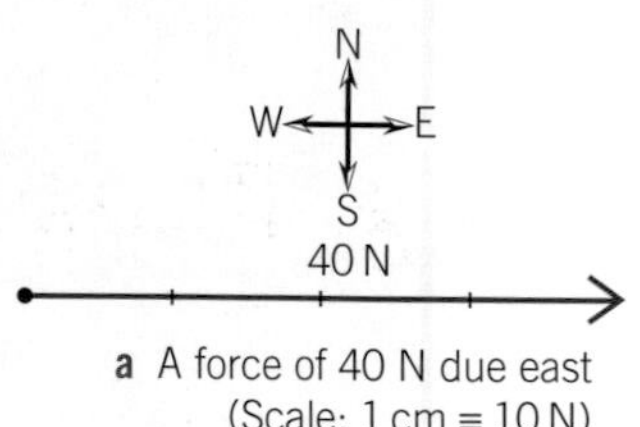


**a** A force of 40 N due east (Scale: 1 cm ≡ 10 N)

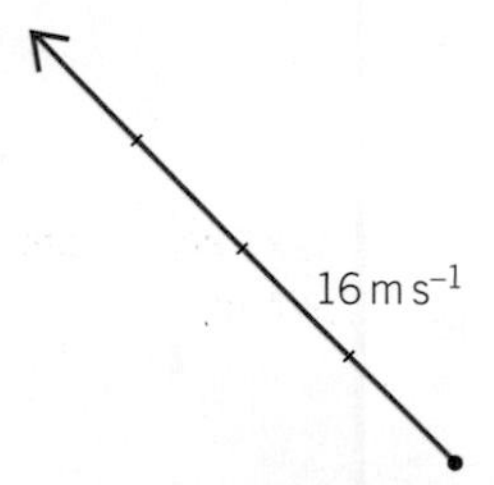


**b** A velocity of 16 m s$^{-1}$ at 45° west of due north (Scale: 1 cm ≡ 4 m s$^{-1}$)

▲ **Figure 2** *Representing vectors*

### Representing a vector

Imagine you are planning to cycle to a friend's home several kilometres away from your home. The **distance** you travel depends on your route. However, the direct distance from your home to your friend's home is the same, whichever route you choose. Distance in a given direction, or **displacement**, is an example of a **vector** quantity because it has magnitude and direction. In Figure 1, suppose your home is at point O on the map and your friend's home is at A. The road distance you would cycle from O to A is greater than the direct distance or displacement from O to A. This is represented by the arrow from O to A.

**A vector is any physical quantity that has a direction as well as a magnitude.**

Further examples of vectors include velocity, acceleration, and force.

**A scalar is any physical quantity that is not directional.**

For example, distance is a scalar because it takes no account of direction. Further examples of scalars include mass, density, volume, and energy.

Any vector can be represented as an arrow. The length of the arrow represents the magnitude of the vector quantity. The direction of the arrow gives the direction of the vector.

- **Displacement** is distance in a given direction. As shown in Figure 1, the displacement from one point to another can be represented on a map or a scale diagram as an arrow from the first point to the second point. The length of the arrow must be in proportion to the least distance between the two points.
- **Velocity** is speed in a given direction. The velocity of an object can be represented by an arrow of length in proportion to the speed pointing in the direction of motion of the object.
- **Force and acceleration** are both vector quantities and therefore can each be represented by an arrow in the appropriate direction and of length in proportion to the magnitude of the quantity.

### Addition of vectors using a scale diagram

Let's go back to the cycle journey in Figure 1. Suppose when you reach your friend's home at A you then go on to another friend's home at B. Your journey is now a two-stage journey.

**Stage 1 from O to A** is represented by the displacement vector **OA**.

**Stage 2 from A to B** is represented by the displacement vector **AB**.

Figure 3 shows how the overall displacement from O to B, represented by vector OB, is the result of adding vector AB to vector OA. The resultant is the third side of a triangle where OA and AB are the other two sides.

$$OB = OA + AB$$

Use Figure 1 to show that the resultant displacement OB is 5.1 km in a direction 11° north of due east.

Any two vectors of the same type can be added together using a scale diagram. For example, Figure 4a shows a ship pulled via cables by two tugboats. The two pull forces $F_1$ and $F_2$ acting on the ship are at 40° to each other. Suppose the forces are both 8.0 kN. Figure 4b shows how you can find the **resultant** (combined effect) of the two forces using a scale diagram.

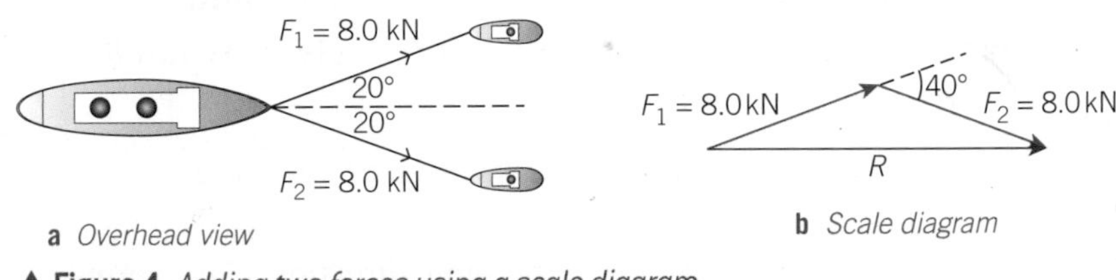


▲ **Figure 4** *Adding two forces using a scale diagram*

## Addition of two perpendicular vectors using a calculator

### 1 *Adding two displacement vectors that are at right angles to each other*

Suppose you walk 10.0 m forward then turn through exactly 90° and walk 7.0 m. At the end, how far will you be from your starting point? The vector diagram to add the two displacements is shown in Figure 5, drawn to a scale of 1 cm to 2.0 m. The two displacements form the two shorter sides of a right-angled triangle with the overall displacement, the resultant, as the hypotenuse. Using a ruler and a protractor, the resultant displacement can be shown to be a distance of 12.2 m at an angle of 35° to the initial direction. You can check this using

- Pythagoras's theorem for the distance (= $(10.0^2 + 7.0^2)^{1/2}$)
- the trigonometry equation, $\tan\theta = 7.0/10.0$ for the angle $\theta$ between the resultant and the initial direction.

### 2 *Two forces acting at right angles to each other*

Figure 6 shows an object O acted on by two forces $F_1$ and $F_2$ at right angles to each other. The vector diagram for this situation is also shown. The two forces in the vector diagram form two of the sides of a right-angled triangle in which the resultant force is represented by the hypotenuse.

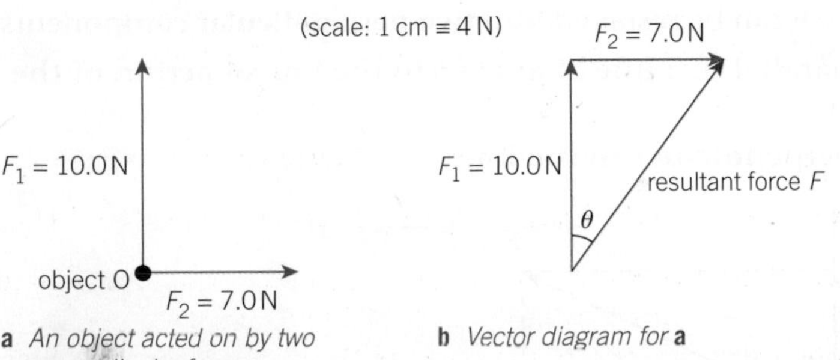


▲ **Figure 6**

**Hint**

Since **OA** and **AB** are vectors, the lengths do not add directly.

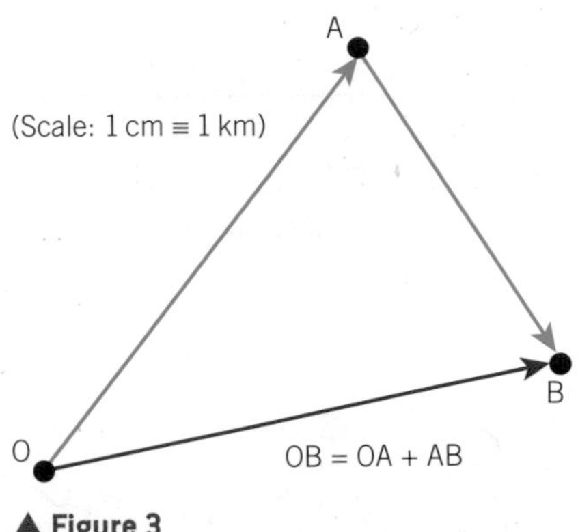


▲ **Figure 3**

**Study tip**

Drawing vector diagrams requires a ruler and careful drawing.

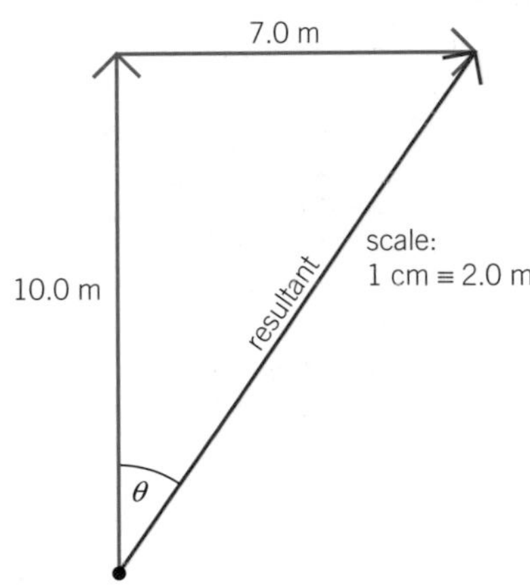


▲ **Figure 5** *Adding two displacements at right angles to each other*

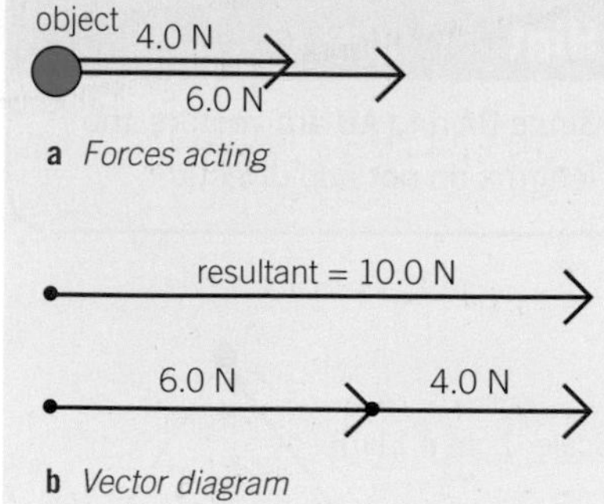


▲ **Figure 7** *Two forces acting in the same direction*

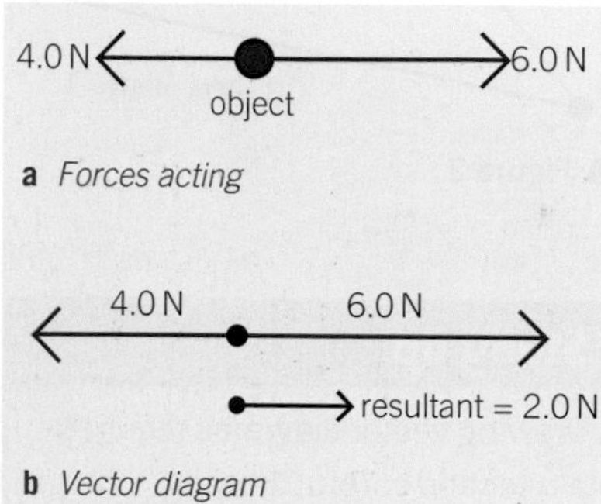


▲ **Figure 8** *Two forces acting in opposite directions*

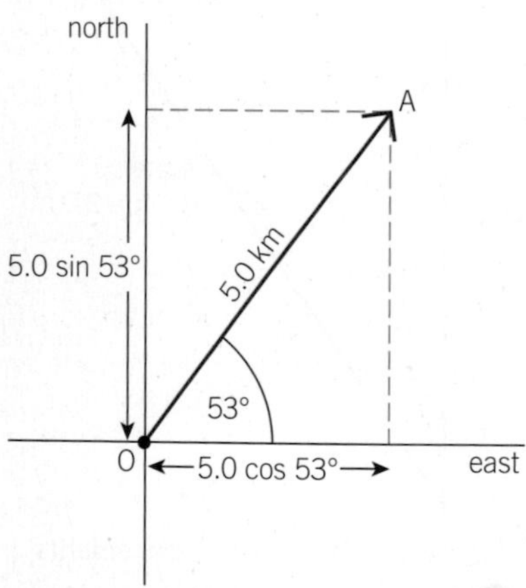


▲ **Figure 9** *Resolving a vector*

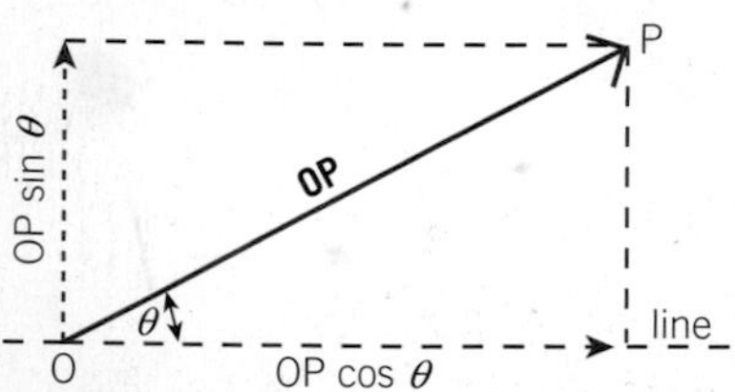


▲ **Figure 10** *The general rule for resolving a vector*

The vector diagram drawn to scale can therefore be used to find the resultant if the magnitudes of $F_1$ and $F_2$ are given. For example, the resultant of a force of 10.0 N acting at right angles to a force of 7.0 N has a magnitude of 12.2 N in a direction of 35° to the direction of the force of 10.0 N. You could check this using Pythagoras's theorem and the appropriate trigonometry equation. You might, however, recognise that you have already done this! See addition of displacement vectors on the previous page.

**In general, if the two perpendicular forces are $F_1$ and $F_2$, then**

- **the magnitude of the resultant $F = (F_1^2 + F_2^2)^{1/2}$ and**
- **the angle $\theta$ between the resultant and $F_1$ is given by $\tan\theta = \dfrac{F_2}{F_1}$**

***Note:***

The resultant of two vectors that act *along the same line* has a magnitude that is

- the sum, if the two vectors are in the same direction. For example, if an object is acted on by a force of 6.0 N and a force of 4.0 N both acting in the same direction, the resultant force is 10.0 N (Figure 7).
- the difference, if the two vectors are in opposite directions. For example, if an object is acted on by a 6.0 N force and a 4.0 N force in opposite directions, the resultant force is 2.0 N in the direction of the 6.0 N force (Figure 8).

## Resolving a vector into two perpendicular components

This is the process of working out the components of a vector in two perpendicular directions from the magnitude and direction of the vector. Figure 9 shows the displacement vector OA represented on a scale diagram that also shows lines due north and due east. The components of this vector along these two lines are 5.0 cos 53° km (= 3.0 km) along the line due east and 5.0 sin 53° km (= 4.0 km) along the line due north.

In general, to resolve any vector into two perpendicular components, draw a diagram showing the two perpendicular directions and an arrow to represent the vector. Figure 10 shows this diagram for a vector OP. The components are represented by the projection of the vector onto each line. If the angle $\theta$ between the vector OP and one of the lines is known,

- the component along that line = OP $\cos\theta$, and
- the component perpendicular to that line (i.e., along the other line) = OP $\sin\theta$

**Thus a force F can be resolved into two perpendicular components:**

- **$F\cos\theta$ parallel to a line at angle $\theta$ to the line of action of the force and**
- **$F\sin\theta$ perpendicular to the line.**

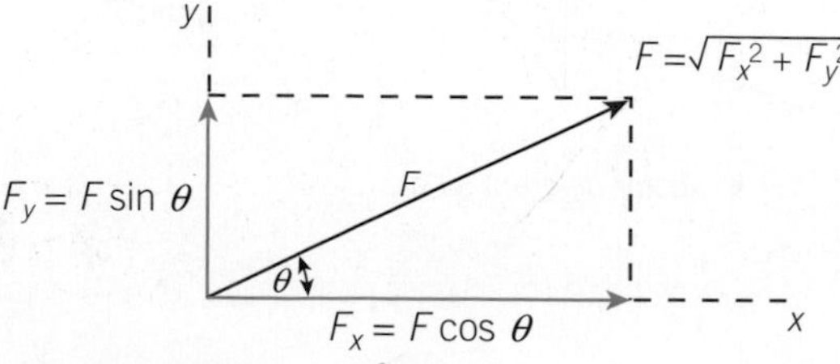


▲ **Figure 11** *Resolving a force*

## Worked example

A paraglider is pulled along at constant height at steady speed by a cable attached to a speedboat as shown in Figure 12. The cable pulls on the paraglider with a force of 500 N at an angle of 35° to the horizontal. Calculate the horizontal and vertical components of this force.

### *Solution*

Because the force on the paraglider is at an angle of 35° below the horizontal, the horizontal and vertical components of this force are

- $500 \cos 35° = 410$ N horizontally to the right
- $500 \sin 35° = 287$ N vertically downwards.

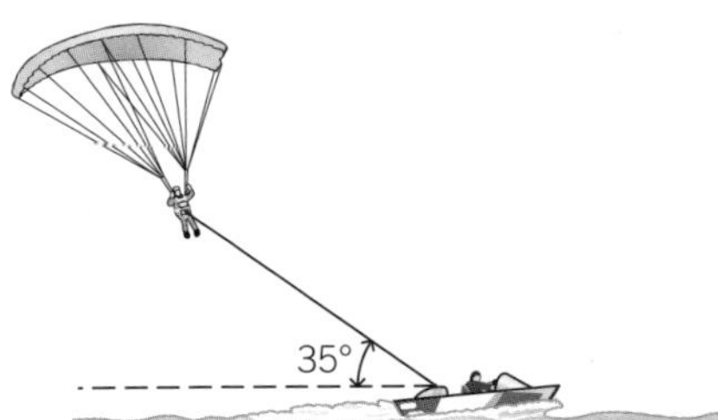


▲ Figure 12

## Summary questions

1 Figure 13 shows three situations a–c where two or more known forces act on an object. For each situation, calculate the magnitude and direction of the resultant force.

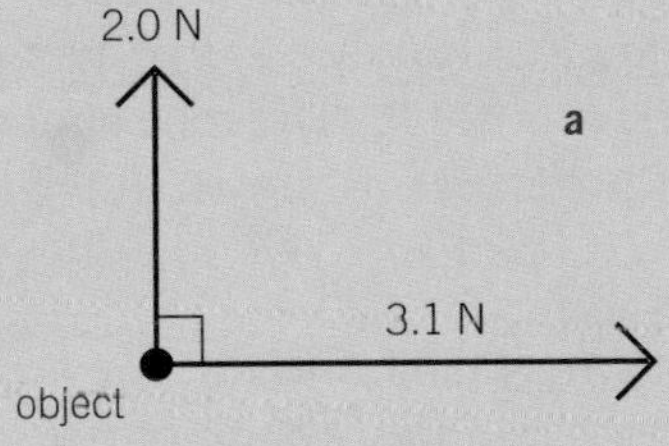


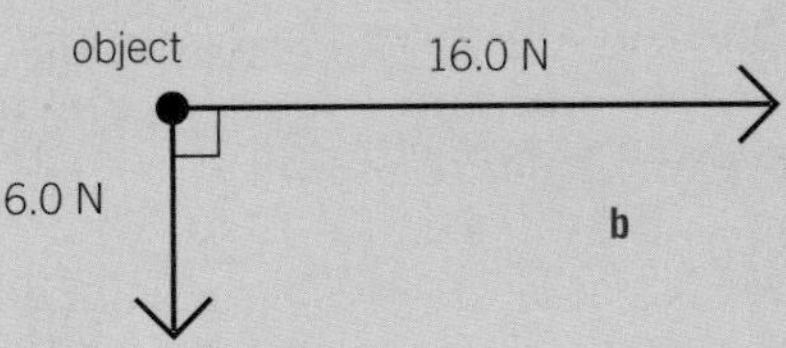


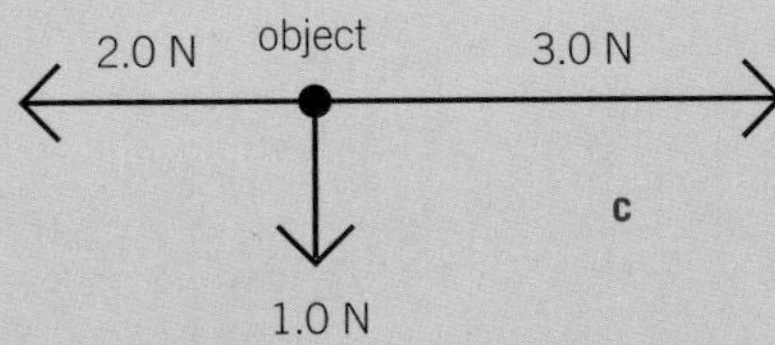


▲ Figure 13

2 Calculate the magnitude and direction of the resultant force on an object which is acted on by a force of 4.0 N and a force of 10 N that are

- **i** in the same direction
- **ii** in opposite directions
- **iii** at right angles to each other.

3 A crane is used to raise one end of a steel girder off the ground, as shown in Figure 14. When the cable attached to the end of the girder is at 20° to the vertical, the force of the cable on the girder is 6.5 kN. Calculate the horizontal and vertical components of this force.

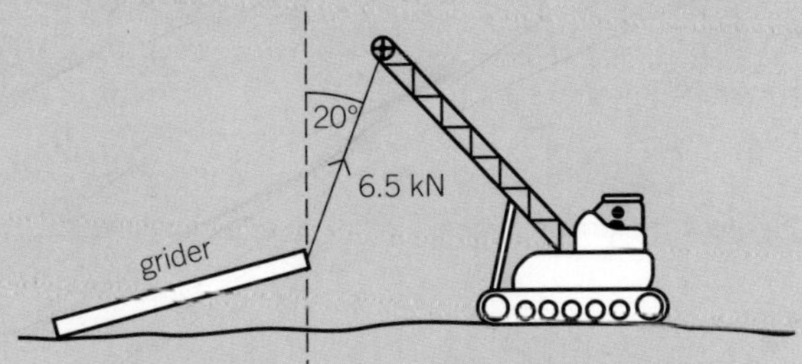


▲ Figure 14

4 A yacht is moving due north as a result of a force, due to the wind, of 350 N in a horizontal direction of 40° east of due north, as shown in Figure 15.

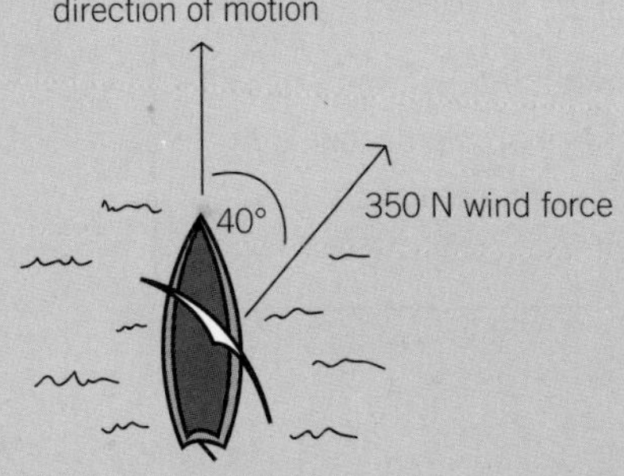


▲ Figure 15

Calculate the component of the force of the wind:

- **a** in the direction the yacht is moving
- **b** perpendicular to the direction in which the yacht is moving.

# 6.2 Balanced forces

## Learning objectives:

- → Explain why we have to consider the direction in which a force acts.
- → Demonstrate when two (or more) forces have no overall effect on a point object.
- → Explain the parallelogram of forces.

*Specification reference: 3.4.1.1*

## Equilibrium of a point object

**When two forces act on a point object**, the object is in **equilibrium** (at rest or moving at constant velocity) only if the two forces are equal and opposite to each other. The resultant of the two forces is therefore zero. The two forces are said to be **balanced**.

For example, an object resting on a horizontal surface is acted on by its weight $W$ (i.e., the force of gravity on it) acting downwards, and a support force $S$ from the surface, acting upwards. Hence $S$ is equal and opposite to $W$, provided the object is at rest (or moving at constant velocity).

$$S = W$$

**When three forces act on a point object**, their combined effect (the resultant) is zero only if the resultant of any two of the forces is equal and opposite to the third force. To check the combined effect of the three forces is zero:

- resolve each force along the same parallel and perpendicular lines
- balance the components along each line.

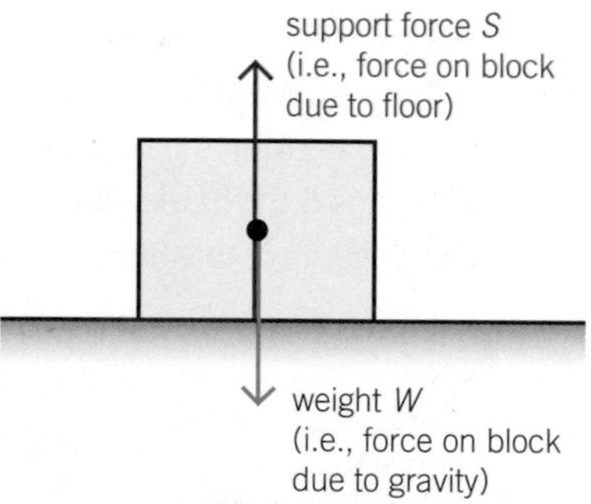


▲ **Figure 1** *Balanced forces*

### Worked example 1

A child of weight $W$ on a swing is at rest due to the swing seat being pulled to the side by a horizontal force $F_1$. The rope supporting the seat is then at an angle $\theta$ to the vertical, as shown in Figure 2.

### *Solution*

Assuming the swing seat is of negligible weight, the swing seat is acted on by three forces, which are: the weight of the child, the horizontal force $F_1$ and the tension $T$ in the rope. Resolving the tension $T$ vertically and horizontally gives $T\cos\theta$ for the vertical component of $T$ (which is upwards) and $T\sin\theta$ for the horizontal component of $T$. Therefore, the balance of forces:

1. horizontally: $F_1 = T\sin\theta$
2. vertically: $W = T\cos\theta$

Because $\sin^2\theta + \cos^2\theta = 1$, then $F_1{}^2 + W^2 = T^2\sin^2\theta + T^2\cos^2\theta = T^2$

$$\therefore T^2 = F_1{}^2 + W^2$$

Also, because $\tan\theta = \dfrac{\sin\theta}{\cos\theta}$, then $\dfrac{F_1}{W} = \dfrac{T\sin\theta}{T\cos\theta} = \tan\theta$

$$\therefore \tan\theta = \frac{F_1}{W}$$

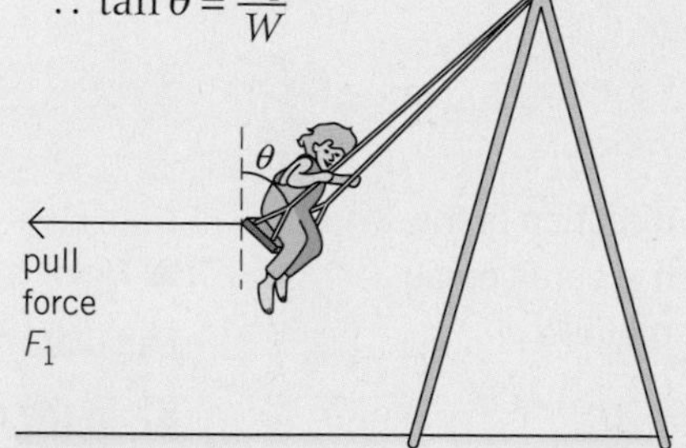


**a** *On a swing*

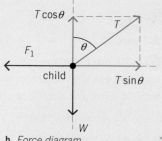

$T\cos\theta$

$T$

$\theta$

$F_1$

child

$T\sin\theta$

$W$

**b** *Force diagram*

▲ **Figure 2**

### Study tip

Forces can often be calculated using several different methods, for example, using trigonometric functions or Pythagoras' theorem.

You will meet another method of solving these problems, by using the triangle of forces, later on in this chapter.

### Synoptic link

Further guidance on trigonometric relationships can be found in Topic 16.2 Trigonometry.

## Worked example 2

An object of weight $W$ is at rest on a rough slope (i.e., a slope with a roughened surface as in Figure 3). The object is acted on by a frictional force $F$, which prevents it sliding down the slope, and a support force $S$ from the slope, perpendicular to the slope.

Resolving the three forces parallel and perpendicular to the slope gives:

1 horizontally: $F = W\sin\theta$

2 vertically: $S = W\cos\theta$

Because $\sin^2\theta + \cos^2\theta = 1$

then $F^2 + S^2 = W^2\sin^2\theta + W^2\cos^2\theta = W^2$

$\therefore W^2 = F^2 + S^2$

Also, because $\tan\theta = \frac{\sin\theta}{\cos\theta}$, then $\frac{F}{S} = \frac{W\sin\theta}{W\cos\theta} = \tan\theta$

$\therefore \tan\theta = \frac{F}{S}$

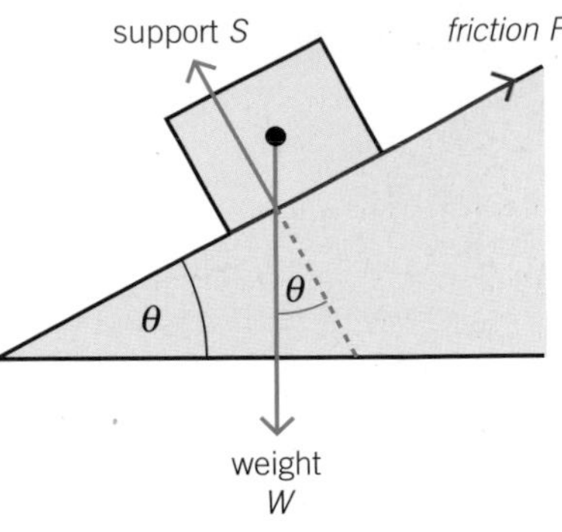


▲ **Figure 3**

## Worked example 3

An object of weight $W$ is supported by a vertical string, which is supported by two strings at different angles $\theta_1$ and $\theta_2$ to the vertical, as shown in Figure 4. Suppose the tension in the string at angle $\theta_1$ to the vertical is $T_1$ and the tension in the other string is $T_2$.

At the point P where the strings meet, the forces $T_1$, $T_2$, and $W$ are in equilibrium.

Resolving $T_1$ and $T_2$ vertically and horizontally gives:

1 horizontally: $T_1\sin\theta_1 = T_2\sin\theta_2$

2 vertically: $T_1\cos\theta_1 + T_2\cos\theta_2 = W$.

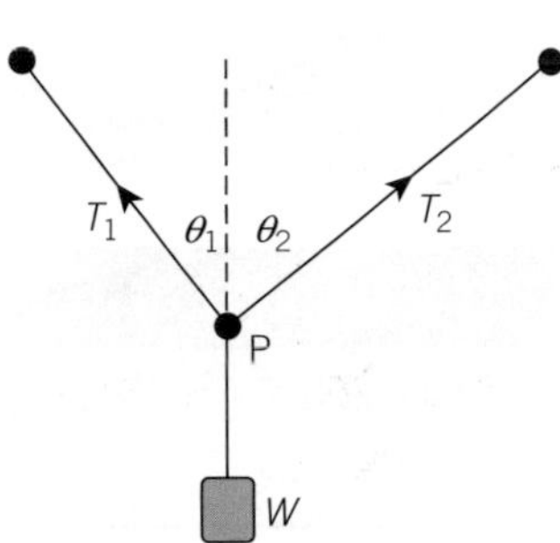


▲ **Figure 4** *A suspended weight*

## Testing three forces in equilibrium

Figure 5 shows a practical arrangement to test three forces acting on a point object P. The tension in each string pulls on P and is due to the weight it supports, either directly or indirectly over a pulley. Provided the pulleys are frictionless, each tension force acting on P is equal to the weight supported by its string.

### Study tip

Always draw force diagrams clearly, representing the magnitude and direction of each force as accurately as possible.

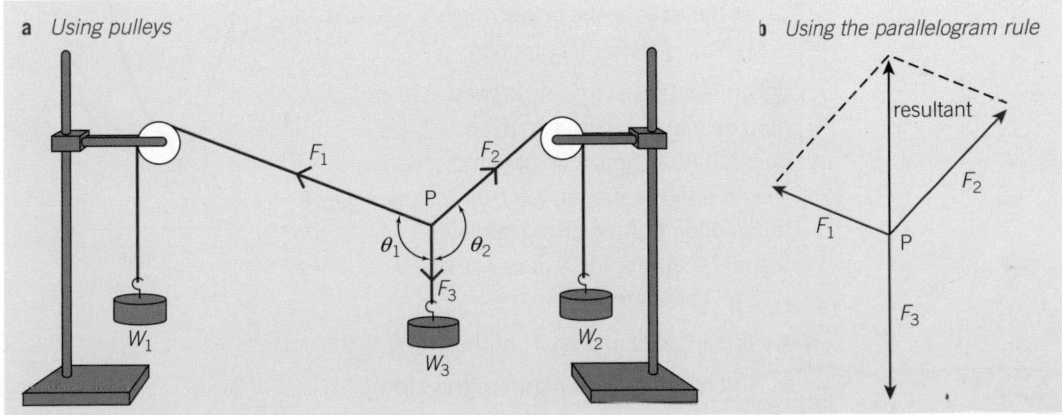


▲ **Figure 5** *Testing three forces*

The three forces $F_1$, $F_2$, and $F_3$ acting on P are in equilibrium, so any two should give a resultant equal and opposite to the third force. For example, the resultant of $F_1$ and $F_2$ is equal and opposite to $F_3$. You can test this by measuring the angle between each of the upper strings and the lower string which is vertical. A scale diagram can then be constructed using the fact that the magnitudes of $F_1$, $F_2$, and $F_3$ are equal to $W_1$, $W_2$, and $W_3$ respectively. Your diagram should show that the resultant of $F_1$ and $F_2$ is equal and opposite to $F_3$.

### *Note:*

Greater accuracy can be obtained by drawing a parallelogram, using the two force vectors $F_1$ and $F_2$ as adjacent sides. The resultant is the diagonal of the parallelogram *between* the two force vectors. This should be equal and opposite to $F_3$.

## Summary questions

1 A point object of weight 6.2 N is acted on by a horizontal force of 3.8 N.

  a Calculate the resultant of these two forces.

  b Determine the magnitude and direction of a third force acting on the object for it to be in equilibrium.

2 A small object of weight 5.4 N is at rest on a rough slope, which is at an angle of 30° to the horizontal.

  a Sketch a diagram and show the three forces acting on the object.

  b Calculate

    i the frictional force on the object

    ii the support force from the slope on the object.

### Hint

Remember that *rough* means friction is involved.

3 An archer pulled a bowstring back until the two halves of the string are at 140° to each other (Figure 6). The force needed to hold the string in this position was 95 N.

  Calculate:

  a the tension in each part of the bowstring in this position

  b the resultant force on an arrow at the instant the bowstring is released from this position.

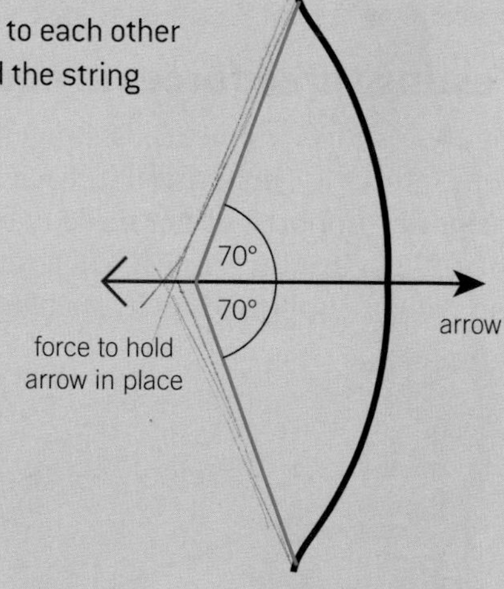


▲ Figure 6

4 An elastic string is stretched horizontally between two fixed points 0.80 m apart. An object of weight 4.0 N is suspended from the midpoint of the string, causing the midpoint to drop a distance of 0.12 m. Calculate:

  a the angle of each part of the string to the vertical

  b the tension in each part of the string.

# 6.3 The principle of moments

## Learning objectives:

- → Describe the conditions under which a force produces a turning effect.
- → Explain how the turning effect of a given force can be increased.
- → Explain what is required to balance a force that produces a turning effect.
- → Explain why the centre of mass is an important idea.

*Specification reference: 3.4.1.2*

## Turning effects

Whenever you use a lever or a spanner, you are using a force to turn an object about a pivot. For example, if you use a spanner to loosen a wheel nut on a bicycle, you need to apply a force to the spanner to make it turn about the wheel axle. The effect of the force depends on how far it is applied from the wheel axle. The longer the spanner, the less force needed to loosen the nut. However, if the spanner is too long and the nut is too tight, the spanner could snap if too much force is applied to it.

**The moment of a force about any point is defined as the force × the perpendicular distance from the line of action of the force to the point.**

The unit of the moment of a force is the newton metre (N m).

For a force $F$ acting along a line of action at perpendicular distance $d$ from a certain point,

**the moment of the force = $F \times d$**

***Notes:***

1 The greater the distance $d$, the greater the moment.
2 The distance $d$ is the **perpendicular** distance from the line of action of the force to the point.

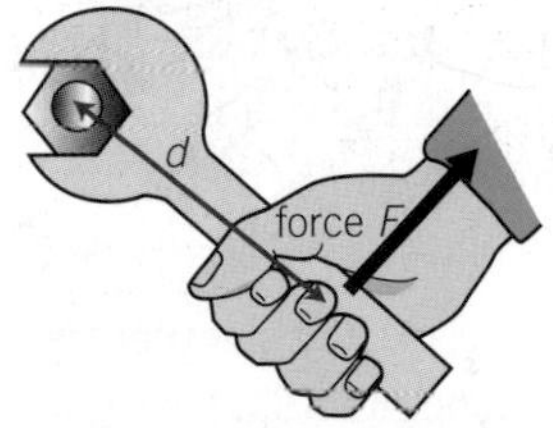


▲ **Figure 1** *A turning force*

## The principle of moments

An object that is not a point object is referred to as a body. Any such object turns if a force is applied to it anywhere other than through its centre of mass. If a body is acted on by more than one force and it is in equilibrium, the turning effects of the forces must balance out. In more formal terms, considering the **moments** of the forces about any point, for equilibrium

**the sum of the clockwise moments = the sum of the anticlockwise moments**

This statement is known as the **principle of moments**.

For example, consider a uniform metre rule balanced on a pivot at its centre, supporting weights $W_1$ and $W_2$ suspended from the rule on either side of the pivot.

- Weight $W_1$ provides an anticlockwise moment about the pivot = $W_1d_1$, where $d_1$ is the distance from the line of action of the weight to the pivot.
- Weight $W_2$ provides a clockwise moment about the pivot = $W_2d_2$, where $d_2$ is the distance from the line of action of the weight to the pivot.

For equilibrium, applying the principle of moments:

$$W_1d_1 = W_2d_2$$

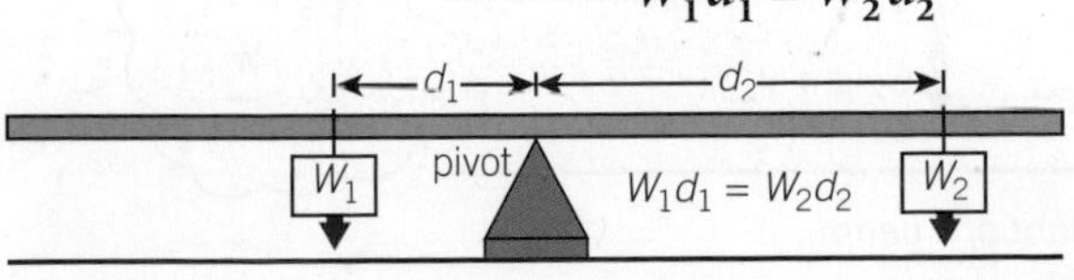


▲ **Figure 2** *The principle of moments*

## Study tip

When there are several unknown forces, take moments about a point through which one of the unknown forces acts. This point, the pivot, will give the force which acts through it a moment of zero, thereby simplifying your calculations.

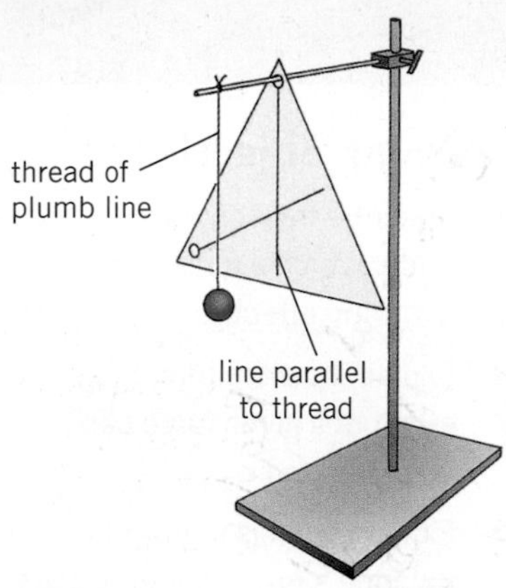


▲ **Figure 3** *A centre of mass test*

## Summary questions

1 A child of weight 200 N sits on a seesaw at a distance of 1.2 m from the pivot at the centre. The seesaw is balanced by a second child sitting on it at a distance of 0.8 m from the centre. Calculate the weight of the second child.

2 A metre rule, pivoted at its centre of mass, supports a 3.0 N weight at its 5.0 cm mark, a 2.0 N weight at its 25 cm mark, and a weight *W* at its 80 cm mark.

   a Sketch a diagram to represent this situation.

   b Calculate the weight *W*.

3 In **2**, the 3.0 N weight and the 2.0 N weight are swapped with each other. Sketch the new arrangement and calculate the new distance of weight *W* from the pivot to balance the metre rule.

4 A uniform metre rule supports a 4.5 N weight at its 100 mm mark. The rule is balanced horizontally on a horizontal knife-edge at its 340 mm mark. Sketch the arrangement and calculate the weight of the rule.

### *Note:*

If a third weight $W_3$ is suspended from the rule on the same side of the pivot as $W_2$ at distance $d_3$ from the pivot, then the rule can be rebalanced by increasing distance $d_1$.

At this new distance $d_1'$ for $W_1$,

$$W_1 d_1' = W_2 d_2 + W_3 d_3$$

## Centre of mass

A tightrope walker knows just how important the centre of mass of an object can be. One slight off-balance movement can be catastrophic. The tightrope walker uses a horizontal pole to ensure his or her overall centre of mass is always directly above the rope. The support force from the rope then acts upwards through the centre of mass of the walker.

**The centre of mass of a body is the point through which a single force on the body has no turning effect.** In effect, it is the point where we consider the weight of the body to act when studying the effect of forces on the body. For a regular solid, for example, a wooden block, the centre of mass is at its centre.

### *Centre of mass tests*

1 Balance a ruler at its centre on the end of your finger. The centre of mass of the ruler is directly above the point of support. Tip the ruler too much and it falls off because the centre of mass is no longer above the point of support.

2 To find the centre of mass of a triangular card, suspend the piece of card on a clamp stand as shown in Figure 3. Draw pencil lines along the plumb line. The centre of mass is where the lines drawn on the card cross.

## Calculating the weight of a metre rule

1 Locate the centre of mass of a metre rule by balancing it horizontally on a horizontal knife-edge. Note the position of the centre of mass. The rule is **uniform** if its centre of mass is exactly at the middle of the rule.

2 Balance the metre rule off-centre on a knife-edge, using a known weight $W_1$ as shown in Figure 4. The position of the known weight needs to be adjusted gradually until the rule is exactly horizontal.

At this position

- the known weight $W_1$ provides an anticlockwise moment about the pivot = $W_1 d_1$, where $d_1$ is the perpendicular distance from the line of action of $W_1$ to the pivot
- the weight of the rule $W_0$ provides a clockwise moment = $W_0 d_0$, where $d_0$ is the distance from the centre of mass of the rule to the pivot.

Applying the principle of moments,

$$W_0 d_0 = W_1 d_1$$

By measuring distance $d_0$ and $d_1$, the weight $W_0$ of the rule can therefore be calculated.

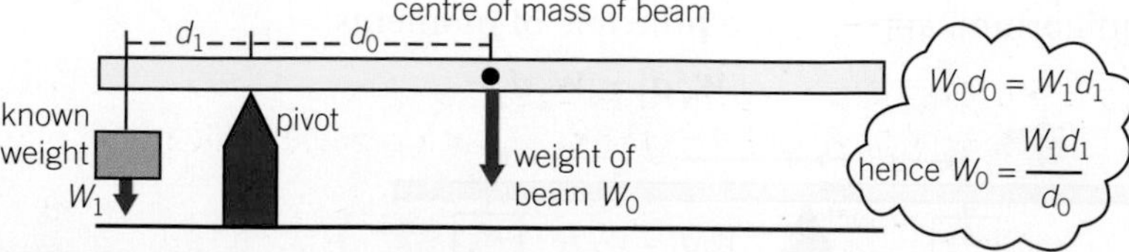


▲ **Figure 4** *Finding the weight of a beam*

# 6.4 More on moments

## Support forces

### Single-support problems

When an object in equilibrium is supported at one point only, the support force on the object is equal and opposite to the total downward force acting on the object. For example, in Figure 1, a uniform rule is balanced on a knife-edge at its centre of mass, with two additional weights $W_1$ and $W_2$ attached to the rule. The support force $S$ on the rule from the knife-edge must be equal to the total downward weight. Therefore

**$S = W_1 + W_2 + W_0$, where $W_0$ is the weight of the rule.**

As explained in Topic 6.3, taking moments about the knife-edge gives

$$W_1 d_1 = W_2 d_2$$

***Note:***

What difference is made by taking moments about a different point? Consider moments about the point where $W_1$ is attached to the rule.

The sum of the clockwise moments $= W_0 d_1 + W_2 (d_1 + d_2)$

The sum of the anticlockwise moments $= S d_1 = (W_1 + W_2 + W_0) d_1$

$\therefore (W_1 + W_2 + W_0) d_1 = W_0 d_1 + W_2 (d_1 + d_2)$

Multiplying out the brackets gives:

$W_1 d_1 + W_2 d_1 + W_0 d_1 = W_0 d_1 + W_2 d_1 + W_2 d_2$

which simplifies to become $W_1 d_1 = W_2 d_2$

This is the same as the equation obtained by taking moments about the original pivot point. So moments can be taken about any point. It makes sense therefore to choose a point through which one or more unknown forces act, as such forces have zero moment about this point.

### Two-support problems

Consider a uniform beam supported on two pillars X and Y, which are at distance $D$ apart. The weight of the beam is shared between the two pillars according to how far the beam's centre of mass is from each pillar. For example:

- If the centre of mass of the beam is midway between the pillars, the weight of the beam is shared equally between the two pillars. In other words, the support force on the beam from each pillar is equal to half the weight of the beam.
- If the centre of mass of the beam is at distance $d_x$ from pillar X and distance $d_y$ from pillar Y, as shown in Figure 2, then taking moments about

  1 where X is in contact with the beam,
  $S_y D = W d_x$, where $S_y$ is the support force from pillar
  gives $S_y = \dfrac{W d_x}{D}$

  2 where Y is in contact with the beam,
  $S_x D = W d_y$, where $S_x$ is the support force from pillar
  gives $S_x = \dfrac{W d_y}{D}$.

Therefore, if the centre of mass is closer to X than to Y, $d_x < d_y$ so $S_y < S_x$.

## Learning objectives:

- → Describe the support force on a pivoted body.
- → When a body in equilibrium is supported at two places, state how much force is exerted on each support.
- → Explain what is meant by a couple.

*Specification reference: 3.4.1.2*

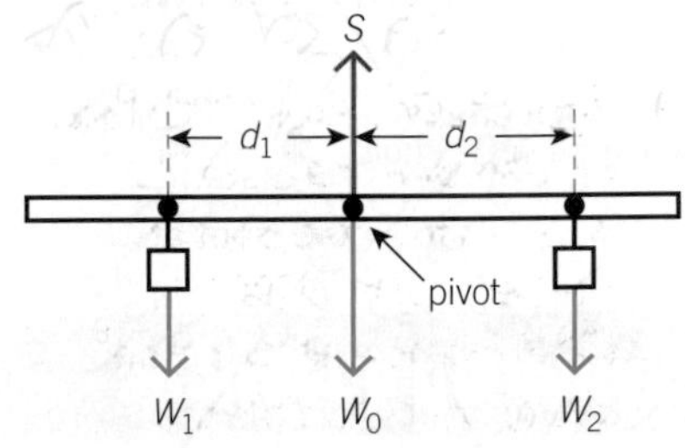


▲ **Figure 1** *Support forces*

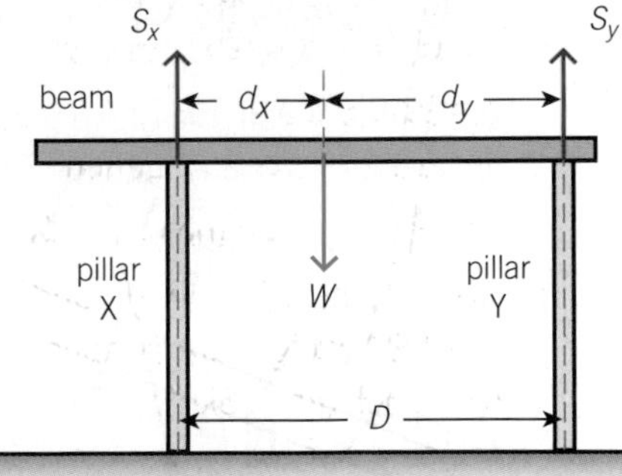


▲ **Figure 2** *A two-support problem*

## Study tip

In calculations, you can always eliminate the turning effect of a force by taking moments about a point through which it acts.

## Couples

A **couple** is a pair of equal and opposite forces acting on a body, but not along the same line. Figure 3 shows a couple acting on a beam. The couple turns or tries to turn the beam.

The moment of a couple = force × perpendicular distance between the lines of action of the forces.

To prove this, consider the couple in Figure 3. Taking moments about an arbitrary point P between the ends at distance $x$ along the beam from one end,

1 the moment due to the force $F$ at that end = $Fx$ clockwise
2 the moment due to the force $F$ at the other end = $F(d - x)$ clockwise, where $d$ is the perpendicular distance between the lines of action of the forces.

▲ **Figure 3** *A couple*

Therefore the total moment = $Fx + F(d - x) = Fx + Fd - Fx = Fd$.

The total moment is the same, regardless of the point about which the moments are taken.

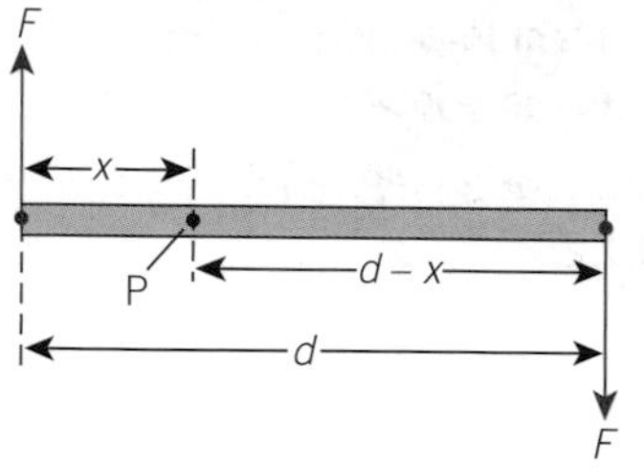

## Summary questions

1 A uniform metre rule of weight 1.2 N rests horizontally on two knife-edges at the 100 mm mark and the 800 mm mark. Sketch the arrangement and calculate the support force on the rule due to each knife-edge.

2 A uniform beam of weight 230 N and of length 10 m rests horizontally on the tops of two brick walls, 8.5 m apart, such that a length of 1.0 m projects beyond one wall and 0.5 m projects beyond the other wall. Figure 4 shows the arrangement.

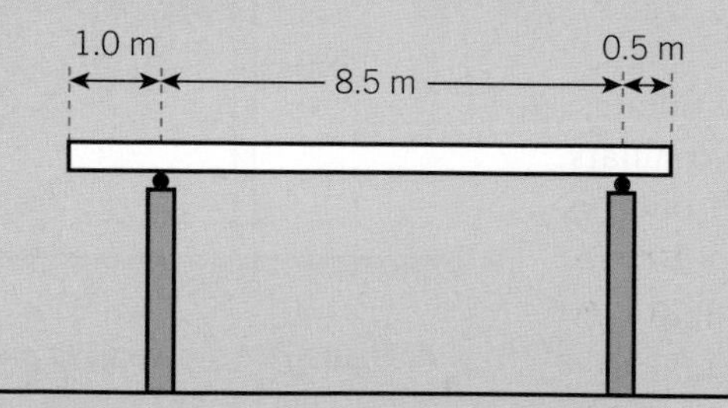


▲ **Figure 4**

Calculate:

a the support force of each wall on the beam

b the force of the beam on each wall.

3 A uniform bridge span of weight 1200 kN and of length 17.0 m rests on a support of width 1.0 m at either end. A stationary lorry of weight 60 kN is the only object on the bridge. Its centre of mass is 3.0 m from the centre of the bridge.

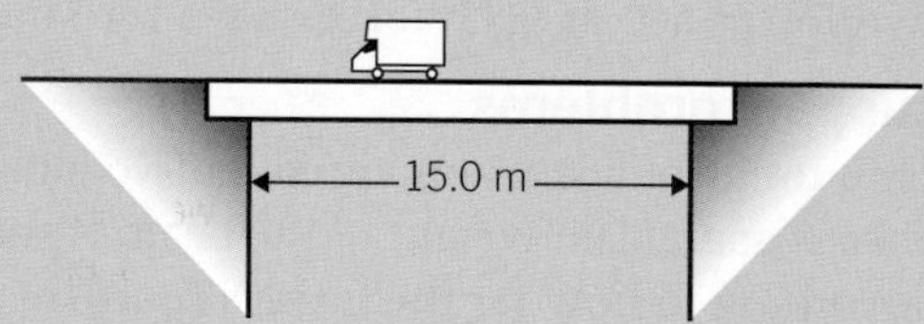


▲ **Figure 5**

Calculate the support force on the bridge at each end. Assume the support forces act where the bridge meets its support.

4 A uniform plank of weight 150 N and of length 4.0 m rests horizontally on two bricks. One of the bricks is at the end of the plank. The other brick is 1.0 m from the other end of the plank.

a Sketch the arrangement and calculate the support force on the plank from each brick.

b A child stands on the free end of the plank and just causes the other end to lift off its support. Sketch this arrangement and calculate the weight of the child.

# 6.5 Stability

## Learning objectives:

- → Explain the difference between stable and unstable equilibrium.
- → Assess when a tilted object will topple over.
- → Explain why a vehicle is more stable when its centre of mass is lower.

*Specification reference: 3.4.1.1; 3.4.1.2*

## Stable and unstable equilibrium

If a body in **stable equilibrium** is displaced then released, it returns to its equilibrium position. For example, if an object such as a coat hanger hanging from a support is displaced slightly, it swings back to its equilibrium position.

Why does an object in stable equilibrium return to equilibrium when it is displaced and then released? The reason is that the centre of mass of the object is directly below the point of support when the object is at rest. The weight of the object is considered to act at its centre of mass. Thus the support force and the weight are directly equal and opposite to each other when the object is in equilibrium. However, when it is displaced, at the instant of release, the line of action of the weight no longer passes through the point of support, so the weight returns the object to equilibrium.

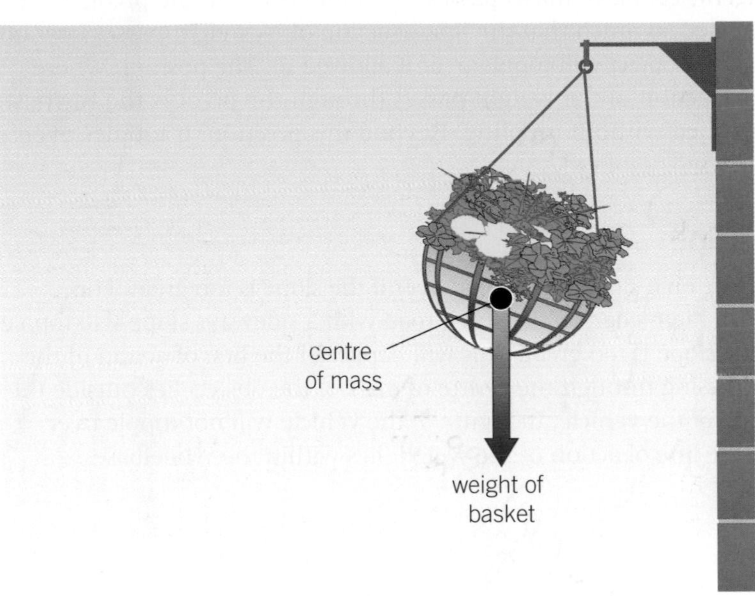


▲ **Figure 1** *Returning to equilbrium*

A plank balanced on a drum is in **unstable equilibrium** (Figure 2). If it is displaced slightly from equilibrium then released, the plank will roll off the drum. The reason is that the centre of mass of the plank is directly above the point of support when it is in equilibrium. The support force is exactly equal and opposite to the weight. If the plank is displaced slightly, the centre of mass is no longer above the point of support. The weight therefore acts to turn the plank further from the equilibrium position.

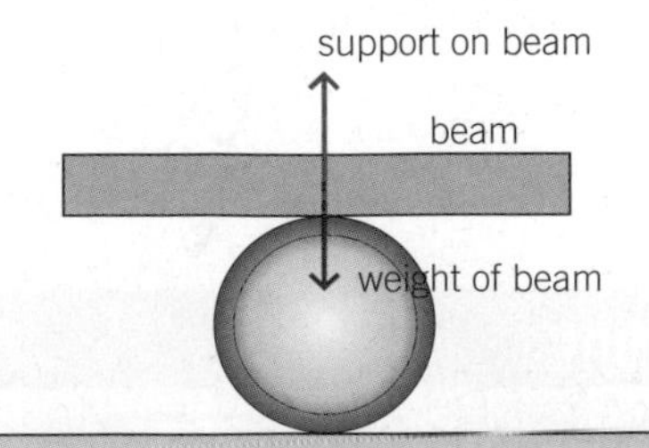


▲ **Figure 2** *Unstable equilbrium*

## Tilting and toppling

Skittles at a bowling alley are easy to knock over because they are tall, so their centre of mass is high and the base is narrow. A slight nudge from a ball causes a skittle to tilt then tip over.

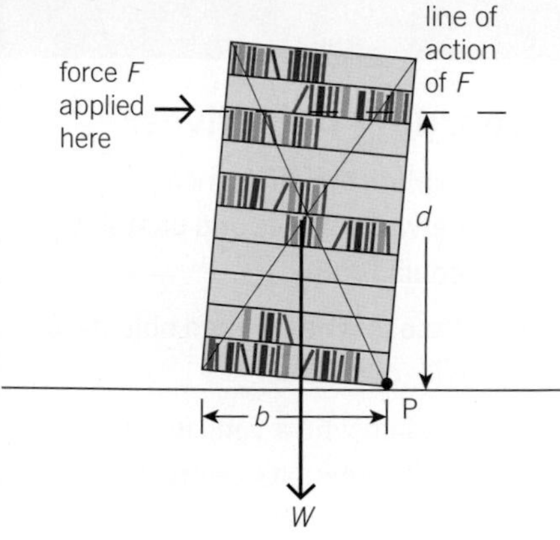


▲ **Figure 3** *Tilting*

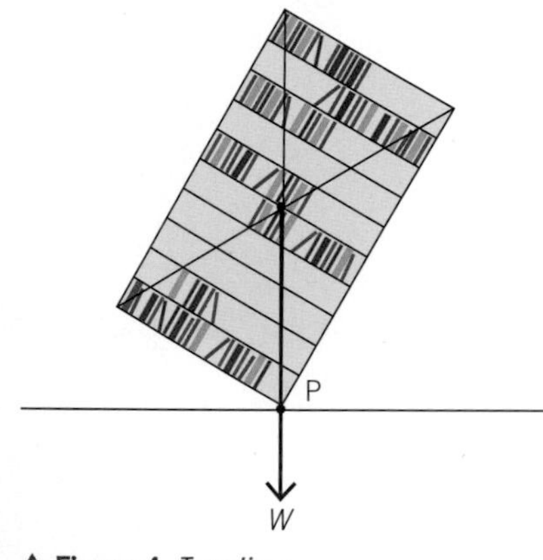


▲ **Figure 4** *Toppling*

## Tilting

This is where an object at rest on a surface is acted on by a force that raises it up on one side. For example, if a horizontal force $F$ is applied to the top of a tall free-standing bookcase, the force can make the bookcase tilt about its base along one edge.

In Figure 3, to make the bookcase tilt, the force must turn it clockwise about point P. The entire support from the floor acts at point P. The weight of the bookcase provides an anticlockwise moment about P.

1 The clockwise moment of $F$ about P = $Fd$, where $d$ is the perpendicular distance from the line of action of $F$ to the pivot.
2 The anticlockwise moment of $W$ about P = $Wb/2$ where $b$ is the width of the base.

Therefore, for tilting to occur $Fd > Wb/2$.

## Toppling

A tilted object will topple over if it is tilted too far. If an object on a flat surface is tilted more and more, the line of action of its weight (which is through its centre of mass) passes closer and closer to the pivot. If the object is tilted so much that the line of action of its weight passes beyond the pivot, the object will topple over if allowed to. The position where the line of action of the weight passes through the pivot is the furthest it can be tilted without toppling. Beyond this position, it topples over if it is released.

# On a slope

A tall object on a slope will topple over if the slope is too great. For example, a high-sided vehicle on a road with a sideways slope will topple over if the slope is too great. This will happen if the line of action of the weight (passing through the centre of mass of the object) lies outside the wheelbase of the vehicle. In Figure 5, the vehicle will not topple over because the line of action of the weight lies within the wheelbase.

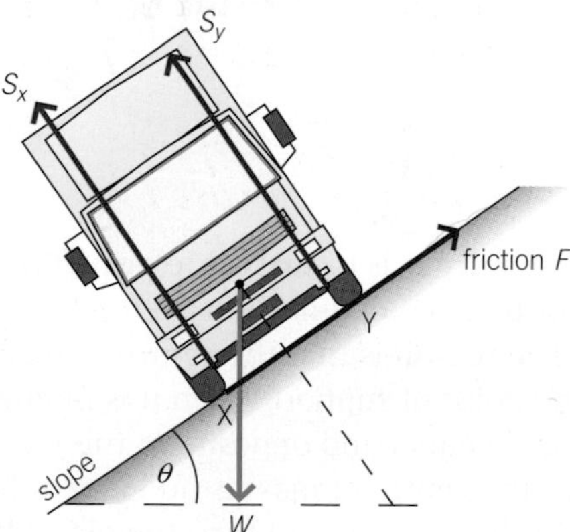


▲ **Figure 5** *A high-sided vehicle on a slope*

### Hint

The lower the centre of mass of an object, the more stable it is.

Consider the forces acting on the vehicle on a sideways slope when it is at rest. The sideways friction $F$, the support forces $S_x$ and $S_y$, and the force of gravity on the vehicle (i.e., its weight) act as shown in Figure 5.

For equilibrium, resolving the forces parallel and perpendicular to the slope gives:

1 parallel to the slope:
$F = W \sin\theta$

2 perpendicular to the slope:
$S_x + S_y = W\cos\theta$.

Note that $S_x$ is greater than $S_y$ because X is lower than Y.

## Summary questions

1 Explain why a bookcase with books on its top shelf only is less stable than if the books were on the bottom shelf.

2 An empty wardrobe of weight 400 N has a square base 0.8 m × 0.8 m and a height of 1.8 m. A horizontal force is applied to the top edge of the wardrobe to make it tilt. Calculate the force needed to lift the wardrobe base off the floor along one side.

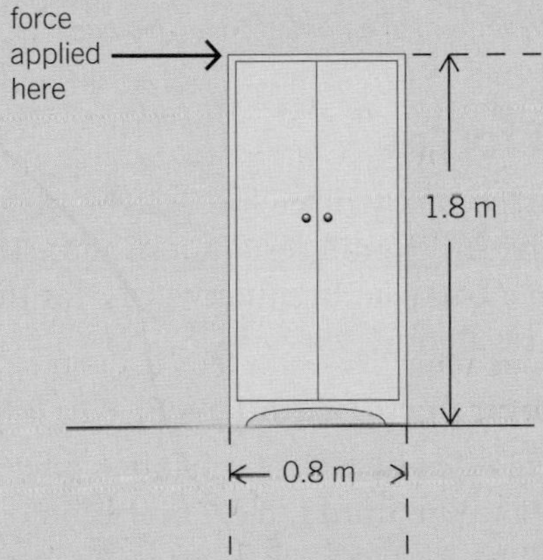


▲ Figure 6

3 A vehicle has a wheelbase of 1.8 m and a centre of mass, when unloaded, which is 0.8 m from the ground.

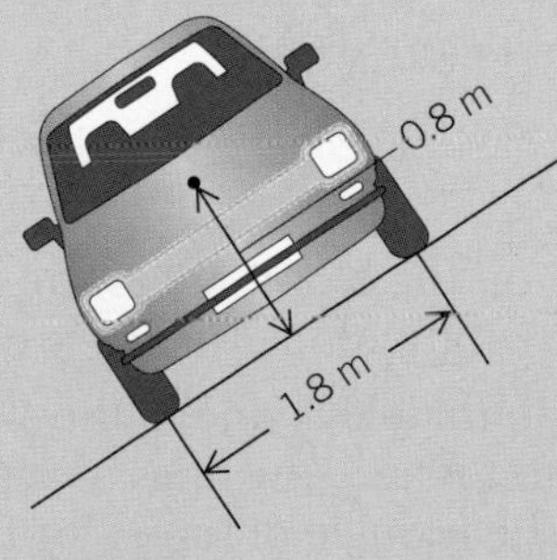


▲ Figure 7

a The vehicle is tested for stability on an adjustable slope. Calculate the maximum angle of the slope to the horizontal if the vehicle is not to topple over.

b If the vehicle carries a full load of people, will it be more or less likely to topple over on a slope? Explain your answer.

# 6.6 Equilibrium rules

## Learning objectives:

- → Explain what condition must apply to the forces on an object in equilibrium.
- → Explain what condition must apply to the turning effects of the forces.
- → Describe how we can apply these conditions to predict the forces acting on a body in equilibrium.

*Specification reference: 3.4.1.1; 3.4.1.2*

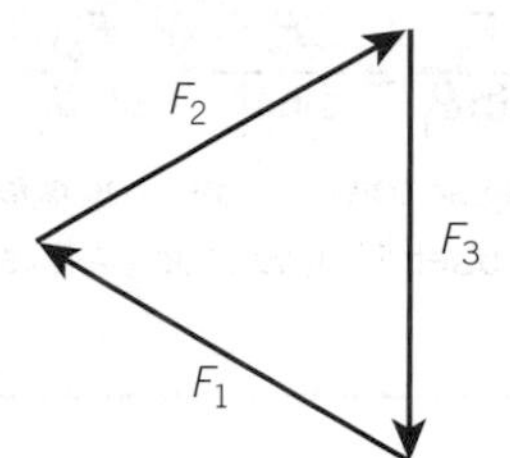


▲ **Figure 2** *The triangle of forces for a point object*

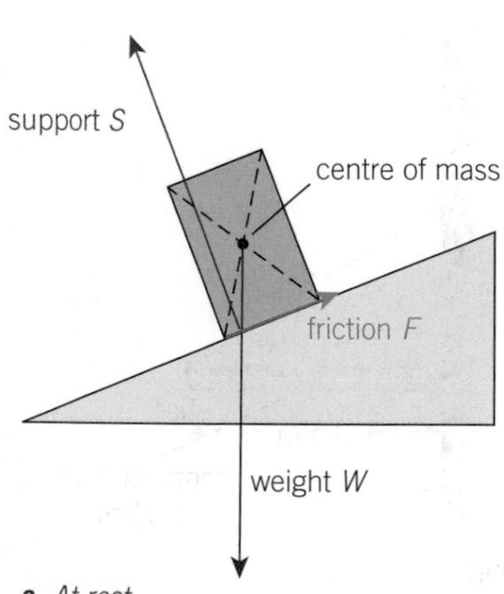


**a** *At rest*

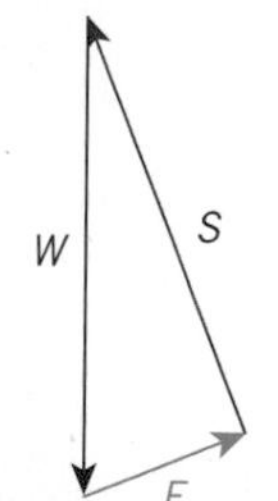


**b** *The triangle of forces*

▲ **Figure 3** *A block on a slope*

## Free body force diagrams

When two objects interact, they always exert equal and opposite forces on one another. A diagram showing the forces acting on an object can become very complicated, if it also shows the forces the object exerts on other objects as well. A **free body force diagram** shows only the forces acting on the object.

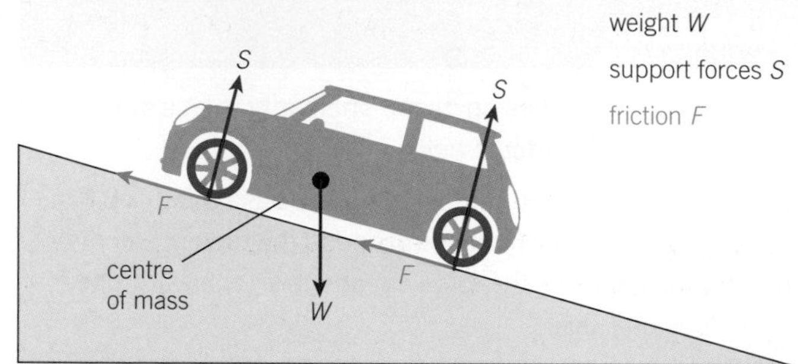


▲ **Figure 1** *A free body diagram of a parked car on a slope*

## The triangle of forces

For a point object acted on by three forces to be in equilibrium, the three forces must give an overall resultant of zero. The three forces as vectors should form a triangle. In other words, for three forces $F_1$, $F_2$ and $F_3$ to give zero resultant,

$$\textbf{their vector sum } F_1 + F_2 + F_3 = 0$$

As explained in Topic 6.1, any two of the forces give a resultant that is equal and opposite to the third force. For example, the resultant of $F_1 + F_2$ is equal and opposite to $F_3$ ($F_1 + F_2 = -F_3$).

The same rule applies to a body in equilibrium acted on by three forces. In addition, their lines of action must intersect at the *same* point, otherwise the body cannot be in equilibrium, as the forces will have a net turning effect. Consider the example of a rectangular block on a rough slope, as shown in Figure 3.

- The weight $W$ of the block acts vertically down through the centre of mass of the block.
- The frictional force $F$ on the block due to the slope prevents the block from sliding down the slope, so it acts up the slope.
- The support force $S$ on the block due to the slope acts normal to the slope through the point where the lines of action of $W$ and $F$ act. Figure 3 also shows the triangle of forces for the three forces $W$, $F$, and $S$ acting on the block.

We can draw a scale diagram of the triangle of forces to find an unknown force or angle, given the other forces and angles in the triangle. For example, to find the unknown force, $F_3$, in the triangle of forces in Figure 4:

1 Draw one of the known force vectors, $F_1$, to scale, as one side of the force triangle.

2 Use a protractor and ruler to draw the other known force (e.g., $F_2$) at the correct angle to $F_1$ as the second side of the triangle. The third side of the triangle can then be drawn in to give the unknown force ($F_3$).

You could calculate the unknown force by resolving it and $F_2$ parallel and perpendicular to the base force $F_1$. Labelling $\theta_2$ as the angle opposite $F_2$ and $\theta_3$ as the angle opposite $F_3$, the perpendicular components of the resolved forces, $F_2 \sin\theta_3$ and $F_3 \sin\theta_2$, correspond to the height of the triangle and are equal and opposite to each other:

$$\boldsymbol{F_2 \sin\theta_3 = F_3 \sin\theta_2}$$

Therefore the unknown force can be calculated if $F_2$, $\theta_2$, and $\theta_3$ are known.

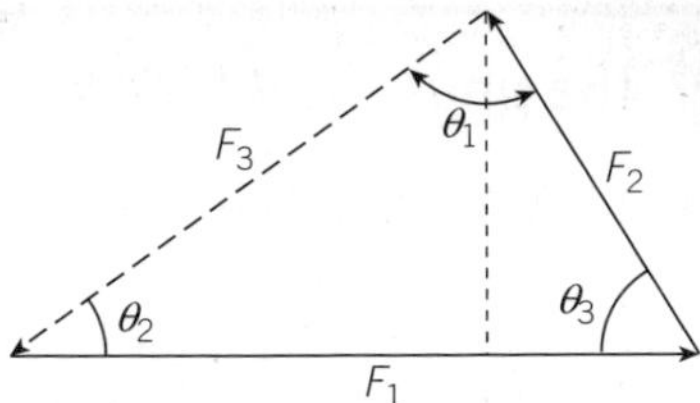


▲ **Figure 4** *Constructing a scale diagram*

## Study tip

Rearranging the equation $F_2 \sin\theta_3 = F_3 \sin\theta_2$ gives

$$\frac{F_2}{\sin\theta_2} = \frac{F_3}{\sin\theta_3}$$

By applying the same theory to $F_1$ and either $F_2$ or $F_3$, it can be shown that

$$\frac{F_1}{\sin\theta_1} = \frac{F_2}{\sin\theta_2} = \frac{F_3}{\sin\theta_3}$$

This rule, known as the *sine rule*, will be useful if you study A level Maths.

## The conditions for equilibrium of a body

An object in equilibrium is either at rest or moving with a constant velocity. The forces acting on it must give zero resultant and their turning effects must balance out as well.

Therefore, for a body in equilibrium:

1 The resultant force must be zero. If there are only three forces, they must form a closed triangle.
2 The principle of moments must apply (i.e., the moments of the forces about the same point must balance out).

See Topics 6.3 and 6.4 again if necessary.

### Worked example

A uniform shelf of width 0.6 m and of weight 12 N is attached to a wall by hinges and is supported horizontally by two parallel cords attached at two corners of the shelf, as shown in Figure 5. The other end of each cord is fixed to the wall 0.4 m above the hinge. Calculate:

**a** the angle between each cord and the shelf
**b** the tension in each cord.

### *Solution*

**a** Let the angle between each cord and the shelf = $\theta$.

From Figure 5, $\tan\theta = \frac{0.4}{0.6}$, so $\theta = 34°$.

**b** Taking moments about the hinge eliminates the force at the hinge (as its moment is zero) to give:

1 The sum of the clockwise moments = weight of shelf × distance from hinge to the centre of mass of the shelf = 12 × 0.3 = 3.6 N m.

2 The sum of the anticlockwise moments = $2Td$, where $T$ is the tension in each cord and $d$ is the perpendicular distance from the hinge to either cord.

From Figure 5 it can be seen that $d = 0.6 \sin\theta = 0.6 \sin 34 = 0.34$ m.

Applying the principle of moments gives:

$$2 \times 0.34 \times T = 3.6$$
$$T = 5.3 \text{ N}$$

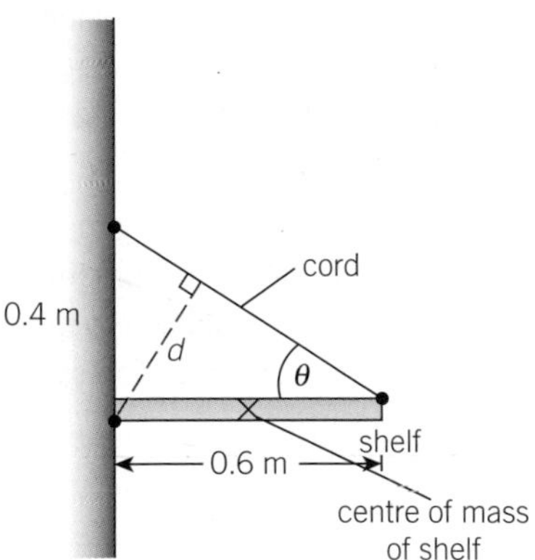


▲ **Figure 5**

## The physics of lifting

Lifting is a process that can damage your spine if it is not done correctly. Figure 6 shows how *not* to lift a heavy suitcase. The spine pivots about the hip joints and is acted upon by the tension $T$ in the back muscles, the weight $W_0$ of the upper part of the body, the weight $W$ of the suitcase, and the reaction force $R$ from the hip.

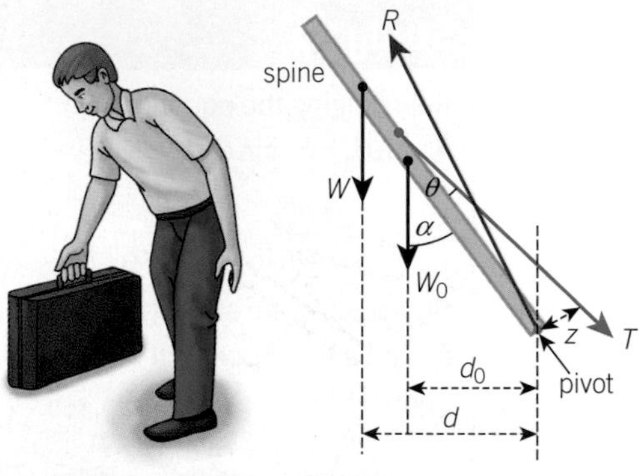


▲ **Figure 6** *Lifting forces*

In Figure 6, the spine is at angle $\alpha$ to the vertical direction, and weight $W$ acts on the spine at distance $d$ from the pivot. Assume that weight $W_0$ acts on the spine at distance $d_0$ from the pivot.

*Before* the suitcase is lifted off the ground:

- weight $W_0$ creates an anticlockwise moment about the hip joints that is equal to $W_0 d_0 \sin\alpha$. This moment is because $W_0$ has a component $W_0 \sin\alpha$ perpendicular to the spine acting on the spine at distance $d_0$ from the pivot.
- tension $T_0$ is necessary to provide an equal and opposite clockwise moment $T_0 \times z$ about the hip joints, where $z$ is the perpendicular distance from the pivot to the line of action of the tension.

Applying the principle of moments to the spine in this position gives $T_0 z = W_0 d_0 \sin\alpha$.

Typically, $d_0 = 10\,z$, and therefore, $\mathbf{T_0 = 10\,W_0 \sin\alpha}$.

*When* the suitcase is lifted off the ground:

- Its weight $W$ creates an extra anticlockwise moment about the hip joints that is equal to $Wd \sin\alpha$. This moment is because $W$ has a component $W \sin\alpha$ perpendicular to the spine acting on the spine at distance $d$ from the pivot.
- Extra tension $\Delta T$ is needed in order to provide an equal and opposite clockwise moment $\Delta T \times z$ to lift the suitcase off the ground, where $z$ is the perpendicular distance from the pivot to the line of action of the tension.

Applying the principle of moments gives $\Delta T z = Wd \sin\alpha$.

Typically, $d = 15\,z$, and therefore, $\mathbf{\Delta T = 15W \sin\alpha}$.

The components of $T$, $W_0$, and $W$ parallel to the spine act down the spine and are opposed by the parallel component of the reaction $R$ acting up the spine. The effect of these forces is to compress the spine. Therefore:

- *Before* the suitcase is lifted off the ground, the compressive force in the spine $= T_0 \cos\theta + W_0 \cos\alpha$, where $\theta$ is the angle between the spine and the line of action of tension $T$. Because $\theta < 10°$, $\cos\theta \approx 1$, the compressive force $\approx T_0 + W_0 \cos\alpha$
  $= \mathbf{10\,W_0 \sin\alpha + W_0 \cos\alpha}$.
- *When* the suitcase is lifted off the ground, the extra compressive force in the spine $= \Delta T \cos\theta + W \cos\alpha$. Because $\theta < 10°$, $\cos\theta \approx 1$, the extra compressive force $\approx \Delta T + W \cos\alpha$
  $= \mathbf{15\,W \sin\alpha + W \cos\alpha}$.

### Questions

1 Estimate the compressive force in the spine of a person leaning forward at angle $\alpha$ equal to **a** 30°, **b** 60° *just before* lifting a 20 kg suitcase off the ground. Assume $W_0 = 400$ N.

2 Estimate the total compressive force in the spine in **1a** and **b** when the suitcase is lifted off the ground.

3 Figure 7 shows a suitcase being lifted by a person who is not leaning over. Explain why the compressive force in the spine is considerably less than if the person had leaned forward to lift the suitcase.

▲ **Figure 7**

## Summary questions

1 A uniform plank of length 5.0 m rests horizontally on two bricks that are 0.5 m from either end. A child of weight 200 N stands on one end of the plank and causes the other end to lift, so it is no longer supported at that end. Calculate:

a the weight of the plank,

b the support force acting on the plank from the supporting brick.

▲ Figure 8

2 A security camera is supported by a frame that is fixed to a wall and ceiling as shown in Figure 9. The support structure must be strong enough to withstand the effect of a downward force of 1500 N acting on the camera (in case the camera is gripped by someone below it). Calculate

a the moment of a force of 1500 N on the camera about the point where the support structure is attached to the wall

b the extra force on the vertical strut supporting the frame, when the camera is pulled with a downward force of 1500 N.

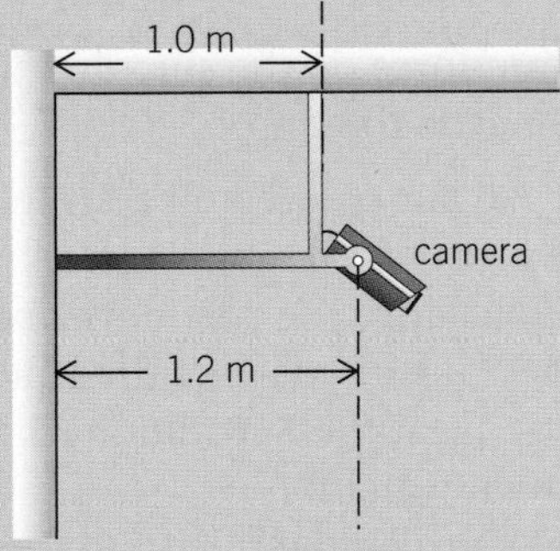


▲ Figure 9

3 A crane is used to raise one end of a 15 kN girder of length 10.0 m off the ground. When the end of the girder is at rest 6.0 m off the ground, the crane cable is perpendicular to the girder, as shown in Figure 10.

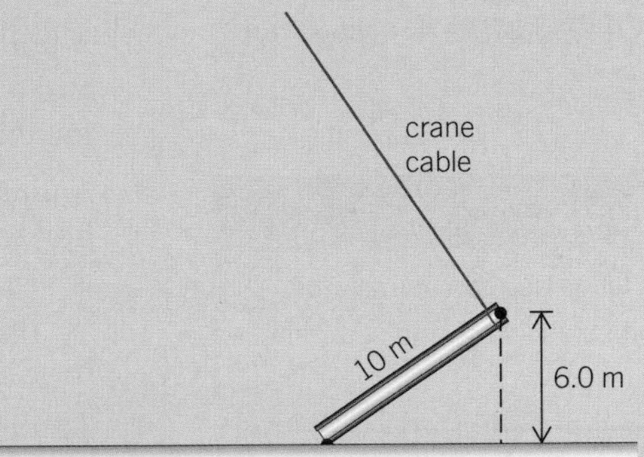


▲ Figure 10

a Calculate the tension in the cable.

b Show that the support force on the girder from the ground has a horizontal component of 3.6 N and a vertical component of 10.2 kN. Hence calculate the magnitude of the support force.

4 A small toy of weight 2.8 N is suspended from a horizontal beam by means of two cords that are attached to the same point on the toy (Figure 11). One cord makes an angle of 60° to the beam and the other makes an angle of 40° to the vertical. Calculate the tension in each cord.

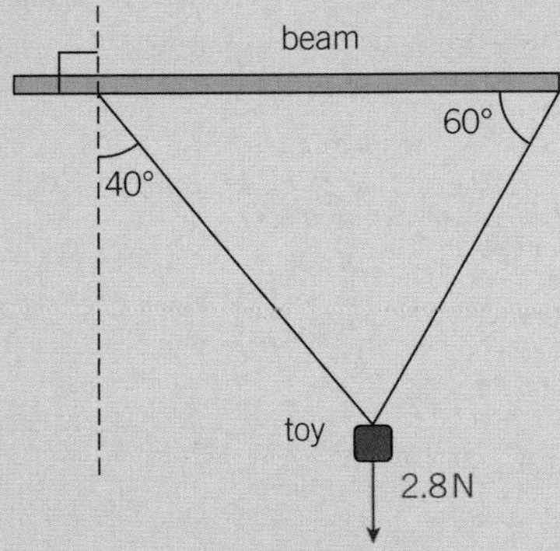


▲ Figure 11

# 6.7 Statics calculations

## Learning objective:

→ State the important principles that always apply to a body in equilibrium.

*Specification reference: 3.4.1.1; 3.4.1.2*

### Study tip

Get as far as you can using force diagrams, resolving, and common sense.

1 Calculate the magnitude of the resultant of a 6.0 N force and a 9.0 N force acting on a point object when the two forces act:
  a in the same direction
  b in opposite directions
  c at 90° to each other.

2 A point object in equilbrium is acted on by a 3 N force, a 6 N force, and a 7 N force. What is the resultant force on the object if the 7 N force is removed?

3 A point object of weight 5.4 N in equilibrium is acted on by a horizontal force of 4.2 N and a second force *F*.
  a Draw a free body force diagram for the object and determine the magnitude of *F*.
  b Calculate the angle between the direction of *F* and the horizontal.

4 An object of weight 7.5 N hangs on the end of a cord, which is attached to the midpoint of a wire stretched between two points on the same horizontal level, as shown in Figure 1. Each half of the wire is at 12° to the horizontal. Calculate the tension in each half of the wire.

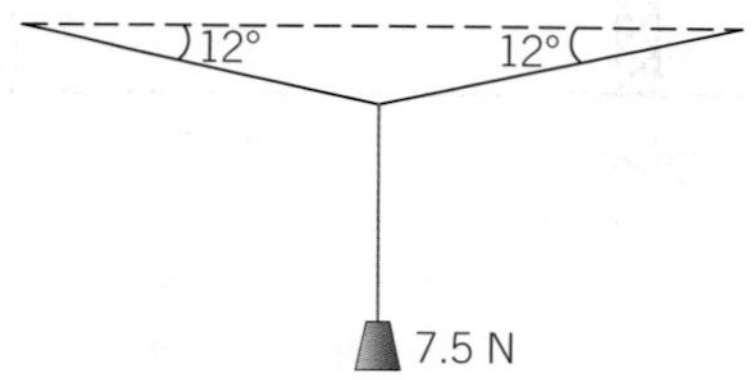


▲ Figure 1

5 A ship is towed at constant speed by two tugboats, each pulling the ship with a force of 9.0 kN. The angle between the tugboat cables is 40°, as shown in Figure 2.

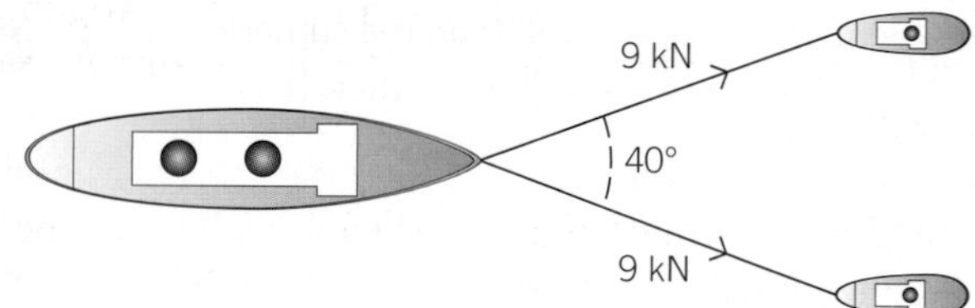


▲ Figure 2

  a Calculate the resultant force on the ship due to the two cables.
  b Calculate the drag force on the ship.

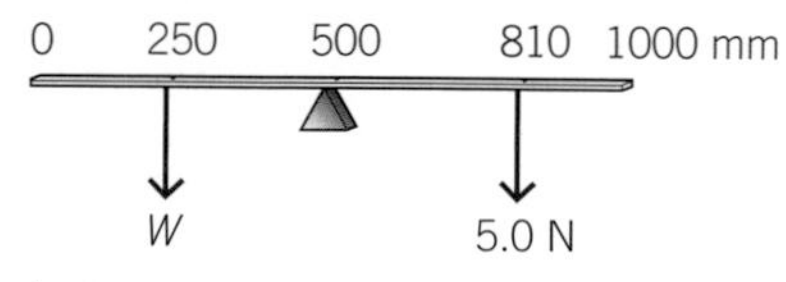


▲ Figure 3

6 A metre rule of weight 1.0 N is pivoted on a knife-edge at its centre of mass, supporting a weight of 5.0 N and an unknown weight *W* as shown in Figure 3. To balance the rule horizontally with the unknown weight on the 250 mm mark of the rule, the position of the 5.0 N weight needs to be at the 810 mm mark.
  a Calculate the unknown weight.
  b Calculate the support force on the rule from the knife-edge.

7 In Figure 3, a 2.5 N weight is also suspended from the rule at its 400 mm mark. What adjustment needs to be made to the position of the 5.0 N weight to rebalance the rule?

8 A uniform metre rule is balanced horizontally on a knife-edge at its 350 mm mark, by placing a 3.0 N weight on the rule at its 10 mm mark.
  a Sketch the arrangement and calculate the weight of the rule.
  b Calculate the support force on the rule from the knife-edge.

9 A uniform diving board has a length 4.0 m and a weight of 250 N, as shown in Figure 4. It is bolted to the ground at one end and projects by a length of 3.0 m beyond the edge of the swimming pool. A person of weight 650 N stands on the free end of the diving board. Calculate:

a the force on the bolts

b the force on the edge of the swimming pool.

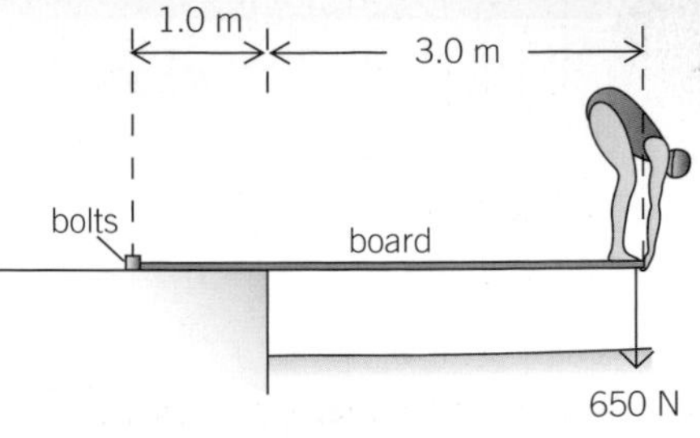


▲ Figure 4

10 A uniform beam XY of weight 1200 N and of length 5.0 m is supported horizontally on a concrete pillar at each end. A person of weight 500 N sits on the beam at a distance of 1.5 m from end X.

a Sketch a free body force diagram of the beam.

b Calculate the support force on the beam from each pillar.

11 A bridge crane used at a freight depot consists of a horizontal span of length 12 m fixed at each end to a vertical pillar, as shown in Figure 5.

a When the bridge crane supports a load of 380 kN at its centre, a force of 1600 kN is exerted on each pillar. Calculate the weight of the horizontal span.

b The same load is moved across a distance of 2.0 m by the bridge crane. Sketch a free body force diagram of the horizontal span and calculate the force exerted on each pillar.

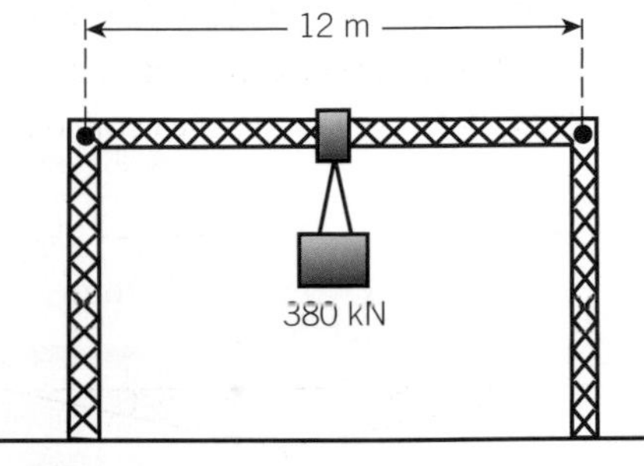


▲ Figure 5

12 A uniform curtain pole of weight 24 N and of length 3.2 m is supported horizontally by two wall-mounted supports X and Y, which are 0.8 m and 1.2 m from each end, respectively.

a Sketch the free body force diagram for this arrangement and calculate the force on each support when there are no curtains on the pole.

b When the pole supports a pair of curtains of total weight 90 N drawn along the full length of the pole, what will be the force on each support?

13 A uniform steel girder of weight 22 kN and of length 14 m is lifted off the ground at one end by means of a crane. When the raised end is 2.0 m above the ground, the cable is vertical.

a Sketch a free body force diagram of the girder in this position.

b Calculate the tension in the cable at this position and the force of the girder on the ground.

14 A rectangular picture 0.80 m deep and 1.0 m wide, of weight 24 N, hangs on a wall, supported by a cord attached to the frame at each of the top corners, as shown in Figure 6. Each section of the cord makes an angle of 25° with the picture, which is horizontal along its width.

a Copy the diagram and mark the forces acting on the picture on your diagram.

b Calculate the tension in each section of the cord.

You will find the answers to these questions at the back of the book.

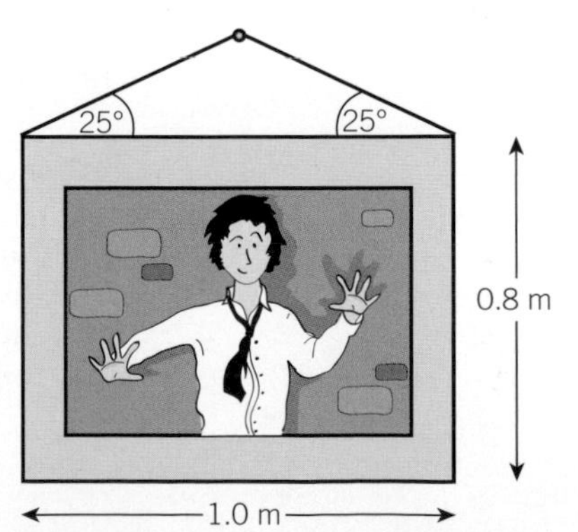


▲ Figure 6

# Practice questions: Chapter 6

1 A student set up a model bridge crane to find out how the support forces change with the position of a load of weight $W$ on the horizontal beam of the crane. With the load at different measured distances $d$ from the fixed support at X, she used a newtonmeter to measure the support force $S$ at Y near the other end, when the beam was horizontal. She repeated the measurements and also measured the distance $D$ from X to Y.

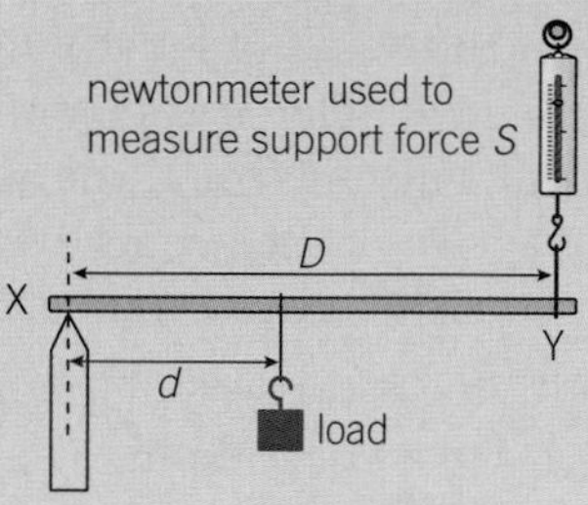


▲ **Figure 1**

The measurements she obtained are in the table below.
Distance XY = 480 mm

| **Distance $d$ / mm** | | 40 | 120 | 200 | 280 | 360 | 440 |
|---|---|---|---|---|---|---|---|
| **Support force $S$ / N** | 1st set | 1.4 | 2.6 | 4.4 | 5.6 | 7.3 | 8.6 |
| | 2nd set | 1.2 | 2.9 | 4.3 | 5.9 | 7.0 | 8.8 |
| | mean | | | | | | |

**(a)** (i) Copy and complete the table.
(ii) Plot a graph of $S$ on the $y$-axis against $d$ on the $x$-axis. *(5 marks)*

**(b)** (i) By taking moments about X, show that $S = \frac{Wd}{D} + 0.5\ W_0$, where $W_0$ is the weight of the beam.
(ii) Use your graph to determine the weight of the load, $W$. *(6 marks)*

**(c)** The distances were measured to an accuracy of ±1 mm and the newtonmeter readings to an accuracy of ±0.1 N.
(i) Show that the percentage uncertainty in $S$ is significantly greater than the percentage uncertainty in $d$.
(ii) Without using additional apparatus, discuss what further measurements you could make to improve the accuracy of your measurements. *(3 marks)*

2 **(a)** An object is acted upon by forces of 9.6 N and 4.8 N, with an angle of 40° between them. Draw a vector diagram of these forces, using a scale of 1 cm representing 1 N. Complete the vector diagram to determine the magnitude of the resultant force acting on the object. Measure the angle between the resultant force and the 9.6 N force. *(3 marks)*

**(b)** Calculate the magnitude of the resultant force when the same two forces act at right angles to each other.
*You must not use a scale diagram for this part.* *(2 marks)*

AQA, 2007

3 Figure 2 shows a uniform steel girder being held horizontally by a crane. Two cables are attached to the ends of the girder and the tension in each of these cables is $T$.

**(a)** If the tension, $T$, in each cable is 850 N, calculate:
(i) the horizontal component of the tension in each cable,
(ii) the vertical component of the tension in each cable,
(iii) the weight of the girder. *(4 marks)*

**(b)** Describe the line of action of the weight of the girder. *(1 mark)*

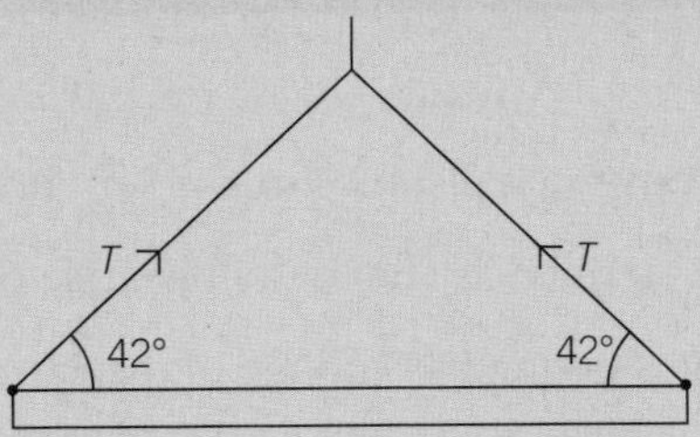


▲ **Figure 2**

AQA, 2005

4 **(a)** Define the moment of a force. *(2 marks)*
**(b)** Figure 3 shows the force, $F$, acting on a bicycle pedal.

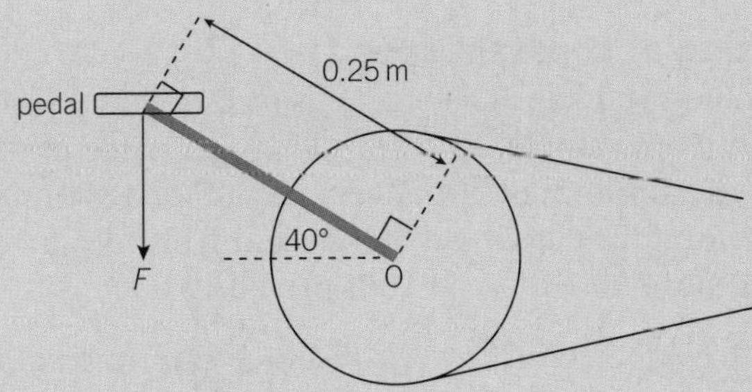


▲ **Figure 3**

(i) The moment of the force about O is 46 N m in the position shown. Calculate the value of the force, $F$.
(ii) Force $F$ is constant in magnitude and direction while the pedal is moving downwards. State and explain how the moment of $F$ changes as the pedal moves through 80°, from the position shown. *(4 marks)*

AQA, 2007

5 Figure 4 shows a student standing on a plank that pivots on a log. The student intends to cross the stream.

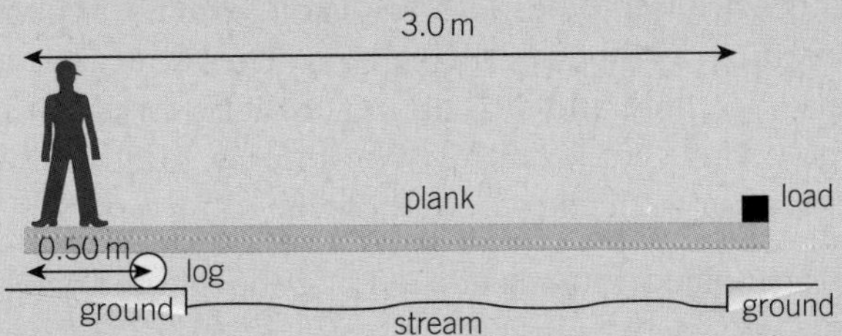


▲ **Figure 4**

**(a)** The plank has a mass of 25 kg and is 3.0 m long with a uniform cross section. The log pivot is 0.50 m from the end of the plank. The student has a mass of 65 kg and stands at the end of the plank. A load is placed on the far end in order to balance the plank horizontally.
Draw on a copy of Figure 4 the forces that act on the plank. *(3 marks)*
**(b)** By taking moments about the log pivot, calculate the load, in N, needed on the right-hand end of the plank in order to balance the plank horizontally. *(3 marks)*
**(c)** Explain why the load will eventually touch the ground as the student walks toward the log. *(2 marks)*

AQA, 2003

# 7 On the move

## 7.1 Speed and velocity

**Learning objectives:**

- → Explain how displacement differs from distance.
- → Explain the difference between instantaneous speed and average speed.
- → Describe when it is necessary to consider velocity rather than speed.

*Specification reference: 3.4.1.3*

**Hint**

1 km = 1000 m, 1 hour = 3600 s, 108 km $h^{-1}$ = 30 m $s^{-1}$

▲ **Figure 1**

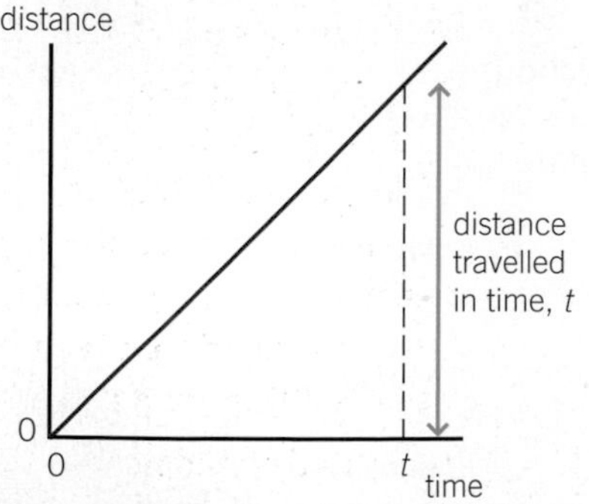


▲ **Figure 2** *Constant speed*

### Speed

**Displacement** is distance in a given direction.

**Speed** is defined as change of distance per unit time.

**Velocity** is defined as change of displacement per unit time. In other words, velocity is speed in a given direction.

Speed and distance are scalar quantities. Velocity and displacement are vector quantities.

The unit of speed and of velocity is the metre per second (m $s^{-1}$).

### Motion at constant speed

An object moving at a constant speed travels equal distances in equal times. For example, a car travelling at a speed of 30 m $s^{-1}$ on a motorway travels a distance of 30 m every second or 1800 m every minute. In 1 hour, the car would therefore travel a distance of 108 000 m or 108 km. So 30 m $s^{-1}$ = 108 km $h^{-1}$.

For an object which travels distance $s$ in time $t$ at constant speed,

$$\textbf{speed } v = \frac{s}{t}$$

$$\textbf{distance travelled } s = vt$$

For an object moving at constant speed on a circle of radius $r$, its speed

$$v = \frac{2\pi r}{T}$$

where $T$ is the time to move round once and $2\pi r$ is the circumference of the circle.

### Motion at changing speed

There are two types of speed cameras. One type measures the speed of a vehicle as it passes the camera. The other type is linked to a second speed camera and a computer, which works out the average speed of the vehicle between the two cameras. This will catch drivers who slow down for a speed camera then speed up again!

For an object moving at changing speed that travels a distance $s$ in time $t$,

$$\textbf{average speed} = \frac{s}{t}$$

In a short time interval $\Delta t$, the distance $\Delta s$ it travels is given by $\Delta s = v\Delta t$, where $v$ is the speed at that time (i.e., its instantaneous speed). Rearranging this equation gives:

$$v = \frac{\Delta s}{\Delta t}$$

In physics and maths, the delta $\Delta$ notation is often used to mean *a change of* something.

### Distance–time graphs

For an object moving at **constant speed**, its graph of distance against time is a straight line with a constant gradient.

**The speed of the object** $= \dfrac{\textbf{distance travelled}}{\textbf{time taken}} =$ **gradient of the line**

For an object moving at **changing speed**, the gradient of the line changes. The gradient of the line at any point can be found by drawing a tangent to the line at that point and then measuring the gradient of the tangent. This is shown in Figure 3 where PR is the tangent at point Y on the line. Show for yourself that the speed at point X on the line is $2.5\,\text{m}\,\text{s}^{-1}$.

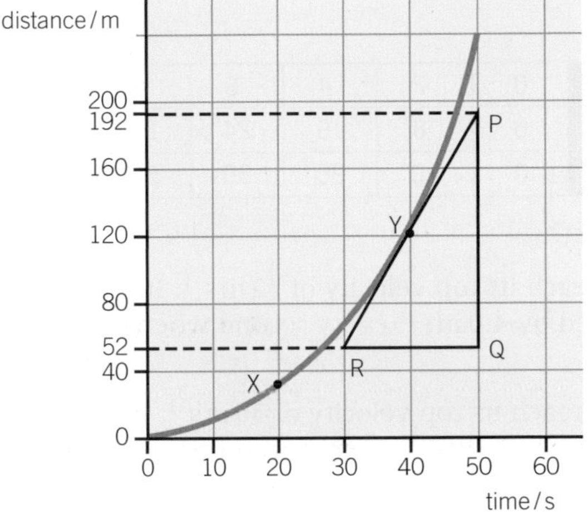


$$\text{speed at Y} = \frac{PQ}{QR} = \frac{192-52}{20}$$
$$= 7\,\text{m}\,\text{s}^{-1}$$

▲ **Figure 3** *Changing speed*

## Velocity

An object moving at **constant velocity** moves at the same speed without changing its direction of motion.

If an object changes its direction of motion or its speed or both, its velocity changes. For example, the velocity of an object moving on a circular path at constant speed changes continuously because its direction of motion changes continuously.

### Displacement–time graphs

An object travelling along a straight line has two possible directions of motion. To distinguish between the two directions, we need a direction code where positive values are in one direction and negative values in the opposite direction.

For example, consider an object thrown vertically into the air. The direction code is + for upwards and – for downwards. Figure 4 shows how its displacement changes with time. The object has an initial positive velocity and a negative return velocity. The displacement and velocity are both positive when the object is ascending. However, when it is descending, its velocity is negative and its displacement is positive until it returns to its initial position. We will consider displacement–time graphs in more detail in Topic 7.5.

displacement / m
5
4
3
2
1
0
0 0.5 1.0 1.5 2.0 2.5
time / s

▲ **Figure 4** *A displacement–time graph*

## Summary questions

**1 kilometre (km) = 1000 m**

1. A car travels a distance of 60 km in 45 min at constant speed. Calculate its speed in:
   **a** $\text{km}\,\text{h}^{-1}$ **b** $\text{m}\,\text{s}^{-1}$.
2. Use Figure 3 to calculate
   **a** the instantaneous speed of the object at X
   **b** the average speed of the object between 0 and Y.
3. A satellite moves round the Earth at constant speed on a circular orbit of radius 8000 km with a time period of 120 min. Calculate its orbital speed in **a** $\text{km}\,\text{h}^{-1}$, **b** $\text{m}\,\text{s}^{-1}$.
4. A vehicle joins a motorway and travels at a steady speed of $25\,\text{m}\,\text{s}^{-1}$ for 30 min then it travels a further distance of 40 km in 20 min before leaving the motorway. Calculate **a** the distance travelled in the first 30 min, **b** its average speed on the motorway.
5. **a** Explain the difference between speed and velocity.
   **b** A police car joins a straight motorway at Junction 4 and travels for 12 km at a constant speed for 400 s. It then leaves at Junction 5 and rejoins on the opposite side and travels for 8 km at a constant speed for 320 s to reach the scene of an accident. Calculate **i** the displacement from Junction 4 to the accident, **ii** the velocity of the car on each side of the motorway.
   **c** Sketch a displacement–time graph for the journey.

# 7.2 Acceleration

**Learning objectives:**

- → Describe what is meant by acceleration and deceleration.
- → Explain why uniform acceleration is a special case.
- → Explain why acceleration is considered to be a vector.

*Specification reference: 3.4.1.3*

▲ **Figure 1** *A racing car on a test track*

## Performance tests

A car maker wants to compare the performance of a new model with the original model. To do this, the velocity of each car is measured on a test track. Each vehicle accelerates as fast as possible to top velocity from a standstill. The results are listed in Table 1.

▼ **Table 1**

| Time from a standing start / s | 0 | 2 | 4 | 6 | 8 | 10 |
|---|---|---|---|---|---|---|
| Velocity of old model / m s$^{-1}$ | 0 | 8 | 16 | 24 | 32 | 32 |
| Velocity of new model / m s$^{-1}$ | 0 | 10 | 20 | 30 | 30 | 30 |

Which car accelerates fastest?

The old model took 8 s to reach its top velocity of 32 m s$^{-1}$. Its velocity must have increased by 4.0 m s$^{-1}$ every second when it was accelerating.

The new model took 6 s to reach its top velocity of 30 m s$^{-1}$. Its velocity must have increased by 5.0 m s$^{-1}$ every second when it was accelerating.

Clearly, the velocity of the new model increases at a faster rate than the old model. In other words, its acceleration is greater.

**Acceleration is defined as change of velocity per unit time.**

- The unit of acceleration is the metre per second per second (m s$^{-2}$).
- Acceleration is a vector quantity.
- Deceleration values are negative and signify that the velocity decreases with respect to time.

For a moving object that does not change direction, its acceleration at any point can be worked out from its rate of change of velocity, because there is no change of direction. For example:

The acceleration of the old model above is 4.0 m s$^{-2}$ because its velocity increased by 4.0 m s$^{-1}$ every second.

The acceleration of the new model above is 5.0 m s$^{-2}$ because its velocity increased by 5.0 m s$^{-1}$ every second.

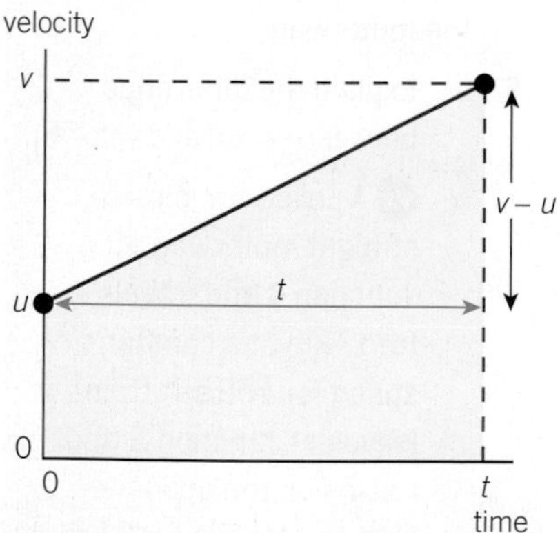


▲ **Figure 2** *Uniform acceleration on a velocity–time graph*

## Uniform acceleration

Uniform acceleration is where the velocity of an object moving along a straight line changes at a constant rate. In other words, the acceleration is constant. Consider an object that accelerates uniformly from velocity $u$ to velocity $v$ in time $t$ along a straight line. Figure 2 shows how its velocity changes with time.

The acceleration, $a$, of the object $= \dfrac{\text{change of velocity}}{\text{time taken}} = \dfrac{v-u}{t}$

$$a = \frac{v-u}{t}$$

To calculate the velocity $v$ at time $t$, rearranging this equation gives

$$at = v - u$$

$$v = u + at$$

**Study tip**

Compare the appearance of uniform acceleration and uniform velocity on velocity–time graphs and on displacement–time graphs.

## Worked example

The driver of a vehicle travelling at $8\,m\,s^{-1}$ applies the brakes for 30 s and reduces the velocity of the vehicle to $2\,m\,s^{-1}$. Calculate the deceleration of the vehicle during this time.

### *Solution*

$u = 8\,m\,s^{-1}$, $v = 2\,m\,s^{-1}$, $t = 30$ s

$$a = \frac{v-u}{t} = \frac{2-8}{30} = \frac{-6}{30} = -0.2\,m\,s^{-2}$$

## Hint

Remember that acceleration is a vector quantity. It is dependent on both speed and direction.

## Non-uniform acceleration

Non-uniform acceleration is where the direction of motion of an object changes, or its speed changes, at a varying rate. Figure 3 shows how the velocity of an object increases for an object moving along a straight line with an increasing acceleration. This can be seen directly from the graph because the gradient increases with time (the graph becomes steeper and steeper) and the gradient represents the acceleration.

The acceleration at any point is the gradient of the tangent to the curve at that point. In Figure 3,

$$\text{the gradient of the tangent at point P} = \frac{\text{height of gradient triangle}}{\text{base of gradient triangle}}$$

$$= \frac{4.0\,m\,s^{-1} - 1.0\,m\,s^{-1}}{2.0\,s} = 1.5\,m\,s^{-2}$$

Therefore the acceleration at P is $1.5\,m\,s^{-2}$.

**Acceleration = gradient of the line on the velocity–time graph.**

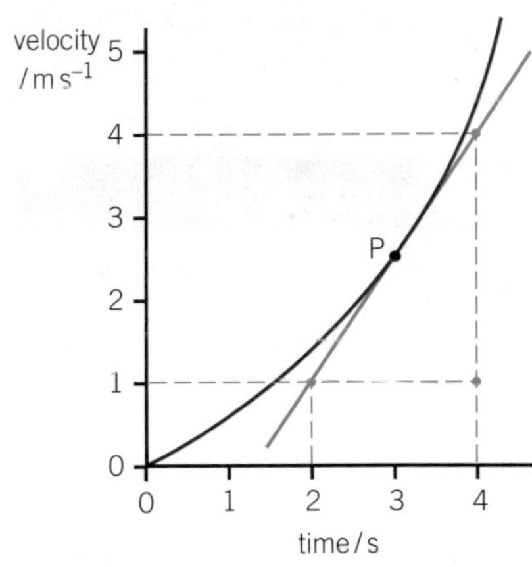


▲ **Figure 3**

## Summary questions

1 a An aeroplane taking off accelerates uniformly on a runway from a velocity of $4\,m\,s^{-1}$ to a velocity of $64\,m\,s^{-1}$ in 40 s. Calculate its acceleration.

b A car travelling at a velocity of $20\,m\,s^{-1}$ brakes sharply to a standstill in 8.0 s. Calculate its deceleration, assuming its velocity decreases uniformly.

2 A cyclist accelerates uniformly from a velocity of $2.5\,m\,s^{-1}$ to a velocity of $7.0\,m\,s^{-1}$ in a time of 10 s. Calculate **a** its acceleration, **b** its velocity 2.0 s later if it continued to accelerate at the same rate.

3 A train on a straight journey between two stations accelerates uniformly from rest for 20 s to a velocity of $12\,m\,s^{-1}$. It then travelled at constant velocity for a further 40 s before decelerating uniformly to rest in 30 s.

**a** Sketch a velocity–time graph to represent its journey.

**b** Calculate its acceleration in each part of the journey.

4 The velocity of an object released in water increased as shown in Figure 4.

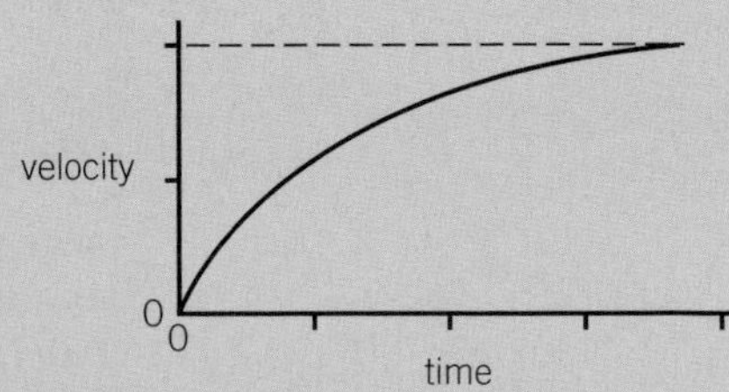


▲ **Figure 4**

Describe how **a** the velocity of the object changed with time, **b** the acceleration of the object changed with time.

# 7.3 Motion along a straight line at constant acceleration

## Learning objectives:

- → Distinguish between $u$ and $v$.
- → Calculate the displacement of an object moving with uniform acceleration.
- → Explain what else we need to know to calculate the acceleration of an object if we know its displacement in a given time.

*Specification reference: 3.4.1.3*

### Maths link 

**Q:** If a space rocket accelerated from rest at $10\,\mathrm{m\,s^{-2}}$, estimate how many years it would take to reach half the speed of light. The speed of light is $3.0 \times 10^8\,\mathrm{m\,s^{-1}}$.

A: 1 year

### Hint

If $a = 0$ (i.e., constant velocity), the equations all reduce to $s = vt$ or $v = u$.

## The dynamics equations for constant acceleration

Consider an object that accelerates uniformly from initial velocity $u$ to final velocity $v$ in time $t$ without change of direction. Figure 2 in 8.2 shows how its velocity changes with time.

1 The acceleration $a = \frac{(v - u)}{t}$, as explained in Topic 7.2, can be rearranged to give

$$v = u + at \qquad \textbf{(Equation 1)}$$

2 The displacement $s$ = average velocity × time taken

Because the acceleration is uniform, average velocity $= \frac{(u + v)}{2}$

Combining this equation with $s = v \times t$ gives

$$s = \frac{(u + v)t}{2} \qquad \textbf{(Equation 2)}$$

3 By combining the two equations above to eliminate $v$, a further useful equation is produced.

To do this, substitute $u + at$ in place of $v$ in equation 2. This gives

$$s = \frac{(u + (u + at))}{2}t = \frac{(u + u + at)t}{2} = \frac{(2ut + at^2)}{2}$$

$$s = ut + \tfrac{1}{2}at^2 \qquad \textbf{(Equation 3)}$$

4 A fourth useful equation is obtained by combining Equations 1 and 2 to eliminate $t$. This can be done by multiplying

$a = \frac{(v - u)}{t}$ and $s = \frac{(u + v)t}{2}$ together to give

$$as = \frac{(v - u)}{t} \times \frac{(u + v)t}{2}$$

This simplifies to become

$$as = \frac{(v - u)\,(v + u)}{2} = \frac{(v^2 - uv + uv - u^2)}{2} = \frac{(v^2 - u^2)}{2}$$

Therefore $2as = v^2 - u^2$ or

$$v^2 = u^2 + 2as \qquad \textbf{(Equation 4)}$$

The four equations, sometimes referred to as the *suvat* equations, are invaluable in any situation where the acceleration is constant.

### Worked example 

A driver of a vehicle travelling at a speed of $30\,\mathrm{m\,s^{-1}}$ on a motorway brakes sharply to a standstill in a distance of 100 m. Calculate the deceleration of the vehicle.

### *Solution*

$u = 30\,\mathrm{m\,s^{-1}}$, $v = 0$, $s = 100$ m, $a = ?$

To find $a$, use $v^2 = u^2 + 2as$ 

Therefore $0 = u^2 + 2as$ because $v = 0$

Rearranging this equation gives

$2as = -u^2$

$$a = -\frac{u^2}{2s} = -\frac{30^2}{2 \times 100} = -4.5\,\text{m}\,\text{s}^{-2}$$

*Note:*

The acceleration is negative which means it is a deceleration as it is in the opposite direction to the velocity.

## Using a velocity–time graph to find the displacement

1 **Consider an object moving at constant velocity, *v*,** as shown in Figure 1. The displacement in time $t$, $s$ = velocity × time taken = $vt$. This displacement is represented on the graph by the area under the line between the start and time $t$. This is a rectangle of height corresponding to velocity $v$ and of base corresponding to the time $t$.

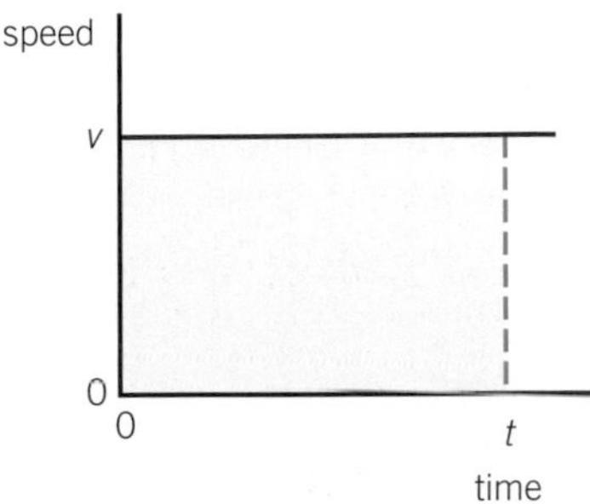


▲ **Figure 1**

2 **Consider an object moving at constant acceleration, *a*,** from initial velocity $u$ to velocity $v$ at time $t$, as shown in Figure 2. From Equation 2, the displacement $s$ moved in this time is given by

$$s = \frac{(u + v)t}{2}$$

This displacement is represented on the graph by the area under the line between the start and time $t$. This is a trapezium which has a base corresponding to time $t$ and an average height corresponding to the average speed $\frac{1}{2}(u + v)$.

Therefore the area of the trapezium (= average height × base) corresponds to $\frac{1}{2}(u + v)t$, and gives the displacement.

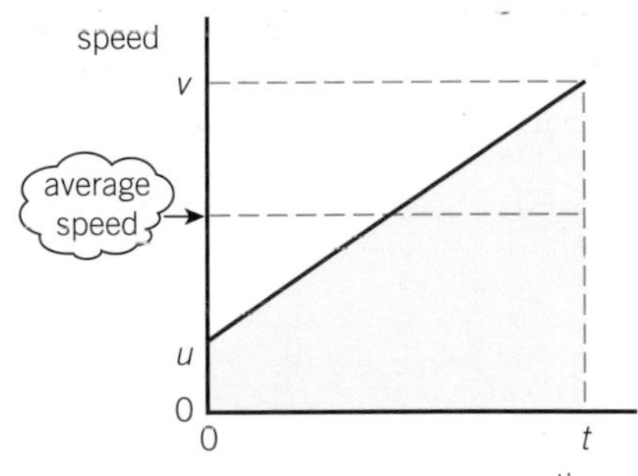


▲ **Figure 2**

3 **Consider an object moving with a changing acceleration,** as shown in Figure 3. Let $v$ represent the velocity at time $t$ and $v + \delta v$ represent the velocity a short time later at $t + \delta t$. ($\delta$ is pronounced 'delta'.)

Because the velocity change $\delta v$ is small compared with the velocity $v$, the dispalcement $\delta s$ in the short time interval $\delta t$ is $v\delta t$.

This is represented on the graph by the area of the shaded strip under the line, which has a base corresponding to $\delta t$ and a height corresponding to $v$. In other words, $\delta s = v\delta t$ is represented by the area of this strip.

By considering the whole area under the line in strips of similar width, the total displacement from the start to time $t$ is represented by the sum of the area of every strip, which is therefore the total area under the line.

Whatever the shape of the line of a velocity–time graph,

**displacement = area under the line of a velocity–time graph.**

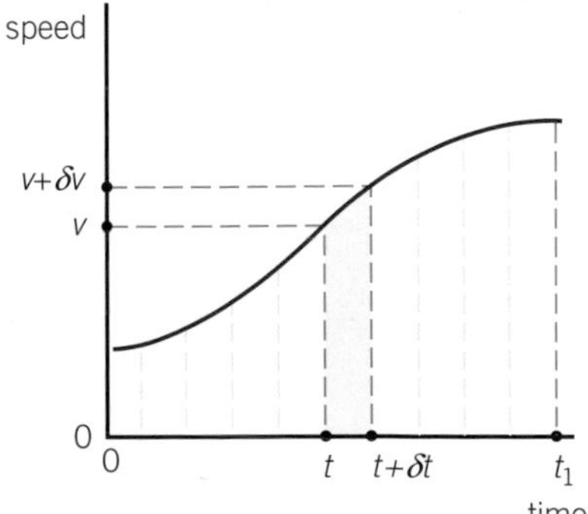


▲ **Figure 3**

## Study tip

Always begin a *suvat* calculation by identifying which three variables are known. Remember the acceleration must be constant for these equations to apply.

The equations for constant acceleration are:

$$v = u + at$$

$$s = \frac{1}{2}(u + v)t$$

$$s = ut + \frac{1}{2}at^2$$

$$v^2 = u^2 + 2as$$

## Summary questions

1 A vehicle accelerates uniformly along a straight road, increasing its velocity from 4.0 m s$^{-1}$ to 30.0 m s$^{-1}$ in 13 s. Calculate:

a its acceleration

b the displacement.

2 An aircraft lands on a runway at a velocity of 40 m s$^{-1}$ and brakes to a halt in a distance of 860 m. Calculate:

a the braking time

b the deceleration of the aircraft.

3 A cyclist accelerates uniformly from rest to a speed of 6.0 m s$^{-1}$ in 30 s then brakes at uniform deceleration to a halt in a distance of 24 m.

a For the first part of the journey, calculate **i** the acceleration, **ii** the displacement.

b For the second part of the journey, calculate **i** the deceleration, **ii** the time taken.

c Sketch a velocity–time graph for this journey.

d Use the graph to determine the average velocity for the journey.

4 The velocity of an athlete for the first 5 s of a sprint race is shown in Figure 4. Use the graph to determine:

a the initial acceleration of the athlete

b the displacement in the first 2 s

c the displacement in the next 2 s

d the average speed over the first 4 s.

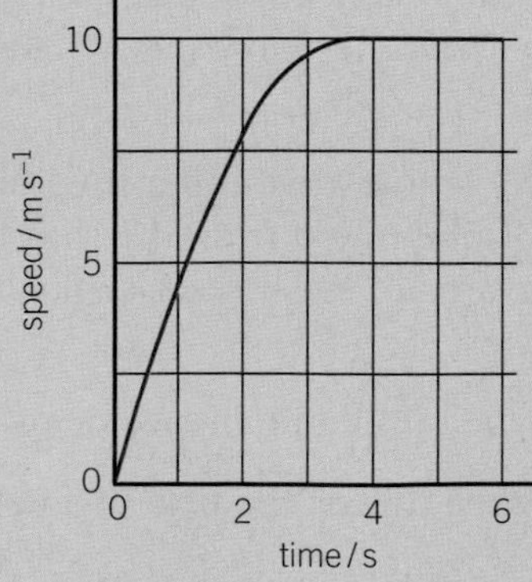


▲ Figure 4

# 7.4 Free fall

## Experimental tests

### Does a heavy object fall faster than a lighter object?

Release a stone and a small coin at the same time. Which one hits the ground first? The answer to this question was first discovered by Galileo Galileli about four centuries ago. He reasoned that because any number of identical objects must fall at the same rate, then any one such object must fall at the same rate as the rest put together. So he concluded that any two objects must fall at the same rate, regardless of their relative weights. He was reported to have demonstrated the correctness of his theory by releasing two different weights from the top of the Leaning Tower of Pisa.

### The inclined plane test

Galileo wanted to know if a falling object speeds up as it falls, but clocks and stopwatches were devices of the future. The simplest test he could think of was to time a ball as it rolled down a plank. He devised a dripping water clock, counting the volume of the drips as a measure of time. He measured how long the ball took to travel equal distances down the slope from rest. His measurements showed that the ball gained speed as it travelled down the slope. In other words, he showed that the ball accelerated as it rolled down the slope. He reasoned that the acceleration would be greater the steeper the slope. So he concluded that an object falling vertically accelerates.

## Acceleration due to gravity

One way to investigate the free fall of a ball is to make a multiflash photo or video clip of the ball's flight as it falls after being released from rest. A vertical metre rule can be used to provide a scale. To obtain a multiflash photo, an ordinary camera with a slow speed shutter may be used to record the ball's descent in a dark room illuminated by a stroboscope (a flashing light). The flashing light needs to flash at a known constant rate of about 20 flashes per second. Figure 2 shows a possible arrangement using a steel ball and a multiflash photo taken with this arrangement. If you make a video clip, you need to be able to rerun the clip at slow speed with the time displayed.

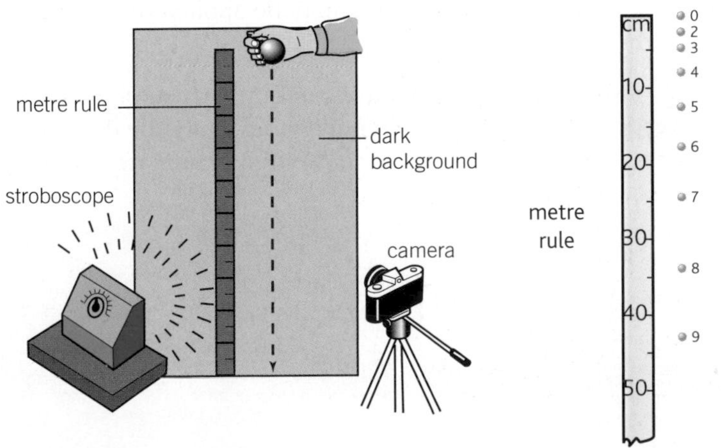


▲ **Figure 2** *Investigating free fall*

## Learning objectives:

- → Define 'free fall'.
- → Explain how the velocity of a freely falling object changes as it falls.
- → Discuss if objects of different masses or sizes all fall with the same acceleration.

*Specification reference: 3.4.1.3*

▲ **Figure 1** *Galileo Galilei 1564–1642*

### Practical link

A digital camera or a similar device can also be used to obtain a video clip that can be analysed to obtain *g*.

### What Galileo did for science

In addition to his discoveries on motion, Galileo made other important scientific discoveries such as the four innermost moons of Jupiter. He became a big supporter of the Copernican, or Sun-centred, model of the Solar System. This upset the Catholic Church, which believed that the Earth was at the centre of the Universe, and Galileo was banned from teaching the Copernican model. He wrote a book about it in protest, which became a bestseller and upset the Church even more. Galileo was condemned to spend the rest of his life under house arrest. However, his trial attracted a lot of publicity all over Europe, and his work was taken up enthusiastically by other scientists as a result. Galileo showed the importance of valid observations and measurements in developing our understanding of the natural world. He also showed that science is powerful enough to change established beliefs.

**Q:** Describe another situation when scientific observations caused a scientific theory to be altered or replaced.

### Synoptic link

You will meet different graphs in more detail in Topic 16.4, Straight line graphs, and Topic 16.5, More on graphs.

For each image of the ball on the photograph, the time of descent of the ball and the distance fallen by the ball from rest can be measured directly. The photograph shows that the ball speeds up as it falls, because it travels further between successive images. Measurements from this photograph are given in Table 1.

▼ **Table 1** *Free fall measurements*

| Number of flashes after start | 0 | 2 | 3 | 4 | 5 | 6 | 7 | 8 | 9 |
|---|---|---|---|---|---|---|---|---|---|
| Time of descent $t$ / s | 0 | 0.06 | 0.10 | 0.13 | 0.16 | 0.19 | 0.23 | 0.26 | 0.29 |
| Distance fallen $s$ / m | 0 | 0.02 | 0.04 | 0.07 | 0.12 | 0.17 | 0.24 | 0.33 | 0.42 |

How can you tell if the acceleration is constant from these results? One way is to consider how the distance fallen, $s$, would vary with time, $t$, for constant acceleration. From Topic 7.3, we know that

$$s = ut + \frac{1}{2}at^2, \text{ where } u = \text{the initial speed and } a = \text{acceleration.}$$

In this experiment, $u = 0$.

Therefore $s = \frac{1}{2}at^2$ for constant acceleration, $a$.

Compare this equation with the general equation for a straight line graph, $y = mx + c$, where $m$ is the gradient and $c$ is the $y$-intercept. If we let $y$ represent $s$ and let $x$ represent $t^2$, then $m = \frac{1}{2}a$ and $c = 0$.

So a graph of $s$ against $t^2$ should therefore give a straight line through the origin. In addition, the gradient of the line $(= \frac{1}{2}a)$ can be measured, enabling the acceleration (= 2 × gradient) to be calculated.

Figure 3 shows this graph. As you can see, it is a straight line through the origin. We can therefore conclude that the equation $s = \frac{1}{2}at^2$ applies here so the acceleration of a falling object is constant. Show for yourself that the gradient of the line is $5.0\,\text{m}\,\text{s}^{-2}$ ($\pm 0.2\,\text{m}\,\text{s}^{-2}$), giving an acceleration of $10\,\text{m}\,\text{s}^{-2}$.

Because there are no external forces acting on the object apart from the force of gravity, this value of acceleration is known as the **acceleration of free fall** and is represented by the symbol $g$. Accurate measurements give a value of $9.8\,\text{m}\,\text{s}^{-2}$ near the Earth's surface.

The 'suvat' equations from Topic 8.3 may be applied to any free fall situation where air resistance is negligible.

The equations can also be applied to situations where objects are thrown vertically upwards. As a general rule, apply the direction code + for *upwards* and – for *downwards* when values are inserted into the suvat equations.

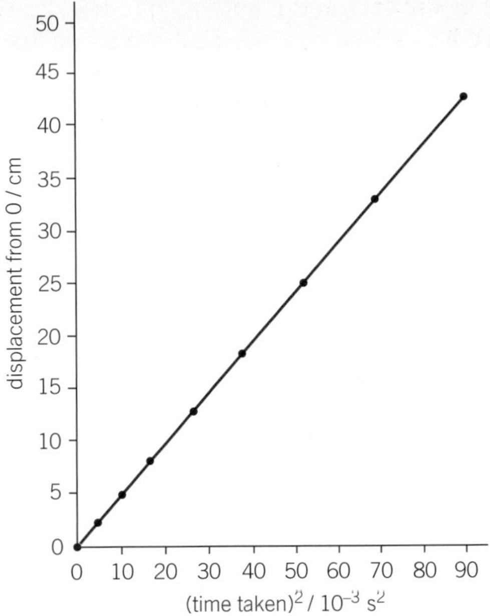


▲ **Figure 3** *A graph of s against $t^2$*

## *Practical*: measuring *g*

Use an electronic timer or motion sensor to make your own measurements of $s$ and $t$. Repeat your measurements to obtain an average timing for each measured distance and plot a suitable graph to find $g$.

### Worked example

$g = 9.8\,\text{m}\,\text{s}^{-2}$

A coin was released at rest at the top of a well. It took 1.6 s to hit the bottom of the well. Calculate **a** the distance fallen by the coin, **b** its speed just before impact.

### *Solution*

$u = 0$, $t = 1.6$ s, $a = -9.8\,\text{m}\,\text{s}^{-2}$ (– as $g$ acts downwards)

**a** To find $s$, use $s = \frac{1}{2}at^2$ as $u = 0$

Therefore $s = \frac{1}{2} \times -9.8 \times 1.6^2$

$= -12.5$ m (– indicates 12.5 m downwards)

**b** To find $v$, use $v = u + at$

$= 0 + (-9.8 \times 1.6) = -15.7\,\text{m}\,\text{s}^{-1}$ (– indicates downward velocity)

## Summary questions

$g = 9.8\,\text{m}\,\text{s}^{-2}$

**1** A pebble, released at rest from a canal bridge, took 0.9 s to hit the water. Calculate:

- **a** the distance it fell before hitting the water
- **b** its speed just before hitting the water.

**2** A spanner was dropped from a hot air balloon when the balloon was at rest 50 m above the ground. Calculate:

- **a** the time taken for the spanner to hit the ground
- **b** the speed of impact of the spanner on hitting the ground.

**3** A bungee jumper jumped off a platform 75 m above a lake, releasing a small object at the instant she jumped off the platform.

- **a** Calculate **i** the time taken by the object to fall to the lake, **ii** the speed of impact of the object on hitting the water, assuming air resistance is negligible.
- **b** Explain why the bungee jumper would take longer to descend than the time taken in **a**.

**4** An astronaut on the Moon threw an object 4.0 m vertically upwards and caught it again 4.5 s later. Calculate:

- **a** the acceleration due to gravity on the Moon
- **b** the speed of projection of the object
- **c** how high the object would have risen on the Earth, for the same speed of projection.

# 7.5 Motion graphs

**Learning objectives:**

- → Distinguish between a distance–time graph and a displacement–time graph.
- → Describe what the gradient of a velocity–time graph represents.
- → Describe what the area under a velocity–time graph represents.

*Specification reference: 3.4.1.3*

## The difference between a distance–time graph and a displacement–time graph

Displacement is distance in a given direction from a certain point. Consider a ball thrown directly upwards and caught when it returns. If the ball rises to a maximum height of 2.0 m, on returning to the thrower its displacement from its initial position is zero. However, the distance it has travelled is 4.0 m.

The displacement of the ball changes with time as shown by Figure 1. The line in this graph fits the equation $s = ut - \frac{1}{2}gt^2$, where $s$ is the displacement, $t$ is the time taken and $u$ is the initial velocity of the object.

The gradient of the line in Figure 1 represents the velocity of the object.

- Immediately after leaving the thrower's hand, the velocity is positive and large so the gradient is positive and large.
- As the ball rises, its velocity decreases so the gradient decreases.
- At maximum height, its velocity is zero so the gradient is zero.
- As the ball descends, its velocity becomes increasingly negative, corresponding to increasing speed in a downward direction. So the gradient becomes increasingly negative.

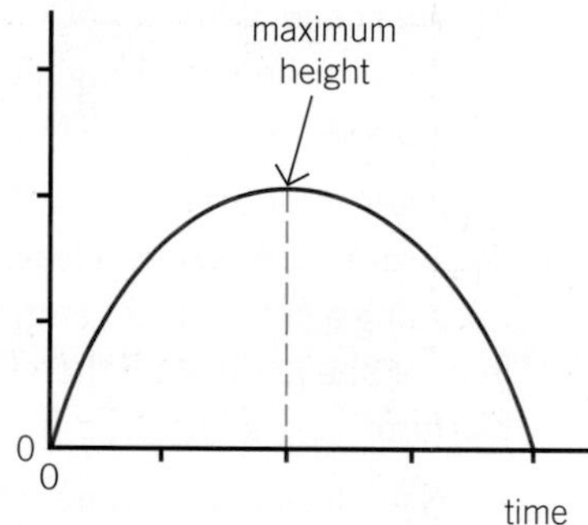


▲ **Figure 1** *Displacement–time graph for an object projected upwards*

The distance travelled by the object changes with time as shown by Figure 2. The gradient of this line represents the speed. From projection to maximum height, the shape is exactly the same as in Figure 1. After maximum height, the distance continues to increase so the line curves up, not down like Figure 1.

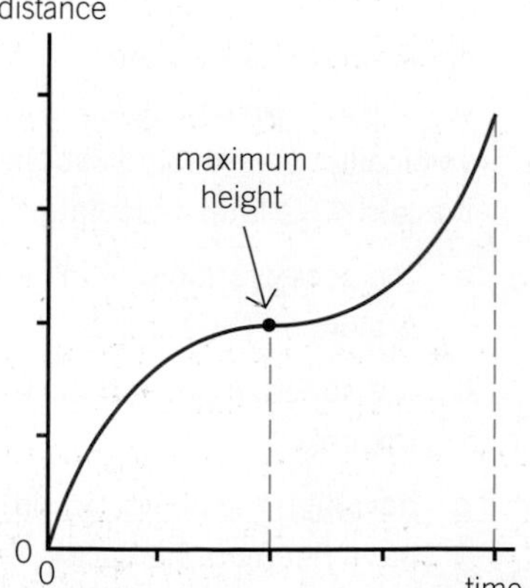


▲ **Figure 2** *Distance–time for an object projected upwards*

## The difference between a speed–time graph and a velocity–time graph

Velocity is speed in a given direction. Consider how the velocity of an object thrown into the air changes with time. The object's velocity decreases from its initial positive (upwards) value to zero at maximum height then increases in the negative (downwards) direction as it falls.

Figure 3 shows how the velocity of the object changes with time.

1. **The gradient of the line represents the object's acceleration.** This is constant and negative, equal to the acceleration of free fall, $g$. The acceleration of the object is the same when it descends as when it ascends so the gradient of the line is always equal to $-9.8\,\text{m}\,\text{s}^{-2}$.
2. **The area under the line represents the displacement of the object from its starting position.**
   - The area between the positive section of the line and the time axis represents the displacement during the ascent.
   - The area under the negative section of the line and the time axis represents the displacement during the descent.

Taking the area for **a** as positive and the area for **b** as negative, the total area is therefore zero. This corresponds to zero for the total displacement.

## Worked example

$g = 9.8\,\mathrm{m\,s^{-2}}$

A ball released from a height of 1.20 m above a concrete floor rebounds to a height of 0.82 m.

**a** Calculate **i** its time of descent, **ii** the speed of the ball immediately before it hits the floor.

**b** Calculate **i** the speed of the ball immediately after it leaves the floor, **ii** its time of ascent.

**c** Sketch a velocity–time graph for the ball. Assume the contact time is negligible compared with the time of descent or ascent and that positive velocity is upwards.

### Solution

**a** $u = 0$, $a = -9.8\,\mathrm{m\,s^{-2}}$, $s = -1.2\,\mathrm{m}$

**i** To find $t$, use $s = ut + \frac{1}{2}at^2$

$-1.2 = 0 + 0.5 \times -9.8 \times t^2$ so $-1.2 = -4.9t^2$

$t^2 = \dfrac{-1.2}{-4.9} = 0.245$ so $t = 0.49\,\mathrm{s}$

**ii** To find $v$, use $v = u + at$

$v = 0 + -9.8 \times 0.49 = -4.8\,\mathrm{m\,s^{-1}}$ (– because downwards)

**b** $v = 0$, $a = -9.8\,\mathrm{m\,s^{-2}}$, $s = +0.82\,\mathrm{m}$

**i** Using $v^2 = u^2 + 2as$ to find $u$ gives $0 = u^2 + 2 \times -9.8 \times 0.82$

$u^2 = 16.1\,\mathrm{m^2\,s^{-2}}$ so $u = +4.0\,\mathrm{m\,s^{-1}}$ (+ because upwards)

**ii** Using $v = u + at$ to find $t$ gives $0 = 4.0 + -9.8 \times t$

$9.8t = 4.0$ so $t = 0.41\,\mathrm{s}$

**c** The sketch graph is given in Figure 4.

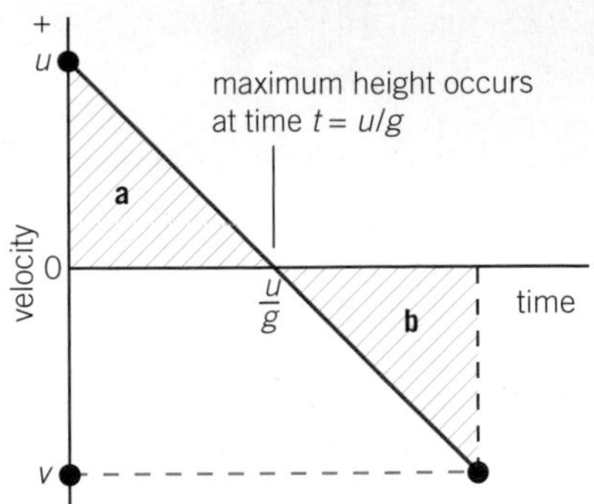


▲ **Figure 3** *Velocity–time graph*

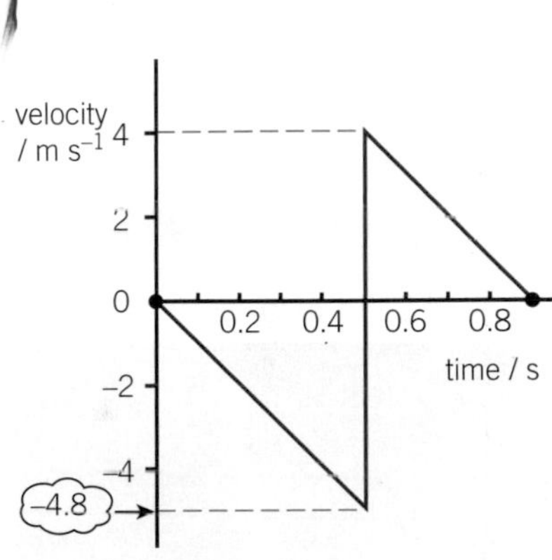


▲ **Figure 4** *Velocity–time graph for a bouncing ball, where gradient = g*

## Summary questions

$g = 9.8\,\mathrm{m\,s^{-2}}$

**1** A swimmer swims 100 m from one end of a swimming pool to the other end at a constant speed of $1.2\,\mathrm{m\,s^{-1}}$, then swims back at constant speed, returning to the starting point 210 s after starting.

**a** Calculate how long the swimmer takes to swim from **i** the starting end to the other end, **ii** back to the start from the other end.

**b** For the swim from start to finish, sketch **i** a displacement–time graph, **ii** a distance–time graph, **iii** a velocity–time graph.

**2** A motorcyclist travelling along a straight road at a constant speed of $8.8\,\mathrm{m\,s^{-1}}$ passes a cyclist travelling in the same direction at a speed of $2.2\,\mathrm{m\,s^{-1}}$. After 200 s, the motorcyclist stops.

**a** Calculate how long the motorcyclist has to wait before the cyclist catches up.

**b** On the same axes, sketch velocity–time graphs for **i** the motorcyclist, **ii** the cyclist.

**3** A hot air balloon is ascending vertically at a constant velocity of $3.5\,\mathrm{m\,s^{-1}}$ when a small metal object falls from its base and hits the ground 3.0 s later.

**a** Sketch a graph to show how the velocity of the object changed with time during its descent.

**b** Show that the balloon base was 33.6 m above the ground when the object fell off the base.

**4** A ball is released from a height of 1.8 m above a level surface and rebounds to a height of 0.90 m.

**a** Given $g = 9.8\,\mathrm{m\,s^{-2}}$, calculate **i** the duration of its descent, **ii** its velocity just before impact, **iii** the duration of its ascent, **iv** its velocity just after impact.

**b** Sketch a graph to show how its velocity changes with time from release to rebound at maximum height.

**c** Sketch a further graph to show how the displacement of the object changes with time.

# 7.6 More calculations on motion along a straight line

**Learning objectives:**

- → Calculate the motion of an object with constant acceleration if its velocity reverses.
- → Deduce whether the overall motion should be broken down into stages.
- → Explain how we use signs to work out if an object is moving forwards or backwards.

*Specification reference: 3.4.1.3*

▲ **Figure 1** *A space vehicle docking*

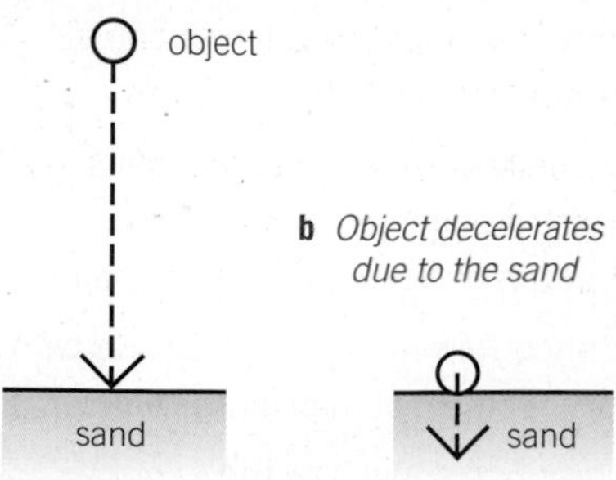

a Object accelerates due to gravity
object
b Object decelerates due to the sand
sand
sand

▲ **Figure 2** *A two-stage problem*

## Motion along a straight line at constant acceleration

- The equations for motion at constant acceleration, $a$, are

$$v = u + at \qquad \textbf{(equation 1)}$$

$$s = \frac{1}{2}(u + v)t \qquad \textbf{(equation 2)}$$

$$s = ut + \frac{1}{2}at^2 \qquad \textbf{(equation 3)}$$

$$v^2 = u^2 + 2as \qquad \textbf{(equation 4)}$$

where $s$ is the displacement in time $t$, $u$ is the initial velocity, and $v$ is the final velocity.

- For motion along a straight line at constant acceleration, one direction along the line is positive and the opposite direction is negative.

### Worked example

A space vehicle moving towards a docking station at a velocity of $2.5\,\text{m s}^{-1}$ is 26 m from the docking station when its reverse thrust motors are switched on to slow it down and stop it when it reaches the station. The vehicle decelerates uniformly until it comes to rest at the docking station when its motors are switched off.

Calculate **a** its deceleration, **b** how long it takes to stop.

### *Solution*

Let the + direction represent motion towards the docking station and – represent motion away from the station.
Initial velocity $u = +2.5\,\text{m s}^{-1}$, final velocity $v = 0$, displacement $s = +26\,\text{m}$.

**a** To find its deceleration, $a$, use $v^2 = u^2 + 2as$

$$0 = 2.5^2 + 2a \times 26 \qquad \text{so } -52a = 2.5^2$$

$$a = \frac{2.5^2}{52} = -0.12\,\text{m s}^{-2}$$

**b** To find the time taken, use $v = u + at$

$$0 = 2.5 - 0.12t \qquad \text{so } 0.12t = 2.5$$

$$t = \frac{2.5}{0.12} = 21\,\text{s}$$

## Two-stage problems

Consider an object released from rest, falling, then hitting a bed of sand. The motion is in two stages:

1. falling motion due to gravity: acceleration = $g$ (downwards)
2. deceleration in the sand: initial velocity = velocity of object just before impact.

The acceleration in each stage is *not* the same. The link between the two stages is that the velocity at the end of the first stage is the same as the velocity at the start of the second stage.

For example, a ball is released from a height of 0.85 m above a bed of sand, and creates an impression in the sand of depth 0.025 m. For directions, let + represent upwards and – represent downwards.

**Stage 1:** $u = 0$, $s = -0.85\,\text{m}$, $a = -9.8\,\text{m}\,\text{s}^{-2}$.

To calculate the speed of impact $v$, use $v^2 = u^2 + 2as$

$$v^2 = 0^2 + 2 \times -9.8 \times -0.85 = 16.7\,\text{m}^2\text{s}^{-2} \qquad v = -4.1\,\text{m}\,\text{s}^{-1}$$

***Note:***
$v^2 = 16.7\,\text{m}^2\text{s}^{-2}$ so $v = -4.1$ or $+\,4.1\,\text{m}\,\text{s}^{-1}$. The negative answer is chosen as the ball is moving downwards.

**Stage 2:** $u = -4.1\,\text{m}\,\text{s}^{-1}$, $v = 0$ (as the ball comes to rest in the sand), $s = -0.025\,\text{m}$

To calculate the deceleration, $a$, use $v^2 = u^2 + 2as$

$$0^2 = (-4.1)^2 + 2a \times -0.025$$

$$2a \times 0.025 = 16.8$$

$$a = \frac{16.8}{2 \times 0.025} = 336\,\text{m}\,\text{s}^{-2}$$

***Note:***
$a > 0$ and therefore in the opposite direction to the direction of motion, which is downwards. Thus the ball slows down in the sand with a deceleration of $336\,\text{m}\,\text{s}^{-2}$.

## Study tip

Applying the sign convention + for forwards (or up), – for backwards (or down), for an object

- moving forwards and accelerating, $v > 0$ and $a > 0$
- moving forwards and decelerating, $v > 0$ and $a < 0$
- moving backwards and accelerating, $v < 0$ and $a < 0$
- moving backwards and decelerating, $v < 0$ and $a > 0$.

## Summary questions

$g = 9.8\,\text{m}\,\text{s}^{-2}$

1 A vehicle on a straight downhill road accelerated uniformly from a speed of $4.0\,\text{m}\,\text{s}^{-1}$ to a speed of $29\,\text{m}\,\text{s}^{-1}$ over a distance of 850 m. The driver then braked and stopped the vehicle in 28 s.

   **a** Calculate **i** the time taken to reach $29\,\text{m}\,\text{s}^{-1}$ from $4\,\text{m}\,\text{s}^{-1}$, **ii** its acceleration during this time.

   **b** Calculate **i** the distance it travelled during deceleration, **ii** its deceleration for the last 28 s.

2 A rail wagon moving at a speed of $2.0\,\text{m}\,\text{s}^{-1}$ on a level track reached a steady incline which slowed it down to rest in 15.0 s and caused it to reverse. Calculate:

   **a** the distance it moved up the incline

   **b** its acceleration on the incline

   **c** its velocity and position on the incline after 20.0 s.

3 A cyclist accelerated from rest at a constant acceleration of $0.4\,\text{m}\,\text{s}^{-2}$ for 20 s, then stopped pedalling and slowed to a standstill at constant deceleration over a distance of 260 m.

   **a** Calculate **i** the distance travelled by the cyclist in the first 20 s, **ii** the speed of the cyclist at the end of this time.

   **b** Calculate **i** the time taken to cover the distance of 260 m after she stopped pedalling, **ii** her deceleration during this time.

4 A rocket was launched directly upwards from rest. Its motors operated for 30 s after it left the launch pad, providing it with a constant vertical acceleration of $6.0\,\text{m}\,\text{s}^{-2}$ during this time. Its motors then switched off.

   **a** Calculate **i** its velocity, **ii** its height above the launch pad when its motors switched off.

   **b** Calculate the maximum height reached after its motors switched off.

   **c** Calculate the velocity with which it would hit the ground if it fell from maximum height without the support of a parachute.

# 7.7 Projectile motion 1

## Learning objectives:

- → Explain why the acceleration of a projectile is always vertically downwards.
- → Identify the horizontal component of a vertical vector.
- → Describe the effect of gravity on horizontal speed.

*Specification reference: 3.4.1.4*

A **projectile** is any object acted upon only by the force of gravity. Three key principles apply to all projectiles:

1. The acceleration of the object is always equal to $g$ and is always downwards because the force of gravity acts downwards. The acceleration therefore only affects the vertical motion of the object.
2. The horizontal velocity of the object is constant because the acceleration of the object does not have a horizontal component.
3. The motions in the horizontal and vertical directions are independent of each other.

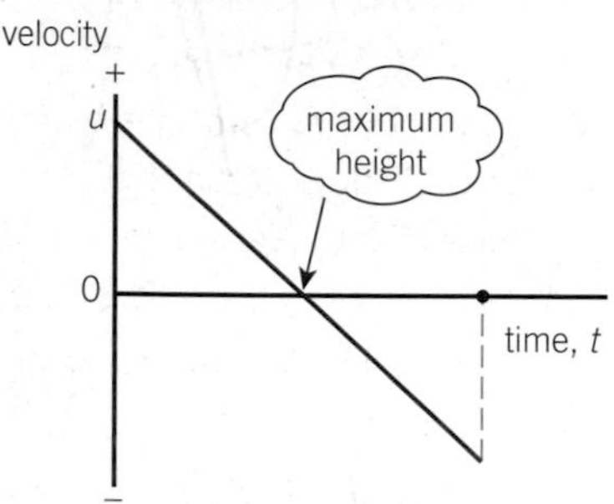


▲ **Figure 1** *Upward projection*

## Vertical projection

Such an object moves vertically as it has no horizontal motion. Its acceleration is $9.8\,\text{m}\,\text{s}^{-2}$ downwards. Using the direction code + is upwards, – is downwards, its displacement, $y$, and velocity, $v$, after time $t$ are given by

$$v = u - gt$$

$$y = ut - \frac{1}{2}gt^2$$

where $u$ is its initial velocity in the upward direction.

See Topic 7.4 for more about vertical projection.

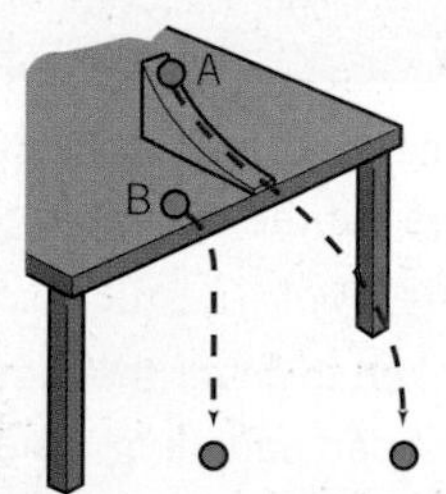


▲ **Figure 2** *Testing horizontal projection*

## Horizontal projection

A stone thrown from a cliff top follows a curved path downwards before it hits the water. If its initial projection is horizontal:

- its path through the air becomes steeper and steeper as it drops
- the faster it is projected, the further away it will fall into the sea
- the time taken for it to fall into the sea does not depend on how fast it is projected.

Suppose two balls are released at the same time above a level floor such that one ball drops vertically and the other is projected horizontally. Which one hits the floor first? In fact, they both hit the floor simultaneously. Try it!

Why should the two balls in Figure 2 hit the ground at the same time? They are both pulled to the ground by the force of gravity, which gives each ball a downward acceleration $g$. The ball that is projected horizontally experiences the same downward acceleration as the other ball. This downward acceleration does not affect the horizontal motion of the ball projected horizontally – only the vertical motion is affected.

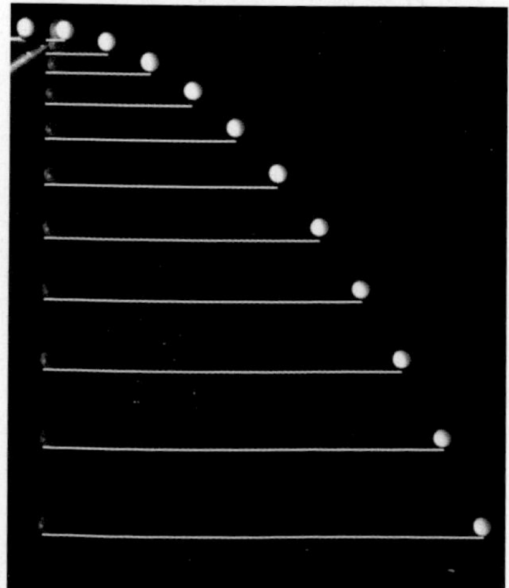

▲ **Figure 3** *Multiflash photo of two falling balls*

### Investigating horizontal projection

A stroboscope and a camera with a slow shutter (or a video camera) may be used to record the motion of a projectile (see Topic 7.4). Figure 3 shows a multiflash photograph of two balls A and B released at the same time. A was released from rest and dropped vertically. B was given an initial horizontal projection so it followed a curved path. The stroboscope flashed at a constant rate so images of both balls were recorded at the same time.

- The **horizontal position** of B changes by equal distances between successive flashes. This shows that the horizontal component of B's velocity is constant.
- The **vertical position** of A and B changes at the same rate. At any instant, A is at the same level as B. This shows that A and B have the same vertical component of velocity at any instant.

### The projectile path of a ball projected horizontally

An object projected horizontally falls in an arc towards the ground as shown in Figure 4. If its initial velocity is $U$, then at time $t$ after projection:

- The horizontal component of its displacement,

$$x = Ut$$

(because it moves horizontally at a constant speed).

- The vertical component of its displacement,

$$y = \frac{1}{2}gt^2$$

(because it has no vertical component of its initial velocity).

- Its velocity has a horizontal component $v_x = U$, and a vertical component $v_y = -gt$.

Note that, at time $t$, the speed $= (v_x^2 + v_y^2)^{1/2}$

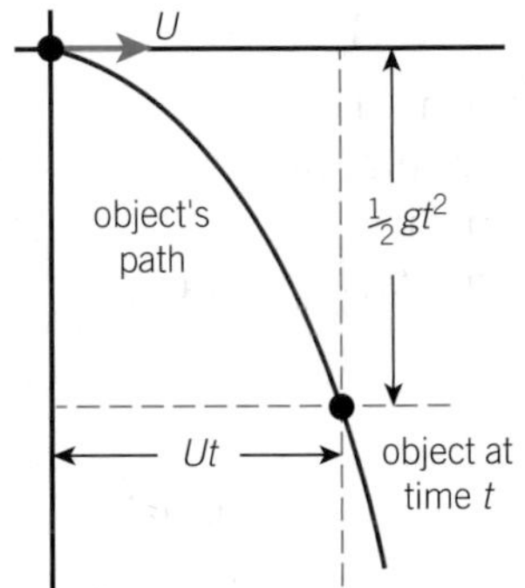

▲ **Figure 4** *Horizontal projection*

## Worked example

$g = 9.8\,\text{m}\,\text{s}^{-2}$

An object is projected horizontally at a speed of $15\,\text{m}\,\text{s}^{-1}$ from the top of a tower of height 35.0 m. Calculate:

a how long it takes to fall to the ground

b how far it travels horizontally

c its speed just before it hits the ground.

### *Solution*

a $y = -35\,\text{m}$, $a = -9.8\,\text{m}\,\text{s}^{-2}$ (– for downwards)

$y = \frac{1}{2}gt^2$

$t^2 = \frac{2y}{a} = 2 \times \frac{-35}{-9.8} = 7.14\,\text{s}^2$

$t = 2.67\,\text{s}$

b $U = 15\,\text{m}\,\text{s}^{-1}$, $t = 2.67\,\text{s}$

$x = Ut = 15 \times 2.67 = 40\,\text{m}$

c Just before impact, $v_X = U = 15\,\text{m}\,\text{s}^{-1}$ and $v_Y = -gt = -9.8 \times 2.67 = 26.2\,\text{m}\,\text{s}^{-1}$

Therefore speed just before impact $v = (v_x^2 + v_y^2)^{1/2} = (15^2 + 26.2^2)^{1/2} = 30.2\,\text{m}\,\text{s}^{-1}$

Remember that the horizontal and vertical motions of a projectile take the same amount of time.

## Summary questions

$g = 9.8\,\text{m}\,\text{s}^{-2}$

1 An object is released, 50 m above the ground, from a hot air balloon, which is descending vertically at a speed of $4.0\,\text{m}\,\text{s}^{-1}$. Calculate:

a the velocity of the object at the ground

b the duration of descent

c the height of the balloon above the ground when the object hits the ground.

2 An object is projected horizontally at a speed of $16\,\text{m}\,\text{s}^{-1}$ into the sea from a cliff top of height 45.0 m. Calculate:

a how long it takes to reach the sea

b how far it travels horizontally

c its impact velocity.

3 A dart is thrown horizontally along a line passing through the centre of a dartboard, 2.3 m away from the point at which the dart was released. The dart hits the dartboard at a point 0.19 m below the centre. Calculate:

a the time of flight of the dart

b its horizontal speed of projection.

4 A parcel is released from an aircraft travelling horizontally at a speed of $120\,\text{m}\,\text{s}^{-1}$ above level ground. The parcel hits the ground 8.5 s later. Calculate:

a the height of the aircraft above the ground

b the horizontal distance travelled in this time by **i** the parcel, **ii** the aircraft

c the speed of impact of the parcel at the ground.

# 7.8 Projectile motion 2

## Learning objectives:

→ Describe where else we meet projectile motion.

→ Describe what would happen if we could switch gravity off.

→ Describe how projectile motion is affected by the air it passes through.

*Specification reference: 3.4.1.4*

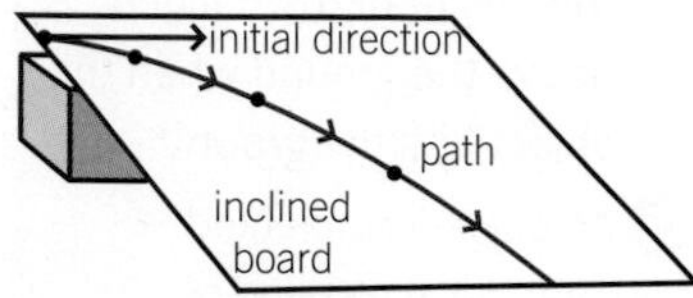


▲ **Figure 1** *Using an inclined board*

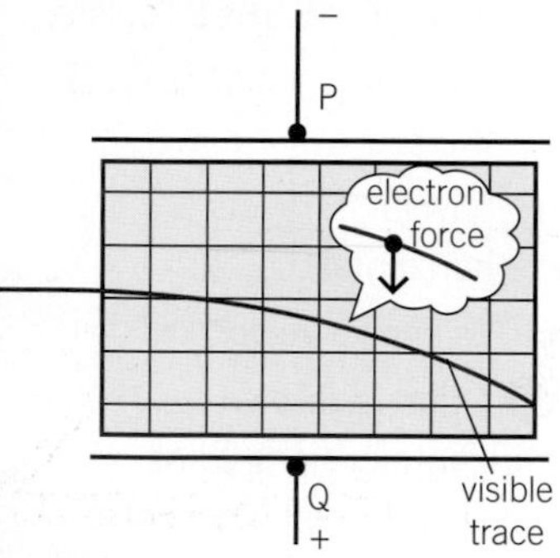


▲ **Figure 2** *An electron beam on a parabolic path*

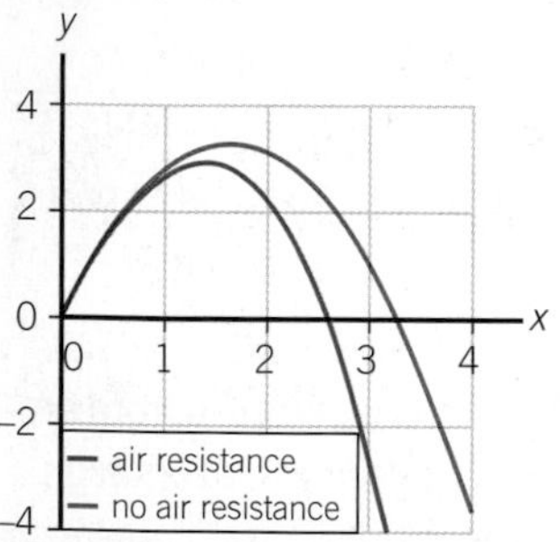


▲ **Figure 3** *Projectile paths*

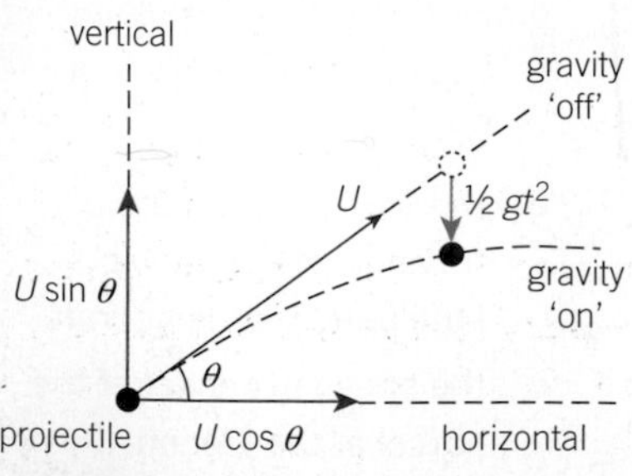


▲ **Figure 4** *Imagination at work*

## Projectile-like motion

Any form of motion where an object experiences a constant acceleration in a different direction to its velocity will be like projectile motion. For example:

- The path of a ball rolling across an inclined board will be a projectile path (Figure 1). The object is projected across the top of the board from the side. Its path curves down the board and is **parabolic**. This is because the object is subjected to constant acceleration acting down the board, and its initial velocity is across the board.
- The path of a beam of electrons directed between two oppositely charged parallel plates is also parabolic (Figure 2). Each electron in the beam is acted on by a constant force towards the positive plate, because the charge of an electron is negative. Therefore, each electron experiences a constant acceleration towards the positive plate. Its path is parabolic because its motion parallel to the plates is at **zero** acceleration whereas its motion perpendicular to the plates is at constant (non-zero) acceleration.

### The effects of air resistance

A projectile moving through air experiences a force that drags on it because of the resistance of the air it passes through. This **drag force** is partly caused by friction between the layers of air near the projectile's surface where the air flows over the surface. The drag force:

- acts in the opposite direction to the direction of motion of the projectile, and it increases as the projectile's speed increases
- has a horizontal component that reduces both the horizontal speed of the projectile and its range
- reduces the maximum height of the projectile if its initial direction is above the horizontal and makes its descent steeper than its ascent.

### *Note:*

The shape of the projectile is important because it affects the drag force and may also cause a **lift force** in the same way as the cross-sectional shape of an aircraft wing creates a lift force. This happens if the shape of the projectile causes the air to flow faster over the top of the object than underneath it. As a result, the pressure of the air on the top surface is less than that on the bottom surface. This pressure difference causes a lift force on the object.

A spinning ball also experiences a force due to the same effect. However, this force can be downwards, upwards, or sideways, depending on how the ball is made to spin.

## Imagination at work

Imagine if gravity could be switched off, with air resistance negligible. An object projected into the air would follow a straight line with no change in its velocity (Figure 4). Suppose it had been launched at speed $U$ in a direction at angle $\theta$ above the horizontal. Its initial velocity would therefore have a horizontal component $U\cos\theta$ and a vertical component $U\sin\theta$.

Without gravity, its velocity would be unchanged, so its displacement at time $t$ after projection would have

- a horizontal component $x = Ut\cos\theta$
- a vertical component $y = Ut\sin\theta$.

Let's now consider the difference made to its displacement and velocity if we bring back gravity. At time $t$ after its launch,

- its vertical position with gravity on would be $\frac{1}{2}gt^2$ lower
- its vertical component of velocity would be changed by $-gt$ from its initial value $U\sin\theta$.

Hence
$$y = Ut\sin\theta - \frac{1}{2}gt^2$$
$$v_y = U\sin\theta - gt$$

**Study tip**

Projections below the horizontal can be analysed by assigning a negative value of $\theta$. Its speed $v$ at time $t$ is given by $v = (v_x^2 + v_y^2)^{1/2}$.

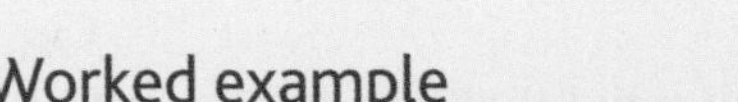

## Worked example

An arrow is fired at a speed of $48\,\mathrm{m\,s^{-1}}$ at an angle of 30° above the horizontal from a height of 1.5 m above a level field. The arrow hits a target at the same height 0.25 s later.

Calculate the maximum height above the field gained by the arrow. Assume air resistance is negligible.

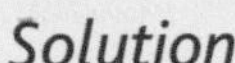

### Solution

At time $t = 0.25\,\mathrm{s}$, vertical displacement $y = Ut\sin\theta - \frac{1}{2}gt^2$.

Therefore, $y = (48 \times 0.25 \times \sin 30) - (0.5 \times 9.8 \times 0.25^2) = 5.7\,\mathrm{m}$.

The maximum height $= 5.7 + 1.5 = 7.2\,\mathrm{m}$.

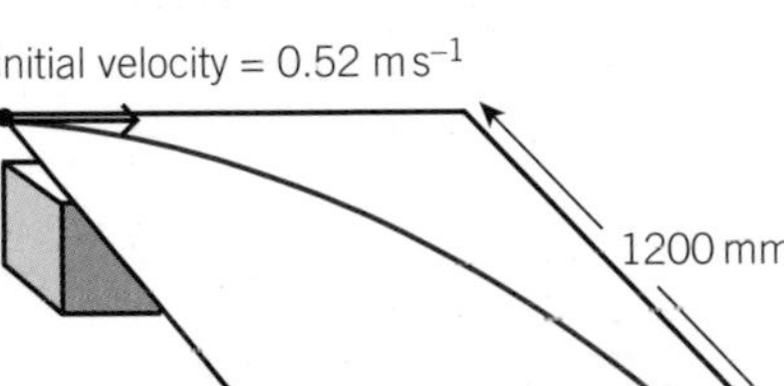


▲ Figure 5

## Summary questions

$g = 9.8\,\mathrm{m\,s^{-2}}$

1 A ball was projected horizontally at a speed of $0.52\,\mathrm{m\,s^{-1}}$ across the top of an inclined board of width 600 mm and length 1200 mm. It reached the bottom of the board 0.9 s later (Figure 5). Calculate:

   **a** the distance travelled by the ball across the board

   **b** its acceleration on the board

   **c** its speed at the bottom of the board.

2 An arrow hits the ground after being fired horizontally at a speed of $25\,\mathrm{m\,s^{-1}}$ from the top of a tower 20 m above the ground. Calculate:

   **a** how long the arrow takes to fall to the ground

   **b** the distance travelled horizontally by the arrow.

3 An object was released from a cable car travelling at a speed of $4.6\,\mathrm{m\,s^{-1}}$ in a direction 40° above the horizontal (Figure 6).

   **a** Calculate the horizontal and vertical components of velocity of the object at the instant it was released.

   **b** The object took 5.8 s to fall to the ground. Calculate **i** the distance fallen by the object from the point of release, **ii** the horizontal distance travelled by the object from the point of release to where it hit the ground.

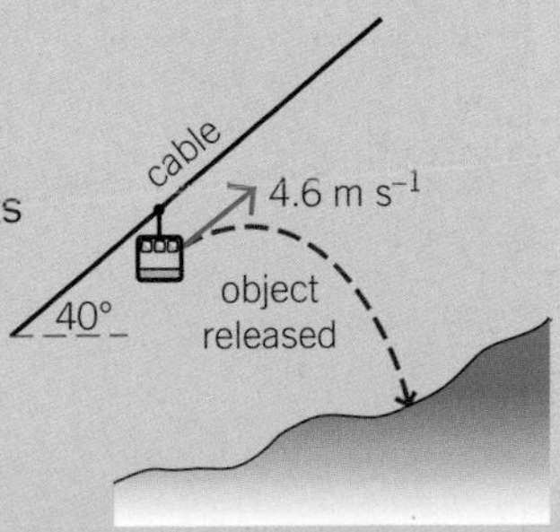

cable
4.6 m s⁻¹
object released
40°

▲ Figure 6

# Practice questions: Chapter 7

1 A car accelerates uniformly from rest to a speed of $100\,\text{km}\,\text{h}^{-1}$ in 5.8 s.

**(a)** Calculate the magnitude of the acceleration of the car in $\text{m}\,\text{s}^{-2}$. *(3 marks)*

**(b)** Calculate the distance travelled by the car while accelerating. *(2 marks)*

AQA, 2005

2 **Figure 1** shows how the velocity of a toy train moving in a straight line varies over a period of time.

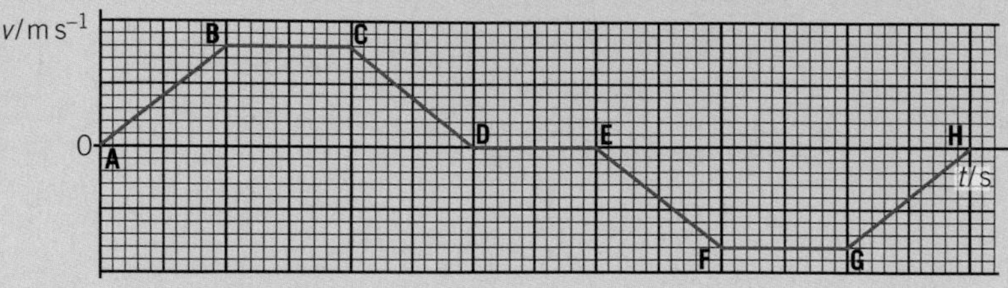


▲ **Figure 1**

**(a)** Describe the motion of the train in the following regions of the graph:

AB BC CD DE EF *(5 marks)*

**(b)** What feature of the graph represents the displacement of the train? *(1 mark)*

**(c)** Explain, with reference to the graph, why the distance travelled by the train is different from its displacement. *(2 marks)*

AQA, 2002

3 A vehicle accelerates uniformly from a speed of $4.0\,\text{m}\,\text{s}^{-1}$ to a speed of $12\,\text{m}\,\text{s}^{-1}$ in 6.0 s.

**(a)** Calculate the vehicle's acceleration. *(2 marks)*

**(b)** Sketch a graph of speed against time for the vehicle covering the 6.0 s period in which it accelerates. *(2 marks)*

**(c)** Calculate the distance travelled by the vehicle during its 6.0 s period of acceleration. *(2 marks)*

AQA, 2002

4 A supertanker, cruising at an initial speed of $4.5\,\text{m}\,\text{s}^{-1}$, takes one hour to come to rest.

**(a)** Assuming that the force slowing the tanker down is constant, calculate

(i) the deceleration of the tanker

(ii) the distance travelled by the tanker while slowing to a stop. *(4 marks)*

**(b)** Sketch a distance–time graph representing the motion of the tanker until it stops. *(2 marks)*

**(c)** Explain the shape of the graph you have sketched in part (b). *(2 marks)*

AQA, 2006

5 **(a)** A cheetah accelerating uniformly from rest reaches a speed of $29\,\text{m}\,\text{s}^{-1}$ in 2.0 s and then maintains this speed for 15 s. Calculate:

(i) its acceleration

(ii) the distance it travels while accelerating

(iii) the distance it travels while it is moving at constant speed. *(4 marks)*

**(b)** The cheetah and an antelope are both at rest and 100 m apart. The cheetah starts to chase the antelope. The antelope takes 0.50 s to react. It then accelerates uniformly for 2.0 s to a speed of $25\,\text{m}\,\text{s}^{-1}$ and then maintains this speed. **Figure 2** shows the speed–time graph for the cheetah.

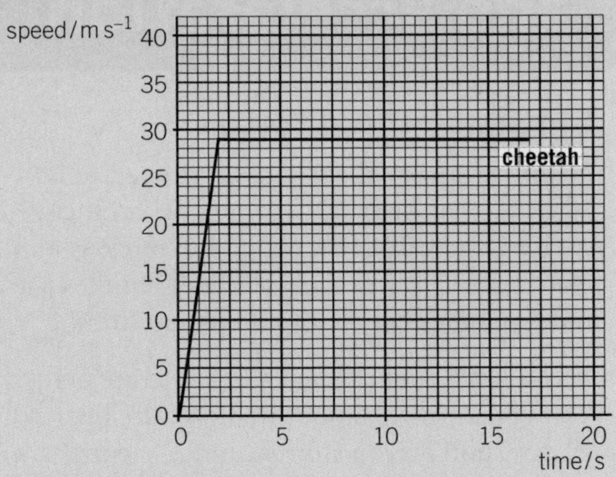


▲ **Figure 2**

(i) Using the same axes plot the speed–time graph for the antelope during the chase.
(ii) Calculate the distance covered by the antelope in the 17 s after the cheetah started to run.
(iii) How far apart are the cheetah and the antelope after 17 s? *(6 marks)*

AQA, 2007

6 The following data were obtained when two students performed an experiment to determine the acceleration of free fall. One student released a lump of lead the size of a tennis ball from a window in a tall building and the other measured the time for it to reach the ground.

distance fallen by lump of lead = 35 m
time to reach the ground = 2.7 s

**(a)** Calculate a value for the acceleration of free fall, $g$, from these observations. *(2 marks)*

**(b)** State and explain the effect on the value of $g$ obtained by the students if a tennis ball were used instead of the lump of lead. *(3 marks)*

**(c)** The graph in **Figure 3** shows how the velocity changes with time for the lump of lead from the time of release until it hits the ground. Sketch on the same axes a graph to show how the velocity would change with time if a tennis ball were used by the students instead of a lump of lead.

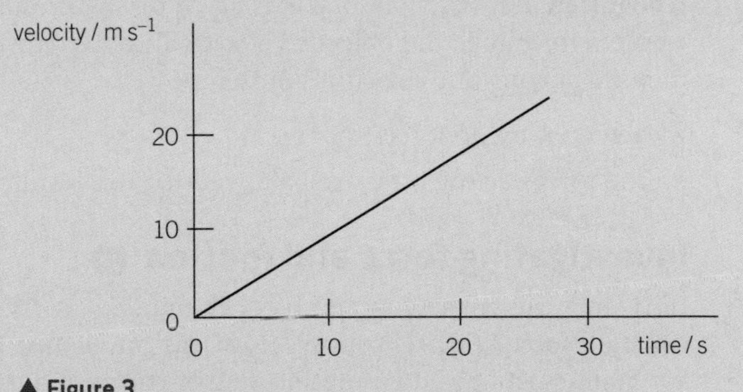


▲ **Figure 3**

*(2 marks)*

AQA, 2004

7 A man jumps from a plane that is travelling horizontally at a speed of $70\,m\,s^{-1}$. If air resistance can be ignored, and before he opens his parachute, determine

**(a)** his horizontal velocity 2.0 s after jumping

**(b)** his vertical velocity 2.0 s after jumping

**(c)** the magnitude and direction of his resultant velocity 2.0 s after jumping. *(5 marks)*

AQA, 2003

# 8 Newton's laws of motion

## 8.1 Force and acceleration

### Learning objectives:

- → Describe what effect a resultant force produces.
- → Describe what would happen to a body that was already in motion if there was no resultant force acting on it.
- → Explain how weight is different from mass.

*Specification reference: 3.4.1.5*

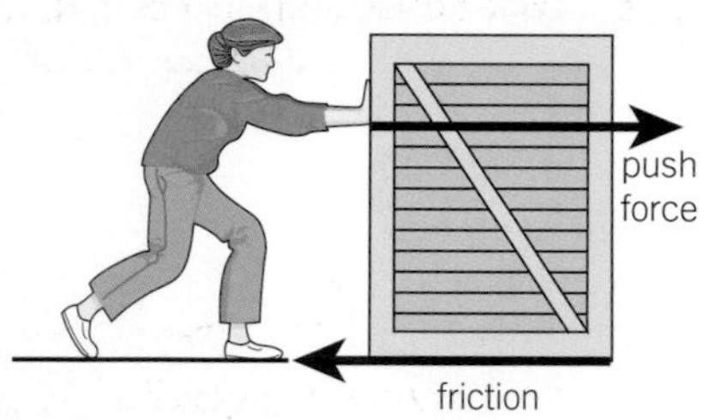


▲ **Figure 1** *Overcoming friction*

### Motion without force

Motorists on icy roads in winter need to be very careful, because the tyres of a car have little or no grip on the ice. Moving from a standstill on ice is very difficult. Stopping on ice is almost impossible, as a car moving on ice will slide when the brakes are applied. **Friction** is a hidden force that we don't usually think about until it is absent!

If you have ever pushed a heavy crate across a rough concrete floor, you will know all about friction. The push force is opposed by friction, and as soon as you stop pushing, friction stops the crate moving. If the crate had been pushed onto a patch of ice, it would have moved across the ice without any further push needed.

Figure 2 shows an air track which allows motion to be observed in the absence of friction. The glider on the air track floats on a cushion of air. Provided the track is level, the glider moves at constant velocity along the track because friction is absent.

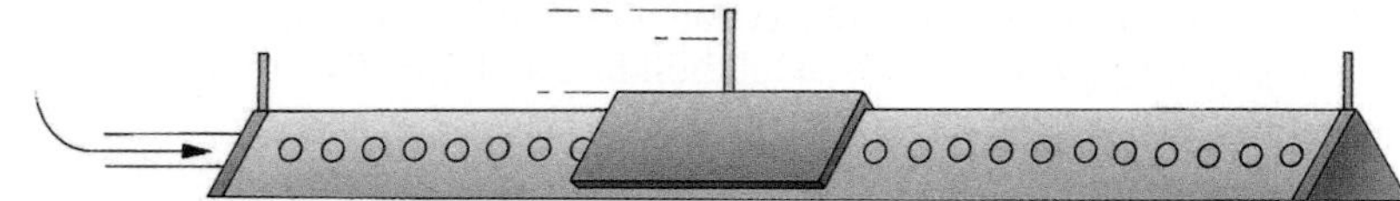

▲ **Figure 2** *The linear air track*

#### Newton's first law of motion

**Objects either stay at rest or moves with constant velocity unless acted on by a force.**

Sir Isaac Newton was the first person to realise that a moving object remains in uniform motion unless acted on by a force. He recognised that when an object is acted on by a resultant force, the result is to change the object's velocity. In other words, an object moving at constant velocity is either

- acted on by no forces, or
- the forces acting on it are balanced (so the resultant force is zero).

### Investigating force and motion

How does the velocity of an object change if it is acted on by a constant force? Figure 3 shows how this can be investigated, using a dynamics trolley and a motion sensor connected to a computer. The computer is used to process a signal from the motion sensor and to display a graph showing how the velocity of the trolley changes with time.

The trolley is pulled along a sloping runway using one or more elastic bands stretched to the same length. The runway is sloped just enough to compensate for friction. To test for the correct slope, the trolley should move down the runway at constant speed after being given a brief push.

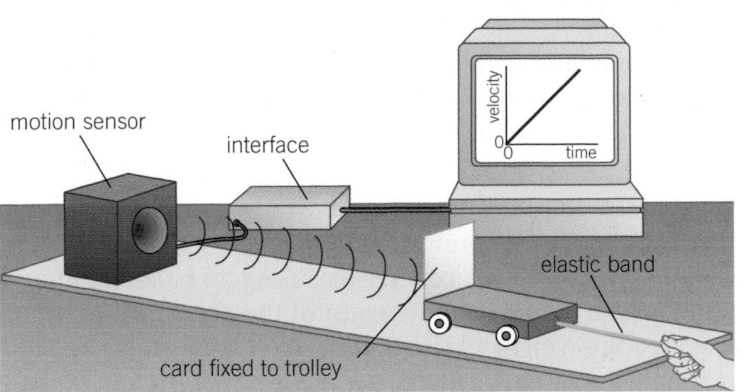


▲ **Figure 3** *Investigating force and motion*

As a result of pulling the trolley with a constant force, the velocity–time graph should show that the velocity increases at a constant rate. The acceleration of the trolley is therefore constant and can be measured from the velocity–time graph. Table 1 below shows typical measurements using different amounts of force (one, two, or three elastic bands in parallel stretched to the same length each time) and different amounts of mass (i.e., a single, double, or triple trolley).

▼ **Table 1**

| Force (no of elastic bands) | 1 | 2 | 3 | 1 | 2 | 3 |
|---|---|---|---|---|---|---|
| Mass (no. of trolleys) | 1 | 1 | 1 | 2 | 2 | 2 |
| Acceleration (in $m\,s^{-2}$) | 12 | 24 | 36 | 6 | 12 | 18 |
| Mass × acceleration | 12 | 24 | 36 | 12 | 24 | 36 |

The results in the table show that the force is proportional to the mass × the acceleration. In other words, if a resultant force $F$ acts on an object of mass $m$, the object undergoes acceleration $a$ such that

**$F$ is proportional to $ma$**

By defining the unit of force, the *newton*, as the amount of force that will give an object of mass 1 kg an acceleration of $1\,m\,s^{-2}$, the above proportionality statement can be expressed as an equation

$$F = ma$$

where $F$ = resultant force (in N), $m$ = mass (in kg), $a$ = acceleration (in $m\,s^{-2}$).

This equation is known as **Newton's second law** for constant mass.

## Worked example

A vehicle of mass 600 kg accelerates uniformly from rest to a velocity of $8.0\,m\,s^{-1}$ in 20 s. Calculate the force needed to produce this acceleration.

### *Solution*

Acceleration $a = \frac{v-u}{t} = \frac{8.0-0}{20} = 0.4\,m\,s^{-2}$

Force $F = ma = 600 \times 0.4 = 240\,N$

## Hint

The acceleration is always in the same direction as the resultant force. For example, a freely moving projectile in motion experiences a force vertically downwards due to gravity. Its acceleration is therefore vertically downwards, no matter what its direction of motion is.

## Study tip

The mass $m$ must be in kg and $a$ in $m\,s^{-2}$ when calculating a force in N.

## Maths link

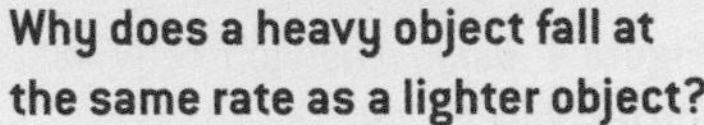

**Why does a heavy object fall at the same rate as a lighter object?**

As outlined in Topic 7.4, Free fall, Galileo proved that objects fall at the same rate, regardless of their weight. He also reasoned that if two objects joined by a string were released and they initially fell at different rates, the faster one would be slowed down by the slower one, which would be speeded up by the faster one until they fell at the same rate. Many years later, Newton explained their identical falling motion because, for an object of mass $m$ in free fall,

$$\text{acceleration} = \frac{\text{force of gravity } mg}{\text{mass } m} = g \text{ which is independent of } m$$

## Synoptic link

You learnt about Galileo in Topic 7.4, Free fall.

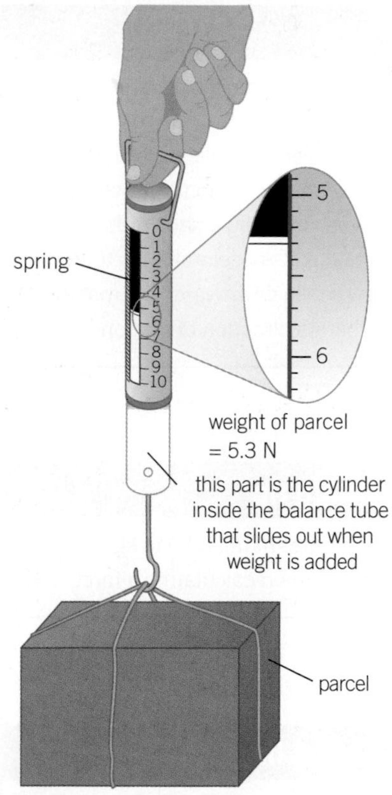


▲ **Figure 4** *Using a newtonmeter to weigh an object*

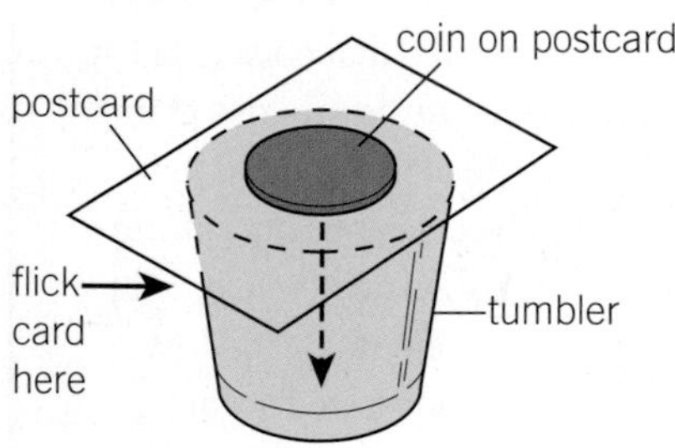


▲ **Figure 5** *An inertia trick*

## Weight

- The acceleration of a falling object acted on by gravity only is $g$. Because the force of gravity on the object is the only force acting on it, its **weight** (in newtons), $W = mg$, where $m$ = the mass of the object (in kg).
- When an object is in equilbrium, the support force on it is equal and opposite to its weight. Therefore, an object placed on a weighing balance (e.g., a newtonmeter or a top pan balance) exerts a force on the balance equal to the weight of the object. Thus the balance measures the weight of the object. See Figure 4.
- $g$ is also referred to as the **gravitational field strength** at a given position, as it is the force of gravity per unit mass on a small object at that position. So the gravitational field strength at the Earth's surface is 9.81 N kg$^{-1}$. Note that the weight of a fixed mass depends on its location. For example, the weight of a 1 kg object is 9.81 N on the Earth's surface and 1.62 N on the Moon's surface.
- The mass of an object is a measure of its **inertia**, which is its resistance to change of motion. More force is needed to give an object a certain acceleration than to give an object with less mass the same acceleration. Figure 5 shows an entertaining demonstration of inertia. When the card is flicked, the coin drops into the glass because the force of friction on it due to the moving card is too small to shift it sideways.
- The scale of a top pan balance is usually calibrated for convenience in grams or kilograms.

## Summary questions

**$g$ = 9.81 N kg$^{-1}$**

**1** A car of mass 800 kg accelerates uniformly along a straight line from rest to a speed of 12 m s$^{-1}$ in 50 s. Calculate:

**a** the acceleration of the car

**b** the force on the car that produced this acceleration

**c** the ratio of the accelerating force to the weight of the car.

**2** An aeroplane of mass 5000 kg lands on a runway at a speed of 60 m s$^{-1}$ and stops 25 s later. Calculate:

**a** the deceleration of the aeroplane

**b** the braking force on the aircraft.

**3** **a** A vehicle of mass 1200 kg on a level road accelerates from rest to a speed of 6.0 m s$^{-1}$ in 20 s, without change of direction. Calculate the force that accelerated the car.

**b** The vehicle in **a** is fitted with a trailer of mass 200 kg. Calculate the time taken to reach a speed of 6.0 m s$^{-1}$ from rest for the same force as in **a**.

**4** A bullet of mass 0.002 kg travelling at a speed of 120 m s$^{-1}$ hit a tree and penetrated a distance of 55 mm into the tree. Calculate:

**a** the deceleration of the bullet

**b** the impact force on the bullet.

# 8.2 Using $F = ma$

## Two forces in opposite directions

When an object is acted on by two unequal forces acting in opposite directions, the object accelerates in the direction of the larger force. If the forces are $F_1$ and $F_2$, where $F_1 > F_2$

$$\textbf{resultant force, } F_1 - F_2 = ma,$$

where $m$ is the mass of the object and $a$ is its acceleration, which is in the same direction as $F_1$.

If the object is on a horizontal surface and $F_1$ and $F_2$ are horizontal and in opposite directions, the above equation still applies. The support force on the object is equal and opposite to its weight.

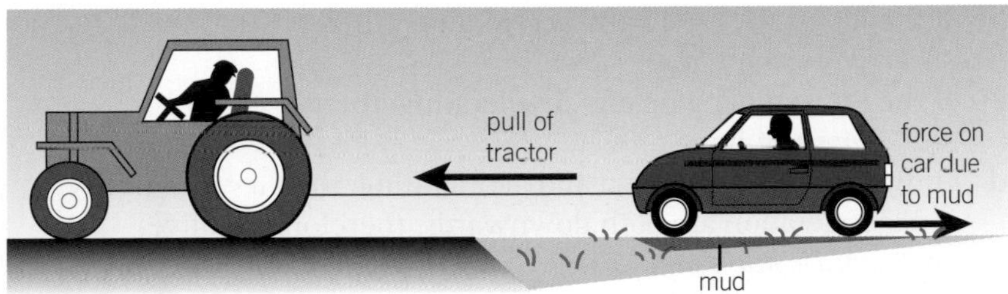


▲ **Figure 1** *Unbalanced forces*

Some examples are given below where two forces act in different directions on an object.

### *Towing a trailer*

Consider the example of a car of mass $M$ fitted with a trailer of mass $m$ on a level road. When the car and the trailer accelerate, the car pulls the trailer forward and the trailer holds the car back. Assume air resistance is negligible.

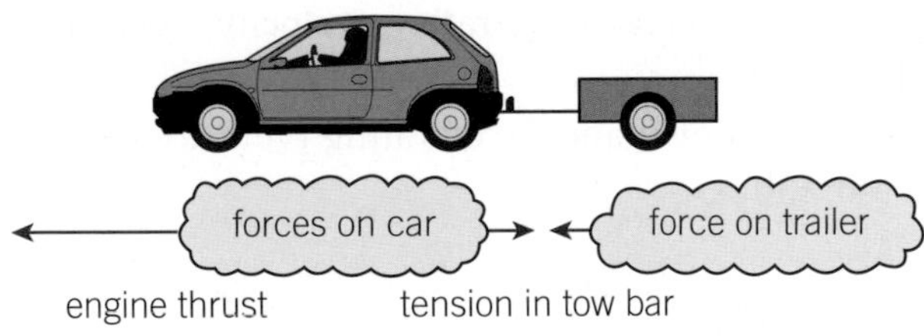


▲ **Figure 2** *Car and trailer*

- The car is subjected to a driving force $F$ pushing it forwards (from its engine thrust) and the tension $T$ in the tow bar holding it back.

  Therefore the resultant force on the car = $F - T = Ma$
- The force on the trailer is due to the tension $T$ in the tow bar pulling it forwards.

  Therefore $T = ma$.

Combining the two equations gives $\mathbf{F = Ma + ma = (M + m)a}$

## Learning objectives:

- → Apply $F = ma$ when the forces on an object are in opposite directions.
- → Explain why you experience less support as an ascending lift stops.
- → Describe any situations in which $F = ma$ cannot be applied.

*Specification reference: 3.4.1.5*

## Study tip

Remember that it is the force of friction that pushes an object. If friction was negligible the object would not be able to move.

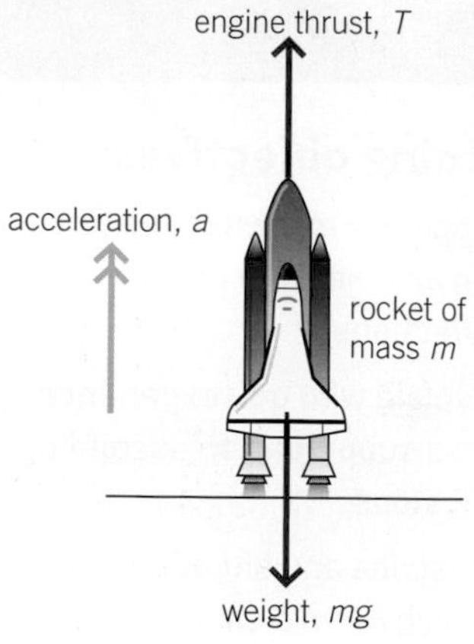


▲ **Figure 3** *Rocket launch*

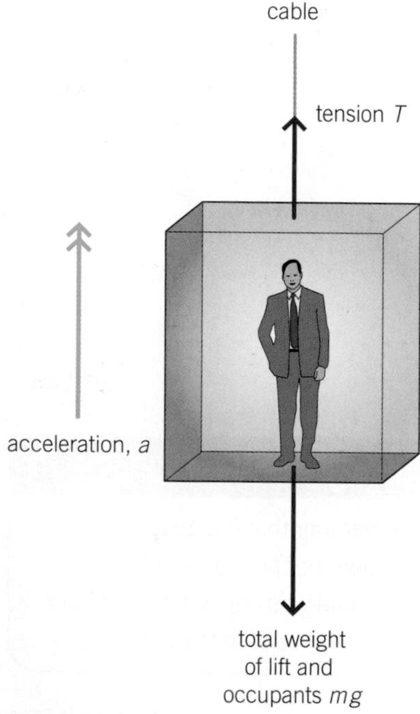


▲ **Figure 4** *In a lift*

## Hint

Downward acceleration = upward deceleration

## Study tip

Identify the separate forces acting, then work out the resultant force – show these steps by clear working (usually starting in algebra, such as $T - mg = ma$).

### *Rocket problems*

If $T$ is the thrust of the rocket engine when its mass is $m$ and the rocket is moving upwards, its acceleration $a$ is given by $\boldsymbol{T - mg = ma}$.

Rocket thrust $\boldsymbol{T = mg + ma}$

The rocket thrust must therefore overcome the weight of the rocket for the rocket to take off (Figure 3).

### *Lift problems*

Using 'upwards is positive' gives the resultant force on the lift as $T - mg$, where $T$ is the tension in the lift cable and $m$ is the total mass of the lift and occupants (Figure 4).

$\boldsymbol{T - mg = ma}$, where $a$ = acceleration.

**a** If the lift is moving at constant velocity, then $a = 0$ so $T = mg$ (tension = weight).

**b** If the lift is moving up and accelerating, then $a > 0$ so $T = mg + ma > mg$.

**c** If the lift is moving up and decelerating, then $a < 0$ so $T = mg + ma < mg$.

**d** If the lift is moving down and accelerating, then $a < 0$ (velocity and acceleration are both downwards, therefore negative) so $T = mg + ma < mg$.

**e** If the lift is moving down and decelerating, then $a > 0$ (velocity downwards and acceleration upwards, therefore positive) so $T = mg + ma > mg$.

**The tension in the cable is less than the weight if**

- the lift is moving up and decelerating (velocity > 0 and acceleration < 0)
- the lift is moving down and accelerating (velocity < 0 and acceleration < 0).

**The tension in the cable is greater than the weight if**

- the lift is moving up and accelerating (velocity > 0 and acceleration > 0)
- the lift is moving down and decelerating (velocity < 0 and acceleration > 0).

## Worked example

$\boldsymbol{g = 9.8\,\mathrm{m\,s^{-2}}}$

A lift of total mass 650 kg moving downwards decelerates at $1.5\,\mathrm{m\,s^{-2}}$ and stops. Calculate the tension in the lift cable during the deceleration.

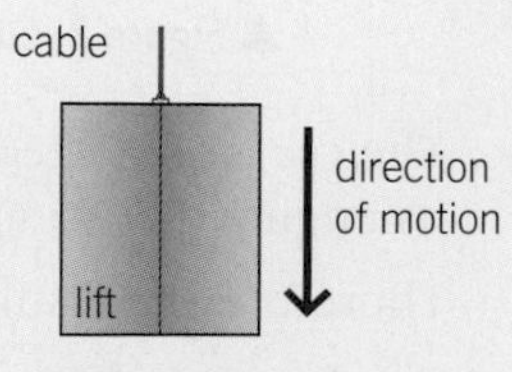


▲ **Figure 5**

### *Solution*

The lift is moving down so its velocity $v < 0$. Since it decelerates, its acceleration $a$ is in the opposite direction to its velocity, so $a > 0$.

Therefore, inserting $a = +1.5\,\mathrm{m\,s^{-2}}$ in the equation $T - mg = ma$ gives

$T = mg + ma = (650 \times 9.8) + (650 \times 1.5) = 7300\,\mathrm{N}$

## Further $F = ma$ problems

### Pulley problems

Consider two masses $M$ and $m$ (where $M > m$) attached to a thread hung over a frictionless pulley, as in Figure 6. When released, mass $M$ accelerates downwards and mass $m$ accelerates upwards. If $a$ is the acceleration and $T$ is the tension in the thread, then

- the resultant force on mass $M$, $Mg - T = Ma$
- the resultant force on mass $m$, $T - mg = ma$.

Therefore, adding the two equations gives

$$Mg - mg = (M + m)a$$

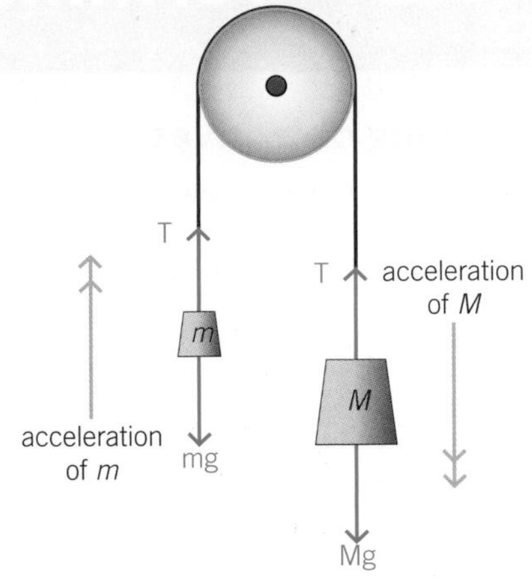

▲ **Figure 6** *Pulley problems*

### Sliding down a slope

Consider a block of mass $m$ sliding down a slope, as in Figure 7.

The component of the block's weight down the slope = $mg\sin\theta$.

If the force of friction on the block = $F_0$, then the resultant force on the block = $mg\sin\theta - F_0$.

Therefore $mg\sin\theta - F_0 = ma$, where $a$ is the acceleration of the block.

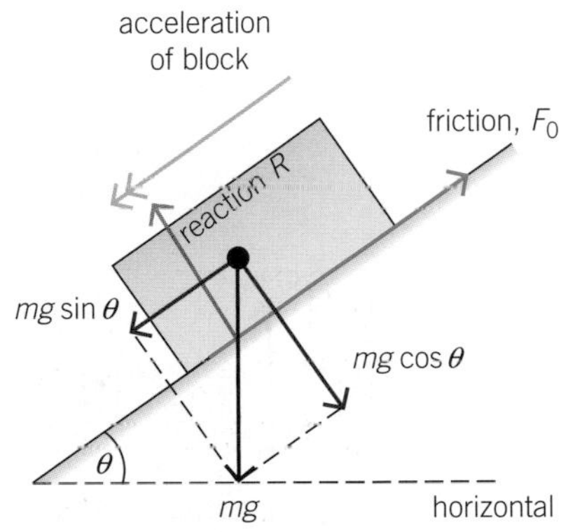

▲ **Figure 7** *Sliding down a slope*

### Note:

With the addition of an engine force $F_E$, the above equation can be applied to a vehicle on a downhill slope of constant gradient. Thus the acceleration is given by $F_E + mg\sin\theta - F_0 = ma$, where $F_0$ is the combined sum of the air resistance and the braking force.

## Summary questions

$g = 9.8\,\text{m s}^{-2}$

1 A rocket of mass 550 kg blasts vertically from the launch pad at an acceleration of 4.2 m s$^{-2}$. Calculate:

  **a** the weight of the rocket

  **b** the thrust of the rocket engines.

2 A car of mass 1400 kg, pulling a trailer of mass 400 kg, accelerates from rest to a speed of 9.0 m s$^{-1}$ in a time of 60 s on a level road. Assuming air resistance is negligible, calculate:

  **a** the tension in the tow bar

  **b** the engine force.

3 A lift and its occupants have a total mass of 1200 kg. Calculate the tension in the lift cable when the lift is:

  **a** stationary

  **b** ascending at constant speed

  **c** ascending at a constant acceleration of 0.4 m s$^{-2}$

  **d** descending at a constant deceleration of 0.4 m s$^{-2}$.

4 A brick of mass 3.2 kg on a sloping flat roof, at 30° to the horizontal, slides at constant acceleration 2.0 m down the roof in 2.0 s from rest. Calculate:

  **a** the acceleration of the brick

  **b** the frictional force on the brick due to the roof.

# 8.3 Terminal speed

## Learning objectives:

- → Explain why the speed reaches a maximum when a driving force is still acting.
- → Explain what we mean by a drag force.
- → Explain what determines the terminal speed of a falling object or a vehicle.

*Specification reference: 3.4.1.5*

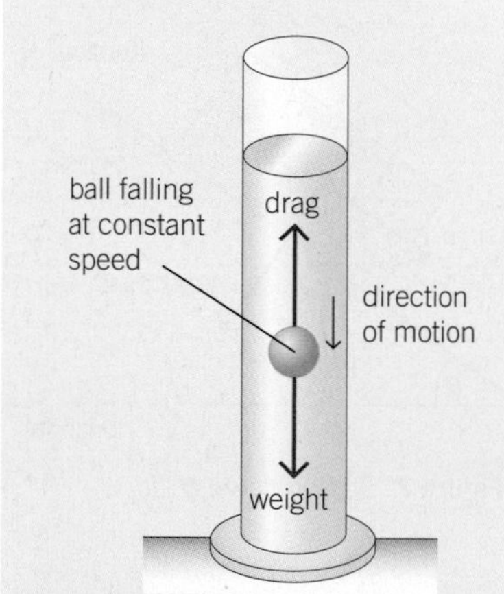


**a** *Falling in a fluid*

**b** *Skydiving*

▲ **Figure 1** *At terminal speed*

Any object moving through a fluid experiences a force that drags on it due to the fluid. The **drag force** depends on

- the shape of the object
- its speed
- the viscosity of the fluid, which is a measure of how easily the fluid flows past a surface.

The faster an object travels in a fluid, the greater the drag force on it.

## Motion of an object falling in a fluid

The speed of an object released from rest in a fluid increases as it falls, so the drag force on it due to the fluid increases. The resultant force on the object is the difference between the force of gravity on it (its weight) and the drag force. As the drag force increases, the resultant force decreases, so the acceleration becomes less as it falls. If it continues falling, it attains **terminal speed**, when the drag force on it is equal and opposite to its weight. Its acceleration is then zero and its speed remains constant as it falls (Figure 1).

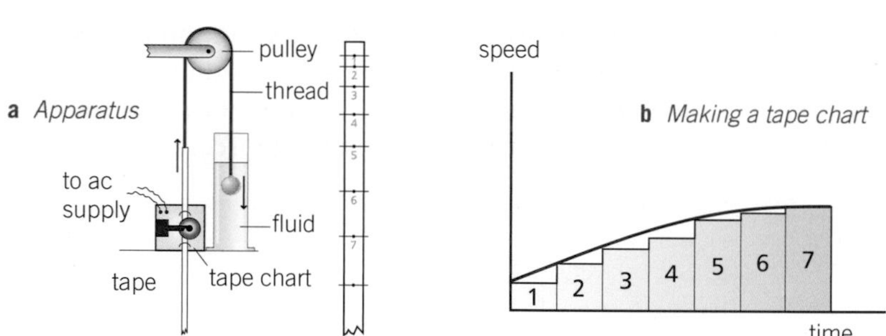


▲ **Figure 2** *Investigating the motion of an object falling in a fluid*

Figure 2a shows how to investigate the motion of an object falling in a fluid. When the object is released, the thread attached to the object pulls a tape through a tickertimer, which prints dots at a constant rate on the tape. The spacing between successive dots is a measure of the speed of the object as the dots are printed on the tape at a constant rate. A tape chart can be made from the tape to show how the speed changes with time. This is done by cutting the tape at each dot and pasting successive lengths side by side on graph paper, as shown in Figure 2b. The results show that:

- the speed increases and reaches a constant value, which is the terminal speed
- the acceleration at any instant is the gradient of the speed–time curve.

At any instant, the resultant force $F = mg - D$, where $m$ is the mass of the object and $D$ is the drag force.

Therefore, the acceleration of the object $= \dfrac{mg - D}{m} = g - \dfrac{D}{m}$

- The initial acceleration $= g$ because the speed is zero, and therefore the drag force is zero, at the instant the object is released.
- At the terminal speed, the potential energy of the object is transferred, as it falls, into internal energy of the fluid by the drag force.

### Study tip

The magnitude of the acceleration at any instant is the gradient of the speed–time curve.

## Motion of a powered vehicle

The top speed of a road vehicle or an aircraft depends on its engine power and its shape. A vehicle with a streamlined shape can reach a higher top speed than a vehicle with the same engine power that is not streamlined.

For a powered vehicle of mass $m$ moving on a level surface, if $F_E$ represents the **motive force** (driving force) provided by the engine, the resultant force on it $= F_E - F_R$, where $F_R$ is the **resistive force** opposing the motion of the vehicle. ($F_R$ = the sum of the drag forces acting on the vehicle.)

Therefore its acceleration $\boldsymbol{a} = \dfrac{\boldsymbol{F_E - F_R}}{\boldsymbol{m}}$

Because the drag force increases with speed, the maximum speed (the terminal speed) of the vehicle $v_{max}$ is reached when the resistive force becomes equal and opposite to the engine force, and $a = 0$.

### Worked example

A car of mass 1200 kg has an engine which provides an engine force of 600 N. Calculate its acceleration when the resistive force is 400 N.

#### *Solution*

When the resistive force = 400 N, the resultant force = engine force – drag force = 600 – 400 = 200 N.

$$\text{Acceleration} = \frac{\text{force}}{\text{mass}} = \frac{200}{1200} = 0.17\,\text{m}\,\text{s}^{-2}$$

### Extension

#### Hydrofoil physics

Some hydrofoil boats travel much faster than an ordinary boat because they have powerful jet engines that enable them to ski on its hydrofoils when the jet engines are switched on.

When the jet engine is switched on and takes over from the less powerful propeller engine, the boat speeds up and the hydrofoils are extended. The boat rides on the hydrofoils, so the drag force is reduced, as its hull is no longer in the water. At top speed, the motive force of the jet engine is equal to the drag force on the hydrofoils.

When the jet engine is switched off, the drag force on the hydrofoils reduces the speed of the boat and the hydrofoils are retracted. The speed drops further until the drag force is equal to the motive force of the propeller engine. The boat then moves at a constant speed, which is much less than its top speed with the jet engine on.

**Q:** Describe how the force of the jet engine compares to the drag force when the boat speeds up.

**Answer:** The driving force is greater than the drag force.

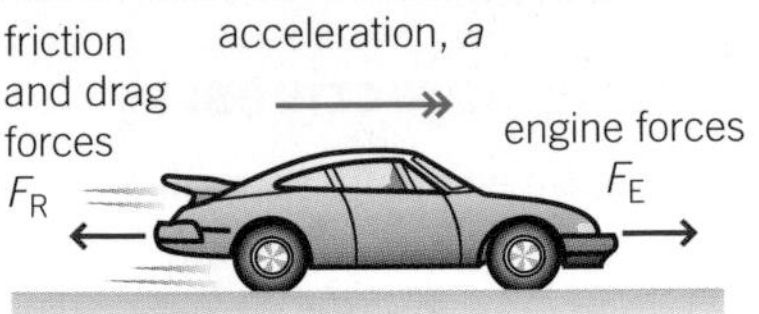


▲ **Figure 3** *Vehicle power*

### Summary questions

$g = 9.8\,\text{m}\,\text{s}^{-2}$

1. **a** A steel ball of mass 0.15 kg, released from rest in a liquid, falls a distance of 0.20 m in 5.0 s. Assuming the ball reaches terminal speed within a fraction of a second, calculate **i** its terminal speed, **ii** the drag force on it when it falls at terminal speed.
   
   **b** State and explain whether or not a smaller steel ball would fall at the same rate in the same liquid.
2. Explain why a cyclist can reach a higher top speed by crouching over the handlebars instead of sitting upright while pedalling.
3. A vehicle of mass 32 000 kg has an engine which has a maximum engine force of 4.4 kN and a top speed of 36 m s$^{-1}$ on a level road.
   
   Calculate **a** its maximum acceleration from rest, **b** the distance it would travel at maximum acceleration to reach a speed of 12 m s$^{-1}$ from rest.
4. Explain why a vehicle has a higher top speed on a downhill stretch of road than on a level road.

# 8.4 On the road

## Learning objectives:

- Describe what is happening to a vehicle's speed while a driver is reacting to a hazard ahead.
- Distinguish between braking distance and stopping distance.
- Describe how road conditions affect these distances.

*Specification reference: 3.4.1.5*

## Stopping distances

Traffic accidents often happen because vehicles are being driven too fast and too close. A driver needs to maintain a safe distance between his or her own vehicle and the vehicle in front. If a vehicle suddenly brakes, the driver of the following vehicle needs to brake as well to avoid a crash.

**Thinking distance** is the distance travelled by a vehicle in the time it takes the driver to react. For a vehicle moving at constant speed $v$, the thinking distance $s_1$ = speed × reaction time = $vt_0$, where $t_0$ is the reaction time of the driver. The reaction time of a driver is affected by distractions, drugs, and alcohol, which is why drivers in the UK are banned from using mobile phones while driving and breathalyser tests are carried out on drivers.

**Braking distance** is the distance travelled by a car in the time it takes to stop safely, from when the brakes are first applied. Assuming constant deceleration, $a$, to zero speed from speed $u$, the braking distance $s_2 = \frac{u^2}{2a}$ since $u^2 = 2as_2$.

**Stopping distance** = thinking distance + braking distance = $ut_0 + \frac{u^2}{2a}$ where $u$ is the speed before the brakes were applied.

Figure 1 shows how thinking distance, braking distance, and stopping distance vary with speed for a reaction time of 0.67 s and a deceleration of 6.75 m s$^{-2}$. Using these values for reaction time and deceleration in the above equations, Figure 1 gives the shortest stopping distances on a dry road as recommended in the Highway Code.

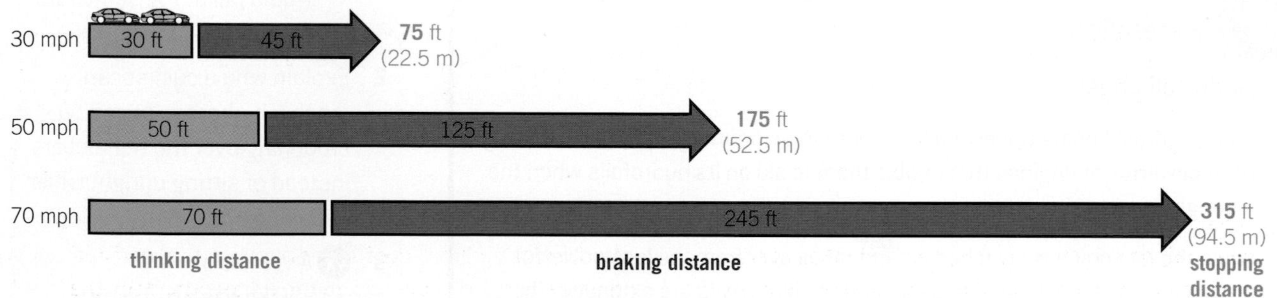


▲ **Figure 1** *Stopping distances*

### Study tip

A car can only move forwards by pushing backwards on the road. The same applies when you take a forward step. When you push backwards on the ground, the ground exerts an equal and opposite force on you that enables you to move forward.

### Skidding

On a front-wheel drive vehicle, the engine turns the front wheels via the transmission system. Friction between the tyres and the road prevent wheel spin (slipping) so the driving wheels therefore roll along the road. If the driver tries to accelerate too fast, the wheels skid. This is because there is an upper limit to the amount of friction between the tyres and the road (Figure 2).

When the brakes are applied, the wheels are slowed down by the brakes. The vehicle therefore slows down, provided the wheels do not skid. If the braking force is increased, the friction force between the tyres and the road increases. However, if the upper limit of friction (usually referred to as the limiting frictional force) between the tyres and the road is reached, the wheels skid. When this happens, the brakes lock and the vehicle slides uncontrollably forward. Most vehicles are now fitted with an anti-lock brake system (ABS), which consists of a speed sensor on each wheel linked to a central electronic control unit that controls hydraulic valves in the brake system. When the brakes are applied, if a wheel starts to lock, the control unit senses that the wheel is rotating much more slowly than the other wheels – as a result, the valves are activated, so the brake pressure on the wheel is reduced to stop it locking.

**Q:** State the direction of the frictional force of the road on the driving wheels when a car accelerates.

**Answer:** In the direction in which the car is moving.

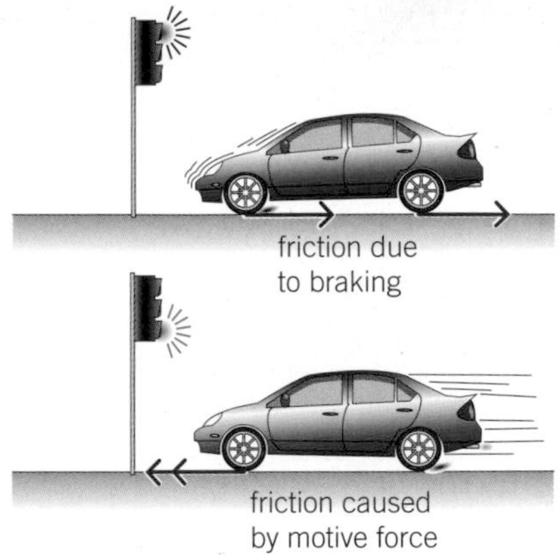

▲ **Figure 2** *Stopping and starting*

### *Practical: Testing friction*

Measure the limiting friction between the underside of a block and the surface it is on. Pull the block with an increasing force until it slides. The limiting frictional force on the block is equal to the pull force on the block just before sliding occurs. Find out how this force depends on the weight of the block.

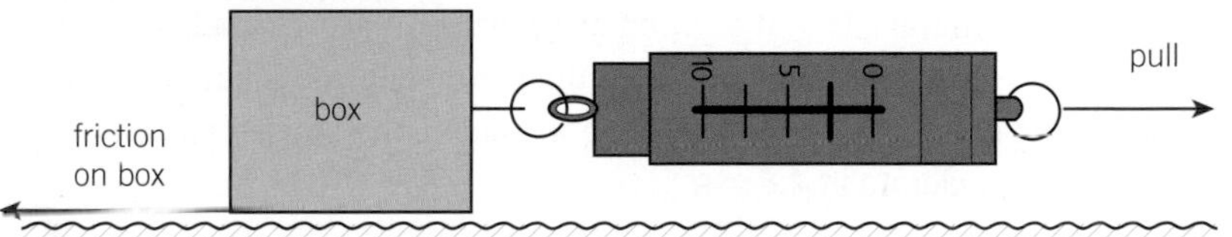

▲ **Figure 3** *Testing friction*

▲ **Figure 4** *Tyre treads*

## More about braking distance

The braking distance for a vehicle depends on the speed of the vehicle at the instant the brakes are applied, on the road conditions, and on the condition of the vehicle tyres.

On a greasy or icy road, skidding is more likely because the limiting frictional force between the road and the tyres is reduced from its dry value. To stop a vehicle safely on a greasy or icy road, the brakes must be applied with less force than on a dry road otherwise skidding will occur. Therefore the braking distance is longer than on a dry road. In fast-moving traffic, a driver must ensure there is a bigger gap to the car in front so as to be able to slow down safely if the vehicle in front slows down.

The condition of the tyres of a vehicle also affects braking distance. The tread of a tyre must not be less than a certain depth, otherwise any grease, oil, or water on the road reduces the friction between the road and the tyre very considerably. In the United Kingdom, the absolute minimum legal requirement is that every tyre of a vehicle must have a tread depth of at least 1.6 mm across three-quarters of the width of the tyre all the way round. If the pressure on a tyre is too small or too great, or the wheel is unbalanced, then the tyre will wear unevenly and will quickly become unsafe. In addition, the braking force may be reduced if the tyre pressure is too high as the tyre area in contact with the road will be reduced. Clearly, a driver must check the condition of the tyres regularly. If you are a driver or hope to become one, remember that the vehicle tyres are the only contacts between the vehicle and the road and must therefore be looked after.

## Worked example

$g = 9.8\,\text{m}\,\text{s}^{-2}$

A vehicle of mass 900 kg, travelling on a level road at a speed of $15\,\text{m}\,\text{s}^{-1}$, can be brought to a standstill without skidding by a braking force equal to 0.5 × its weight. Calculate **a** the deceleration of the vehicle, **b** the braking distance.

### Solution

**a** Weight = 900 × 9.8 = 8800 N

Braking force = 0.5 × 8800 = 4400 N

$$\text{Deceleration} = \frac{\text{braking force}}{\text{mass}} = \frac{4400}{900} = 4.9\,\text{m}\,\text{s}^{-2}$$

**b** $u = 15\,\text{m}\,\text{s}^{-1}$, $v = 0$, $a = -4.9\,\text{m}\,\text{s}^{-2}$

To calculate $s$, use $v^2 = u^2 + 2as$

$$s = \frac{-u^2}{2a} = \frac{-15^2}{2 \times -4.9} = 23\,\text{m}$$

## Summary questions

$g = 9.8\,\text{m}\,\text{s}^{-2}$

**1** A vehicle is travelling at a speed of $18\,\text{m}\,\text{s}^{-1}$ on a level road, when the driver sees a pedestrian stepping off the pavement into the road 45 m ahead. The driver reacts within 0.4 s and applies the brakes, causing the car to decelerate at $4.8\,\text{m}\,\text{s}^{-2}$.

- **a** Calculate **i** the thinking distance, **ii** the braking distance.
- **b** How far does the driver stop from where the pedestrian stepped into the road?

**2** The braking distance of a vehicle for a speed of $18\,\text{m}\,\text{s}^{-1}$ on a dry level road is 24 m. Calculate:

- **a** the deceleration of the vehicle from this speed to a standstill over this distance
- **b** the frictional force on a vehicle of mass 1000 kg on this road as it stops.

**3**
- **a** What is meant by the braking distance of a vehicle?
- **b** Explain, in terms of the forces acting on the wheels of a car, why a vehicle slows down when the brakes are applied.

**4** The frictional force on a vehicle travelling on a certain type of level road surface is 0.6 × the vehicle's weight. For a vehicle of mass 1200 kg,

- **a** show that the maximum deceleration on this road is $5.9\,\text{m}\,\text{s}^{-2}$
- **b** calculate the braking distance on this road for a speed of $30\,\text{m}\,\text{s}^{-1}$.

# 8.5 Vehicle safety

## Impact forces

▲ **Figure 1** *Head-on collision*

## Learning objectives:

- → Describe the force on a moving body if it is suddenly stopped, for example in a road accident.
- → Explain what should be increased to give a smaller deceleration from a given speed.
- → State which design features attempt to achieve this in a modern vehicle.

*Specification reference: 3.4.1.5*

### Measuring impacts

The effect of a collision on a vehicle can be measured in terms of the acceleration or deceleration of the vehicle. By expressing an acceleration or deceleration in terms of $g$, the acceleration due to gravity, the force of the impact can then easily be related to the weight of the vehicle. For example, suppose a vehicle hits a wall and its acceleration is $-30\,\text{m}\,\text{s}^{-2}$. In terms of $g$, the acceleration = $-3g$. So the **impact force** of the wall on the vehicle must have been three times its weight (= $3mg$, where $m$ is the mass of the vehicle). Such an impact is sometimes described as being equal to $3g$. This statement, although technically wrong because the deceleration not the impact force is equal to $3g$, is a convenient way of expressing the effect of an impact on a vehicle or a person.

### Study tip

Remember that a negative acceleration is equal to a positive deceleration.

### Application

#### How much acceleration or deceleration can a person withstand?

The duration of the impact as well as the magnitude of the acceleration affect the person. A person who is sitting or upright can survive a deceleration of $20\,g$ for a time of a few milliseconds, although not without severe injury. A deceleration of over $5\,g$ lasting for a few seconds can cause injuries. Car designers carry out tests using dummies in remote-control vehicles to measure the change of motion of different parts of a vehicle or a dummy. Sensors linked to data recorders and computers are used, as well as video cameras that record the motion to allow video clips to be analysed.

▲ **Figure 2** *A side-on impact*

## Contact time and impact time

When objects collide and bounce off each other, they are in contact with each other for a certain time, which is the same for both objects. The shorter the contact time, the greater the impact force for the same initial velocities of the two objects. When two vehicles collide, they may or may not separate from each other after the collision. If they remain tangled together, they exert forces on each other until they are moving at the same velocity. The duration of the impact force is not the same as the contact time in this situation.

The **impact time** $t$ (the duration of the impact force) can be worked out by applying the equation $s = \frac{1}{2}(u + v)t$ to one of the vehicles, where $s$ is the distance moved by that vehicle during the impact, $u$ is its initial velocity, and $v$ is its final velocity. If the vehicle mass is known, the impact force can also be calculated.

For a vehicle of mass $m$ in time $t$,

$$\textbf{the impact time } t = \frac{2s}{u + v}$$

$$\textbf{the acceleration } a = \frac{v - u}{t}$$

$$\textbf{the impact force } F = ma$$

### Study tip

When writing answers and when you are driving (especially just after passing your driving test), don't rush – it's safer to take longer over what you are doing!

### Worked example

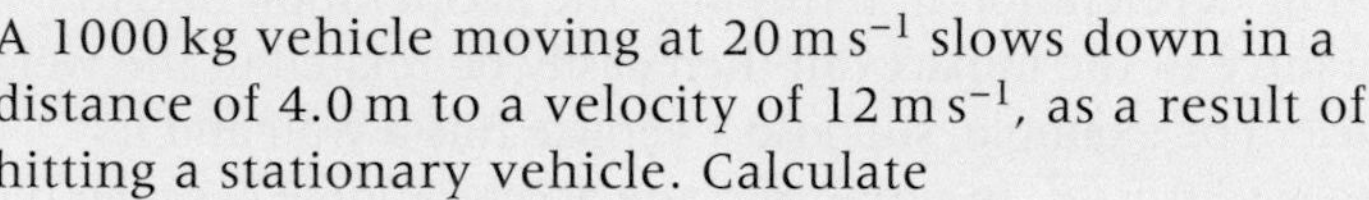

A 1000 kg vehicle moving at $20\,\text{m}\,\text{s}^{-1}$ slows down in a distance of 4.0 m to a velocity of $12\,\text{m}\,\text{s}^{-1}$, as a result of hitting a stationary vehicle. Calculate

**a** the vehicle's acceleration in terms of $g$

**b** the impact force for this collision.

#### Solution

**a** $t = \dfrac{2s}{(u + v)} = \dfrac{2 \times 4.0}{(20 + 12)} = 0.25\,\text{s}$

The acceleration, $a = \dfrac{(v - u)}{t}$

$= \dfrac{12 - 20}{0.25}$

$= -32\,\text{m}\,\text{s}^{-2}$

$= -3.3g$ (where $g = 9.8\,\text{m}\,\text{s}^{-2}$).

**b** The impact force $F = ma = 1000 \times -32 = -32\,000\,\text{N}$.

### Study tip

The work done by the impact force $F$ over an impact distance $s$ $(= Fs)$ is equal to the change of kinetic energy of the vehicle. The impact force can also be worked out using the equation

$$F = \frac{\text{change of kinetic energy}}{\text{impact distance}}$$

### Synoptic link

You will learn in Topic 10.2, Kinetic energy and potential energy, that the kinetic energy of a vehicle of mass $m$ moving at speed $v = \frac{1}{2}mv^2$. Prove for yourself that the vehicle in the example above has 200 kJ of kinetic energy at $20\,\text{m}\,\text{s}^{-1}$ and 72 kJ at $12\,\text{m}\,\text{s}^{-1}$. You can then show the change of kinetic energy ÷ impact distance gives −32 000 N for the impact force.

## Extension

### Car safety features

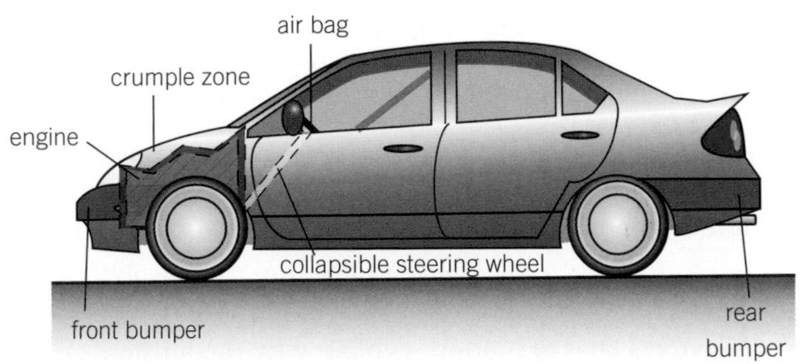


▲ **Figure 3** *Vehicle safety features*

In the example on the previous page, the vehicle's deceleration and hence the impact force would be lessened if the impact time was greater. So design features that would increase the impact time reduce the impact force. With a reduced impact force, the vehicle occupants are less affected. The following vehicle safety features are designed to increase the impact time and so reduce the impact force.

- **Vehicle bumpers** give way a little in a low-speed impact and so increase the impact time. The impact force is therefore reduced as a result. If the initial speed of impact is too high, the bumper and/or the vehicle chassis are likely to be damaged.
- **Crumple zones** The engine compartment of a car is designed to give way in a front-end impact. If the engine compartment were rigid, the impact time would be very short, so the impact force would be very large. By designing the engine compartment so it crumples in a front-end impact, the impact time is increased and the impact force is therefore reduced.
- **Seat belts** In a front-end impact, a correctly fitted seat belt restrains the wearer from crashing into the vehicle frame after the vehicle suddenly stops. The restraining force on the wearer is therefore much less than the impact force would be if the wearer hit the vehicle frame. With the seat belt on, the wearer is stopped more gradually than without it.
- **Collapsible steering wheel** In a front-end impact, the seat belt restrains the driver without holding the driver rigidly. If the driver makes contact with the steering wheel, the impact force is lessened as a result of the steering wheel collapsing in the impact.
- **Airbags** An airbag reduces the force on a person, because the airbag acts as a cushion and increases the impact time on the person. More significantly, the force of the impact is spread over the contact area, which is greater than the contact area with a seat belt. So the pressure on the body is less.

**Q:** Explain why an air bag cushions an impact.

## Summary questions

$g = 9.8\,\text{m}\,\text{s}^{-2}$

1 A car of mass 1200 kg travelling at a speed of $15\,\text{m}\,\text{s}^{-1}$ is struck from behind by another vehicle, causing its speed to increase to $19\,\text{m}\,\text{s}^{-1}$ in a time of 0.20 s. Calculate:

   **a** the acceleration of the car, in terms of $g$

   **b** the impact force on the car.

2 The front end of a certain type of car of mass 1500 kg travelling at a speed of $20\,\text{m}\,\text{s}^{-1}$ is designed to crumple in a distance of 0.8 m, if the car hits a wall. Calculate:

   **a** the impact time if it hits a wall at $20\,\text{m}\,\text{s}^{-1}$

   **b** the impact force.

3 The front bumper of a car of mass 900 kg is capable of withstanding an impact with a stationary object, provided the car is not moving faster than $3.0\,\text{m}\,\text{s}^{-1}$, when the impact occurs. The impact time at this speed is 0.40 s. Calculate:

   **a** the deceleration of the car from $3.0\,\text{m}\,\text{s}^{-1}$ to rest in 0.40 s

   **b** the impact force on the car.

4 In a crash, a vehicle travelling at a speed of $25\,\text{m}\,\text{s}^{-1}$ stops after 4.5 m. A passenger of mass 68 kg is wearing a seat belt, which restrains her forward movement relative to the car to a distance of 0.5 m. Calculate **a** the deceleration of the passenger in terms of $g$, **b** the resultant force on the passenger.

# Practice questions: Chapter 8

1 A car moving at a velocity of $20\,m\,s^{-1}$ brakes to a standstill in a distance of 40 m. A child of mass 15 kg is sitting in a forward-facing child car seat fitted to the back seat of the car.

**(a)** Calculate (i) the deceleration of the car, (ii) the force on the child. *(3 marks)*

**(b)** In 2006, the UK government passed a law requiring that children in cars must travel in child car seats. If the child had not been in the child car seat, explain why the force on her would have been much greater than the value calculated in (a)(ii). *(2 marks)*

**(c)** Road accidents involving children are being reduced by reducing the speed limit near schools to 20 mph ($9\,m\,s^{-1}$). Discuss the effect this has on road safety where the speed limit is currently 30 mph. *(3 marks)*

2 A packing case is being lifted vertically at a constant speed by a cable attached to a crane.

**(a)** With reference to one of Newton's laws of motion, explain why the tension, $T$, in the cable must be equal to the weight of the packing case. *(3 marks)*

**(b)** A 12.0 N force and a 8.0 N force act on a body of mass 6.5 kg at the same time. For this body, calculate

(i) the maximum resultant acceleration that it could experience

(ii) the minimum resultant acceleration that it could experience. *(4 marks)*

AQA, 2005

3 **Figure 1** shows how the velocity of a motor car increases with time as it accelerates from rest along a straight horizontal road.

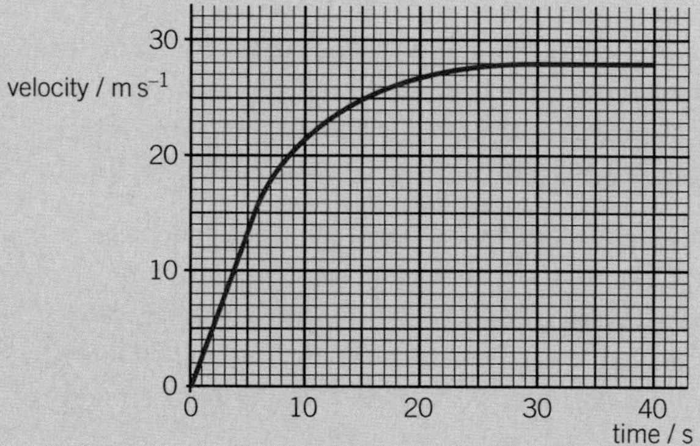


▲ **Figure 1**

**(a)** The acceleration is approximately constant for the first five seconds of the motion. Show that, over the first five seconds of the motion, the acceleration is approximately $2.7\,m\,s^{-2}$. *(3 marks)*

**(b)** Throughout the motion shown in **Figure 1** there is a constant driving force of 2.0 kN acting on the car.

(i) Calculate the mass of the car and its contents.

(ii) What is the magnitude of the resistive force acting on the car after 40 s? *(3 marks)*

**(c)** Find the distance travelled by the car during the first 40 s of the motion. *(3 marks)*

AQA, 2006

4 A constant resultant horizontal force of $1.8 \times 10^3$ N acts on a car of mass 900 kg, initially at rest on a level road.

**(a)** Calculate:

(i) the acceleration of the car

(ii) the speed of the car after 8.0 s

(iii) the distance travelled by the car in the first 8.0 s of its motion. *(5 marks)*

**(b)** In practice the resultant force on the car changes with time. Air resistance is one factor that affects the resultant force acting on the vehicle.

(i) Suggest, with a reason, how the resultant force on the car changes as its speed increases.

(ii) Explain, using Newton's laws of motion, why the vehicle has a maximum speed. *(5 marks)*

AQA, 2004

**5** $\sqrt{x}$ A passenger aircraft has a mass of $3.2 \times 10^5$ kg when fully laden. It is powered by four jet engines each producing a maximum thrust of 270 kN.

**(a)** (i) Calculate the total force of the engines acting on the aircraft.

(ii) Show that the initial acceleration of the aircraft with the engines set to full thrust is about $3.4\,\mathrm{m\,s^{-2}}$. Ignore any frictional forces. *(3 marks)*

**(b)** The aircraft starts from rest at the beginning of its take-off and has a take-off speed of $90\,\mathrm{m\,s^{-1}}$.

(i) Calculate the time taken for the aircraft to reach its take-off speed if frictional forces are ignored.

(ii) Frictional forces reduce the actual acceleration of the aircraft to $2.0\,\mathrm{m\,s^{-2}}$. Calculate the mean total frictional force acting against the aircraft during the time taken to reach its take-off speed. *(4 marks)*

**(c)** Calculate the minimum runway length required by this aircraft for take-off when the acceleration is $2.0\,\mathrm{m\,s^{-2}}$. *(2 marks)*

**(d)** The cruising speed of the aircraft in level flight with the engines at maximum thrust is a constant $260\,\mathrm{m\,s^{-1}}$. The pilot adjusts the thrust so that the *horizontal* acceleration is always $2.0\,\mathrm{m\,s^{-2}}$.
Calculate the time taken from take-off to reach its cruising speed. *(2 marks)*

**(e)** When it is at cruising speed, the aircraft travels at a constant velocity and at a constant height. Explain, in terms of the horizontal and vertical forces acting on the aircraft, how this is achieved. *(2 marks)*

AQA, 2002

**6** $\sqrt{x}$ A fairground ride ends with a car moving up a ramp at a slope of 30° to the horizontal.

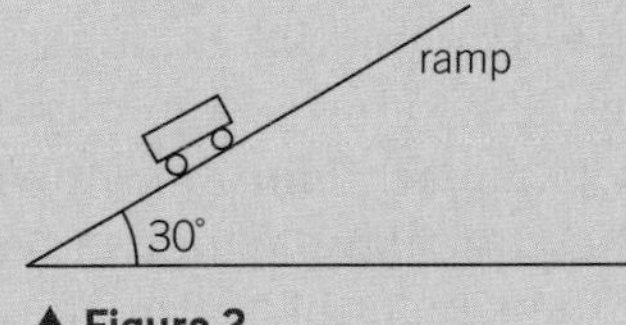


▲ **Figure 2**

**(a)** The car and its passengers have a total weight of $7.2 \times 10^3$ N. Show that the component of the weight parallel to the ramp is $3.6 \times 10^3$ N. *(1 mark)*

**(b)** Calculate the deceleration of the car assuming the only force causing the car to decelerate is that calculated in part (a). *(2 marks)*

**(c)** The car enters at the bottom of the ramp at $18\,\mathrm{m\,s^{-1}}$. Calculate the minimum length of the ramp for the car to stop before it reaches the end. The length of the car should be neglected. *(2 marks)*

**(d)** Explain why the stopping distance is, in practice, shorter than the value calculated in part (c). *(2 marks)*

AQA, 2005

# 9 Force and momentum
## 9.1 Momentum and impulse

**Learning objectives:**

- → Calculate momentum.
- → Describe the connection between Newton's first and second laws of motion and momentum.
- → Define the impulse of a force, and calculate it from a force versus time graph.

*Specification reference: 3.4.1.6*

### Momentum

If you have ever run into someone on the sports field, you will know something about momentum. If the person you ran into was more massive than you, then you probably came off worse than the other person. When two bodies collide, the effect they have on each other depends not only on their initial velocities but also on the mass of each object. You can easily test the idea using coins, as shown in Figure 1. You might already have developed your skill in this area! It is not too difficult to show that when a large coin and a small coin collide, the motion of the small coin is affected more.

Sir Isaac Newton was the first person to realise that a **force** was needed to change the velocity of an object. He realised that the effect of a force on an object depended on its mass as well as on the amount of force. He defined the **momentum** of a moving object as its mass × its velocity and showed how the momentum of an object changes when a force acts on it. In this chapter, you will consider the ideas that Newton established in full.

▲ **Figure 1** *Momentum games*

**Synoptic link**

You have met Newton's laws of motion in Chapter 8, and will study Newton's laws of gravitation later on if you're studying A Level Physics.

Although Newton put forward his ideas over 300 years ago, his laws continue to provide the essential mathematical rules for predicting the motion of objects in any situation except inside the atom (where the rules of quantum physics apply) and at speeds approaching the speed of light or in very strong gravitational fields (where Einstein's theories of relativity apply). For example, the launch of a rocket is carefully planned using **Newton's laws of motion** and his law of gravitation. However, the laws do not, for example, predict the existence of black holes, which were a confirmed prediction of Einstein's theory of general relativity. In fact, Einstein showed that his theories of relativity simplify into Newton's laws where gravity is weak and the speed of objects is much less than the speed of light.

**The momentum of an object is defined as its mass × its velocity.**

- The unit of momentum is $\mathrm{kg\,m\,s^{-1}}$. The symbol for momentum is $p$.
- Momentum is a vector quantity. Its direction is the same as the direction of the object's velocity.
- For an object of mass $m$ moving at velocity $v$, its momentum $p = mv$.

For example, a ball of mass 2.0 kg moving at a velocity of $10\,\mathrm{m\,s^{-1}}$ has the same amount of momentum as a person of mass 50 kg moving at a velocity of $0.4\,\mathrm{m\,s^{-1}}$.

### Momentum and Newton's laws of motion

**Newton's first law of motion: An object remains at rest or in uniform motion unless acted on by a force.**

In effect, Newton's first law tells us that a force is needed to change the momentum of an object. If the momentum of an object is constant, there is no resultant force acting on it. Clearly, if the mass of an object is constant and the object has constant momentum, it follows that the velocity of the object is also constant. If a moving object with constant momentum gains or loses mass, however, its velocity would change to keep its momentum constant. For example, a cyclist in a race who collects a water bottle as he or she speeds past a service point gains mass (i.e., the water bottle) and therefore loses velocity.

**Newton's second law of motion: The rate of change of momentum of an object is proportional to the resultant force on it. In other words, the resultant force is proportional to the change of momentum per second.**

## Synoptic link

You have met the force equation in Topic 8.2, Using $F = ma$.

You can write Newton's second law in the form force = mass × acceleration. Now, you will look at how this equation is derived from Newton's second law in its general form as stated above.

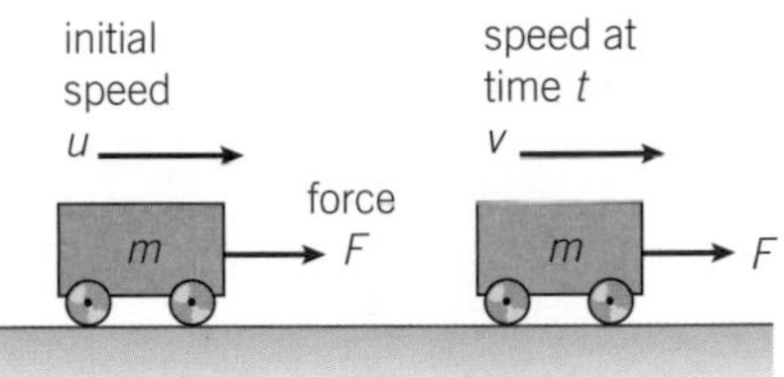


▲ **Figure 2** *Force and momentum*

**Consider an object of constant mass $m$ acted on by a constant force $F$.** Its acceleration causes a change of its speed from initial speed $u$ to speed $v$ in time $t$ without change of direction:

- its initial momentum = $mu$, and its final momentum = $mv$
- its change of momentum = its final momentum $(mv)$ – its initial momentum $(mu)$.

According to Newton's second law, the force is proportional to the change of momentum per second.

Therefore, force $F \propto \frac{\text{change of momentum}}{\text{time taken}} = \frac{mv - mu}{t} = \frac{m(v-u)}{t} = ma$

where $a = \frac{v-u}{t}$ = the acceleration of the object.

This proportionality relationship (i.e., $F \propto ma$) can be written as $F = kma$, where $k$ is a constant of proportionality.

## Study tip

The unit of momentum may be either $kg\,m\,s^{-1}$ or (more neatly) N s. The equation $F = \frac{\Delta(mv)}{\Delta t}$ always applies but $F = ma$ applies only to objects of constant mass.

The value of $k$ is made equal to 1 by defining the unit of force, **the newton**, as the amount of force that gives an object of mass 1 kg an acceleration of $1\,m\,s^{-2}$ (i.e., force $F = 1$ N, mass $m = 1$ kg, acceleration $a = 1\,m\,s^{-2}$ so $k = 1$).

Therefore, with $k = 1$, the equation $\boldsymbol{F = ma}$ follows from Newton's second law provided the mass of the object is constant.

*In general*, the change of momentum of an object may be written as $\Delta(mv)$, where the symbol $\Delta$ means change of. Therefore, if the momentum of an object changes by $\Delta(mv)$ in time $\Delta t$, the force $F$ on the object is given by the equation

$$F = \frac{\Delta(mv)}{\Delta t}$$

1 **If $m$ is constant,** then $\Delta(mv) = m\Delta v$, where $\Delta v$ is the change of velocity of the object.

$\therefore F = \frac{m\Delta v}{\Delta t} = ma$ where acceleration $a = \frac{\Delta v}{\Delta t}$

**2** **If $m$ changes at a constant rate** as a result of mass being transferred at constant velocity, then $\Delta(mv) = v\Delta m$, where $\Delta m$ is the change of mass of the object.

$$\therefore F = \frac{v\Delta m}{\Delta t} \text{ where } \frac{\Delta m}{\Delta t} = \text{change of mass per second.}$$

This form of Newton's second law is used in any situation where an object gains or loses mass continuously.

For example, if a rocket ejects burnt fuel as hot gas from its engine at speed $v$, the force $F$ exerted by the engine to eject the hot gas is given by

$$F = \frac{v\Delta m}{\Delta t} \text{ where } \frac{\Delta m}{\Delta t} = \text{mass of hot gas lost per second.}$$

An equal and opposite reaction force acts on the jet engine due to the hot gas, propelling the rocket forwards.

The **impulse** of a force is defined as the force × the time for which the force acts. Therefore, for a force $F$ which acts for time $\Delta t$,

**the impulse = $F\Delta t = \Delta(mv)$**

Hence the impulse of a force acting on an object is equal to the change of momentum of the object.

### Synoptic link

The area rule is especially useful when the force changes with time. This rule can also be used for a velocity–time graph where the velocity changes. You have met this in Topic 7.3, Motion along a straight line at constant acceleration.

## Force–time graphs

Suppose an object of constant mass $m$ is acted on by a constant force $F$ which changes its velocity from initial velocity $u$ to velocity $v$ in time $t$. As explained earlier in this topic, Newton's second law gives

$$F = \frac{mv - mu}{t}$$

Rearranging this equation gives $Ft = mv - mu$.

Figure 3 is a graph of force versus time for this situation. Because force $F$ is constant for time $t$, the area under the line represents the impulse of the force $Ft$, which is equal to $mv - mu$. In other words,

**the area under the line of a force–time graph represents the change of momentum or the impulse of the force**

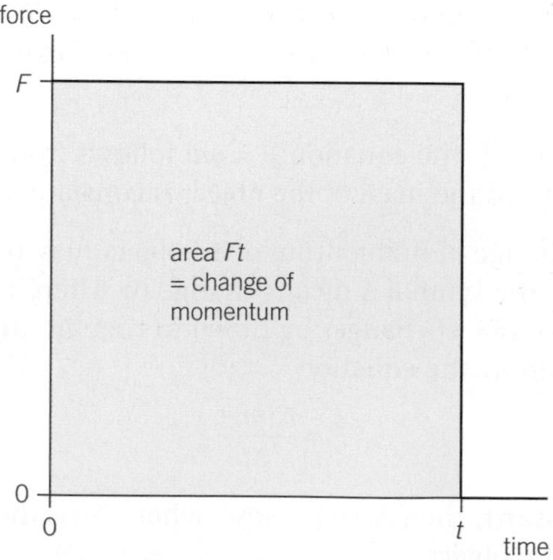


▲ **Figure 3** *Force against time for constant force*

***Note:***
The unit of momentum can be given as the newton second (N s) or the kilogram metre per second (kg m s$^{-1}$). The unit of impulse is usually given as the newton second.

## Worked example

A force of 10 N acts for 20 s on an object of mass 50 kg which is initially at rest.

Calculate:

**a** the change of momentum of the object
**b** the velocity of the object at 20 s.

### *Solution*

**a** Change of momentum = impulse of the force = $Ft = 10 \times 20$ = 200 N s

**b** Momentum at 20 s = 200 N s as the object was initially at rest.

$$\therefore \text{Velocity} = \frac{\text{momentum}}{\text{mass}} = \frac{200}{50} = 4.0\,\text{m s}^{-1}$$

## Summary questions

**1** **a** Calculate the momentum of:

  **i** an atom of mass $4.0 \times 10^{-25}$ kg moving at a velocity of $3.0 \times 10^{6}$ m s$^{-1}$
  **ii** a pellet of mass $4.2 \times 10^{-4}$ kg moving at a velocity of 120 m s$^{-1}$
  **iii** a bird of mass 0.56 kg moving at a velocity of 25 m s$^{-1}$.

**b** Calculate:

  **i** the mass of an object moving at a velocity of 16 m s$^{-1}$ with momentum of 96 kg m s$^{-1}$
  **ii** the velocity of an object of mass 6.4 kg that has a momentum of 128 kg m s$^{-1}$.

**2** A train of mass 24 000 kg moving at a velocity of 15.0 m s$^{-1}$ is brought to rest by a braking force of 6000 N. Calculate

**a** the initial momentum of the train
**b** the time taken for the brakes to stop the train.

**3** An aircraft of total mass 45 000 kg accelerates on a runway from rest to a velocity of 120 m s$^{-1}$ when it takes off. During this time, its engines provide a constant driving force of 120 kN. Calculate:

**a** the gain of momentum of the aircraft
**b** the take-off time.

**4** The velocity of a vehicle of mass 600 kg was reduced from 15 m s$^{-1}$ by a constant force of 400 N which acted for 20 s then by a constant force of 20 N for a further 20 s.

**a** Sketch the force versus time graph for this situation.

**b** **i** Calculate the initial momentum of the vehicle.
  **ii** Use the force versus time graph to determine the total change of momentum.
  **iii** Hence show that the final velocity of the vehicle is 1 m s$^{-1}$.

# 9.2 Impact forces

## Learning objectives:

- → Describe what happens to the impact force (and why) if the duration of impact is reduced.
- → Calculate $\Delta(mv)$ for a moving object that stops or reverses.
- → Describe what happens to the momentum of a ball when it bounces off a wall.

*Specification reference: 3.4.1.6*

## Synoptic link

Use of $F = ma$ and $a = \frac{v-u}{t}$ is another way to calculate an impact force.

You have met impact force in road accidents in Topic 8.5, Vehicle safety.

## Application

### Vehicle safety reminders

In Topic 8.5, you looked at the physics of vehicle safety features such as crumple zones. These and other features such as side-impact bars are all designed to lessen the effect of an impact on passengers in the vehicle. The essential idea is to increase the time taken by an impact so the acceleration or deceleration is less and therefore reducing the impact force. The result is the same using the idea of momentum – for a given change of momentum, the force is reduced if the impact time is increased. You will meet how these ideas can be developed further by using the concept of momentum in Topic 9.3, Conservation of momentum.

▲ **Figure 1** *A golf ball impact*

A sports person knows that the harder a ball is hit, the further it travels. The impact changes the momentum of the ball in a very short time when the object exerting the impact force is in contact with the ball.

- If the ball is initially stationary and the impact causes it to accelerate to speed $v$ in time $t$, the gain of momentum of the ball due to the impact $= mv$, where $m$ is the mass of the ball.

  Therefore, the force of the impact $F = \frac{\text{change of momentum}}{\text{contact time}} = \frac{mv}{t}$

- If the ball is moving with an initial velocity $u$, and the impact changes its velocity to $v$ in time $t$, the change of momentum of the ball $= mv - mu$.

  Therefore, the force of impact $F = \frac{\text{change of momentum}}{\text{contact time}}$

$$F = \frac{mv - mu}{t}$$

### Worked example

A ball of mass 0.63 kg initially at rest was struck by a bat which gave it a velocity of $35\,\text{m}\,\text{s}^{-1}$. The contact time between the bat and ball was 25 ms. Calculate:

**a** the momentum gained by the ball

**b** the average force of impact on the ball.

(**Note** The 'm' in ms stands for milli.)

### *Solution*

**a** Momentum gained $= 0.63 \times 35 = 22\,\text{kg}\,\text{m}\,\text{s}^{-1}$

**b** Impact force $= \frac{\text{gain of momentum}}{\text{contact time}} = \frac{22}{0.025} = 880\,\text{N}$

## Force–time graphs for impacts

The variation of an impact force with time on a ball can be recorded using a force sensor connected using suitably long wires or a radio link to a computer. The force sensor is attached to the object (e.g., a bat) that causes the impact. Because the force on the bat is equal and opposite to the force on the ball during impact, the force on the ball due to the bat varies in exactly the same way as the force on the bat due to the ball. The variation of force with time is displayed on the computer screen.

Figure 3 shows a typical force–time graph for an impact. The graph shows that the impact force increases then decreases during the impact. As explained in Topic 9.1, the area under the graph is equal to the change of momentum. The average force of impact can be worked out from the change of momentum divided by the contact time.

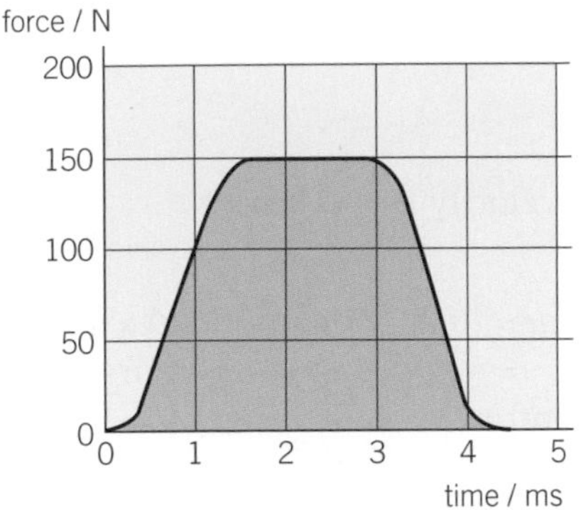


area under curve = 9 blocks

*Ft* for 1 block = 50 N × 1 ms
= $5.0 \times 10^{-2}$ N s

change of momentum
= $9 \times 5.0 \times 10^{-2}$
= 0.45 N s

▲ **Figure 3** *Force against time for an impact*

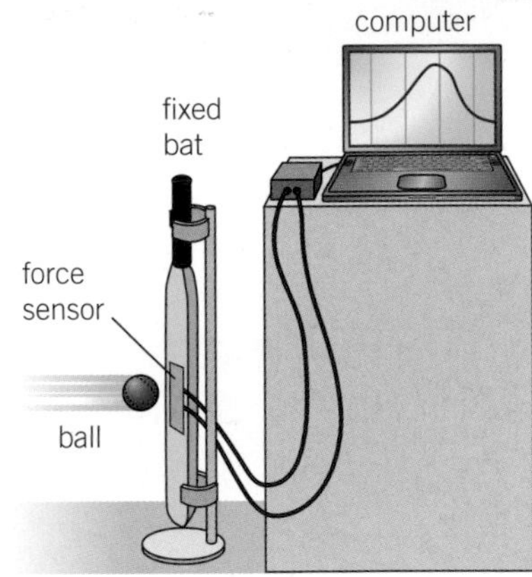


▲ **Figure 2** *Investigating an impact force on a ball*

## Rebound impacts

When a ball hits a wall and rebounds, its momentum changes direction due to the impact. If the ball hits the wall normally, it rebounds normally so the direction of its momentum is reversed. The velocity and therefore the momentum after the impact is in the opposite direction to the velocity before the impact and therefore has the opposite sign. Figure 4 shows the idea.

Suppose the ball hits the wall normally with an initial speed $u$ and it rebounds at speed $v$ in the opposite direction. Since its direction of motion reverses on impact, a sign convention is necessary to represent the two directions. Using + for 'towards the wall' and – for 'away from the wall', its initial momentum = $+mu$, and its final momentum = $-mv$.

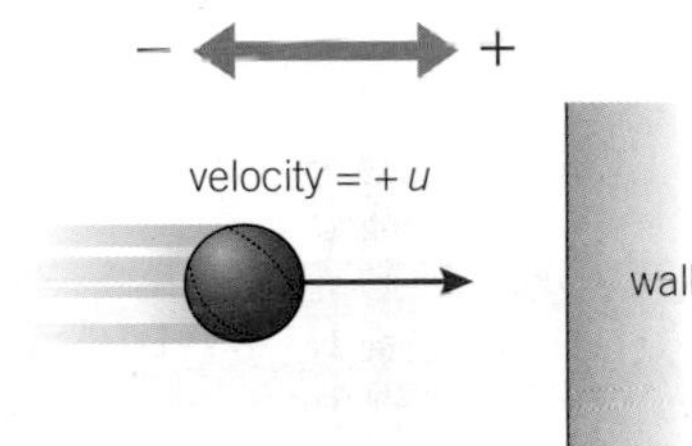


**a** *Before impact*

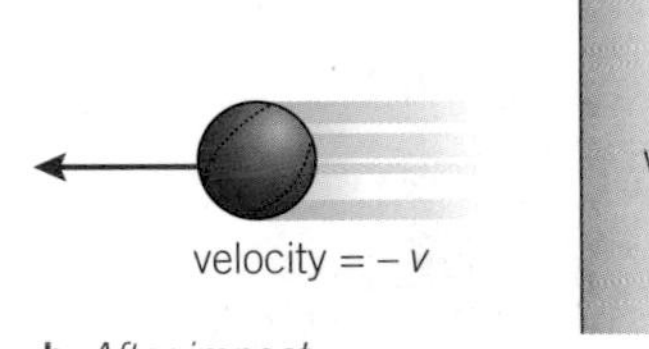


**b** *After impact*

▲ **Figure 4** *A rebound*

Therefore,

its change of momentum = final momentum – initial momentum

$$= (-mv) - (mu)$$

$$\text{The impact force } F = \frac{\text{change of momentum}}{\text{contact time}} = \frac{(-mv) - (mu)}{t}$$

### *Notes:*

1 If there is no loss of speed on impact, then $v = u$ so the impact force

$$F = \frac{(-mv) - (mu)}{t} = \frac{-2mu}{t}$$

2 If the impact is oblique (i.e., the initial direction of the ball is not perpendicular to the wall, as in Figure 5), the normal components of the velocity must be used. For an impact in which the initial and final direction of the ball are at the same angle $\theta$ to the normal and there is no loss of speed (i.e., $u = v$), the normal component of the initial velocity is $+u\cos\theta$ and the normal component of the final velocity is $-u\cos\theta$. The change of momentum of the ball is therefore $-2\,mu\cos\theta$.

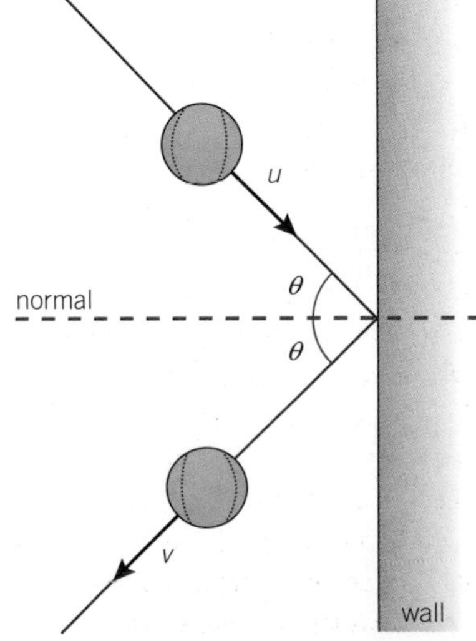


▲ **Figure 5** *An oblique impact*

## Worked example

A tennis ball of mass 0.20 kg moving at a speed of $18\,\mathrm{m\,s^{-1}}$ was hit by a bat, causing the ball to go back in the direction it came from at a speed of $15\,\mathrm{m\,s^{-1}}$. The contact time was 0.12 s. Calculate:

**a** the change of momentum of the ball

**b** the impact force on the ball.

### *Solution*

**a** Mass of ball $m = 0.20\,\mathrm{kg}$, initial velocity $u = +18\,\mathrm{m\,s^{-1}}$, final velocity $= -15\,\mathrm{m\,s^{-1}}$.

Change of momentum $= mv - mu = (0.20 \times -15) - (0.20 \times 18)$
$= -3.0 - 3.6 = -6.6\,\mathrm{kg\,m\,s^{-1}}$

**b** Impact force $= \dfrac{\text{change of momentum}}{\text{time taken}} = \dfrac{-6.6}{0.12} = -55\,\mathrm{N}$

***Note:*** The minus sign indicates the force on the ball is in the same direction as the velocity after the impact.

## Summary questions

**1** A 2000 kg lorry reversing at a speed of $0.80\,\mathrm{m\,s^{-1}}$ backs accidentally into a steel fence. The fence stops the lorry 0.5 s after the lorry first makes contact with the fence. Calculate:

**a** the initial momentum of the lorry

**b** the force of the impact.

**2** A car of mass 600 kg travelling at a speed of $3.0\,\mathrm{m\,s^{-1}}$ is struck from behind by another vehicle. The impact lasts for 0.40 s and causes the speed of the car to increase to $8.0\,\mathrm{m\,s^{-1}}$. Calculate:

**a** the change of momentum of the car due to the impact

**b** the impact force.

**3** A molecule of mass $5.0 \times 10^{-26}$ kg moving at a speed of $420\,\mathrm{m\,s^{-1}}$ hits a surface at right angles to the surface and rebounds at the same speed in the opposite direction in an impact lasting 0.22 ns. Calculate:

**a** the change of momentum

**b** the force on the molecule.

**4** Repeat the calculation in Q3 for a molecule of the same mass at the same speed which hits the surface at 60° to the normal and rebounds without loss of speed at 60° to the normal, as shown in Figure 6. You will need to work out the components of the molecule's velocity parallel to the normal before and after the impact. Assume the contact time is the same.

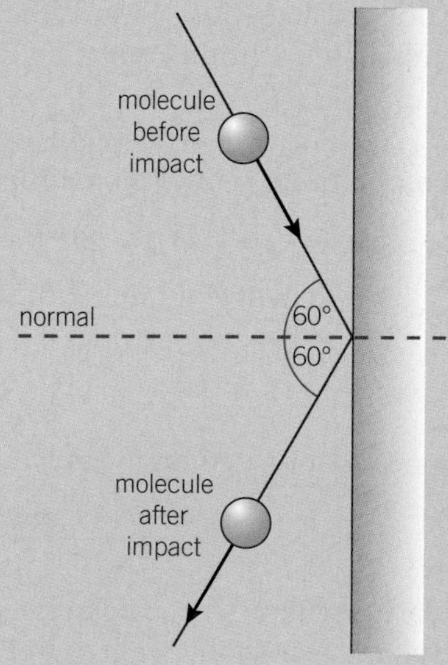


▲ **Figure 6**

# 9.3 Conservation of momentum

## Newton's third law of motion

**When two objects interact, they exert equal and opposite forces on each other.**

In other words, if object A exerts a force on object B, there must be an equal and opposite force acting on object A due to object B. For example,

- the Earth exerts a force due to gravity on an object which exerts an equal and opposite force on the Earth
- a jet engine exerts a force on hot gas in the engine to expel the gas – the gas being expelled exerts an equal and opposite force on the engine.

However, it is worth remembering that two forces must be of the same type, and acting on different objects, for the forces to be considered a force pair. For example, weight and normal reaction would not constitute a force pair in this case. (See Figure 1 for further examples.)

## The principle of conservation of momentum

When an object is acted on by a resultant force, its momentum changes. If there is no change of its momentum, there can be no resultant force on the object. Now consider several objects which interact with each other. If no external resultant force acts on the objects, the total momentum does not change. However, interactions between the objects can transfer momentum between them. But the total momentum does not change.

**The principle of conservation of momentum states that for a system of interacting objects, the total momentum remains constant, provided no external resultant force acts on the system.**

Consider two objects that collide with each other then separate. As a result, the momentum of each object changes. They exert equal and opposite forces on each other when they are in contact. So the change of momentum of one object is equal and opposite to the change of momentum of the other object. In other words, if one object gains momentum, the other object loses an equal amount of momentum. So the total amount of momentum is unchanged.

Let's look in detail at the example of two snooker balls A and B in collision, as shown in Figure 2 on the next page.

The impact force $F_1$ on ball A due to ball B changes the velocity of A from $u_A$ to $v_A$ in time $t$

Therefore, $F_1 = \frac{m_A v_A - m_A u_A}{t}$, where $t$ = the time of contact between A and B, and $m_A$ = the mass of ball A.

The impact force $F_2$ on ball B due to ball A changes the velocity of B from $u_B$ to $v_B$ in time $t$.

## Learning objectives:

- → Consider whether momentum is ever lost in a collision.
- → Define conservation of momentum.
- → State the condition that must be satisfied if the momentum of a system is conserved.

*Specification reference: 3.4.1.6*

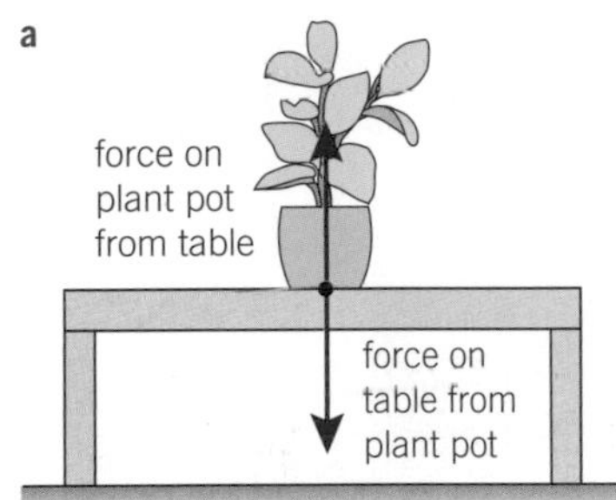


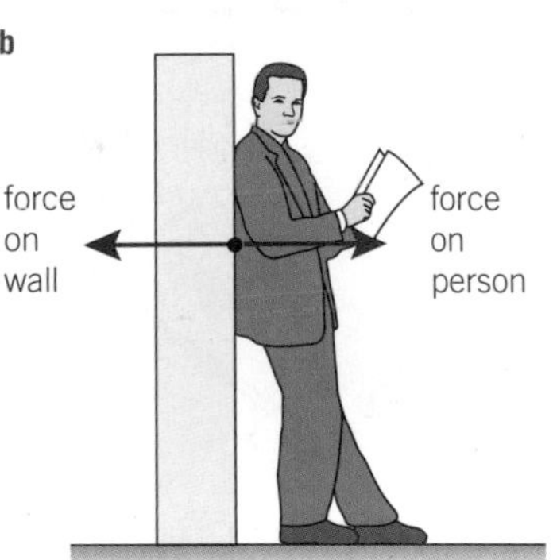


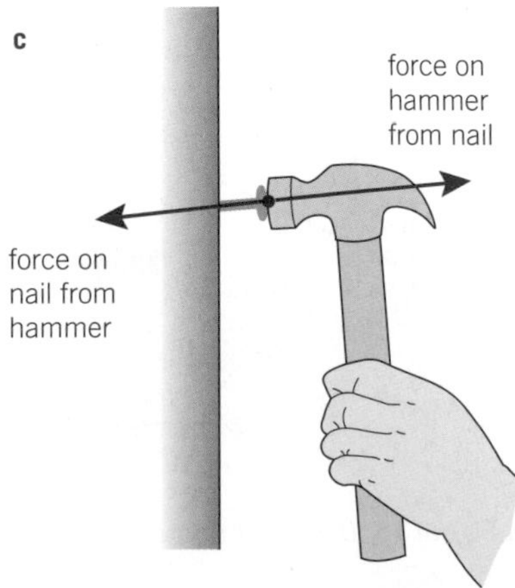


▲ **Figure 1** *Examples of Newton's third law*

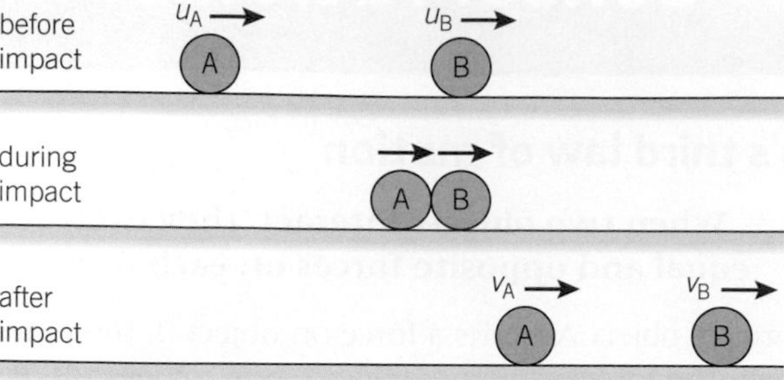


▲ **Figure 2** *Conservation of momentum*

Therefore, $F_2 = \frac{m_B v_B - m_B u_B}{t}$, where $t$ = the time of contact between A and B, and $m_B$ = the mass of ball B.

Because the two forces are equal and opposite to each other, $F_2 = -F_1$.

Therefore, $\frac{m_B v_B - m_B u_B}{t} = -\frac{(m_A v_A - m_A u_A)}{t}$.

Cancelling $t$ on both sides gives

$$m_B v_B - m_B u_B = -(m_A v_A - m_A u_A)$$

Rearranging this equation gives

$$m_B v_B + m_A v_A = m_A u_A + m_B u_B$$

Therefore,

**the total final momentum = the total initial momentum**

Hence the total momentum is unchanged by this collision, that is, it is conserved.

***Note:***

If the colliding objects stick together as a result of the collision, they have the same final velocity. The above equation with $V$ as the final velocity may therefore be written

$$(m_B + m_A)V = m_A u_A + m_B u_B$$

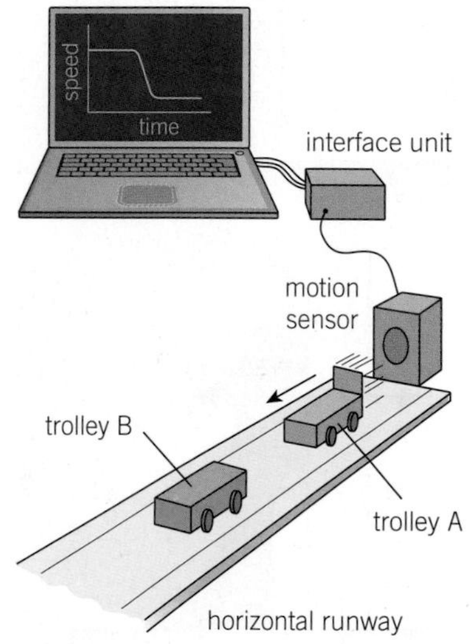


▲ **Figure 3a** *Using a motion sensor*

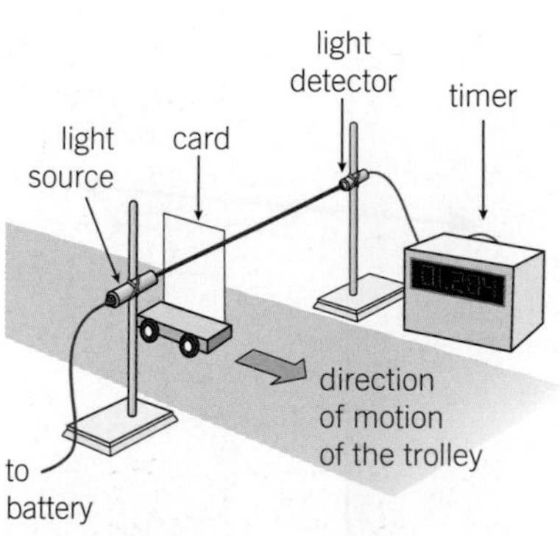


▲ **Figure 3b** *Using light gates*

## Testing conservation of momentum

Figure 3a shows an arrangement that can be used to test **conservation of momentum** using a motion sensor linked to a computer. The mass of each trolley is measured before the test. With trolley B at rest, trolley A is given a push so it moves towards trolley B at constant velocity. The two trolleys stick together on impact. The computer records and displays the velocity of trolley A throughout this time.

The computer display shows that the velocity of trolley A dropped suddenly when the impact took place. The velocity of trolley A immediately before the collision, $u_A$, and after the collision, $V$, can be measured. The measurements should show that the total momentum of both trolleys after the collision is equal to the momentum of trolley A before the collision. In other words,

$$(m_B + m_A)V = m_A u_A$$

## Worked example

A rail wagon C of mass 4500 kg moving along a level track at a speed of 3.0 m s$^{-1}$ collides with and couples to a second rail wagon D of mass 3000 kg which is initially stationary. Calculate the speed of the two wagons immediately after the collision.

### *Solution*

Total initial momentum = initial momentum of C + initial momentum of D

$= (4500 \times 3.0) + (3000 \times 0) = 13\,500\,\text{kg m s}^{-1}$

Total final momentum = total mass of C and D × velocity $v$ after the collision

$= (4500 + 3000)\,v = 7500\,v$

Using the principle of conservation of momentum,

$$7500\,v = 13\,500$$

$$v = \frac{13\,500}{7500} = 1.8\,\text{m s}^{-1}$$

Instead of a motion sensor, we may choose to use light gates linked to a computer or a data logger as shown in Figure 3b. A card attached to trolley A passes through the first of two light gates. Trolley B is positioned so that A collides with it just after passing through the first light gate. The velocity of A before the collision, $u_A$, is calculated by dividing the card length by the time taken to pass through the first gate. The velocity of A after the collision, $V$, is calculated by dividing the card length by the time taken to pass through the second gate.

## Head-on collisions

Consider two objects moving in opposite directions that collide with each other. Depending on the masses and initial velocities of the two objects, the collision could cause them both to stop. The momentum of the two objects after the collision would then be zero. This could only happen if the initial momentum of one object was exactly equal and opposite to that of the other object. In general, if two objects move in opposite directions before a collision, then the vector nature of momentum needs to be taken into account by assigning numerical values of velocity + or – according to the direction.

For example, if a car of mass 600 kg travelling at a velocity of 25 m s$^{-1}$ collides head-on with a lorry of mass 2400 kg travelling at a velocity of 10 m s$^{-1}$ in the opposite direction, the total momentum before the collision is 9000 kg m s$^{-1}$ in the direction the lorry was moving. As momentum is conserved in a collision, the total momentum after the collision is the same as the total momentum before the collision. Prove for yourself that if the two vehicles were to stick together after the collision, they must have had a velocity of 3.0 m s$^{-1}$ in the direction the lorry was moving immediately after the impact.

## Summary questions

1 A rail wagon of mass 3000 kg moving at a velocity of 1.2 m s$^{-1}$ collides with a stationary wagon of mass 2000 kg. After the collision, the two wagons couple together. Calculate their speed immediately after the collision.

2 A rail wagon of mass 5000 kg moving at a velocity of 1.6 m s$^{-1}$ collides with a stationary wagon of mass 3000 kg (see Figure 4). After the collision, the 3000 kg wagon moves away at a velocity of 1.5 m s$^{-1}$. Calculate the speed and direction of the 5000 kg wagon after the collision.

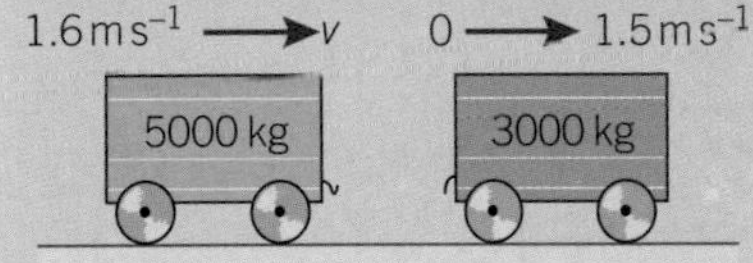


▲ **Figure 4**

3 In a laboratory experiment, a trolley of mass 0.50 kg moving at a speed of 0.25 m s$^{-1}$ collides with a trolley of mass 1.0 kg moving in the opposite direction at a speed of 0.20 m s$^{-1}$. The two trolleys couple together on collision. Calculate their speed and direction immediately after the collision.

4 A ball of mass 0.80 kg moving at a speed of 2.5 m s$^{-1}$ along a straight line collides with a ball of mass 2.5 kg which was initially stationary. As a result of the collision, the 2.5 kg ball has a velocity of 1.0 m s$^{-1}$ along the same line. Calculate the speed and direction of the 0.80 kg ball immediately after the collision.

# 9.4 Elastic and inelastic collisions

## Learning objectives:

- → Distinguish between an elastic collision and an inelastic collision.
- → Describe what is conserved in a perfectly elastic collision.
- → Discuss whether any real collisions are ever perfectly elastic.

*Specification reference: 3.4.1.6*

Drop a bouncy rubber ball from a measured height onto a hard floor. The ball should bounce back almost to the same height. Try the same with a cricket ball and there will be very little bounce! An elastic ball would be one that bounces back to exactly the same height. Its kinetic energy just after impact must equal its **kinetic energy** just before impact. Otherwise, it cannot regain its initial height. There is no loss of kinetic energy in an **elastic collision**.

**An elastic collision is one where there is no loss of kinetic energy.**

A very low speed impact between two cars is almost perfectly elastic, provided no damage is done. However, if the collision causes damage to the vehicles, some of the initial kinetic energy is transferred to the surroundings. This collision may be described as **inelastic**.

**An inelastic collision occurs where the colliding objects have less kinetic energy after the collision than before the collision.**

Objects that collide and couple together undergo inelastic collisions, as some of the initial kinetic energy is transferred to the surroundings.

To work out whether a collision is elastic or inelastic, the kinetic energy of each object before and after the collision must be calculated.

## Examples

1 For a ball of mass $m$ falling in air from a measured height $H$ above the floor and rebounding to a height $h$,

- i the kinetic energy immediately before impact = loss of potential energy through height $H = mgH$
- ii the kinetic energy immediately after impact = gain of potential energy through height $h = mgh$.

So the height ratio $h/H$ gives the fraction of the initial kinetic energy that is recovered as kinetic energy after the collision.

2 For a collision between two objects, the kinetic energy of each object can be worked out using the kinetic energy equation $E_K = \frac{1}{2}mv^2$, where $m$ is the mass of the object and $v$ is its speed (see Topic 10.2). Using this equation, the total initial kinetic energy and the total final kinetic energy can be worked out if the mass, initial speed, and speed after collision of each object is known.

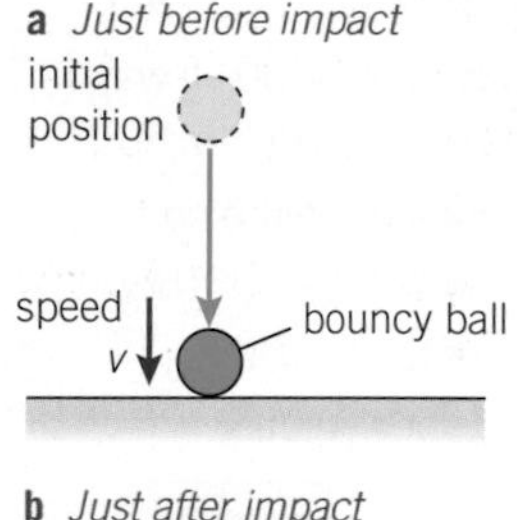


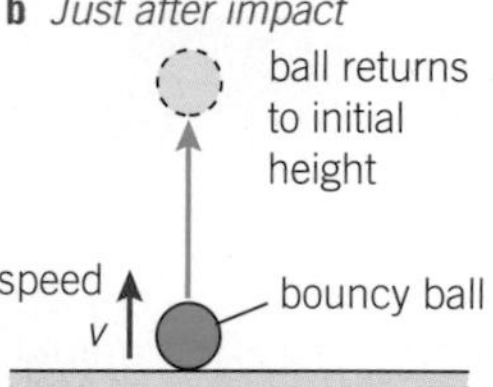


▲ **Figure 1** *An elastic impact*

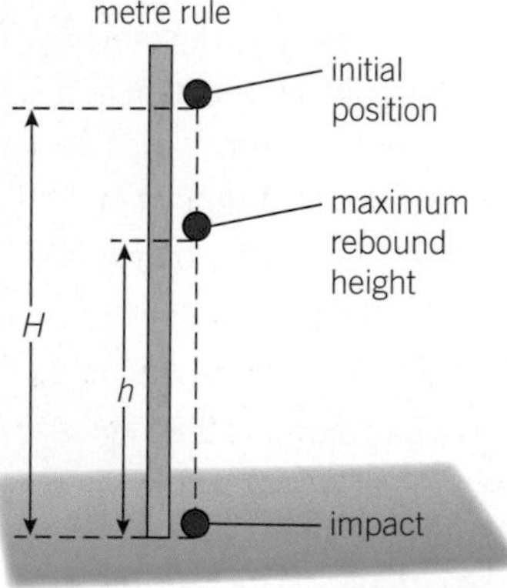


▲ **Figure 2** *Testing an impact*

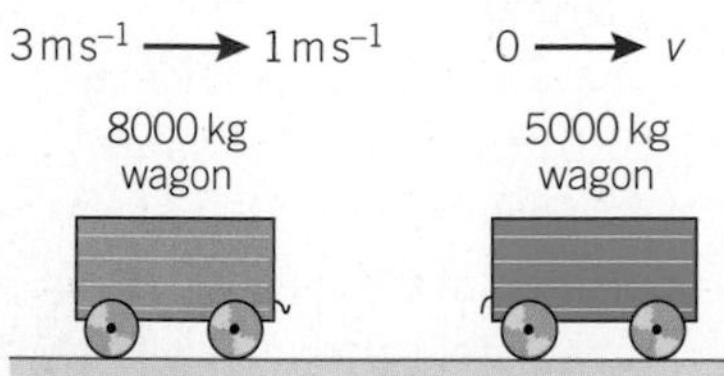


▲ **Figure 3**

## Worked example

A railway wagon of mass 8000 kg moving at 3.0 m s$^{-1}$ collides with an initially stationary wagon of mass 5000 kg. The two wagons separate after the collision. The 8000 kg wagon moves at a speed of 1.0 m s$^{-1}$ without change of direction after the collision. Calculate:

- a the speed and direction of the 5000 kg wagon after the collision
- b the loss of kinetic energy due to the collision.

## Solution

**a** The total initial momentum = $8000 \times 3 = 24\,000\,\text{kg m s}^{-1}$.

The total final momentum = $(8000 \times 1.0) + 5000V$, where $V$ is the speed of the 5000 kg wagon after the collision.

Using the principle of conservation of momentum

$$8000 + 5000V = 24\,000$$

$$5000V = 24\,000 - 8000 = 16\,000$$

$$V = \frac{16\,000}{5000} = 3.2\,\text{m s}^{-1}$$

**b** Kinetic energy of the 8000 kg wagon before the collision

$$= \frac{1}{2} \times 8000 \times 3.0^2 = 36\,000\,\text{J}$$

Kinetic energy of the 8000 kg wagon after the collision

$$= \frac{1}{2} \times 8000 \times 1.0^2 = 4000\,\text{J}$$

Kinetic energy of the 5000 kg wagon after the collision

$$= \frac{1}{2} \times 5000 \times 3.2^2 = 25\,600\,\text{J}$$

∴ loss of kinetic energy due to the collision

$$= 36\,000 - (4000 + 25\,600)$$
$$= 6400\,\text{J}$$

### Study tip

Momentum is always conserved in collisions provided no external forces act. Total energy is always conserved, but kinetic energy may be transferred into other energy stores.

### Synoptic link

Excitation by collision in a gas occurs when the gas molecules undergo collisions and are excited to higher energy states. You have met particle collisions in Topic 3.3, Collisions of electrons with atoms. You will also learn more about kinetic energy in Topic 10.2, Kinetic energy and potential energy.

## Summary questions

**1** **a** A squash ball is released from rest above a flat surface. Describe how its energy changes if:

- **i** it rebounds to the same height
- **ii** it rebounds to a lesser height.

**b** In **a ii**, the ball is released from a height of 1.20 m above the surface and it rebounds to a height of 0.90 m above the surface. Show that 25% of its kinetic energy is lost in the impact.

**2** A vehicle of mass 800 kg moving at a speed of $15.0\,\text{m s}^{-1}$ collided with a vehicle of mass 1200 kg moving in the same direction at a speed of $5.0\,\text{m s}^{-1}$. The two vehicles locked together on impact. Calculate:

**a** the velocity of the two vehicles immediately after impact

**b** the loss of kinetic energy due to the impact.

**3** An ice puck of mass 1.5 kg moving at a speed of $4.2\,\text{m s}^{-1}$ collides head-on with a second ice puck of mass 1.0 kg moving in the opposite direction at a speed of $4.0\,\text{m s}^{-1}$. After the impact, the 1.5 kg ice puck continues in the same direction at a speed of $0.80\,\text{m s}^{-1}$. Calculate:

**a** the speed and direction of the 1.0 kg ice puck after the collision

**b** the loss of kinetic energy due to the collision.

**4** The bumper cars at a fairground are designed to withstand low-speed impacts without damage. A bumper car of mass 250 kg moving at a velocity of $0.90\,\text{m s}^{-1}$ collides elastically with a stationary car of mass 200 kg. Immediately after the impact, the 250 kg car has a velocity of $0.10\,\text{m s}^{-1}$ in the same direction as it was initially moving in.

**a** **i** Calculate the velocity of the 200 kg car immediately after the impact.

**ii** Show that the collision was an elastic collision.

**b** Without further calculations, discuss the effect of the impact on the driver of each car.

# 9.5 Explosions

**Learning objectives:**

- → Describe the energy changes that take place in an explosion.
- → State what can always be said about the total momentum of a system that has exploded.
- → Describe the consequences when, after the explosion, only two bodies move apart.

*Specification reference: 3.4.1.6*

▲ **Figure 1** *The gun barrel recoils when the shell is fired. Large springs fitted to the barrel take away and store the kinetic energy of the barrel as it recoils*

When two objects fly apart after being initially at rest, they recoil from each other with equal and opposite amounts of momentum. So they move away from each other in opposite directions. Consider Figure 2, where two spring-loaded trolleys of mass $m_A$ and $m_B$ respectively are initially positioned at rest and in contact. These trolleys move apart at speeds $v_A$ and $v_B$ respectively when the trigger is tapped to release the spring in trolley A.

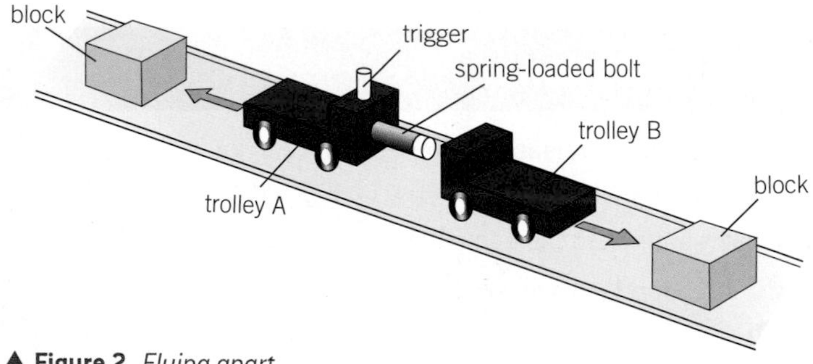


▲ **Figure 2** *Flying apart*

The total initial momentum = 0

The total momentum immediately after the explosion

$$= \text{momentum of A} + \text{momentum of B}$$

$$= m_A v_A + m_B v_B$$

Using the principle of the conservation of momentum, $m_A v_A + m_B v_B = 0$

$$\therefore m_B v_B = -m_A v_A$$

The minus sign means that the two masses move apart from each other in opposite directions.

For example, if $m_A = 1.0\,\text{kg}$, $v_A = 2.0\,\text{m}\,\text{s}^{-1}$ and $m_B = 0.5\,\text{kg}$,

$$\text{then } v_B = -\frac{m_A v_A}{m_B} = -4.0\,\text{m}\,\text{s}^{-1}$$

So A and B move apart at speeds of $2.0\,\text{m}\,\text{s}^{-1}$ and $4.0\,\text{m}\,\text{s}^{-1}$ in opposite directions.

**Synoptic link**

The $\alpha$ particles from a given isotope are always emitted with the same kinetic energy because they are emitted with the same speed, provided no external forces act. This is because each $\alpha$ particle and the nucleus that emits it move apart with equal and opposite amounts of momentum. This isn't the case with $\beta$ particles because a neutrino or antineutrino is emitted as well. You have met these ideas in more detail in Topic 1.2, Stable and unstable nuclei.

## Testing a model explosion

In Figure 2, when the spring is released from one of the trolleys, the two trolleys, A and B, push each other apart. The blocks are positioned so that the trolleys hit the blocks at the same moment. The distance travelled by each trolley to the point of impact with the block is equal to its speed × the time taken to travel that distance. As the time taken is the same for the two trolleys, the distance ratio is the same as the speed ratio. Because the trolleys have equal (and opposite) amounts of momentum, the ratio of their speeds is the inverse of the mass ratio. The distance ratio should therefore be equal to the inverse of the mass ratio. In other words, if trolley A travels twice as far as trolley B, then the mass of A must be half the mass of B (so they carry away equal amounts of momentum).

### *Note:*

In this experiment, the kinetic energy of the two trolleys immediately after they separate from each other is equal to the energy stored in the spring when it was originally compressed. For two or more objects that fly apart due to an explosion, their total kinetic energy immediately after the explosion is less than the total chemical energy released in the explosion because heat, light, and sound all carry away energy.

## Summary questions

1. A shell of mass 2.0 kg is fired at a speed of $140\ \text{m s}^{-1}$ from an artillery gun of mass 800 kg. Calculate the recoil velocity of the gun.

2. In a laboratory experiment to measure the mass of an object X, two identical trolleys A and B, each of mass 0.50 kg, were initially stationary on a track. Object X was fixed to trolley A. When a trigger was pressed, the two trolleys moved apart in opposite directions at speeds of $0.30\ \text{m s}^{-1}$ and $0.25\ \text{m s}^{-1}$.

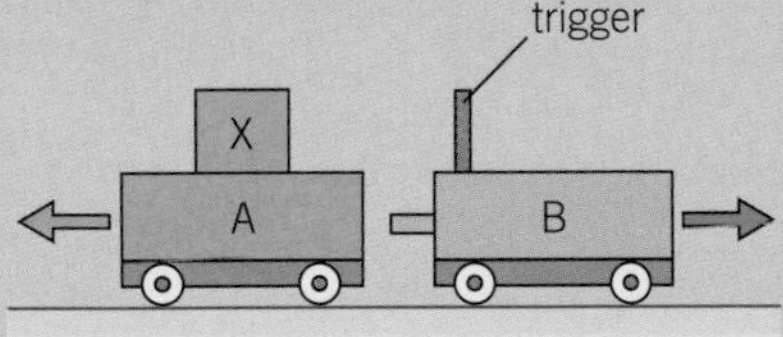


▲ Figure 3

   **a** Which of the two speeds given above was the speed of trolley A? Give a reason for your answer.

   **b** Show that the mass of X must have been 0.10 kg.

3. Two trolleys, X of mass 1.20 kg and Y of mass 0.80 kg, are initially stationary on a level track.

   **a** When a trigger is pressed on one of the trolleys, a spring pushes the two trolleys apart. Trolley Y moves away at a velocity of $0.15\ \text{m s}^{-1}$.

      **i** Calculate the velocity of X.

      **ii** Calculate the total kinetic energy of the two trolleys immediately after the explosion.

   **b** In part **a**, if the test had been carried out with trolley X held firmly, calculate the speed at which Y would have recoiled, assuming the energy stored in the spring before release is equal to the total kinetic energy calculated in **a ii**.

4. A person in a stationary boat of total mass 150 kg throws a rock of mass 2.0 kg out of the boat. As a result, the boat recoils at a speed of $0.12\ \text{m s}^{-1}$. Calculate

   **a** the speed at which the rock was thrown from the boat

   **b** the kinetic energy gained by

      **i** the boat

      **ii** the rock.

# Practice questions: Chapter 9

1 (a) Collisions can be described as *elastic* or *inelastic*.
State what is meant by an inelastic collision. *(1 mark)*

(b) A ball of mass 0.12 kg strikes a stationary cricket bat with a speed of $18\,m\,s^{-1}$. The ball is in contact with the bat for 0.14 s and returns along its original path with a speed of $15\,m\,s^{-1}$. Calculate:

(i) the momentum of the ball before the collision
(ii) the momentum of the ball after the collision
(iii) the total change of momentum of the ball
(iv) the average force acting on the ball during contact with the bat
(v) the kinetic energy lost by the ball as a result of the collision. *(6 marks)*

AQA, 2001

2 Two carts A and B, with a compressed spring between them, are pushed together and held at rest, as shown in **Figure 1**. The spring is not attached to either cart. The carts are then released.

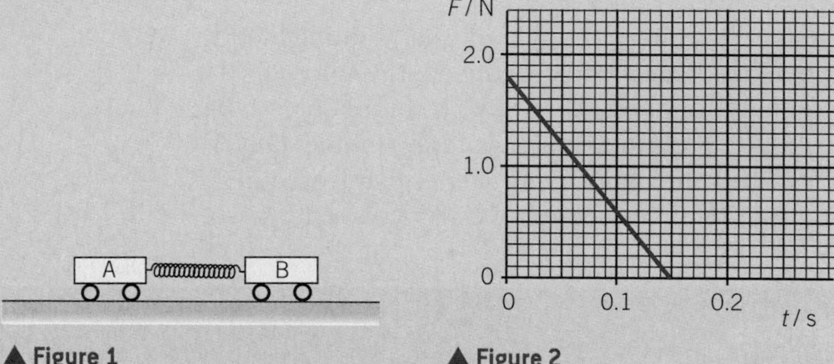


▲ Figure 1 ▲ Figure 2

**Figure 2** shows how the force, $F$, exerted by the spring on the carts varies with time, $t$, after release.
When the spring returns to its unstretched length and drops away, cart A is moving at $0.60\,m\,s^{-1}$.

(a) Calculate the impulse given to each cart by the spring as it expands.
(b) Calculate the mass of cart A.
(c) State the final total momentum of the system at the instant the spring drops away. *(5 marks)*

AQA, 2004

3 A railway engine is about to couple with a stationary carriage of mass $4.0 \times 10^4$ kg. When they have joined up, the engine and the carriage move at a constant speed. The engine has a mass of $6.2 \times 10^4$ kg and is moving at $0.35\,m\,s^{-1}$ just before coupling.

(a) Calculate the momentum of the engine.
(b) Calculate the speed of the engine and carriage after coupling. *(5 marks)*

AQA, 2007

4 (a) State two quantities that are conserved in an elastic collision. *(2 marks)*

(b) A gas molecule makes an elastic collision with the walls of a gas cylinder. The molecule is travelling at $450\,m\,s^{-1}$ at right angles towards the wall before the collision.

(i) What is the magnitude and direction of its velocity after the collision?
(ii) Calculate the change in momentum of the molecule during the collision if it has a mass of $8.0 \times 10^{-26}$ kg. *(4 marks)*

(c) Use Newton's laws of motion to explain how the molecules of a gas exert a force on the wall of a container. *(4 marks)*

AQA, 2006

5 **(a)** When an α particle is emitted from a nucleus of the polonium isotope $^{210}_{84}Po$, a nucleus of lead (Pb) is formed. Complete the equation below:

$$^{210}_{84}Po \rightarrow \alpha + Pb$$ *(2 marks)*

**(b)** √x The α particle in part (a) is emitted at a speed of $1.6 \times 10^7\,m\,s^{-1}$.

(i) The mass of the α particle is 4.0 u (where $u$ = atomic mass unit). Calculate the kinetic energy, in MeV, of the α particle immediately after it has been emitted. Ignore relativistic effects.

(ii) Calculate the speed of recoil of the daughter nucleus immediately after the α particle has been emitted. Assume the parent nucleus is initially at rest. *(6 marks)*

AQA, 2006

6 **Figure 3** shows how the force, $F$, on a steel ball varies with time, $t$, when the ball is dropped onto a thick steel plate and rebounds. The kinetic energy of the ball after the collision is the same as it was before the collision.

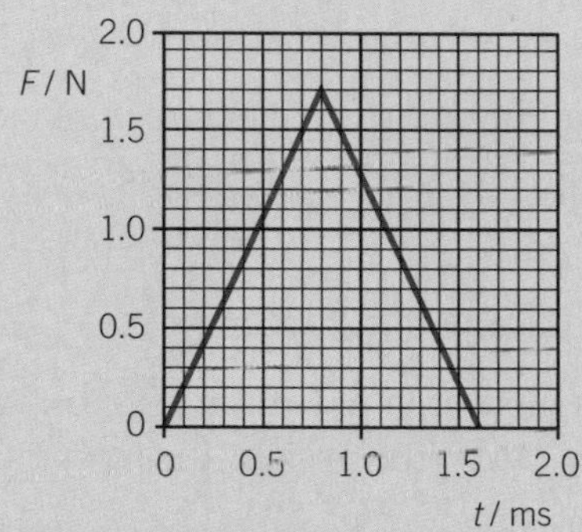


▲ **Figure 3**

**(a)** State the name of the quantity that is obtained by determining the shaded area.

**(b)** √x Use the graph **Figure 3** to determine the initial momentum of the ball.

**(c)** √x Sketch a graph to show how the momentum of the ball varies between times $t = 0$ and $t = 2.0\,ms$. *(6 marks)*

AQA, 2006

7 **(a)** Explain what is meant by the principle of conservation of momentum. *(2 marks)*

**(b)** √x A hose pipe is used to water a garden. The supply delivers water at a rate of $0.31\,kg\,s^{-1}$ to the nozzle which has a cross-sectional area of $7.3 \times 10^{-5}\,m^2$.

(i) Show that water leaves the nozzle at a speed of about $4\,m\,s^{-1}$.
density of water = $1000\,kg\,m^{-3}$

(ii) Before it leaves the hose, the water has a speed of $0.68\,m\,s^{-1}$. Calculate the force on the hose.

(iii) The water from the hose is sprayed onto a brick wall the base of which is firmly embedded in the ground. Explain why there is no overall effect on the rotation of the Earth. *(7 marks)*

AQA, 2005

# 10 Work, energy, and power

## 10.1 Work and energy

**Learning objectives:**

- → Define energy and describe how we measure it.
- → Discuss whether energy ever disappears.
- → Define work (in the scientific sense).

*Specification reference: 3.4.1.7*

### Energy rules

**Energy** is needed to make stationary objects move, to change their shape, or to warm them up. When you lift an object, you transfer energy from your muscles to the object.

Objects can possess energy in different types of stores, including:

- gravitational potential stores (the position of objects in a gravitational field)
- kinetic stores (moving objects)
- thermal stores (hot objects)
- elastic stores (objects compressed or stretched).

Energy can be transferred between objects in different ways, including:

- by radiation (e.g., light)
- electrically
- mechanically (e.g., by sound).

Energy is measured in **joules (J)**. One joule is equal to the energy needed to raise a 1 N weight through a vertical height of 1 m.

Whenever energy is transferred, the total amount of energy after the transfer is always equal to the total amount of energy before the transfer. The total amount of energy is unchanged.

**Energy cannot be created or destroyed.**

This statement is known as the **principle of conservation of energy.**

### Forces at work

**Work** is done on an object when a force acting on it makes it move. As a result, energy is transferred to the object. The amount of work done depends on the force and the distance the object moved. The greater the force or the further the distance, the greater the work done.

**Work done = force × distance moved in the direction of the force.**

The unit of work is the joule (J), equal to the work done when a force of 1 N moves its point of application by a distance of 1 m in the direction of the force. For example, as shown in Figure 1:

- A force of 1 N is required to raise an object of weight 1 N steadily. If it is raised by 1 m, the work done by the force is 1 J (= 1 N × 1 m). The gain potential energy of the raised object is 1 J.
- For a 2 N object raised to a height of 1 m, the work done, and hence potential energy of the raised object, is 2 J (= 2 N × 1 m).

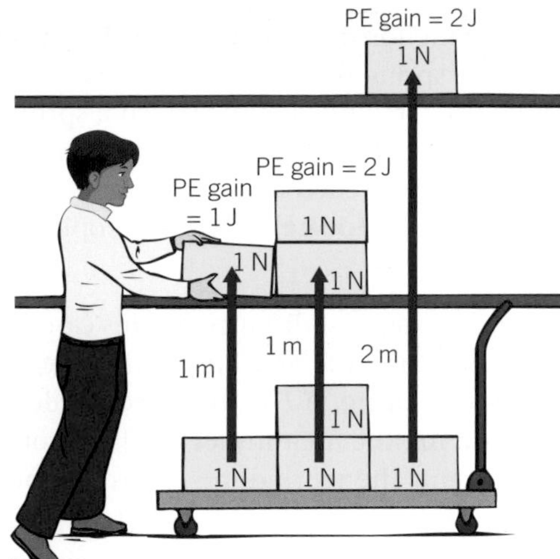


▲ **Figure 1** *Using joules*

## Force and displacement

Imagine a yacht acted on by a wind force $F$ at an angle $\theta$ to the direction in which the yacht moves. The wind force has a component $F\cos\theta$ in the direction of motion of the yacht and a component $F\sin\theta$ at right angles to the direction of motion. If the yacht is moved a distance $s$ by the wind, the work done on it, $W$, is equal to the component of force in the direction of motion × the distance moved.

$$W = Fs\cos\theta$$

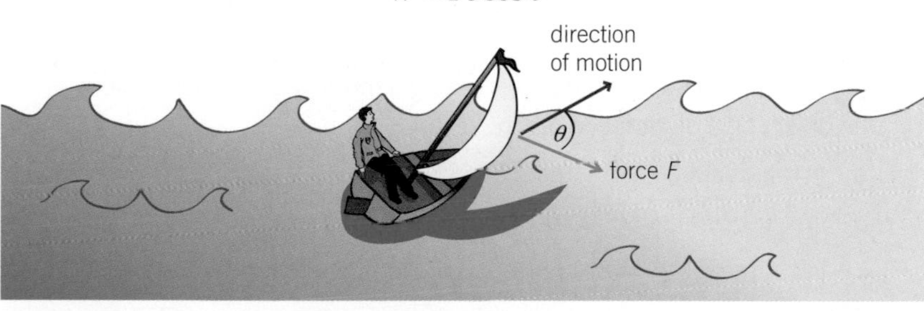


▲ **Figure 2** *Force and displacement*

Note that if $\theta = 90°$ (which means that the force is perpendicular to the direction of motion) then, because $\cos 90° = 0$, the work done is zero.

### Hint

No work is done when $F$ and $s$ are at right angles to each other.

## Force–distance graphs

- If a constant force $F$ acts on an object and makes it move a distance $s$ in the direction of the force, the work done on the object $W = Fs$. Figure 3 shows a graph of force against distance in this situation. The area under the line is a rectangle of height representing the force and of base length representing the distance moved. **Therefore the area under the line represents the work done.**
- If a variable force $F$ acts on an object and causes it to move in the direction of the force, the work done for a small amount of distance $\Delta s$, $\Delta W = F\Delta s$. This is represented on a graph of the force $F$ against distance $s$ by the area of a strip under the line of width $\Delta s$ and height $F$ (Figure 4). The total work done is therefore the sum of the areas of all the strips (i.e., the total area under the line).

**The area under the line of a force–distance graph represents the total work done.**

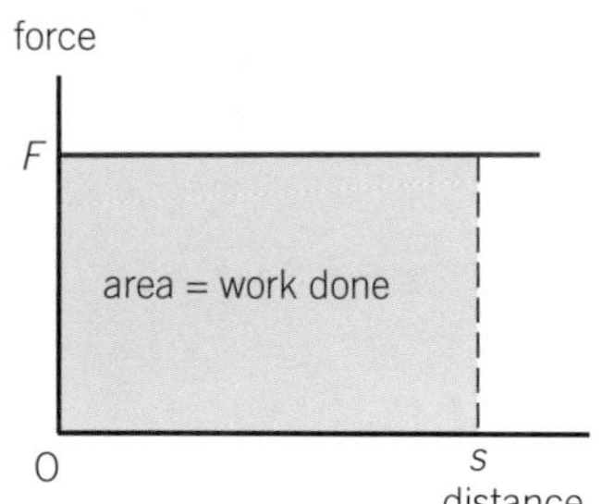


▲ **Figure 3** *Force–distance graph for a constant force*

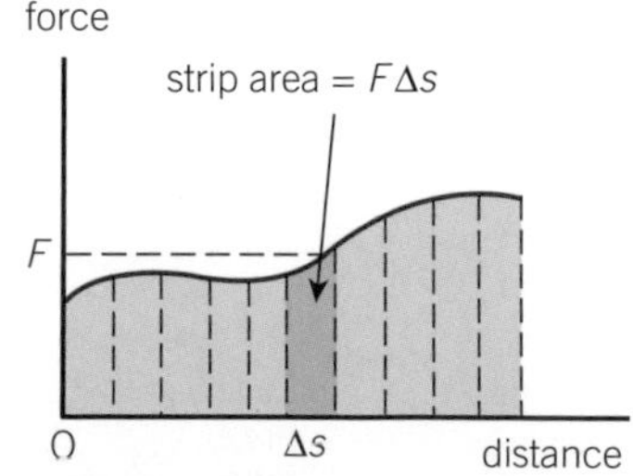


▲ **Figure 4** *Force–distance graph for a variable force*

For example, consider the force needed to stretch a spring. The greater the force, the more the spring is extended from its unstretched length. Figure 5 shows how the force needed to stretch a spring changes with the extension of the spring. The graph is a straight line through the origin. Therefore, the force needed is proportional to the extension of the spring. This is known as **Hooke's law**. See Topic 11.2 for more about springs.

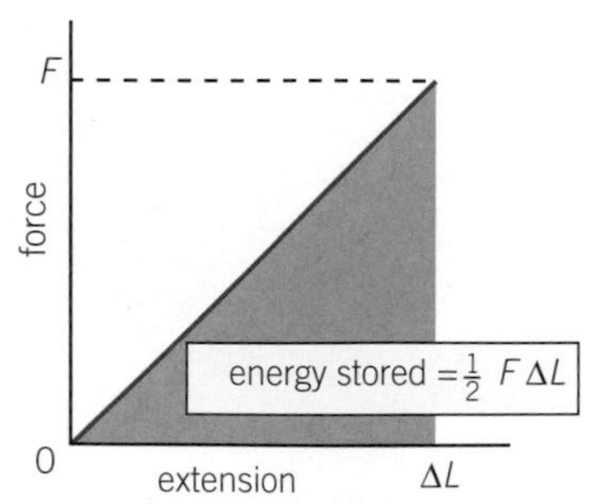


▲ **Figure 5** *Force against extension for a spring*

Figure 5 is a graph of force against distance, in this case the distance the spring is extended. Therefore, the area under the line represents the work done to stretch the spring. If $F$ is the force needed to extend the spring to extension $\Delta L$, the area under the line from the origin to extension $\Delta L$ represents the work done to stretch the spring to extension $\Delta L$. As this area is a triangle, the work done = $\frac{1}{2}$ × height × base = $\frac{1}{2} F \Delta L$.

**To stretch the spring to extension $\Delta L$,** work done = $\frac{1}{2} F \Delta L$

## Summary questions

$g = 9.8\,\text{m}\,\text{s}^{-2}$

1 Calculate the work done when:

   a a weight of 40 N is raised by a height of 5.0 m

   b a spring is stretched to an extension of 0.45 m by a force that increases to 20 N.

2 Calculate the energy transferred by a force of 12 N when it moves an object by a distance of 4.0 m:

   a in the direction of the force

   b in a direction at 60° to the direction of the force

   c in a direction at right angles to the direction of the force.

3 A luggage trolley of total weight 400 N is pushed at steady speed 20 m up a slope, by a force of 50 N acting in the same direction as the object moves in. At the end of this distance, the trolley is 1.5 m higher than at the start. Calculate:

   a the work done pushing the trolley up the slope

   b the gain of potential energy of the trolley

   c the energy wasted due to friction.

4 A spring that obeys Hooke's law requires a force of 1.2 N to extend it to an extension of 50 mm. Calculate:

   a the force needed to extend it to an extension of 100 mm

   b the work done when the spring is stretched to an extension of 100 mm from zero extension.

# 10.2 Kinetic energy and potential energy

## Learning objectives:

- → Describe what happens to the work done on an object when it is lifted.
- → Describe what energy change takes place when an object is allowed to fall.
- → Describe the effect on the kinetic energy of a car if its speed is doubled.

*Specification reference: 3.4.1.8*

## Kinetic energy

**Kinetic energy** is the energy of an object due to its motion. The faster an object moves, the more kinetic energy it has. To see the exact link between kinetic energy and speed, consider an object of mass $m$, initially at rest, acted on by a constant force $F$ for a time $t$.

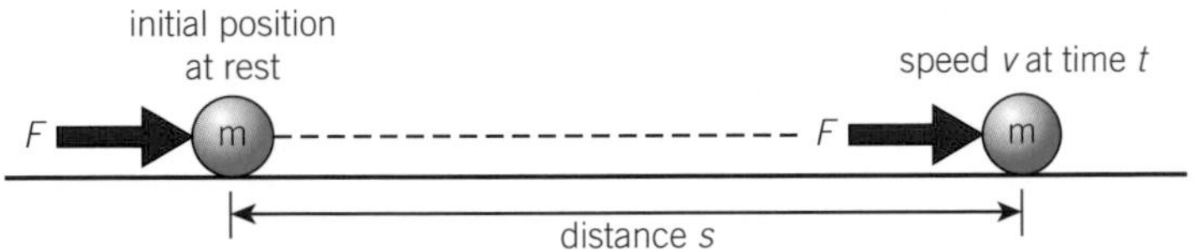


▲ **Figure 1** *Gaining kinetic energy*

Let the speed of the object at time $t$ be $v$.

Therefore, distance travelled, $s = \frac{1}{2}(u + v)t = \frac{1}{2}vt$ because $u = 0$

acceleration $a = \frac{(v - u)}{t} = \frac{v}{t}$

Using Newton's second law, $F = ma = \frac{mv}{t}$

Therefore, the work done $W$ by force $F$ to move the object through distance $s$,

$$W = Fs = \frac{mv}{t} \times \frac{vt}{2} = \frac{1}{2}mv^2$$

Because the gain of kinetic energy is due to the work done, then

$$\textbf{kinetic energy, } E_K = \frac{1}{2}mv^2$$

### Hint

The equation for kinetic energy does not hold at speeds approaching the speed of light. Einstein's theory of special relativity tells us that the mass of an object increases with speed and that the energy $E$ of an object can be worked out from the equation $E = mc^2$, where $c$ is the speed of light in free space and $m$ is the mass of the object.

## Potential energy

**Potential energy** is the energy of an object due to its position.

If an object of mass $m$ is raised through a vertical height $\Delta h$ at steady speed, the force needed to raise it is equal and opposite to its weight $mg$. Therefore,

**the work done to raise the object = force × distance moved**
$$= mg\Delta h$$

The work done on the object increases its gravitational potential energy.

**Change of gravitational potential energy** $\Delta E_P = mg\Delta h$

At the Earth's surface, $g = 9.8\,\mathrm{m\,s^{-2}}$

### Hint

The equation only holds if the change of height $h$ is much smaller than the Earth's radius. If height $h$ is not insignificant compared with the Earth's radius, the value of $g$ is not the same over height $h$. The force of gravity on an object decreases with increased distance from the Earth.

## Energy changes involving kinetic and potential energy

Consider an object of mass $m$ released above the ground. If air resistance is negligible, the object gains speed as it falls. Its potential energy therefore decreases and its kinetic energy increases.

After falling through a vertical height $\Delta h$, its kinetic energy is equal to its loss of potential energy:

$$\frac{1}{2}mv^2 = mg\Delta h$$

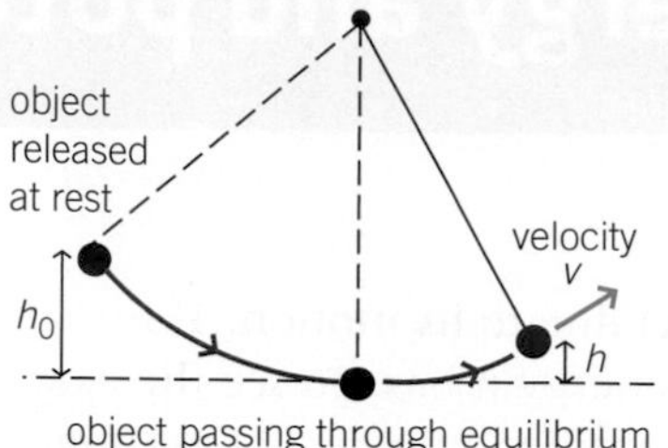


▲ **Figure 2** *A pendulum in motion*

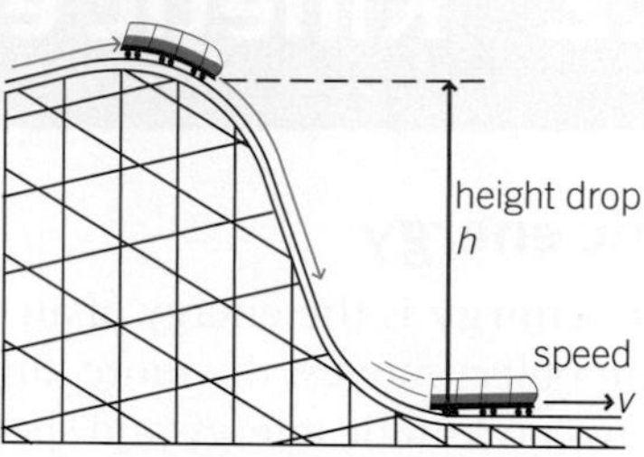


▲ **Figure 3** *At the fairground*

## A pendulum bob

A pendulum bob is displaced from its equilibrium position and then released with the thread taut. The bob passes through the equilibrium position at maximum speed then slows down to reach maximum height on the other side of the equilibrium position. If its initial height above equilibrium position $= h_0$, then whenever its height above the equilibrium position $= h$, its speed $v$ at this height is such that

kinetic energy = loss of potential energy from maximum height

$$\frac{1}{2}mv^2 = mg\,(h_0 - h)$$

## A fairground vehicle of mass *m* on a downward track

If a fairground vehicle was initially at rest at the top of the track and its speed is $v$ at the bottom of the track, then at the bottom of the track

- its kinetic energy $= \frac{1}{2}mv^2$
- its loss of potential energy $= mgh$, where $h$ is the vertical distance between the top and the bottom of the track
- the work done to overcome friction and air resistance $= mgh - \frac{1}{2}mv^2$

### Worked example

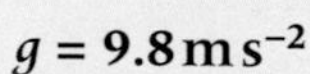

$g = 9.8\,\text{m}\,\text{s}^{-2}$

On a fairground ride, the track descends by a vertical drop of 55 m over a distance of 120 m along the track. A train of mass 2500 kg on the track reaches a speed of $30\,\text{m}\,\text{s}^{-1}$ at the bottom of the descent after being at rest at the top. Calculate **a** the loss of potential energy of the train, **b** its gain of kinetic energy, **c** the average frictional force on the train during the descent.

### *Solution*

**a** Loss of potential energy $= mg\Delta h = 2500 \times 9.8 \times 55 = 1.35 \times 10^6\,\text{J}$

**b** Its gain of kinetic energy $= \frac{1}{2}mv^2 = 0.5 \times 2500 \times 30^2 = 1.13 \times 10^6\,\text{J}$

**c** Work done to overcome friction $= mg\Delta h - \frac{1}{2}mv^2 = 1.35 \times 10^6 - 1.13 \times 10^6 = 2.2 \times 10^5\,\text{J}$

Because the work done to overcome friction = frictional force × distance moved along track,

$$\text{the frictional force} = \frac{\text{work done to overcome friction}}{\text{distance moved}} = \frac{2.2 \times 10^5}{120} = 1830\,\text{N}$$

## Summary questions

$g = 9.8\,\text{m}\,\text{s}^{-2}$

**1** A ball of mass 0.50 kg was thrown directly up at a speed of $6.0\,\text{m}\,\text{s}^{-1}$. Calculate

- **a** its kinetic energy at $6\,\text{m}\,\text{s}^{-1}$
- **b** its maximum gain of potential energy
- **c** its maximum height gain.

**2** A cyclist of mass 80 kg (including the bicycle) freewheels from rest 500 m down a hill. The foot of the hill is 20 m lower than the cyclist's starting point and the cyclist reaches a speed of $12\,\text{m}\,\text{s}^{-1}$ at the foot of the hill. Calculate:

- **a** the loss of potential energy
- **b** the gain of kinetic energy of the cyclist and cycle
- **c** the work done against friction and air resistance during the descent
- **d** the average resistive force during the descent.

**3** A fairground vehicle of total mass 1200 kg moving at a speed of $2\,\text{m}\,\text{s}^{-1}$ descends through a height of 50 m to reach a speed of $28\,\text{m}\,\text{s}^{-1}$ after travelling a distance of 75 m along the track. Calculate:

- **a** its loss of potential energy
- **b** its initial kinetic energy
- **c** its kinetic energy after the descent
- **d** the work done against friction
- **e** the average frictional force on it during the descent.

## Power and energy

Energy can be transferred from one object to another by means of

- **work done** by a force due to one object making the other object move
- **heat transfer** from a hot object to a cold object. Heat transfer can be due to conduction or convection or radiation.

In addition, electricity, sound waves and **electromagnetic radiation** such as light or radio waves transfer energy.

In any energy transfer process, the more energy transferred per second, the greater the power of the transfer process. For example, in a tall building where there are two elevators of the same total weight, the more powerful elevator is the one that can reach the top floor fastest. In other words, its motor transfers energy from electricity at a faster rate than the motor of the other elevator. The energy transferred per second is the **power** of the motor.

**Power is defined as the rate of transfer of energy.**

The unit of power is the watt (W), equal to an energy transfer rate of 1 joule per second. Note that 1 kilowatt (kW) = 1000 W, and 1 megawatt (MW) = $10^6$ W.

If energy $\Delta E$ is transferred steadily in time $\Delta t$,

$$\textbf{power } P = \frac{\Delta E}{\Delta t}$$

Where energy is transferred by a force doing work, the energy transferred is equal to the work done by the force. Therefore, the rate of transfer of energy is equal to the work done per second. In other words, if the force does work $\Delta W$ in time $\Delta t$,

$$\textbf{power } P = \frac{\Delta W}{\Delta t}$$

## Learning objectives:

- → State which physical quantities are involved in power.
- → Explain how you could develop more power as you go up a flight of stairs.
- → Explain why a 100 W light bulb is more powerful than a 40 W light bulb when each works at the same mains voltage.

*Specification reference: 3.4.1.7*

## Power measurements

### Muscle power

Test your own muscle power by timing how long it takes you to walk up a flight of steps. To calculate your muscle power, you will also need to know your weight and the total height gain of the flight of steps.

- Your gain of potential energy = your weight × total height gain.
- Your muscle power = $\frac{\text{energy transferred}}{\text{time taken}} = \frac{\text{weight} \times \text{height gain}}{\text{time taken}}$.

For example, a person of weight 480 N who climbs a flight of stairs of height 10 m in 12 s has leg muscles of power 400 W (= 480 N × 10 m/12 s). Each leg would therefore have muscles of power 200 W.

### Electrical power

The power of a 12 V light bulb can be measured using a joulemeter, as shown in Figure 2. The joulemeter is read before and after the light bulb is switched on. The difference between the readings is the energy supplied to the light bulb. If the light bulb is switched on for a measured time, the power of the light bulb can be calculated from the energy supplied to it ÷ the time taken.

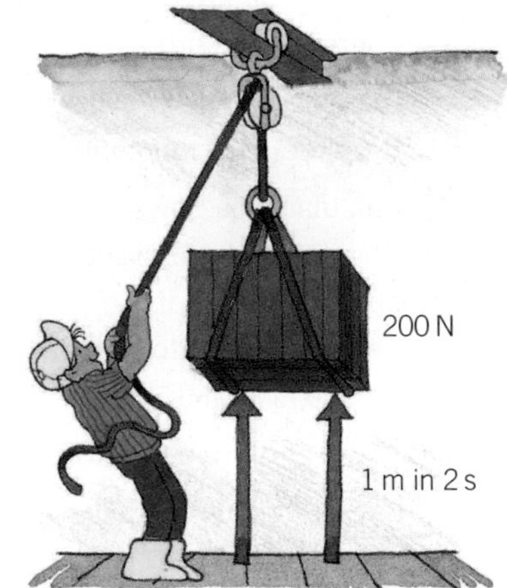

▲ **Figure 1** *A 100 watt worker*

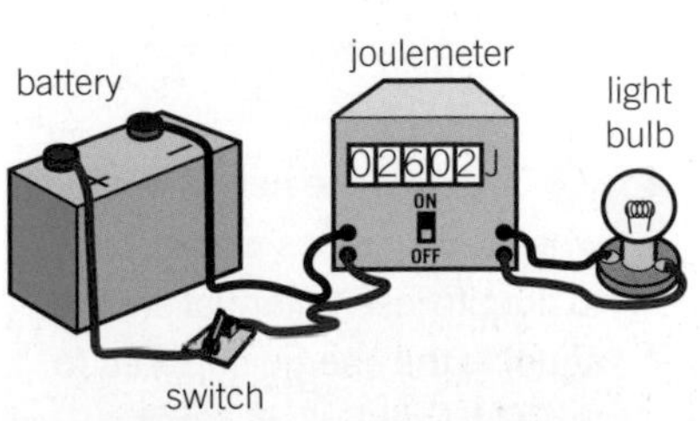

▲ **Figure 2** *Using a joulemeter*

## Engine power

Vehicle engines, marine engines, and aircraft engines are all designed to make objects move. The output power of an engine is sometimes called its motive power.

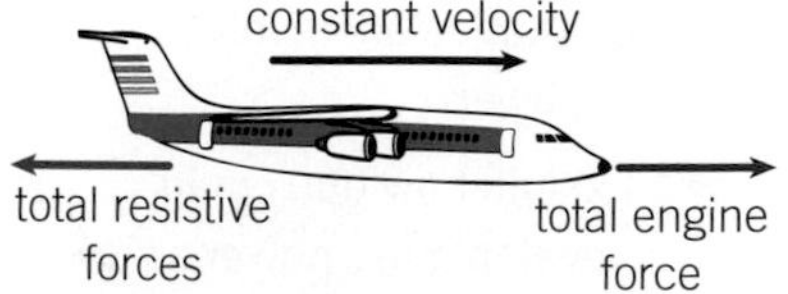


▲ **Figure 3** *Engine power*

**When a powered object moves at constant velocity at constant height**, the resistive forces (e.g., friction, air resistance, drag) are equal and opposite to the motive force.

The work done by the engine is transferred into the internal energy of the surroundings by the resistive forces.

For a powered vehicle driven by a constant force $F$ moving at speed $v$,

**the work done per second = force × distance moved per second**

Therefore, the output power of the engine $\boldsymbol{P = Fv}$.

### Worked example

An aircraft powered by engines that exert a force of 40 kN is in level flight at a constant velocity of $80\,\mathrm{m\,s^{-1}}$. Calculate the output power of the engine at this speed.

#### *Solution*

$F = 40\,\mathrm{kN} = 40\,000\,\mathrm{N}$

Power = force × velocity = $40\,000 \times 80 = 3.2 \times 10^6\,\mathrm{W}$

**When a powered object gains speed, the output force exceeds the resistive forces on it.**

Consider a vehicle that speeds up on a level road. The output power of its engine is the work done by the engine per second. The work done by the engine increases the kinetic energy of the vehicle and enables the vehicle to overcome the resistive forces acting on it. Because the resistive forces increase the internal energy of the surroundings,

**the motive power = energy per second wasted due to the resistive force + the gain of kinetic energy per second**

### Juggernaut physics

The maximum weight of a truck with six or more axles on UK roads must not exceed 44 tonnes, which corresponds to a total mass of 44 000 kg. This limit is set so as to prevent damage to roads and bridges. European Union regulations limit the output power of a large truck to a maximum of 6 kW per tonne. Therefore, the maximum output power of a 44 tonne truck is 264 kW. Prove for yourself that a truck with an output power of 264 kW moving at a constant speed of $31\,\mathrm{m\,s^{-1}}$ (= 70 miles per hour) along a level road experiences a drag force of 8.5 kN.

## Summary questions

$g = 9.8\,\mathrm{m\,s^{-2}}$

1. A student of weight 450 N climbed 2.5 m up a rope in 18 s. Calculate:
   a. the gain of potential energy of the student
   b. the useful energy transferred per second.
2. Calculate the power of the engines of an aircraft at a speed of $250\,\mathrm{m\,s^{-1}}$ if the total engine thrust to maintain this speed is 2.0 MN.
3. A rocket of mass 5800 kg accelerates vertically from rest to a speed of $220\,\mathrm{m\,s^{-1}}$ in 25 s. Calculate:
   a. its gain of potential energy
   b. its gain of kinetic energy
   c. the power output of its engine, assuming no energy is wasted due to air resistance.
4. Calculate the height through which a 5 kg mass would need to drop to lose potential energy equal to the energy supplied to a 100 W light bulb in 1 min.

# 10.4 Energy and efficiency

## Machines at work

A machine that lifts or moves an object applies a force to the object to move it. If the machine exerts a force $F$ on an object to make it move through a distance s in the direction of the force, the work done $W$ on the object by the machine can be calculated using the equation

**work done, $W = Fs$**

If the object moves at a constant velocity $v$ due to this force being opposed by an equal and opposite force caused by friction, the object moves a distance $s = vt$ in time $t$.

Therefore, the output power of the machine

$$P_{OUT} = \frac{\text{work done by the machine}}{\text{time taken}} = \frac{Fvt}{t} = Fv$$

**output power, $P_{OUT} = Fv$**

where $F$ = output force of the machine and $v$ = speed of the object.

### Examples

1 An electric motor operating a sliding door exerts a force of 125 N on the door, causing it to open at a constant speed of $0.40\,m\,s^{-1}$. The output power of the motor is $125\,N \times 0.40\,m\,s^{-1} = 50\,W$. The motor must therefore transfer 50 J every second to the sliding door while the door is being opened.

Friction in the motor bearings and also electrical resistance of the motor wires means that some of the electrical energy supplied to the motor is wasted. For example, if the motor is supplied with electrical energy at a rate of $150\,J\,s^{-1}$ and it transfers $50\,J\,s^{-1}$ to the door, the difference of $100\,J\,s^{-1}$ is wasted as a result of friction and electrical resistance in the motor.

2 A pulley system is used to raise a **load** of 80 N at a speed of $0.15\,m\,s^{-1}$ by means of a constant **effort** of 30 N applied to the system. Figure 1 shows the arrangement. Note that for every metre the load rises, the effort needs to act over a distance of 3 m because the load is supported by three sections of rope. The effort must therefore act at a speed of $0.45\,m\,s^{-1}$ ($= 3 \times 0.15\,m\,s^{-1}$).

- The work done on the load each second = load × distance raised per second = $80\,N \times 0.15\,m\,s^{-1} = 12\,J\,s^{-1}$
- The work done by the effort each second = effort × distance moved by the effort each second = $30\,N \times 0.45\,m\,s^{-1} = 13.5\,J\,s^{-1}$.

The difference of 1.5 W is the energy wasted each second in the pulley system. This is due to friction in the bearings and also because energy must be supplied to raise the lower pulley.

## Learning objectives:

- → State the force that is mainly responsible for energy dissipation when mechanical energy is transferred from one store to another.
- → State the energy store that wasted energy is almost always transferred into.
- → Discuss whether any device can ever achieve 100% efficiency.

*Specification reference: 3.4.1.7*

▲ **Figure 1** *Using pulleys*

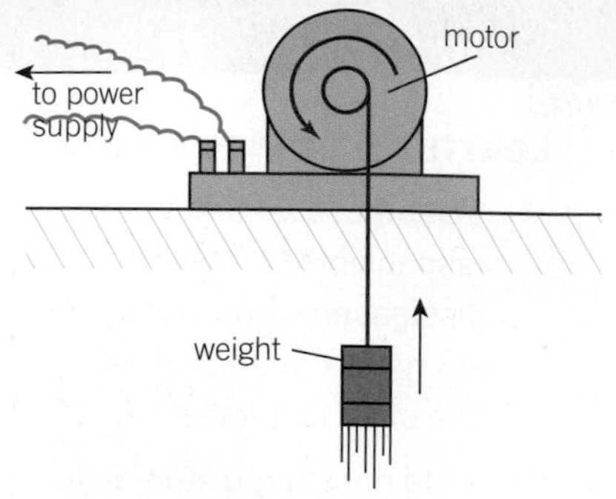


▲ **Figure 2** *Efficiency*

## Efficiency measures

**Useful energy** is energy transferred for a purpose. In any machine where friction is present, some of the energy transferred by the machine is wasted. In other words, not all the energy supplied to the machine is transferred for the intended purpose. For example, suppose a 500 W electric winch raises a weight of 150 N by 6.0 m in 10 s.

- The electrical energy supplied to the winch = 500 W × 10 s = 5000 J.
- The useful energy transferred by the machine = potential energy gain of the load = 150 N × 6 m = 900 J.

Therefore, in this example, 4100 J of energy is wasted.

$$\textbf{The efficiency of a machine} = \frac{\textbf{useful energy transferred by the machine}}{\textbf{energy supplied to the machine}}$$

$$= \frac{\textbf{work done by the machine}}{\textbf{energy supplied to the machine}}$$

*Notes:*

- **Efficiency** can be expressed as $\frac{\text{the output power of a machine}}{\text{the input power to the machine}}$.
- **Percentage efficiency** = efficiency × 100%. In the above example, the efficiency of the machine is therefore 0.18 or 18%.

### Improving efficiency

In any process or device where energy is transferred for a purpose, the efficiency of the transfer process or the device is the fraction of the energy supplied that is used for the intended purpose. The rest of the energy supplied is wasted, usually as heat and/or sound. The devices we use could be made more efficient. For example:

- A 100 W filament light bulb that is 12% efficient emits 12 J of energy as light for every 100 J of energy supplied to it by electricity. It therefore wastes 88 J of energy per second as heat.
- A fluorescent lamp with the same light output that is 80% efficient wastes just 3 J per second as heat. It gives the same light output for only 15 J of electrical energy supplied each second.

Is it possible to stop energy being wasted as heat? If the petrol engine of a car were insulated to stop heat loss, the engine would overheat. Less energy is wasted in an electric motor because it doesn't burn fuel so an electric car would be more efficient than a petrol car. However, the power stations where our electricity is generated are typically less than 40% efficient. This is partly because we need to burn fuel to produce the steam or hot gases used to drive turbines which turn the electricity generators. If the turbines were not kept cool, they would stop working because the pressure inside would build up and prevent the steam or hot gas from entering. Stopping the heat transfer to the cooling system would stop the generators working.

## Renewable energy

▲ **Figure 1** *A wind farm*

Renewable energy sources will contribute increasingly to the world's energy supplies in the future. Most of the energy we use at present is obtained from fossil fuels. But scientists think that the use of fossil fuels is causing increased climate change, due to the increasing amount of carbon dioxide in the atmosphere. Increased climate change could have disastrous consequences (e.g., rising sea levels, changing weather patterns, etc.). To cut carbon emissions, many countries are now planning to build new nuclear power stations and to develop more renewable energy resources. The UK's demand for power is about 500 000 MW. Our nuclear power stations currently provide about a quarter of our electricity, which is about 7% of our total energy supply. In terms of physics, let's consider how much energy could be supplied from typical renewable energy resources.

### Wind power

A wind turbine is an electricity generator on a tall tower, driven by large blades pushed round by the force of the wind. A typical modern wind turbine on a suitable site can generate about 2 MW of electrical power. Let's consider how this estimate is obtained.

- The kinetic energy of a mass $m$ of wind moving at speed $v = \frac{1}{2}mv^2$.
- Therefore, the kinetic energy per unit volume of air from the wind at speed $v = \frac{1}{2}\rho v^2$, where $\rho$ is the density (i.e., mass per unit volume) of air.
- Suppose the blades of a wind turbine sweep out an area $A$ when they rotate. For wind at speed $v$, a cylinder of air of area $A$ and length $v$ passes every second through the area swept out by the blades. So the volume of air passing per second = $vA$.

Therefore, the kinetic energy per second of the wind passing through a wind turbine $= \frac{1}{2}\rho v^2\, vA = \frac{1}{2}\rho v^3 A$.

For a wind turbine with blades of length 20 m, $A = \pi\,(20)^2 = 1300\,\text{m}^2$.

The density of air, $\rho = 1.2\,\text{kg}\,\text{m}^{-3}$.

The power of the wind at $v = 15\,\text{m}\,\text{s}^{-1}$

$$= \frac{1}{2} \times 1.2 \times 15^3 \times 1300$$

$$= 2.6 \times 10^6\,\text{W} = 2.6\,\text{MW}$$

The calculation shows that the maximum power output of a large wind turbine at a windy site could not be more than about 2 MW. To generate the same power as a 5000 MW power station, about 2500 wind turbines would need to be constructed and connected to the electricity network. Britain has many coastal sites where wind turbines could be located. Thousands of wind turbines would be needed to make a significant contribution to UK energy needs.

### Water power

Hydroelectricity and tidal power stations both make use of the potential energy released by water when it runs to a lower level.

- A tidal power station covering an area of 100 km$^2$ could trap a depth of 6 m of sea water twice per day. This would mean releasing a volume of $6 \times 10^8$ m$^2$ of sea water over a few hours. The mass of such a volume is about $6 \times 10^{11}$ kg and if its centre of mass drops through an average height of about 1 m, the potential energy released would be about $6 \times 10^{12}$ J. Prove for yourself this would give an energy transfer rate of more than 250 MW if released over about 6 hours. Several tidal power stations would make a significant contribution to UK energy needs.

▲ **Figure 2** *A tidal power station*

- A hydroelectric power station releases less water per second than a tidal power station but the water drops through a much greater height. However, even a large height drop of 500 m with rainfall over an area of 1000 km$^2$ at a depth of about 10 mm per day would transfer no more than about 500 MW. Lots of hydroelectric power stations would be needed to make a big contribution to UK energy needs.

### Solar power

A solar panel of area 1 m$^2$ in space would absorb solar energy at a rate of about 1400 J s$^{-1}$ if it absorbed all the incident solar energy. At the Earth's surface, the incident energy would be less, because some would be absorbed in the atmosphere. In addition, some of the Sun's energy would be reflected by the panel itself.

- A solar heating panel can heat water running though it to 70 °C on a hot sunny day. In Britain in summer, a typical solar heating panel can absorb up to 500 J s$^{-1}$ of solar energy.
- A solar cell panel produces electricity directly. A potential difference is produced across each solar cell when light is incident on the cell. A large array of solar cells and plenty of sunshine are necessary to produce useful amounts of power.

To make a significant contribution to UK energy needs, millions of homes and buildings would need to be fitted with solar panels.

▲ **Figure 3** **a** *A solar heating panel* **b** *A solar cell panel*

### Renewable energy overview

The UK demand for electricity uses about 30% of our total energy supply. If nuclear energy continues to meet a quarter of this electricity demand, could renewable energy meet the rest of the demand – approximately 100 000 MW?

- 10 million buildings each with a 0.5 kW solar panel could produce 5000 MW.

- 100 off-shore wind farms, each producing 100 MW, could produce 10 000 MW.
- 100 tidal and hydroelectric power stations could produce 25 000 MW.

Other renewable resources such as ground source heat (geothermal power) could contribute in some areas. But the simplified analysis above fails to take account of the unreliable nature of renewables. Supply and demand could be smoothed out by using renewable energy to pump water into upland reservoirs (pumped storage). Hundreds more reservoirs would be needed for this purpose.

Better insulation in homes and buildings and more efficient machines would help to reduce demand and so help to cut carbon emissions. Carbon capture of carbon emissions from fossil-fuel power stations could cut overall carbon emissions significantly. Road transport would need to switch from fossil fuels if overall carbon emissions are to be cut even more and the UK is to avoid overdependence on imported fuel. Biofuels for transport could contribute and more electricity will be needed as more people use electric vehicles.

**Questions**

1 $\sqrt{x}$ A solar cell panel of area 1 $m^2$ can produce 200 W of electrical power on a sunny day. Calculate the area of panels that would be needed to produce 2000 MW of electrical power.

2 $\sqrt{x}$ The maximum power that can be obtained from a wind turbine is proportional to the cube of the wind speed. When the wind speed is 10 m $s^{-1}$, the power output of a certain wind turbine is 1.2 MW. Calculate the power output of this wind turbine when the wind speed is 15 m $s^{-1}$.

3 $\sqrt{x}$ A hydroelectric power station produces electrical power at an overall efficiency of 25%. The power station is driven by water that has descended from an upland reservoir 650 m above the power station. Calculate the volume of water passing through the power station per second when it produces 200 MW of electrical power.

The density of water = 1000 kg $m^{-3}$

4 $\sqrt{x}$ At a tidal power station, water is trapped over an area of 200 $km^2$ when the tide is 3.0 m above the power station turbines. The trapped water is released gradually over a period of 6 hours. Calculate:

a the mass of trapped water

b the average loss of potential energy per second of this trapped water when it is released over a period of 6 hours.

The density of sea water = 1050 kg $m^{-3}$

## Summary questions

1 $\sqrt{x}$ In a test of muscle efficiency, an athlete on an exercise bicycle pedals against a brake force of 30 N at a speed of 15 m $s^{-1}$.

a Calculate the useful energy supplied per second by the athlete's muscles.

b If the efficiency of the muscles is 25%, calculate the energy per second supplied to the athlete's muscles.

2 $\sqrt{x}$ A 60 W electric motor raises a weight of 20 N through a height of 2.5 m in 8.0 s. Calculate:

a the electrical energy supplied to the motor

b the useful energy transferred by the motor

c the efficiency of the motor.

3 $\sqrt{x}$ A power station has an overall efficiency of 35% and it produces 200 MW of electrical power. The fuel used in the power station releases 80 MJ per kilogram of fuel burned. Calculate:

a the energy per second supplied by the fuel

b the mass of fuel burned per day.

4 $\sqrt{x}$ A vehicle engine has a power output of 6.2 kW and uses fuel which releases 45 MJ per kilogram when burned. At a speed of 30 m $s^{-1}$ on a level road, the fuel usage of the vehicle is 18 km per kilogram. Calculate:

a the time taken by the vehicle to travel 18 km at 30 m $s^{-1}$

b the useful energy supplied by the engine in this time

c the overall efficiency of the engine.

# Practice questions: Chapter 10

1 (a) An electric car is fitted with a battery, which is used to drive an electric motor that drives the car. The battery has a maximum power output of 12 kW, which gives a maximum driving force of 600 N.
Calculate (i) the top speed of the car, (ii) the car's maximum range, if the battery lasts for 90 minutes at maximum power output without being recharged. *(4 marks)*

(b) A hybrid vehicle has a battery-driven electric motor and a petrol engine, which takes over from the electric motor when the vehicle has reached a certain speed. Above this speed, the petrol engine also recharges the battery. The vehicle has an overall fuel efficiency of 18 km per litre, compared with 10 km per litre for an equivalent petrol-only car, which has a carbon emission of 180 grams per km. Discuss the benefits of the use of such hybrid cars instead of petrol-only cars, in terms of carbon emissions, given the average annual distance travelled by a driver in the UK is about 20 000 km. The average annual carbon emission per UK household, including driving, is about 10 000 kg. *(5 marks)*

2 **Figure 1** shows apparatus that can be used to investigate energy changes.

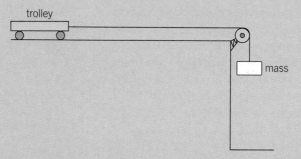

▲ **Figure 1**

The trolley and the mass are joined by an inextensible string. In an experiment to investigate energy changes, the trolley is initially held at rest, and is then released so that the mass falls vertically to the ground.

(a) (i) State the energy changes of the falling mass.
(ii) Describe the energy changes that take place in this system. *(4 marks)*

(b) State what measurements would need to be made to investigate the conservation of energy. *(2 marks)*

(c) Describe how the measurements in part (b) would be used to investigate the conservation of energy. *(4 marks)*

AQA, 2006

3 A small hydroelectric power station uses water which falls through a height of 4.8 m.

(a) Calculate the change in potential energy of a 1.0 kg mass of water falling through a vertical height of 4.8 m. *(2 marks)*

(b) Calculate the maximum power available from the water passing through this power station when water flows through it at a rate of $6.7 \times 10^7$ kg per hour. *(3 marks)*

(c) State *two* factors that affect the usefulness of hydroelectric power stations for electricity production. *(2 marks)*

AQA, 2002

4 **Figure 2** represents the motion of a car of mass $1.4 \times 10^3$ kg, travelling in a straight line.

(a) Describe, without calculation, how the *resultant* force acting on the car varies over this 10 second interval. *(2 marks)*

(b) Calculate the maximum kinetic energy of the car. *(2 marks)*

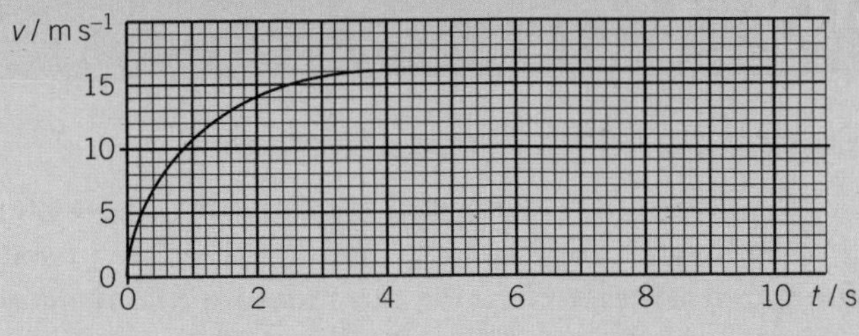


▲ **Figure 2**

**(c)** At some time later, when the car is travelling at a steady speed of $30\,\mathrm{m\,s^{-1}}$, the useful power developed by the engine is 20 kW. Calculate the driving force required to maintain this speed. *(2 marks)*

AQA, 2002

**5** A skydiver of mass 70 kg, jumps from a stationary balloon and reaches a speed of $45\,\mathrm{m\,s^{-1}}$ after falling a distance of 150 m.

**(a)** Calculate the skydiver's:
(i) loss of gravitational potential energy
(ii) gain in kinetic energy. *(4 marks)*

**(b)** The difference between the loss of gravitational potential energy and the gain in kinetic energy is equal to the work done against air resistance. Use this fact to calculate:
(i) the work done against air resistance
(ii) the average force due to air resistance acting on the skydiver. *(3 marks)*

AQA, 2004

**6** A car travels at constant velocity along a horizontal road.

**(a)** The car has an effective power output of 18 kW and is travelling at a constant velocity of $10\,\mathrm{m\,s^{-1}}$. Show that the total resistive force acting is 1800 N. *(1 mark)*

**(b)** The total resistive force consists of two components. One of these is a constant frictional force of 250 N and the other is the force of air resistance, which is proportional to the square of the car's speed. Calculate:
(i) the force of air resistance when the car is travelling at $10\,\mathrm{m\,s^{-1}}$
(ii) the force of air resistance when the car is travelling at $20\,\mathrm{m\,s^{-1}}$
(iii) the effective output power of the car required to maintain a constant speed of $20\,\mathrm{m\,s^{-1}}$ on a horizontal road. *(4 marks)*

AQA, 2001

**7** **(a)** A ball bearing is released when near the top of a tall cylinder containing oil. Discuss the energy changes which take place when the ball bearing (i) accelerates from rest, (ii) travels at constant velocity. *(6 marks)*

**(b)** A pump-operated hydraulic jack is used to raise a large object of mass 470 kg. The jack is used by pushing down on the handle of a lever connected to the pump. During each stroke, a force of 150 N is applied downward on the handle, which is moved through a distance of 0.42 m. 52 strokes of the handle are needed to raise the object through a vertical height of 0.58 m.
Calculate the efficiency of this process. *(5 marks)*

# 11 Materials

## 11.1 Density

**Learning objectives:**

- → Define density.
- → State the unit of density.
- → Measure the density of an object.

*Specification reference: 3.4.2.1*

## Density and its measurement

Lead is much more dense than aluminium. Sea water is more dense than tap water. To find how much more dense one substance is compared with another, we can measure the mass of equal volumes of the two substances. The substance with the greater mass in the same volume is more dense. For example, a lead sphere of volume $1\,cm^3$ has a mass of 11.3 g whereas an aluminium sphere of the same volume has a mass of 2.7 g.

**The density of a substance is defined as its mass per unit volume.**

For a certain amount of a substance of mass m and volume $V$, its density $\rho$ (pronounced 'rho') may be calculated using the equation

$$\textbf{density} = \frac{m}{V}$$

The unit of density is the kilogram per cubic metre ($kg\,m^{-3}$).

Rearranging the above equation gives $m = \rho V$ or $V = \dfrac{m}{\rho}$.

Table 1 shows the density of some common substances in $kg\,m^{-3}$. You can see that gases are much less dense than solids or liquids. This is because the average separation between the molecules in a gas is much greater than in a liquid or solid.

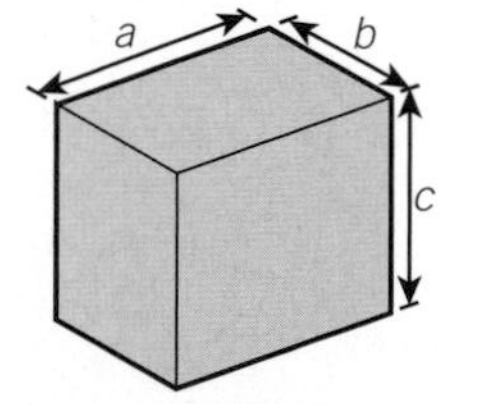


**a** *Volume of cuboid* = $a \times b \times c$

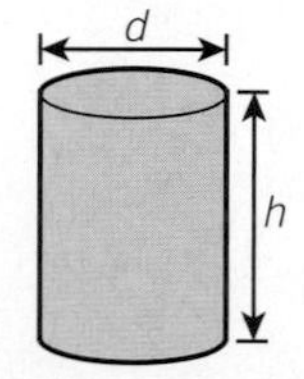


**b** *Volume of cylinder* = $\dfrac{\pi d^2}{4} \times h$

▲ **Figure 1** *Volume equations*

## Density measurements

An unknown substance can often be identified if its density is measured and compared with the density of known substances. The following procedures could be used to measure the density of a substance.

### 1 A regular solid

- Measure its mass using a top pan balance.
- Measure its dimensions using vernier calipers or a micrometer and calculate its volume using the appropriate equation (e.g., for a sphere of radius $r$, volume = $\frac{4}{3}\pi r^3$ – see Figure 1 for other volume equations). Calculate the density from mass/volume.

### 2 A liquid

- Measure the mass of an empty measuring cylinder. Pour some of the liquid into the measuring cylinder and measure the volume of the liquid directly. Use as much liquid as possible to reduce the percentage error in your measurement.
- Measure the mass of the cylinder and liquid to enable the mass of the liquid to be calculated. Calculate the density from mass/volume.

### 3 An irregular solid

- Measure the mass of the object.
- Immerse the object on a thread in liquid in a measuring cylinder, observe the increase in the liquid level. This is the volume of the object.
- Calculate the density of the object from its mass/volume.

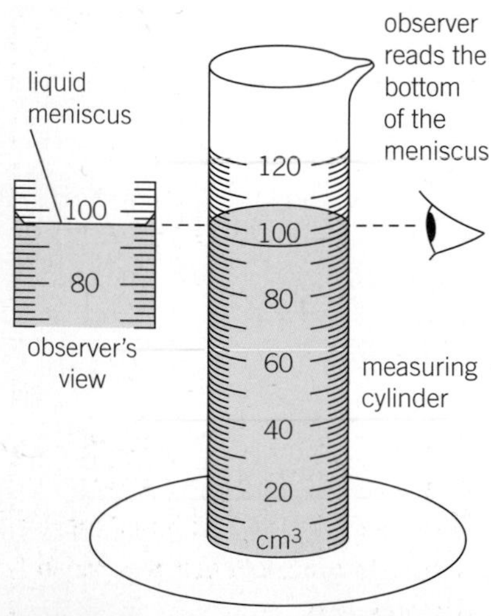


▲ **Figure 2** *Using a measuring cylinder*

## Density of alloys

An alloy is a solid mixture of two or more metals. For example, brass is an alloy of copper and zinc that has good resistance to corrosion and wear.

For an alloy, of volume $V$, that consists of two metals A and B,

- if the volume of metal A = $V_A$, the mass of metal A = $\rho_A V_A$, where $\rho_A$ is the density of metal A
- if the volume of metal B = $V_B$, the mass of metal B = $\rho_B V_B$, where $\rho_B$ is the density of metal B.

Therefore, the mass of the alloy, $m = \rho_A V_A + \rho_B V_B$.

Hence the density of the alloy $\rho = \frac{m}{V} = \frac{\rho_A V_A + \rho_B V_B}{V} = \frac{\rho_A V_A}{V} + \frac{\rho_B V_B}{V}$.

▼ **Table 1** *Densities of common substances*

| Substance | Density / kg m$^{-3}$ |
|---|---|
| air | 1.2 |
| aluminium | 2700 |
| copper | 8900 |
| gold | 19 300 |
| hydrogen | 0.083 |
| iron | 7900 |
| lead | 11 300 |
| oxygen | 1.3 |
| silver | 10 500 |
| water | 1000 |

### Worked example

A brass object consists of $3.3 \times 10^{-5}$ m$^3$ of copper and $1.7 \times 10^{-5}$ m$^3$ of zinc. Calculate the mass and the density of this object. The density of copper = 8900 kg m$^{-3}$. The density of zinc = 7100 kg m$^{-3}$.

*Solution*

Mass of copper = density of copper × volume of copper

$= 8900 \times 3.3 \times 10^{-5}\,\text{m}^3 = 0.294\,\text{kg}$

Mass of zinc = density of zinc × volume of zinc

$= 7100 \times 1.7 \times 10^{-5}\,\text{m}^3 = 0.121\,\text{kg}$

Total mass, $m = 0.294 + 0.121 = 0.415\,\text{kg}$

Total volume, $V = 5.0 \times 10^{-5}\,\text{m}^3$

Density of alloy $\rho = \frac{m}{V} = \frac{0.415\,\text{kg}}{5.0 \times 10^{-5}\,\text{m}^3} = 8300\,\text{kg m}^{-3}$

### Study tip

**More about units**

mass: 1 kg = 1000 g

length: 1 m = 100 cm = 1000 mm

volume: 1 m$^3$ = 10$^6$ cm$^3$

density: $1000\,\text{kg m}^{-3} = \frac{10^6\,\text{g}}{10^6\,\text{cm}^3} = 1\,\text{g cm}^{-3}$

## Summary questions

1. A rectangular brick of dimensions 5.0 cm × 8.0 cm × 20.0 cm has a mass of 2.5 kg. Calculate **a** its volume, **b** its density.
2. An empty paint tin of diameter 0.150 m and of height 0.120 m has a mass of 0.22 kg. It is filled with paint to within 7 mm of the top. Its total mass is then 6.50 kg. Calculate **a** the mass, **b** the volume, **c** the density of the paint in the tin.
3. A solid steel cylinder has a diameter of 12 mm and a length of 85 mm. Calculate **a** its volume in m$^3$, **b** its mass in kg. The density of steel = 7800 kg m$^{-3}$.
4. An alloy tube of volume $1.8 \times 10^{-4}$ m$^3$ consists of 60% aluminium and 40% magnesium by volume. Calculate **a** the mass of **i** aluminium, **ii** magnesium in the tube, **b** the density of the alloy. The density of aluminium = 2700 kg m$^{-3}$. The density of magnesium = 1700 kg m$^{-3}$.

### Study tip

Unit errors are commonplace in density calculations. Avoid such errors by writing the unit and the numerical value of each quantity in your working.

# 11.2 Springs

## Learning objectives:

- → Discuss whether there is any limit to the linear graph of force against extension for a spring.
- → Define the spring constant, and state its unit of measurement.
- → If the extension of a spring is doubled, calculate how much more energy it stores.

*Specification reference: 3.4.2.1*

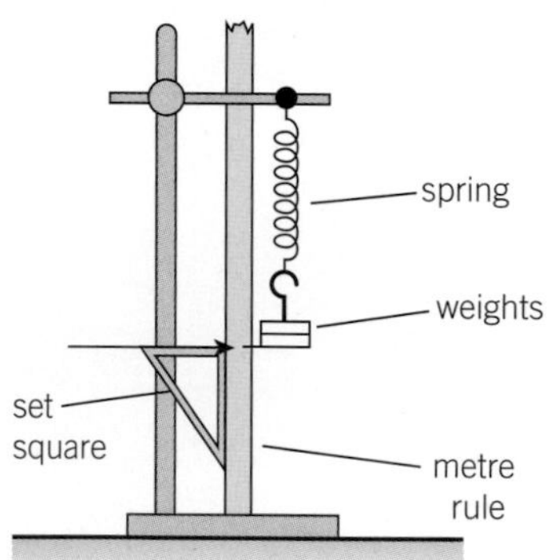


▲ **Figure 1** *Testing the extension of a spring*

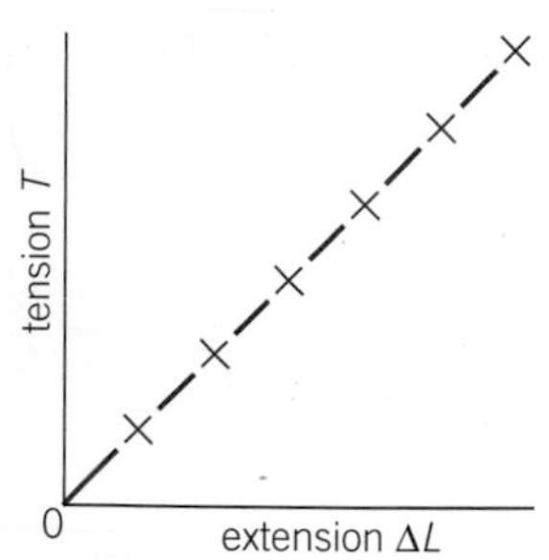


▲ **Figure 2** *Hooke's law*

## Hooke's law

A stretched spring exerts a pull on the object holding each end of the spring. This pull, referred to as the *tension* in the spring, is equal and opposite to the force needed to stretch the spring. The more a spring is stretched, the greater the tension in it. Figure 1 shows a stretched spring supporting a weight at rest. This arrangement may be used to investigate how the tension in a spring depends on its extension from its unstretched length. The measurements may be plotted on a graph of tension against extension, as shown in Figure 2. The graph shows that the force needed to stretch a spring is proportional to the extension of the spring. This is known as **Hooke's law**, after its discoverer, Robert Hooke, a seventeenth century scientist.

**Hooke's law states that the force needed to stretch a spring is directly proportional to the extension of the spring from its natural length.**

Hooke's law may be written as

$$\textbf{Force } F = k\Delta L$$

where $k$ is the spring constant (sometimes referred to as the stiffness constant) and $\Delta L$ is the extension from its natural length $L$.

- The greater the value of $k$, the stiffer the spring is. The unit of $k$ is $\text{N m}^{-1}$.
- The graph of $F$ against $\Delta L$ is a straight line of gradient $k$ through the origin.
- If a spring is stretched beyond its **elastic limit**, it does not regain its initial length when the force applied to it is removed.
- In AS/A Level Maths students may meet Hooke's law in the form $F = \lambda\, \Delta L/L$, where $L$ is the unstretched length of the spring and $\lambda$ (= $kL$) is the spring modulus. Note that $\lambda$ is not needed on this course.

### Worked example

A vertical steel spring fixed at its upper end has an unstretched length of 300 mm. Its length is increased to 385 mm when a 5.0 N weight attached to the lower end is at rest. Calculate:

**a** the spring constant

**b** the length of the spring when it supports an 8.0 N weight at rest.

### *Solution*

**a** Use $F = k\Delta L$ with $F = 5.0\,\text{N}$ and $\Delta L = 385 - 300\,\text{mm} = 85\,\text{mm}$ $= 0.085\,\text{m}$.

Therefore $k = \dfrac{F}{\Delta L} = \dfrac{5.0\,\text{N}}{0.085\,\text{m}} = 59\,\text{N m}^{-1}$.

**b** Use $F = k\Delta L$ with $F = 8.0\,\text{N}$ and $k = 59\,\text{N m}^{-1}$ to calculate $\Delta L$.

$$\Delta L = \frac{F}{k} = \frac{8.0\,\text{N}}{59\,\text{N m}^{-1}} = 0.136\,\text{m}$$

Therefore the length of the spring = 0.300 m + 0.136 m = 0.436 m.

## Spring combinations

### Springs in parallel

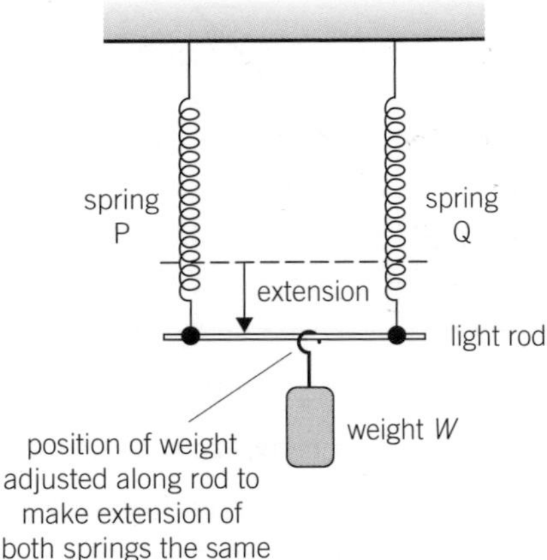


▲ **Figure 3** *Two springs in parallel*

Figure 3 shows a weight supported by means of two springs P and Q in parallel with each other. The extension, $\Delta L$, of each spring is the same. Therefore

- the force needed to stretch P, $F_P = k_P \Delta L$
- the force needed to stretch Q, $F_Q = k_Q \Delta L$

where $k_P$ and $k_Q$ are the spring constants of P and Q, respectively.

Since the weight $W$ is supported by both springs,

$W = F_P + F_Q = k_P \Delta L + k_Q \Delta L = k \Delta L$

where the effective spring constant, $k = k_P + k_Q$.

### Springs in series

Figure 4 shows a weight supported by means of two springs joined end-on in series with each other. The tension in each spring is the same and is equal to the weight $W$.

Therefore

- the extension of spring P, $\Delta L_P = \dfrac{W}{k_P}$
- the extension of spring Q, $\Delta L_Q = \dfrac{W}{k_Q}$

where $k_P$ and $k_Q$ are the spring constants of P and Q, respectively.

Therefore the total extension, $\Delta L = \Delta L_P + \Delta L_Q = \dfrac{W}{k_P} + \dfrac{W}{k_Q} = \dfrac{W}{k}$

where $k$, the effective spring constant, is given by the equation

$$\frac{1}{k} = \frac{1}{k_P} + \frac{1}{k_Q}$$

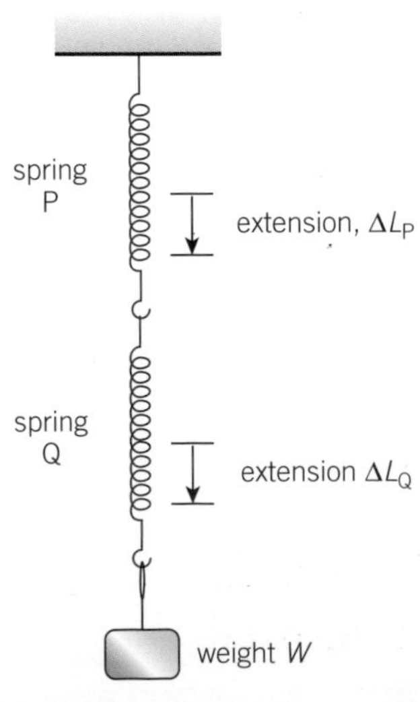


▲ **Figure 4** *Two springs in series*

## The energy stored in a stretched spring

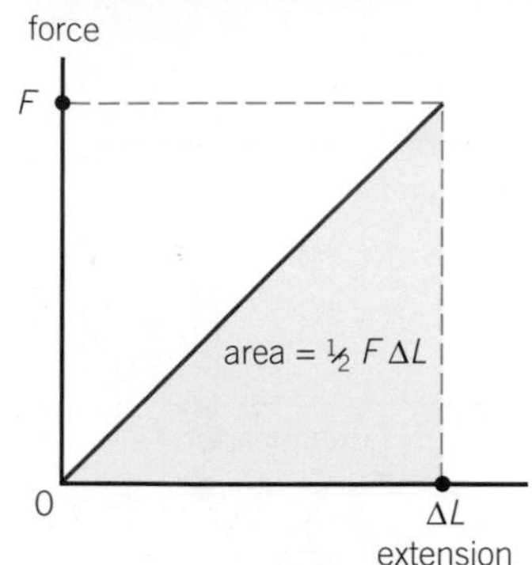


▲ **Figure 4** *Energy stored in a stretched spring*

### Study tip

Energy stored by a spring is $\frac{1}{2}F\,\Delta L$, not $F\,\Delta L$.

### Synoptic link

You have met work done in Topic 10.1, Work and energy.

Elastic potential energy is stored in a stretched spring. If the spring is suddenly released, the elastic energy stored in it is suddenly transferred into kinetic energy of the spring. The work done to stretch a spring by extension $\Delta L$ from its unstretched length $= \frac{1}{2} F\,\Delta L$, where $F$ is the force needed to stretch the spring to extension $\Delta L$. The work done on the spring is stored as elastic potential energy. Therefore, the elastic potential energy $E_p$ in the spring $= \frac{1}{2} F\,\Delta L$. Also, since $F = k\,\Delta L$, where $k$ is the spring constant, then $E_p = \frac{1}{2} k\,\Delta L^2$.

**Elastic potential energy stored in a stretched spring,**

$$E_p = \frac{1}{2} F\,\Delta L = \frac{1}{2} k\,\Delta L^2$$

## Summary questions

$g = 9.8\,\mathrm{m\,s^{-2}}$

**1** A steel spring has a spring constant of $25\,\mathrm{N\,m^{-1}}$. Calculate:

- **a** the extension of the spring when the tension in it is equal to 10 N
- **b** the tension in the spring when it is extended by 0.50 m from its unstretched length.

**2** Two identical steel springs of length 250 mm are suspended vertically side by side from a fixed point. A 40 N weight is attached to the ends of the two springs. The length of each spring is then 350 mm. Calculate:

- **a** the tension in each spring
- **b** the extension of each spring
- **c** the spring constant of each spring.

**3** Repeat **2a** and **b** for the two springs in series and vertical.

**4** An object of mass 0.150 kg is attached to the lower end of a vertical spring of unstretched length 300 mm, which is fixed at its upper end. With the object at rest, the length of the spring becomes 420 mm as a result. Calculate:

- **a** the spring constant
- **b** the energy stored in the spring
- **c** the weight that needs to be added to extend the spring to 600 mm.

# 11.3 Deformation of solids

## Force and solid materials

Look around at different materials and think about the effect of force on each material. To stretch or twist or compress the material, a pair of forces is needed. For example, stretching a rubber band requires the rubber band to be pulled by a force at either end. Some materials, such as rubber, bend or stretch easily. The **elasticity** of a solid material is its ability to regain its shape after it has been deformed or distorted and the forces that deformed it have been released. Deformation that stretches an object is **tensile**, whereas deformation that compresses an object is **compressive**.

Figure 1 in Topic 11.2 shows how to test different materials to see how easily they stretch. In each case, the material is held at its upper end and loaded by hanging weights at its lower end. A set square or pointer attached to the bottom of the weights may be used to measure the extension of the material, as the weight of the load is increased in steps then decreased to zero. The extension of the strip of material at each step is its increase of length from its unloaded length. The tension in the material is equal to the weight. The measurements may be plotted as a tension–extension graph, as shown in Figure 1.

- A steel spring gives a straight line (blue line in Figure 1), in accordance with Hooke's law (Topic 11.2).
- A rubber band at first extends easily when it is stretched. However, it becomes fully stretched and very difficult to stretch further when it has been lengthened considerably (red line in Figure 1).
- A polythene strip 'gives' and stretches easily after its initial stiffness is overcome. However, after 'giving' easily, it extends little and becomes difficult to stretch (green line in Figure 1).

## Tensile stress and tensile strain

The extension of a wire under tension may be measured using Searle's apparatus, as shown in Figure 2 on the next page (or similar apparatus with a vernier scale). A micrometer attached to the control wire is adjusted so the spirit level between the control and test wire is horizontal. When the test wire is loaded, it extends slightly, causing the spirit level to drop on one side. The micrometer is then readjusted to make the spirit level horizontal again. The change of the micrometer reading is therefore equal to the extension. The extension may be measured for different values of tension by increasing the test weight in steps.

For a wire of length $L$ and area of cross section $A$ under tension,

- the **tensile stress** in the wire, $\sigma = T/A$, where $T$ is the tension. The unit of stress is the **pascal** (Pa) equal to $1\,\mathrm{N\,m^{-2}}$.
- the **tensile strain** in the wire, $\varepsilon = \Delta L/L$, where $\Delta L$ is the extension (increase in length) of the wire. Strain is a ratio and therefore has no unit.

## Learning objectives:

- → Relate stress to force, and strain to extension.
- → Describe what is meant by the Young modulus.
- → Define 'tensile'.
- → Explain why we bother with stress and strain, when force and extension are more easily measured.

*Specification reference 3.4.2.2*

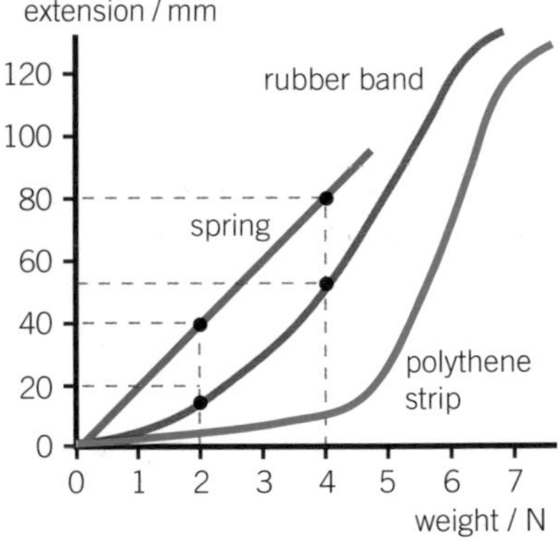


▲ **Figure 1** *Typical curves*

### Study tip

Remember that the extension is always measured from the original (unstretched) length of the object.

## Hint

Some materials can be strong and brittle, while others can be plastic and tough. Toughness is a measure of the energy needed to break a material.

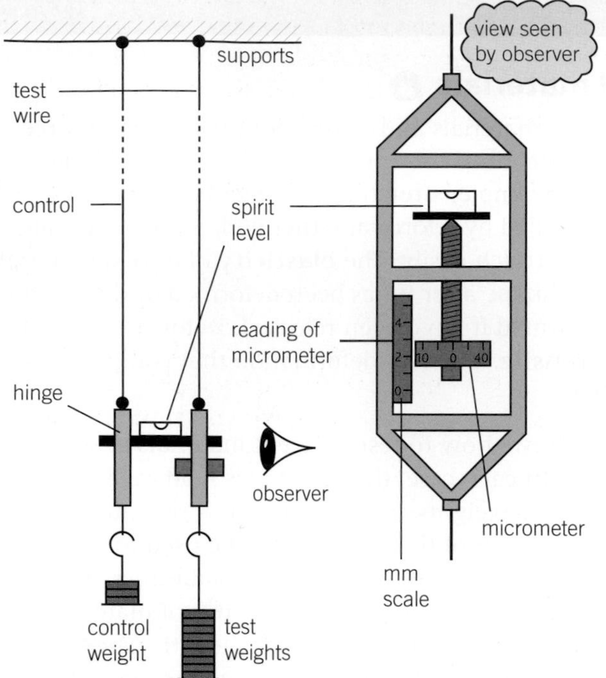


▲ **Figure 2** *Searle's apparatus*

Figure 3 shows how the tensile stress in a wire varies with tensile strain.

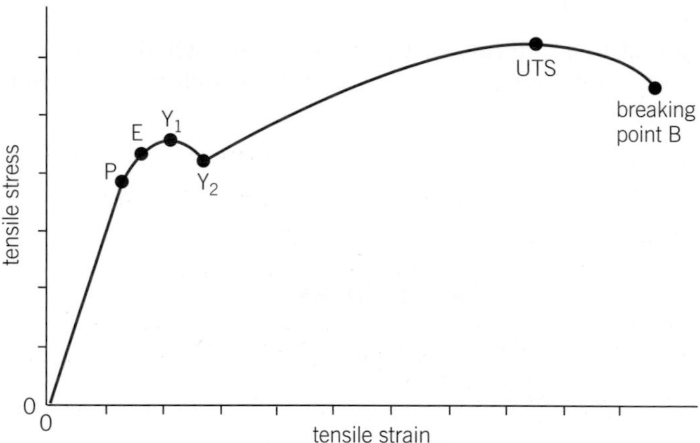


▲ **Figure 3** *Tensile stress versus tensile strain for a metal wire*

- From 0 to the limit of proportionality P, the tensile stress is proportional to the tensile strain.
  The value of stress/strain is a constant, known as the **Young modulus** of the material.

$$\textbf{Young modulus } E = \frac{\textbf{tensile stress, } \sigma}{\textbf{tensile strain, } \varepsilon} = \frac{T}{A} \div \frac{\Delta L}{L} = \frac{TL}{A \Delta L}$$

For a wire of uniform diameter $d$, the area of cross section

$$A = \frac{\pi d^2}{4}$$

- Beyond P, the line curves and continues beyond the elastic limit E to the **yield point** $Y_1$, which is where the wire weakens temporarily. The **elastic limit** is the point beyond which the wire is permanently stretched and suffers **plastic deformation**.

## Study tip

Don't forget the italicised words in the definitions below:

**Stress** is tension per unit *cross-sectional area*.

**Strain** is extension per *unit length*.

- Beyond $Y_2$, a small increase in the tensile stress causes a large increase in tensile strain as the material of the wire undergoes plastic flow. Beyond maximum tensile stress, the **ultimate tensile stress** (UTS), the wire loses its strength, extends, and becomes narrower at its weakest point. Increase of tensile stress occurs due to the reduced area of cross section at this point until the wire breaks at point B. The ultimate tensile stress is sometimes called the **breaking stress**.

## Worked example

A crane fitted with a steel cable of uniform diameter 2.3 mm and length 28 m is used to lift an iron girder of weight 3200 N off the ground. Calculate the extension of the cable when it supports the girder at rest.

The Young modulus for steel = $2.1 \times 10^{11}$ Pa

### *Solution*

Tension $T = 3200\,\text{N}$, $L = 28\,\text{m}$

Area of cross section of wire $= \dfrac{\pi(2.3 \times 10^{-3})^2}{4} = 4.15 \times 10^{-6}\,\text{m}^2$

To find the extension, rearranging the Young modulus equation

$E = \dfrac{TL}{A\Delta L}$ gives

$$\Delta L = \frac{TL}{AE} = \frac{3200 \times 28}{4.15 \times 10^{-6} \times 2.1 \times 10^{11}} = 0.103\,\text{m}$$

## Stress–strain curves for different materials

The **stiffness** of different materials can be compared using the **gradient** of the stress–strain line, which is equal to the Young modulus of the material. Thus steel is stiffer than copper.

The **strength** of a material is its ultimate tensile stress (UTS), which is its maximum tensile stress. Steel is stronger than copper because its maximum tensile stress is greater.

A **brittle** material snaps without any noticeable yield. For example, glass breaks without any give.

A **ductile** material can be drawn into a wire. Copper is more ductile than steel.

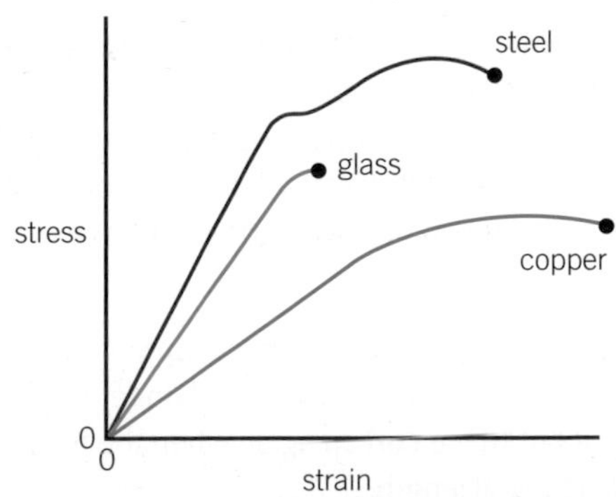


▲ **Figure 4** *Stress–strain curves*

## Summary questions

1. Calculate the tensile stress in a wire of diameter 0.25 mm when the tension in the wire is 50 N.
2. A metal wire of diameter 0.23 mm and of unstretched length 1.405 m was suspended vertically from a fixed point. When a 40 N weight was suspended from the lower end of the wire, the wire stretched by an extension of 10.5 mm. Calculate the Young modulus of the wire material.
3. A vertical steel wire of length 2.5 m and diameter 0.35 mm supports a weight of 90 N. Calculate:
   - **a** the tensile stress in the wire
   - **b** the extension of the wire. The Young modulus of steel = $2.1 \times 10^{11}$ Pa.
4. Compare the two stress–strain curves in Figure 5. Use the curves to identify
   - **a** the material, X or Y, that is **i** stiffest, **ii** strongest
   - **b** the material, X or Y, that is **i** brittle, **ii** ductile.

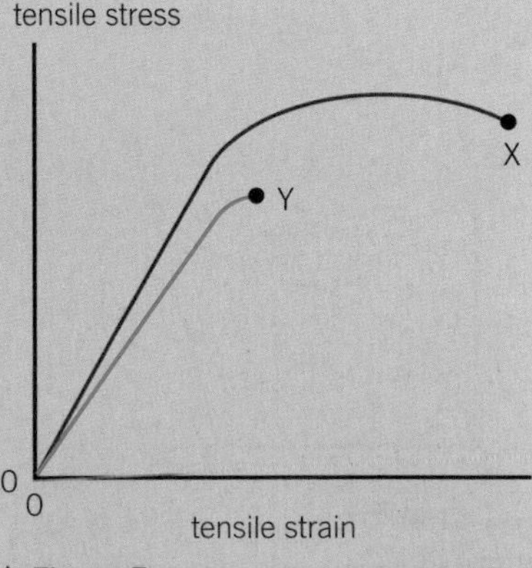


▲ **Figure 5**

# 11.4 More about stress and strain

**Learning objectives:**

- ➔ Predict whether a metal wire stretched below its elastic limit will return to its original length.
- ➔ Describe what happens when a metal wire is stretched beyond its elastic limit and then unloaded.
- ➔ Compare the deformation of other materials such as rubber and polythene with a metal wire.

*Specification reference: 3.4.2.2*

## Loading and unloading of different materials

How does the strength of a material change as a result of being stretched? Figure 1 in Topic 11.2 may be used to investigate this question. The tension in a strip of material is increased by increasing the weight it supports in steps. At each step, the extension of the material is measured. Typical results for different materials are shown in Figure 1. For each material, the loading curve and the subsequent unloading curve are shown.

- For a metal wire, its loading and unloading curves are the same, provided its elastic limit is not exceeded. This means the wire returns to its original length when unloaded. However, beyond its elastic limit, the unloading line is parallel to the loading line. In this case, the wire is slightly longer when unloaded – it has a **permanent extension**.
- For a rubber band, the change of length during unloading is greater than during loading for a given change in tension. The rubber band returns to the same unstretched length, but the unloading curve is below the loading curve except at zero and maximum extensions. The rubber band remains elastic as it regains its initial length, but it has a low **limit of proportionality**.
- For a polythene strip, the extension during unloading is also greater than during loading. However, the strip does not return to the same initial length when it is completely unloaded. The polythene strip has a low limit of proportionality *and* suffers **plastic deformation**.

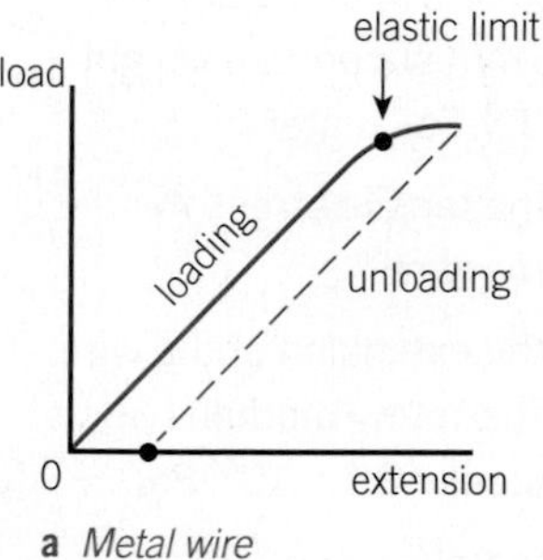


**a** *Metal wire*

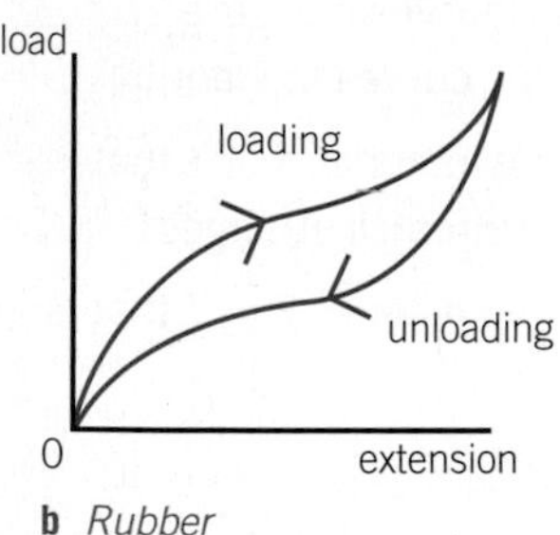


**b** *Rubber*

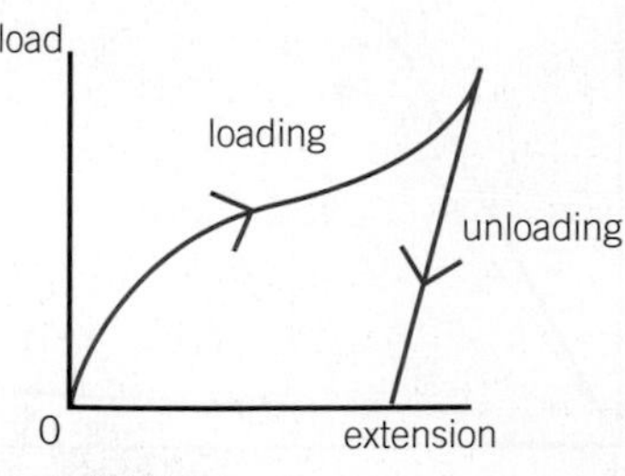


**c** *Polythene*

▲ **Figure 1** *Loading and unloading curves*

## Strain energy

As explained in Topic 11.2, the area under the line of a force–extension graph is equal to the work done to stretch the wire. The work done to deform an object is referred to as **strain energy**. Consider the energy transfers for each of the three materials in Figure 1 when each material is loaded then unloaded.

### Polythene and rubber

Polythene is an example of a polymer, which means that its molecules are long chains of atoms. Before a strip of polythene is stretched, the molecules are tangled together. Weak bonds, or cross-links, form between the molecules. When polythene is under tension, it easily stretches as the weak cross-links break. In this stretched state, new weak cross-links form, and, when the tension is removed, the polythene strip stays stretched.

Rubber is also a polymer but its molecules are curled and tangled together when it is in an unstretched state. When placed under tension, its molecules are straightened out but these curl up again when the tension is removed – the rubber regains its initial length.

### 1 Metal wire (or spring)

Provided the limit of proportionality is not exceeded, to stretch a wire to an extension $\Delta L$, the work done $= \frac{1}{2} T\Delta L$, where $T$ is the tension in the wire at this extension. Because the elastic limit is not reached, the work done is stored as elastic energy in the wire.

$$\textbf{the elastic energy stored in a stretched wire} = \frac{1}{2} T\Delta L$$

Because the graph of tension against extension is the same for unloading as for loading, all the energy stored in the wire can be recovered when the wire is unloaded.

## Worked example

A steel wire of uniform diameter 0.35 mm and of length 810 mm is stretched to an extension of 2.5 mm. Calculate **a** the tension in the wire, **b** the elastic energy stored in the wire.

The Young modulus for steel = $2.1 \times 10^{11}$ Pa

### Solution

**a** Extension, $\Delta L = 2.5\,\text{mm} = 2.5 \times 10^{-3}$ m

Area of cross section of wire $= \dfrac{\pi(0.35 \times 10^{-3})^2}{4} = 9.6 \times 10^{-8}\,\text{m}^2$

To find the tension, rearranging the Young modulus equation

$E = \dfrac{TL}{A\,\Delta L}$ gives

$$T = \frac{EA\,\Delta L}{L} = \frac{2.1 \times 10^{11} \times 9.6 \times 10^{-8} \times 2.5 \times 10^{-3}}{0.810} = 62\,\text{N}$$

**b** Elastic energy stored in the wire $= \frac{1}{2} T\Delta L = 0.5 \times 62 \times 2.5 \times 10^{-3}$ $= 7.8 \times 10^{-2}$ J.

### 2 Rubber band

The work done to stretch the rubber band is represented by the area under the loading curve. The work done by the rubber band, when it is unloaded, is represented by the area under the unloading curve. The area between the loading curve and the unloading curve therefore represents the difference between energy stored in the rubber band when it is stretched and the useful energy recovered from it when it is unstretched. The difference occurs because some of the energy stored in the rubber band becomes the internal energy of the molecules when the rubber band unstretches.

### 3 Polythene

As it does not regain its initial length, the area between the loading and unloading curves represents work done to deform the material permanently, as well as internal energy retained by the polythene when it unstretches.

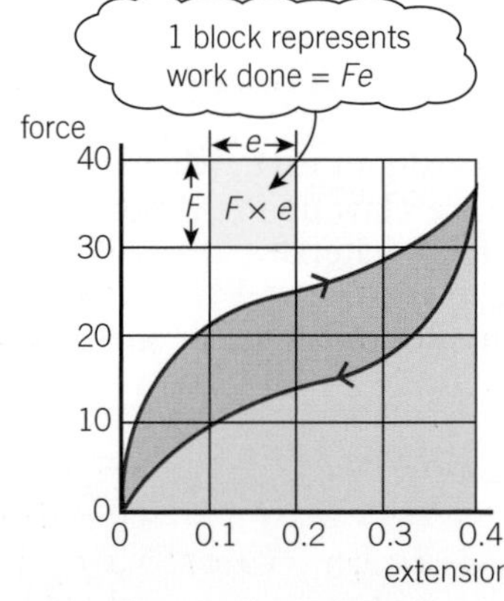


▲ **Figure 2** *Energy changes when loading and unloading rubber*

## Summary questions

**The Young modulus for**

**steel = $1.3 \times 10^{11}$ Pa**

**copper = $1.3 \times 10^{11}$ Pa**

1. A vertical steel cable of diameter 24 mm and of length 18 m supports a weight of 1500 N attached to its lower end. Calculate **a** the tensile stress in the cable, **b** the extension of the cable, **c** the elastic energy stored in the cable, assuming its elastic limit has not been reached.
2. A vertical steel wire of diameter 0.28 mm and of length 2.0 m is fixed at its upper end, and has a weight of 15 N suspended from its lower end. Calculate **a** the extension of the wire, **b** the elastic energy stored in the wire.
3. A steel bar of length 40 mm and cross-sectional area $4.5 \times 10^{-4}\,\text{m}^2$ is placed in a vice and compressed by 0.20 mm when the vice is tightened. Calculate **a** the compressive force exerted on the bar, **b** the work done to compress it.
4. Figure 2 shows a force against extension curve for rubber band. Use the graph to determine **a** the work done to stretch the rubber band to an extension of 0.40 m, **b** the internal energy retained by the rubber band when it unstretches.

# Practice questions: Chapter 11

1 (a) Define the *density* of a material. (*1 mark*)

(b) Brass, an alloy of copper and zinc, consists of 70% *by volume* of copper and 30% *by volume* of zinc.

density of copper = $8.9 \times 10^3 \text{ kg m}^{-3}$

density of zinc = $7.1 \times 10^3 \text{ kg m}^{-3}$

(i) Determine the mass of copper and the mass of zinc required to make a rod of brass of volume $0.80 \times 10^{-3} \text{ m}^3$.

(ii) Calculate the density of brass. (*5 marks*)

AQA, 2004

2 **Figure 1** shows a lorry of mass $1.2 \times 10^3$ kg parked on a platform used to weigh vehicles. The lorry compresses the spring that supports the platform by 0.030 m.

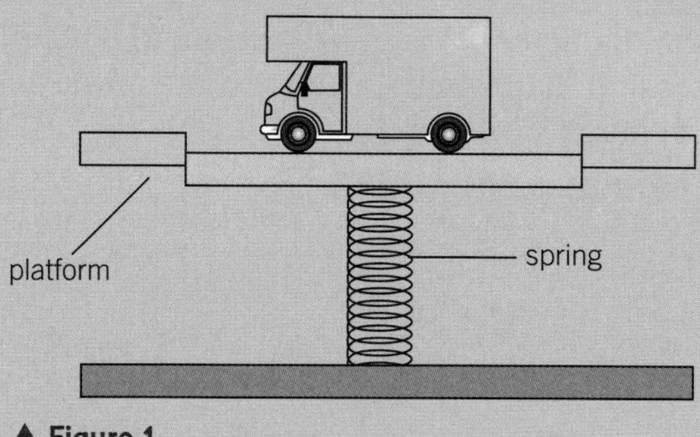


▲ **Figure 1**

Calculate the energy stored in the spring. (*3 marks*)

AQA, 2002

3 (a) **Figure 2** shows the variation of tensile stress with tensile strain for two wires **X** and **Y**, having the same dimensions, but made of different materials. The materials fracture at the points $\mathbf{F_X}$ and $\mathbf{F_Y}$ respectively.

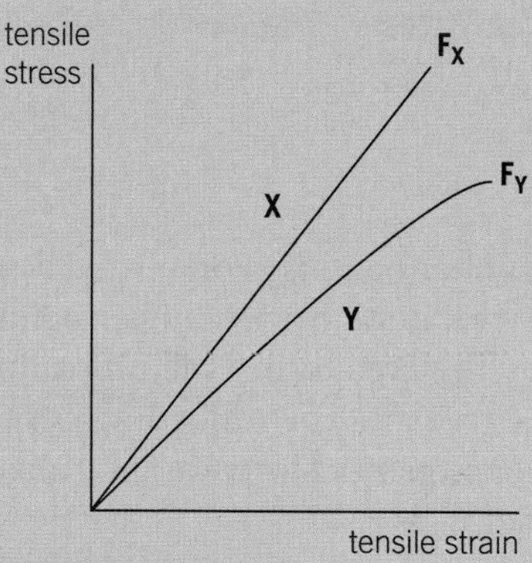


▲ **Figure 2**

State, with a reason for each, which material, **X** or **Y**,

(i) obeys Hooke's law up to the point of fracture

(ii) is the weaker material

(iii) is ductile

(iv) has the greater elastic strain energy for a given tensile stress. (*8 marks*)

**(b)** An elastic cord of unstretched length 160 mm has a cross-sectional area of 0.64 mm$^2$. The cord is stretched to a length of 190 mm. Assume that Hooke's law is obeyed for this range and that the cross-sectional area remains constant.

the Young modulus for the material of the cord = $2.0 \times 10^7$ Pa

(i) Calculate the tension in the cord at this extension.

(ii) Calculate the energy stored in the cord at this extension. *(5 marks)*

AQA, 2003

4 A material in the form of a wire, 3.0 m long and with cross-sectional area of $2.8 \times 10^{-7}$ m$^2$, is suspended from a support so that it hangs vertically. Different masses may be suspended from its lower end. The table shows the extension of the wire when it is subjected to an increasing load and then a decreasing load.

| Load / N | 0 | 24 | 52 | 70 | 82 | 88 | 94 | 101 | 71 | 50 | 16 | 0 |
|---|---|---|---|---|---|---|---|---|---|---|---|---|
| Extension / mm | 0 | 2.2 | 4.6 | 6.4 | 7.4 | 8.2 | 9.6 | 13.0 | 10.2 | 8.0 | 4.8 | 3.2 |

**(a)** Plot a graph of load (on the *y*-axis) against extension (on the *x*-axis) for both increasing and decreasing loads. *(4 marks)*

**(b)** Explain what the shape of the graph tells us about the behaviour of the material in the wire. *(4 marks)*

**(c)** Using the graph, determine a value of the Young modulus for the material of the wire. *(3 marks)*

AQA, 2003

5 **Figure 3** shows two wires, one made of steel and the other of brass, firmly clamped together at their ends. The wires have the same unstretched length and the same cross-sectional area. One of the clamped ends is fixed to a horizontal support and a mass M is suspended from the other end, so that the wires hang vertically.

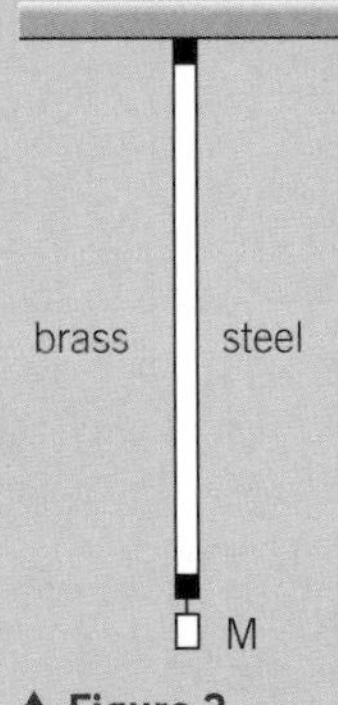


▲ **Figure 3**

(i) Since the wires are clamped together the extension of each wire will be the same. If $E_S$ is the Young modulus for steel and $E_B$ the Young modulus for brass, show that

$$\frac{E_S}{E_B} = \frac{F_S}{F_B}$$

where $F_S$ and $F_B$ are the respective forces in the steel and brass wires.

(ii) The mass M produces a total force of 15 N. Show that the magnitude of the force $F_S = 10$ N.

the Young modulus for steel = $2.0 \times 10^{11}$ Pa
the Young modulus for brass = $1.0 \times 10^{11}$ Pa

(iii) The cross-sectional area of each wire is $1.4 \times 10^{-6}$ m$^2$ and the unstretched length is 1.5 m. Determine the extension produced in either wire. *(3 marks)*

AQA, 2005

# Section 3 Summary

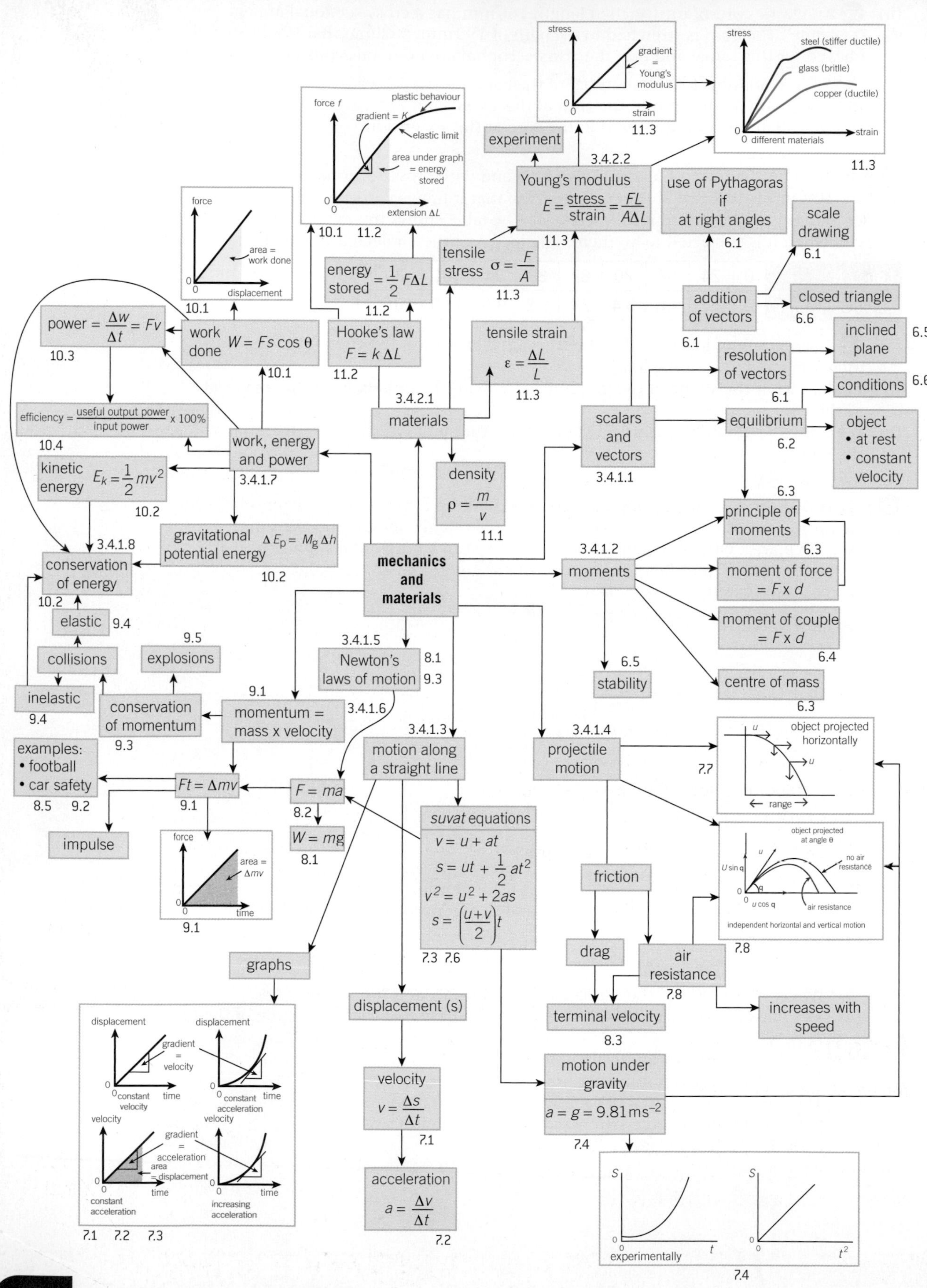

## Practical skills

In this section you have met the following practical skills:

- use of a micrometer to find the diameter of wire (e.g. Young's modulus experiment)
- use of vernier calipers to measure the diameter of a cylinder
- use of a balance to measure mass and a measuring cylinder to measure volume
- use of a forcemeter to measure force
- use of a plumb line to find a vertical line and with a set square to find a horizontal line
- use a protractor to measure angles
- use of pulleys and pivots to test equilibrium
- use of weights and springs in stretch tests
- use of light gates and a data logger to measure velocity or acceleration or to investigate collisions
- use of a stop watch for timing when investigating speed/acceleration/free fall/terminal velocity
- use of a metre rule to measure distance.

In all practical experiments:

- present experimental data in a table with headings and units using significant figures correctly
- repeat results to improve reliability
- record the precision of any instrument used
- plot a graph from data, labelling axes correctly with quantity and units
- consider the accuracy of your results/conclusion, compared to a known value and consider reasons for any types of error
- identify uncertainties in measurements and combine them to find the uncertainty in the overall experiment.

## Maths skills

In this section you have met the following maths skills:

- use of standard form and conversion to standard form from units with prefixes
- use calculators to solve calculations involving powers of ten
- use of an appropriate number of significant figures in all answers to calculations
- use of appropriate units in all answers to calculations
- calculation of a mean from a set of repeat readings (e.g. diameter of wire)
- calculate the area of cross section of a wire
- use fractions and percentages
- change the subject of an equation in order to solve calculations
- plot a graph from data provided or found experimentally (e.g. Hooke's law, free fall)
- relate $y = mx + c$ to a linear graph with physics variables to find the gradient and intercept and understand what the gradient and intercept represent in physics.

## Extension task

You are about to buy a small car. Research different manufacturers' specifications on the internet and choose the one that is the best from a safety point of view. Present your findings in a suitable way (for example, as a presentation) including a comparison relating to:

- air bags
- crumple zones
- seat belts
- collapsible steering wheel
- bumpers and any other features a particular model has that will improve your safety in the event of an accident.

# Practice questions: Mechanics and materials

1 A uniform heavy metal bar of weight 250 N is suspended by two vertical wires, supported at their upper ends from a horizontal surface, as shown in **Figure 1**.

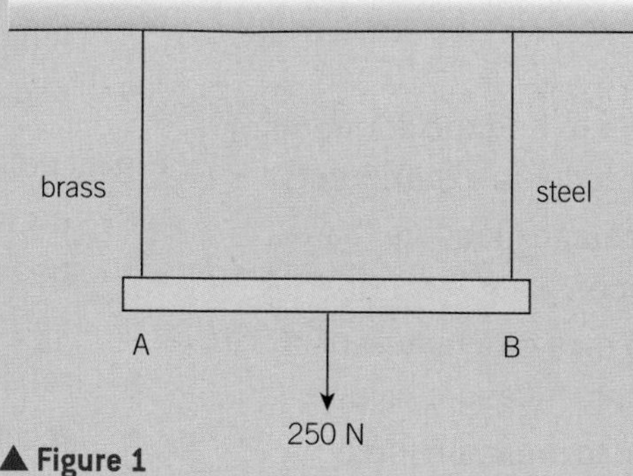


▲ **Figure 1**

One wire is made of brass and the other of steel. The cross-sectional area of each wire is $2.5 \times 10^{-7}$ m$^2$ and the unstretched length of each wire is 2.0 m.

the Young modulus for brass = $1.0 \times 10^{11}$ Pa
the Young modulus for steel = $2.0 \times 10^{11}$ Pa

(i) Calculate the extension of the steel wire, if the tension in it is 125 N.
(ii) Estimate how much lower end A will be than end B. *(5 marks)*

AQA, 2002

2 Spectacle lenses can be tested by dropping a small steel ball onto the lens, as shown in **Figure 2**, and then checking the lens for damage.

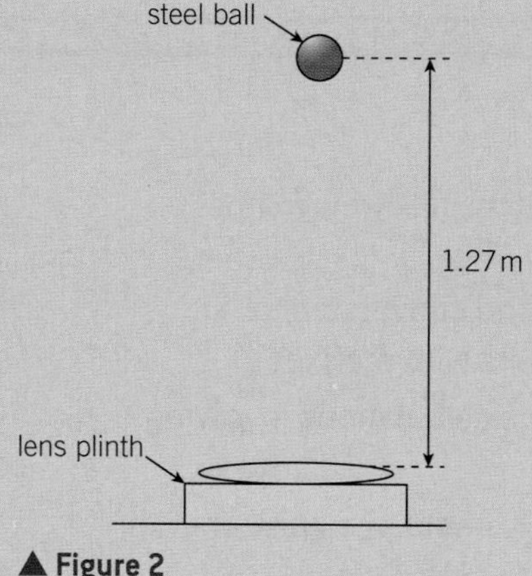


▲ **Figure 2**

A test requires the following specifications:
diameter of the ball = 16 mm
mass of ball = 16 g
height of drop = 1.27 m

**(a)** Calculate the density of the steel used for the ball. *(3 marks)*
**(b)** In a test the ball bounced back to a height of 0.85 m. Calculate the speed of the ball just before impact. *(2 marks)*
**(c)** Calculate the speed of the ball just after impact. *(2 marks)*
**(d)** Calculate the change in momentum of the ball due to the impact. *(2 marks)*
**(e)** The time of contact was 40 ms. Calculate the average force of the ball on the lens during the impact. *(2 marks)*
**(f)** Explain, with reference to momentum, why the test should also specify the material of the plinth the lens sits on. *(2 marks)*

AQA Specimen paper 1, 2015

3 Figure 3 shows a spacecraft that initially moves at a constant velocity of 890 m$^{-1}$ towards A. To change course, a sideways force is produced by firing thrusters. This increases the velocity towards B from 0 to 60 m s$^{-1}$ in 25 s.

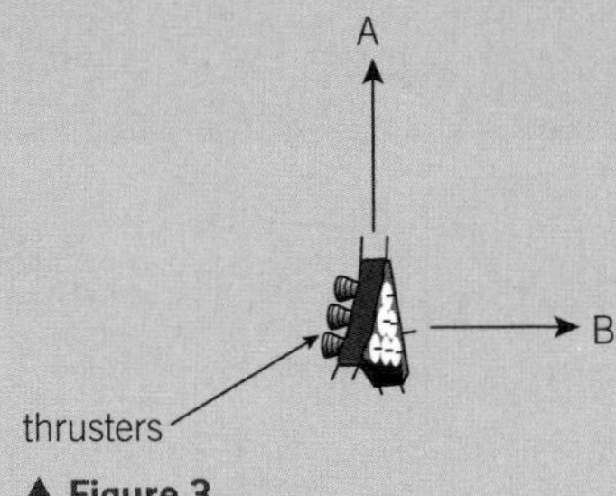


▲ Figure 3

(a) The spacecraft has a mass of $5.5 \times 10^4$ kg. Calculate:
(i) the acceleration of the spacecraft towards B
(ii) the force on the spacecraft produced by the thrusters. *(3 marks)*

(b) Calculate the magnitude of the resultant velocity after 25 s. *(2 marks)*

(c) Calculate the angle between the initial and final directions of travel. *(1 mark)*

AQA, 2004

4 (a) Explain why a raindrop falling vertically through still air reaches a constant velocity. *(4 marks)*

(b) A raindrop falls at a constant vertical velocity of 1.8 m s$^{-1}$ in still air. The mass of the raindrop is $7.2 \times 10^{-9}$ kg.
Calculate:
(i) the kinetic energy of the raindrop
(ii) the work done on the raindrop as it falls through a vertical distance of 4.5 m. *(4 marks)*

(c) The raindrop in part (b) now falls through air in which a horizontal wind is blowing. If the velocity of the wind is 1.4 m s$^{-1}$, use a scale diagram or calculation to determine the magnitude and direction of the resultant velocity of the raindrop. *(3 marks)*

AQA, 2005

5 Tidal power could make a significant contribution to UK energy requirements. This question is about a tidal power station which traps sea water behind a tidal barrier at high tide and then releases the water through turbines 10.0 m below the high tide mark.

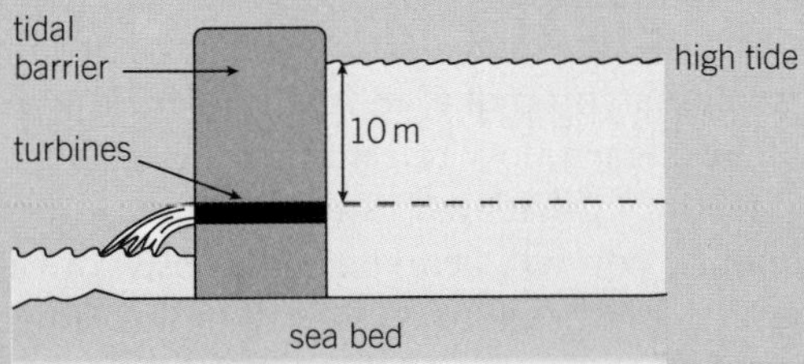


▲ Figure 4

(i) Calculate the mass of sea water covering an area of 120 km$^2$ and depth 10.0 m.
density of sea water = 1100 kg m$^{-3}$

(ii) Calculate the maximum loss of potential energy of the sea water in part (i) when it is released through the turbines.

(iii) The potential energy of the sea water released through the turbines, calculated in part (ii), is lost over a period of 6.0 hours. Estimate the average power output of the power station over this time period. Assume the power station efficiency is 40%. *(7 marks)*

AQA, 2003

# Section 4
## Electricity

**Chapters in this section:**

**12** Electric current

**13** Direct current circuits

## Introduction

Our electricity supplies have improved our lives in many ways. About 200 years ago, scientists like André-Marie Ampère and Alessandro Volta discovered how to produce and measure electricity. They carried out many experiments that helped them to discover the principles and laws of electricity, which engineers today use to design the circuits in all of our electrical devices. After Michael Faraday discovered how to generate electricity and transform it to different voltages, electricity became part of everyday life. Engineers have designed and installed electric circuits to supply electricity from power stations to our homes, offices, and factories. Today, we expect electricity to be available to us day and night, throughout the year. All electrical engineers start their training by studying the basic principles of electricity, which you will meet in this section. You will deepen your understanding of these principles and gain experience of practical work in electricity, which will help you to design and construct low-voltage electric circuits and to analyse circuits that have different components.

In studying electricity, you will also look at materials such as *semiconductors*, which were unknown to scientists before the twentieth century, and are used today in many areas of the electronics industry. You will explore properties such as *superconductivity*, which could in the future lead to entirely new industries. Michael Faraday would have been amazed at just how important electricity has become today!

## Working scientifically

In this part of the course, you will carry out calculations from the measurements you make, to further develop your calculation skills. You will also learn how to estimate the uncertainty of a measurement or of a quantity calculated from your measurements. In some experiments, you will be expected to use your measurements to plot a graph that is predicted to be a straight line. In these experiments, you may be asked to determine the graph gradient (and/or intercept) and then relate the values you get to a physical quantity. Make good use of the notes and exercises in Chapter 16, including the section on straight line graphs, to help you with your maths skills.

Practical work in this part of the course involves using low-voltage circuits. Electrical safety and good practice is *very* important whenever you are carrying out practical work in electricity. Here are some things you need to remember to help you stay safe:

- If you are using a power supply unit instead of a battery as the voltage source in your circuit, always make sure the supply cable of a power supply unit is out of the way of the low-voltage circuit wires and any other apparatus you are using.

- Make sure you have a switch in the circuit or on the power supply unit that you can use to switch the current on or off, and *never* touch the switch, the circuit, or the power supply unit with damp or wet hands.
- Before switching a circuit on, make sure ammeters and voltmeters are correctly connected in the circuit and that the range setting of each meter is correct.
- Ask your teacher to check your circuit *before* you switch it on. Switch off the circuit if any component becomes hot, and switch the circuit off when it is not in use.

## What you already know

From your GCSE studies on electricity, you should know that:

- ☐ An electric current is a flow of charge around a circuit due to the movement of electrons around the circuit.
- ☐ Electric current is measured in amperes (A) by using an ammeter. Electric potential difference, or voltage, is measured in volts (V) by using a voltmeter.
- ☐ For components in series, the current is the same in each component, and the sum of the voltages across the components is equal to the total voltage.
- ☐ For components in parallel, the voltage is the same across each component, and the sum of the currents through the components is equal to the total current.
- ☐ For resistors in series, the total resistance is equal to the sum of the resistances of each resistor.
- ☐ For resistors in parallel, the current is greatest through the resistor that has least resistance.

# 12 Electric current

## 12.1 Current and charge

**Learning objectives:**

- → Define an electric current.
- → Calculate the charge flow in a circuit.
- → Define charge carriers.

*Specification reference: 3.5.1.1*

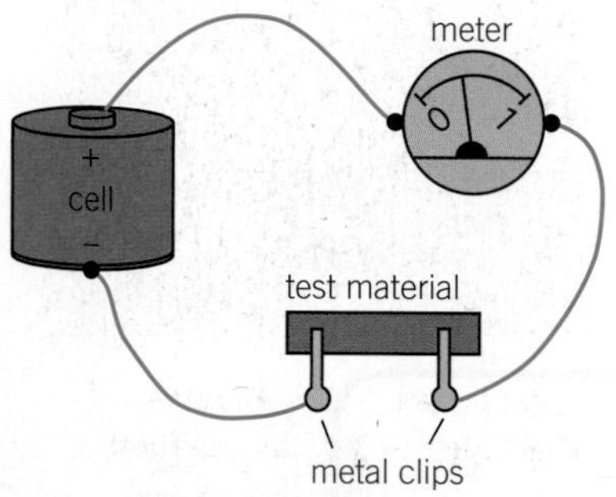


▲ **Figure 1** *Testing for conduction*

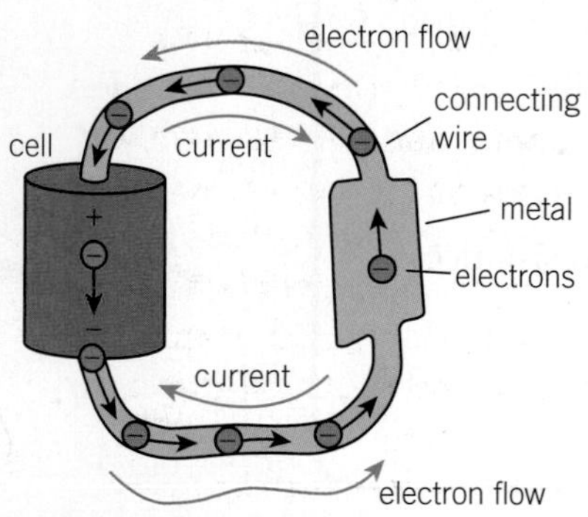


▲ **Figure 2** *Convention for current*

### Electrical conduction

To make an electric current pass round a circuit, the circuit must be complete and there must be a source of potential difference, such as a battery, in the circuit. The electric current is the rate of flow of charge in the wire or component. The current is due to the passage of charged particles. These charged particles are referred to as **charge carriers**.

- In metals, the charge carriers are conduction electrons. They move about inside the metal, repeatedly colliding with each other and the fixed positive ions in the metal.
- In comparison, when an electric current is passed through a salt solution, the charge is carried by ions, which are charged atoms or molecules.

A simple test for conduction of electricity is shown in Figure 1. The meter shows a non-zero reading whenever any conducting material is connected into the circuit. The battery forces the charge carriers through the conducting material and causes them to pass through the battery and the meter. If the test material is a metal, the charge carriers in all parts of the circuit are electrons. These electrons enter the battery at its positive terminal after passing through the metal and the ammeter, and leave at the negative terminal to continue to cycle again.

The convention for the direction of current in a circuit is from positive (+) to negative (−), as shown in Figure 2. The convention was agreed long before the discovery of electrons. When it was set up, it was known that an electric current is a flow of charge one way round a circuit. However, it was not known whether the current was due to positive charge flowing round the circuit from + to −, or if it was due to negative charge flowing from − to +.

The unit of current is the *ampere* (A), which is defined in terms of the magnetic force between two parallel wires when they carry the same current. The symbol for current is $I$.

The unit of charge is the *coulomb* (C), equal to the charge flow in one second when the current is one ampere. The symbol for charge is $Q$.

For a current $I$, the charge flow $\Delta Q$ in time $\Delta t$ is given by

$$\Delta Q = I\Delta t$$

For example, the charge flow for a current of:

- 1 A in 10 s is 10 C
- 5 A in 200 s is 1000 C
- 10 mA in 500 s is 5 C.

For charge flow $\Delta Q$ in a time interval $\Delta t$, the current $I$ is given by

$$I = \frac{\Delta Q}{\Delta t}$$

The equation shows that a current of 1 A is due to a flow of charge of 1 coulomb per second. As the magnitude of the charge of the electron is $1.6 \times 10^{-19}$ C, a current of 1 A along a wire must be due to $6.25 \times 10^{18}$ electrons passing along the wire each second.

## More about charge carriers

Materials can be classified in electrical terms as conductors, insulators, or semiconductors.

- In an insulator, each electron is attached to an atom and cannot move away from the atom. When a voltage is applied across an insulator, no current passes through the insulator, because no electrons can move through the insulator.
- In a metallic conductor, most electrons are attached to atoms but some are delocalised – the delocalised electrons are the charge carriers in the metal. When a voltage is applied across the metal, these conduction electrons are attracted towards the positive terminal of the metal.
- In a semiconductor, the number of charge carriers increases with an increase of temperature. The resistance of a semiconductor therefore decreases as its temperature is raised. A pure semiconducting material is referred to as an intrinsic semiconductor because conduction is due to electrons that break free from the atoms of the semiconductor.

### Rechargeable batteries

A car battery is a 12 V rechargeable battery designed to supply a very large current to start the engine. The battery is recharged when the car engine is running. Smaller rechargeable batteries are used in portable electronic equipment, for example, in mobile phones. Such a battery supplies a much smaller current than a car battery. Disposable batteries can't be recharged. Once a disposable battery has run down, it is no longer of any use – not as environmentally friendly as a rechargeable battery!

## Summary questions

$e = 1.6 \times 10^{-19}$ C

1 a The current in a certain wire is 0.35 A. Calculate the charge passing a point in the wire **i** in 10 s, **ii** in 10 min.

b Calculate the average current in a wire through which a charge of 15 C passes in **i** 5 s, **ii** 100 s.

2 Calculate the number of electrons passing a point in the wire in 10 min when the current is **a** 1.0 μA **b** 5.0 A.

3 In an electron beam experiment, the beam current is 1.2 mA. Calculate

**a** the charge flowing along the beam each minute

**b** the number of electrons that pass along the beam each minute.

4 A certain type of rechargeable battery is capable of delivering a current of 0.2 A for 4000 s before its voltage drops and it needs to be recharged. Calculate the maximum time it could be used for without being recharged if the current through it was **a** 0.5 A **b** 0.1 A.

### Hint

We use Δ (the Greek capital letter delta) where a change takes place.

So $\Delta Q/\Delta t$ means 'change per second' of charge (i.e., rate of change of charge).

### Physics and the human genome project

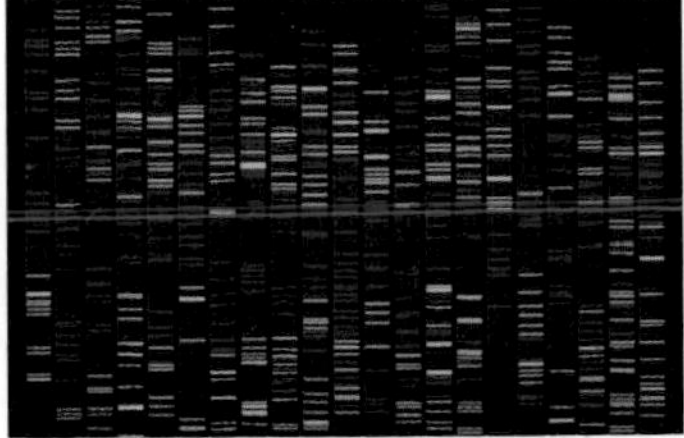

To map the human genome, fragments of DNA are tagged with amino acid bases. Each tagged fragment carries a negative charge. A voltage is applied across a strip of gel with a spot of liquid on it containing tagged fragments. The fragments are attracted to the positive electrode. The smaller the fragment, the faster it moves, so the fragments separate out according to size as they move to the positive electrode. The fragments pass through a spot of laser light, which causes a dye attached to each tag to fluoresce as it passes through the laser spot. Light sensors linked to a computer detect the glow from each tag. The computer is programmed to work out and display the sequence of bases in the DNA fragments.

**Q:** Why are the tagged fragments attracted to the positive electrode?

**Answer:** Because they carry a negative charge, and opposite charges attract.

# 12.2 Potential difference and power

## Learning objectives:

- → Define potential difference.
- → Calculate electrical power.
- → Explain how energy transfers take place in electrical devices.

*Specification reference: 3.5.1.1*

## Energy and potential difference

When a torch bulb is connected to a battery, electrons deliver energy from the battery to the torch bulb. Each electron moves around the circuit and takes a fixed amount of energy from the battery as it passes through it. The electrons then deliver energy to the bulb as they pass through it. After delivering energy to the bulb, each electron re-enters the battery via the positive terminal to be resupplied with more energy to deliver to the bulb.

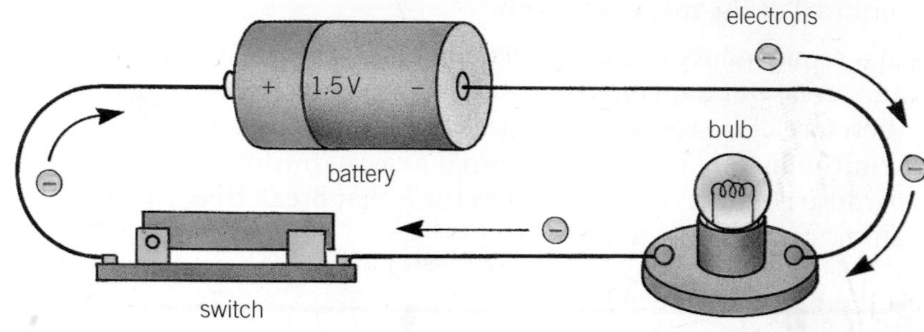


▲ **Figure 1** *Energy transfer by electrons*

A battery has the potential to transfer energy from its chemical store if the battery is not part of a complete circuit. When the battery is in a circuit, each electron passing through a circuit component does work to pass through the component and therefore transfers some or all of its energy. The work done by an electron is equal to its loss of energy. The work done per unit charge is defined as the **potential difference** (abbreviated as pd) or *voltage* across the component.

▲ **Figure 2** *Sources of emf*

**Potential difference is defined as the work done (or energy transfer) per unit charge.** The unit of pd is the volt, which is equal to 1 joule per coulomb.

If work $W$ is done when charge $Q$ flows through the component, the pd across the component, $V$, is given by

$$V = \frac{W}{Q}$$

Rearranging this equation gives $W = QV$ for the work done or energy transfer when charge $Q$ passes through a component which has a pd $V$ across its terminals.

For example:

- If 30 J of work is done when 5 C of charge passes through a component, the pd across the component must be 6 V (= 30 J/5 C).
- If the pd across a component in a circuit is 12 V, then 3 C of charge passing through the component would transfer 36 J of energy from the battery to the component.

**The emf of a source of electricity is defined as the electrical energy produced per unit charge passing through the source.** The unit of emf is the volt, the same as the unit of pd.

### Study tip

Remember 1 volt = 1 joule per coulomb.

For a source of emf $\varepsilon$ in a circuit, the electrical energy produced when charge $Q$ passes through the source = $Q\varepsilon$. This energy is transferred to other parts of the circuit and some may be dissipated in the source itself due to the source's internal resistance.

## Synoptic link

You will meet emf in more detail in Topic 13.3, Electromotive force and internal resistance.

### Energy transfer in different devices

▲ **Figure 3** *Electrical devices*

An electric current has a heating effect when it passes through a component with resistance. It also has a magnetic effect, which is made use of in electric motors and loudspeakers.

1 In a device that has resistance, such as an electrical heater, the work done on the device is transferred as thermal energy. This happens because the charge carriers repeatedly collide with atoms in the device and transfer energy to them, so the atoms vibrate more and the resistor becomes hotter.
2 In an electric motor turning at a constant speed, the work done on the motor is equal to the energy transferred to the load and surroundings by the motor, so the kinetic energy of the motor remains constant. The charge carriers are electrons that need to be forced through the wires of the spinning motor coil against the opposing force on the electrons due to the motor's magnetic field.
3 For a loudspeaker, the work done on the loudspeaker is transferred as sound energy. Electrons need to be forced through the wires of the vibrating loudspeaker coil against the force on them due to the loudspeaker magnet.

## Electrical power and current

Consider a component or device that has a potential difference $V$ across its terminals and a current $I$ passing through it. In time $\Delta t$:

- the charge flow through it, $Q = I\,\Delta t$
- the work done by the charge carriers, $W = QV = (I\,\Delta t)\,V = IV\,\Delta t$.

**Work done $W = IV\Delta t$**

The energy transfer $\Delta E$ in the component or device is equal to the work done $W$.

Because power = $\frac{\text{energy}}{\text{time}}$, the electrical power $P$ supplied to the device is

$$\frac{IV\,\Delta t}{\Delta t} = IV$$

**Electrical power $P = IV$**

***Notes:***

1 This equation can be rearranged to give $I = \frac{P}{V}$ or $V = \frac{P}{I}$.
2 The unit of power is the watt (W). Therefore one volt is equal to one watt per ampere. For example, if the pd across a component is 4 V, then the power delivered to the component is 4 W per ampere of current.

## Summary questions

1 Calculate the energy transfer in 1200 s in a component when the pd across it is 12 V and the current is:
 a 2 A
 b 0.05 A.
2 A 6 V, 12 W light bulb is connected to a 6 V battery. Calculate:
 a the current through the light bulb
 b the energy transfer to the light bulb in 1800 s.
3 A 230 V electrical appliance has a power rating of 800 W. Calculate **i** the energy transfer in the appliance in 1 min, **ii** the current taken by the appliance.
4 A battery has an emf of 9 V and negligible internal resistance. It is capable of delivering a total charge of 1350 C. Calculate:
 a the maximum energy the battery could deliver
 b the power it would deliver to the components of a circuit if the current through it was 0.5 A
 c how long the battery would last for, if it were to supply power at the rate calculated in **b**.

# 12.3 Resistance

**Learning objectives:**

- → Describe what causes electrical resistance.
- → Discuss when Ohm's law can be used.
- → Explain what a superconductor is.

*Specification reference: 3.5.1.1; 3.5.1.3*

## Definitions and laws

The resistance of a component in a circuit is a measure of the difficulty of making current pass through the component. Resistance is caused by the repeated collisions between the charge carriers in the material with each other and with the fixed positive ions of the material.

**The resistance of any component is defined as** $\dfrac{\textbf{the pd across the component}}{\textbf{the current through it}}$

For a component which passes current $I$ when the pd across it is $V$, its resistance $R$ is given by the equation

$$R = \frac{V}{I}$$

The unit of resistance is the ohm (Ω), which is equal to 1 volt per ampere.

Rearranging the above equation gives $V = IR$ or $I = \dfrac{V}{R}$

### *Reminder about prefixes*

▼ **Table 1** *Prefixes*

| Prefix | nano | micro | milli | kilo | mega | giga |
|---|---|---|---|---|---|---|
| Symbol | n | μ | m | k | M | G |
| Value | $10^{-9}$ | $10^{-6}$ | $10^{-3}$ | $10^{3}$ | $10^{6}$ | $10^{9}$ |

### Worked example

The current through a component is 2.0 mA when the pd across it is 12 V. Calculate:

**a** its resistance at this current

**b** the pd across the component when the current is 50 μA, assuming its resistance is unchanged.

*Solution*

**a** $R = \dfrac{V}{I} = \dfrac{12}{2.0 \times 10^{-3}} = 6000\ \Omega$

**b** $V = IR = 50 \times 10^{-6} \times 6000 = 0.30\ \text{V}$

## Measurement of resistance

A resistor is a component designed to have a certain resistance, which is the same regardless of the current through it. The resistance of a resistor can be measured using the circuit shown in Figure 1.

- The ammeter is used to measure the current through the resistor. The ammeter must be in series with the resistor, so the same current passes through both the resistor and the ammeter.
- The voltmeter is used to measure the pd across the resistor. The voltmeter must be in parallel with the resistor so that they

have the same pd. Also, no current should pass through the voltmeter, otherwise the ammeter will not record the exact current through the resistor. In theory, the voltmeter should have infinite resistance. In practice, a voltmeter with a sufficiently high resistance would be satisfactory.

- The variable resistor is used to adjust the current and pd as necessary. To investigate the variation of current with pd, the variable resistor is adjusted in steps. At each step, the current and pd are recorded from the ammeter and voltmeter, respectively. The measurements can then be plotted on a graph of pd against current, as shown in Figure 2 (or for current against pd – see Topic 12.4).

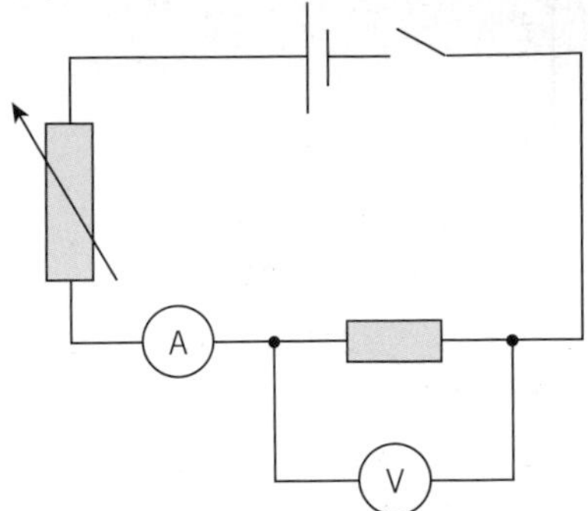


▲ **Figure 1** *Measuring resistance*

The graph for a resistor is a straight line through the origin. The resistance is the same, regardless of the current. The resistance is equal to the gradient of the graph because the gradient is constant and at any point is equal to the pd divided by the current. In other words, the pd across the resistor is proportional to the current. The discovery that the pd across a metal wire is proportional to the current through it was made by Georg Ohm in 1826, and is known as Ohm's law.

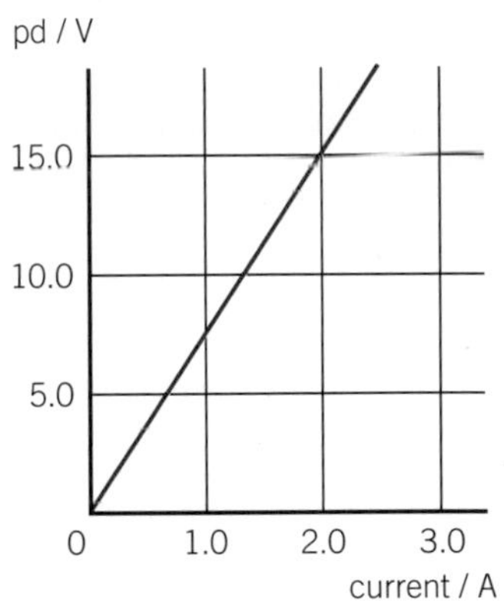


▲ **Figure 2** *Graph of pd versus current for a resistor*

**Ohm's law states that the pd across a metallic conductor is proportional to the current through it, provided the physical conditions do not change.**

***Notes:***

1 Ohm's law is equivalent to the statement that the resistance of a metallic conductor under constant physical conditions (e.g., temperature) is constant.

2 For an ohmic conductor, $V = IR$, where $R$ is constant. A resistor is a component designed to have a certain resistance.

3 If the current and pd measurements for an ohmic conductor are plotted with current on the $y$-axis and pd on the $x$-axis, the gradient of this graph gives $1/R$.

## Resistivity

For a conductor of length $L$ and uniform cross-sectional area $A$, as shown in Figure 3, its resistance $R$ is

- proportional to $L$
- inversely proportional to $A$.

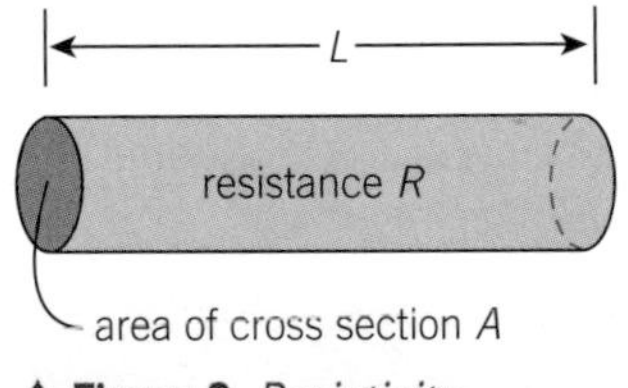


▲ **Figure 3** *Resistivity*

Hence $R = \frac{\rho L}{A}$, where $\rho$ is a constant for that material, known as its resistivity.

Rearranging this equation gives the following equation, which can be used to calculate the resistivity of a sample of material of length $L$ and uniform cross-sectional area $A$:

$$\textbf{resistivity, } \rho = \frac{RA}{L}$$

***Notes:***

1 The unit of resistivity is the ohm metre ($\Omega$m).

2 For a conductor with a circular cross section of diameter $d$, $A = \frac{\pi d^2}{4}$ ($= \pi r^2$ where radius $r = d/2$).

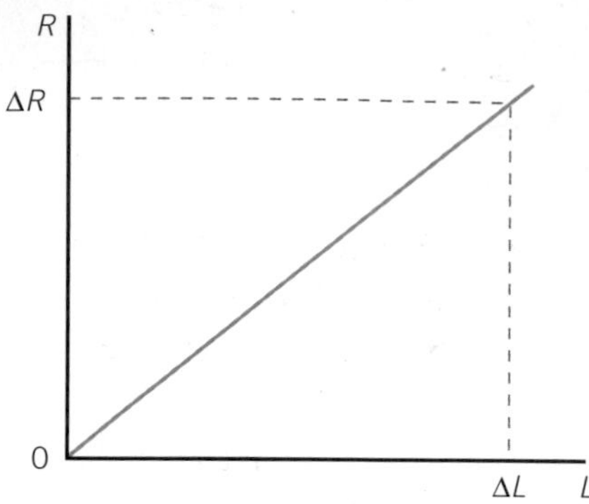


▲ **Figure 4** *Graph of resistance against length for a wire*

▼ **Table 2** *Resistivity values of different materials at room temperature*

| Material | Resistivity / Ω m |
|---|---|
| copper | $1.7 \times 10^{-8}$ |
| constantan | $5.0 \times 10^{-7}$ |
| carbon | $3 \times 10^{-5}$ |
| silicon | 2300 |
| PVC | about $10^{14}$ |

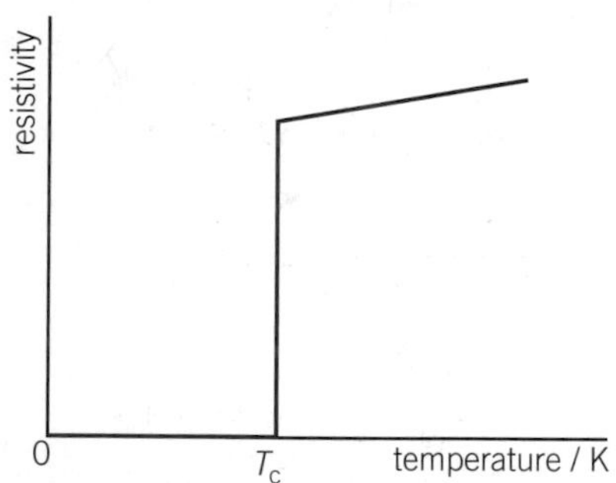


▲ **Figure 5** *Resistivity of a superconductor versus temperature near the critical temperature*

## Synoptic link

For more information about MRI scanners and particle accelerators, see Topic 3.6, Wave–particle duality, and Topic 2.1, The particle zoo, respectively.

## Study tip

Resistivity is a property of a material. Don't confuse resistivity and resistance.

To determine the resistivity of a wire

- Measure the diameter of the wire $d$ using a micrometer at several different points along the wire, to give a mean value for $d$ to calculate its cross-sectional area $A$.
- Measure the resistance $R$ of different lengths $L$ of wire to plot a graph of $R$ against $L$ (Figure 4).

The resistivity of the wire is given by the graph gradient × $A$.

## Superconductivity

A **superconductor** is a wire or a device made of material that has zero resistivity at and below a **critical temperature** that depends on the material. This property of the material is called **superconductivity**. The wire or device has zero resistance below the critical temperature of the material. When a current passes through it, there is no pd across it because its resistance is zero. So the current has no heating effect.

A superconductor material loses its superconductivity if its temperature is raised above its critical temperature. In 2014, the highest critical temperature claimed and awaiting verification is 150 K (−123 °C) for a compound containing mercury, barium, calcium, copper, and oxygen. Any material with a critical temperature above 77 K (−196 °C), the boiling point of liquid nitrogen, is referred to as a high-temperature superconductor.

Superconductors are used to make high-power electromagnets that generate very strong magnetic fields in devices such as MRI scanners and particle accelerators. These strong magnetic fields are also used in the development of new applications such as lightweight electric motors and power cables that transfer electrical energy without energy dissipation.

## Summary questions

**1** **a** Complete the table below by calculating the missing value for each resistor.

| | 1 | 2 | 3 | 4 | 5 |
|---|---|---|---|---|---|
| Current / A | 2.0 | 0.45 | | $5.0 \times 10^{-3}$ | |
| Pd / V | 12.0 | | 5.0 | 0.80 | $5.0 \times 10^{4}$ |
| Resistance / Ω | | 22 | $4.0 \times 10^{4}$ | | $2.0 \times 10^{7}$ |

**b** Find the resistance of the resistor that gave the results shown in Figure 2.

**2** Calculate the resistance of a uniform wire of diameter 0.32 mm and length 5.0 m. The resistivity of the material = $5.0 \times 10^{-7}$ Ω m.

**3** Calculate the resistance of a rectangular strip of copper of length 0.08 m, thickness 15 mm, and width 0.80 mm. The resistivity of copper = $1.7 \times 10^{-8}$ Ω m.

**4** A wire of uniform diameter 0.28mm and length 1.50 m has a resistance of 45 Ω. Calculate:

**a** its resistivity

**b** the length of this wire that has a resistance of 1.0 Ω.

# 12.4 Components and their characteristics

## Learning objectives:

→ Describe how the current through a filament lamp varies with pd.

→ State the characteristics of a diode.

→ Describe what we use a thermistor for.

*Specification reference: 3.5.1.2; 3.5.1.3*

## Circuit diagrams

Each type of component has its own symbol, which is used to represent the component in a circuit diagram. You need to recognise the symbols for different types of components to make progress – just like a motorist needs to know what different road signs mean. Note that on a circuit diagram, the direction of the current is always shown from + to – round the circuit.

You should be able to recognise the component symbols shown in Figure 1. Here are some notes about some of the components.

- A cell is a source of electrical energy. Note that a battery is a combination of cells.
- The symbol for an indicator is the same as that for a light source (including a filament lamp, but not a light-emitting diode).
- A diode allows current in one direction only. A light-emitting diode (or LED) emits light when it conducts. The direction in which the diode conducts is referred to as its forward direction. The opposite direction is referred to as its reverse direction. Examples of the use of diodes include the protection of dc circuits (in case the voltage supply is connected the wrong way round).
- A resistor is a component designed to have a certain resistance.
- The resistance of a thermistor decreases with increasing temperature, if the thermistor is an intrinsic semiconductor such as silicon.
- The resistance of a **light-dependent resistor** (LDR) decreases with increasing light intensity.

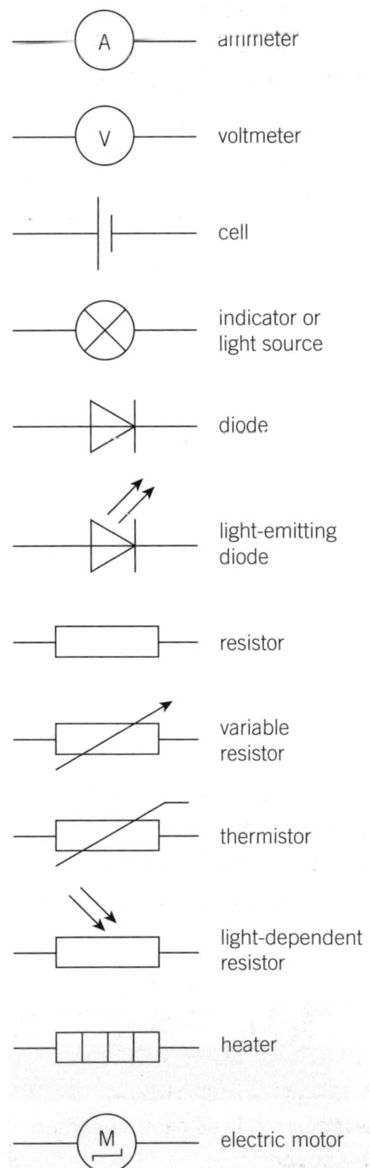


▲ **Figure 1** *Circuit components*

## Investigating the characteristics of different components

To measure the variation of current with pd for a component, use either

- a potential divider to vary the pd from zero, or
- a variable resistor to vary the current to a minimum.

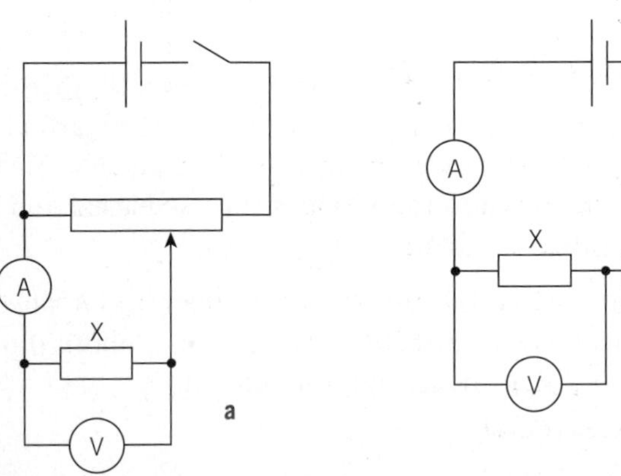


▲ **Figure 2** *Investigating component characteristics* **a** *Using a potential divider,* **b** *Using a variable resistor*

The advantage of using a potential divider is that the current through the component and the pd across it can be reduced to zero. This is not possible with a variable resistor circuit.

The measurements for each type of component are usually plotted as a graph of current (on the $y$-axis) against pd (on the $x$-axis). Typical graphs for a wire, a filament lamp, and a thermistor are shown in Figure 3. Note that the measurements are the same, regardless of which way the current passes through each of these components.

In both circuits, an ammeter sensor and a voltmeter sensor connected to a data logger could be used to capture data (i.e., to measure and record the readings) which could then be displayed directly on an oscilloscope or a computer.

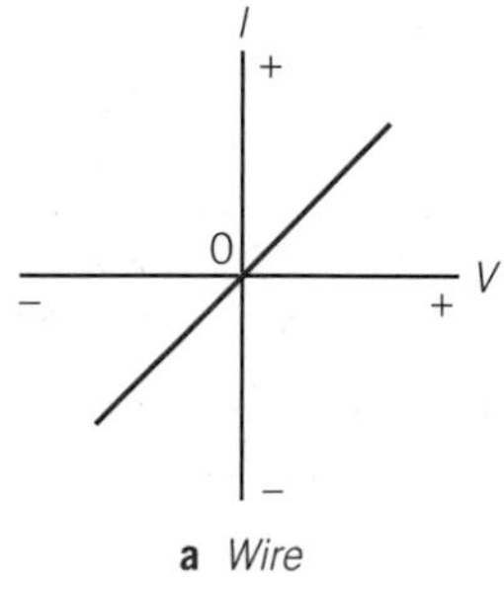


**a** *Wire*

**b** *Lamp*

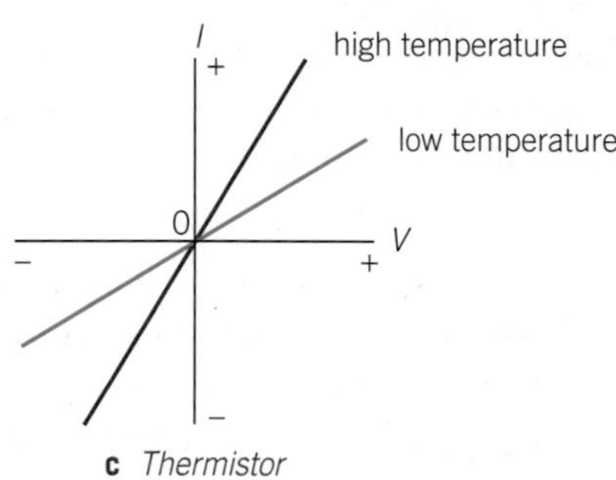


**c** *Thermistor*

▲ **Figure 3** *Current versus pd for different components*

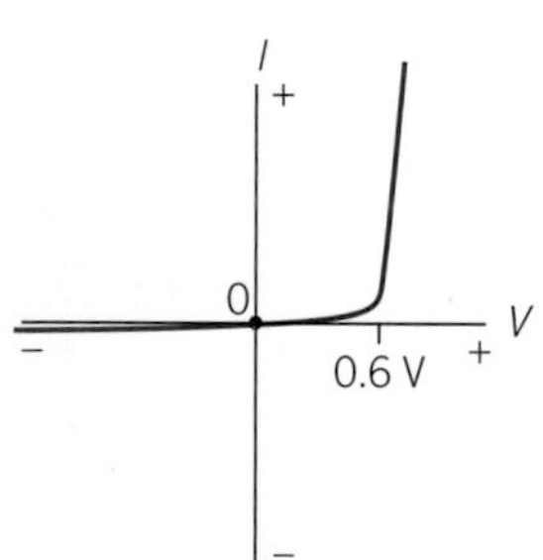


▲ **Figure 4** *Current versus pd for a diode*

- A wire gives a straight line through the origin. This means that at any point on the line, the value of $V/I$ is the same. In other words, the resistance of the wire ($= V/I$) does not change when the current changes. The gradient of the line is equal to $1/$resistance $R$ of the wire. Any resistor at constant temperature would give a straight line. The majority of components do not follow this proportional (straight-line) relationship.
- A filament bulb gives a curve with a decreasing gradient because its resistance increases as it becomes hotter.
- A thermistor at constant temperature gives a straight line. The higher the temperature, the greater the gradient of the line, as the resistance falls with increase of temperature. The same result is obtained for a light-dependent resistor in respect of light intensity.

## The diode

To investigate the characteristics of the diode, one set of measurements is made with the diode in its forward direction (i.e., forward biased) and another set with it in its reverse direction (i.e., reverse biased). The current is very small when the diode is reverse biased and can only be measured using a milliammeter.

Typical results for a silicon diode are shown in Figure 4. A silicon diode conducts easily in its forward direction above a pd of about 0.6 V and hardly at all below 0.6 V or in the opposite direction.

**Hint**

Remember, a diode needs a certain pd to conduct.

## Resistance and temperature

The resistance of a metal increases with increase of temperature. This is because the positive ions in the conductor vibrate more when its temperature is increased. The charge carriers (conduction electrons) therefore cannot pass through the metal as easily when a pd is applied across the conductor. A metal is said to have a **positive temperature coefficient** because its resistance increases with increase of temperature.

The resistance of an intrinsic semiconductor decreases with increase of temperature. This is because the number of charge carriers (conduction electrons) increases when the temperature is increased. A thermistor made from an intrinsic semiconductor therefore has a **negative temperature coefficient**. Its percentage change of resistance per kelvin change of temperature is much greater than for a metal. For this reason, thermistors are often used as the temperature-sensitive component in a temperature sensor.

Figure 5 shows how the resistance of a thermistor and a metal wire vary with temperature. Both components have the same resistance at 0 °C. The resistance of the thermistor decreases non-linearly with increase of temperature, whereas the resistance of the metal wire increases much less over the same temperature range.

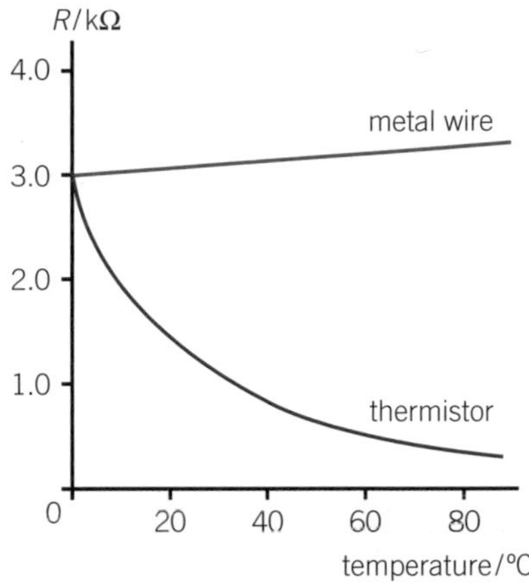


▲ **Figure 5** *Resistance variation with temperature for a thermistor and a metal wire*

### Summary questions

**1** A filament bulb is labelled '3.0 V, 0.75 W'.

- **a** Calculate its current and its resistance at 3.0 V.
- **b** State and explain what would happen to the filament bulb if the current was increased from the value in **a**.

**2** A certain thermistor has a resistance of 50 000 Ω at 20 °C and a resistance of 4000 Ω at 60 °C.

It is connected in series with an ammeter and a 1.5 V cell. Calculate the ammeter reading when the thermistor is:

- **a** at 20 °C
- **b** at 60 °C.

**3** A silicon diode is connected in series with a cell and a torch bulb.

- **a** Sketch the circuit diagram showing the diode in its forward direction.
- **b** Explain why the torch bulb would not light if the polarity of the cell were reversed in the circuit.

**4** The resistance of a certain metal wire increased from 25.3 Ω at 0 °C to 35.5 Ω at 100 °C. Assuming the resistance over this range varies linearly with temperature, calculate:

- **a** the resistance at 50 °C
- **b** the temperature when the resistance is 30.0 Ω.

1 The following measurements were made in an investigation to measure the resistivity of the material of a certain wire.

| Pd across the wire / V | 0.0 | 2.0 | 4.0 | 6.0 | 8.0 | 10.0 |
|---|---|---|---|---|---|---|
| Current through the wire / A | 0.00 | 0.15 | 0.31 | 0.44 | 0.62 | 0.74 |

Length of wire = 1.60 m.
Diameter of wire = 0.28 mm.

**(a)** Plot a graph of the pd against the current. *(4 marks)*

**(b)** Show that the pd, $V$, across the wire varies with the current $I$ according to the equation

$$V = \frac{\rho L I}{A}$$

where $\rho$ is the resistivity of the wire, $L$ is its length, and $A$ is its area of cross section. *(2 marks)*

**(c)** Use the graph to calculate the resistivity of the material of the wire. *(6 marks)*

2 A battery is connected across a uniform conductor. The current in the conductor is 40 mA.

**(a)** Calculate the total charge that flows past a point in the conductor in 3 minutes.

**(b)** Using data from the Data Booklet calculate the number of electron charge carriers passing the same point in the conductor in this time.

**(c)** If 8.6 J of energy are transferred to the conductor in this time, calculate the potential difference across the conductor.

**(d)** Calculate the resistance of the conductor. *(6 marks)*

AQA, 2004

3 **(a)** **Figure 1** shows a graph of $V$ against $I$ for a filament lamp. Calculate the maximum resistance of the lamp over the range shown by the graph. *(3 marks)*

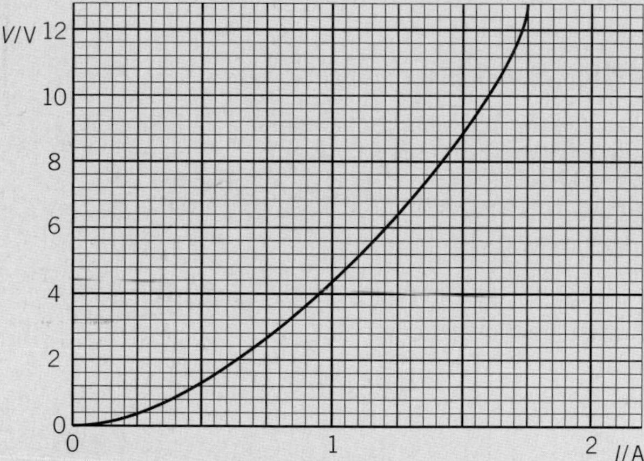


▲ **Figure 1**

**(b)** Sketch on a copy of the axes below, a graph of current against potential difference for a diode. *(2 marks)*

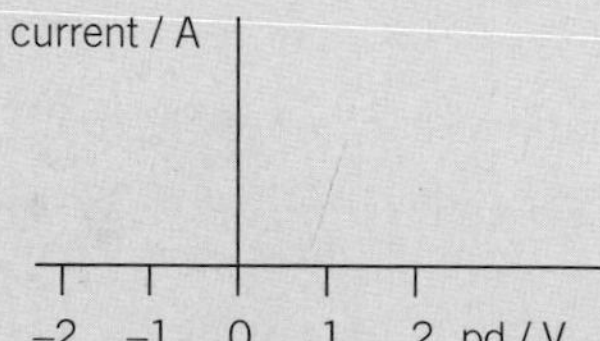


AQA, 2002

4 **Figure 2** shows the general shape of the current–time graph during the 2 seconds after a 12 V filament lamp is switched on.

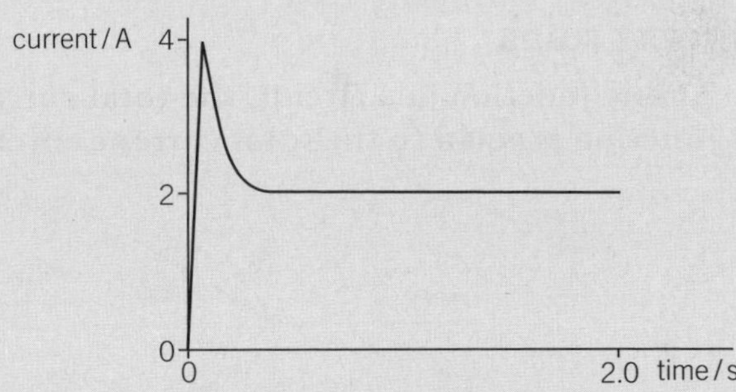


▲ **Figure 2**

**(a)** A student wishes to perform an experiment to obtain this graph.
(i) Explain why sampling data using a sensor and a computer is a sensible option.
(ii) Suggest a suitable sampling rate for such an experiment, giving a reason for your answer. *(3 marks)*

**(b)** Explain why the current rises to a high value before falling to a steady value and why a filament is more likely to fail when being switched on than at other times. *(6 marks)*

AQA, 2005

5 **(a)** A metal wire of length 1.4 m has a uniform cross-sectional area = $7.8 \times 10^{-7}\,\text{m}^2$.
Calculate the resistance, $R$, of the wire.
resistivity of the metal = $1.7 \times 10^{-8}\,\Omega\,\text{m}$ *(2 marks)*

**(b)** The wire is now stretched to twice its original length by a process that keeps its volume constant.
If the resistivity of the metal of the wire remains constant, show that the resistance increases to $4R$. *(2 marks)*

AQA, 2003

6 **(a)** The resistivity of a material in the form of a uniform resistance wire is to be measured. The area of cross section of the wire is known.
The apparatus available includes a battery, a switch, a variable resistor, an ammeter, and a voltmeter.
(i) Draw a circuit diagram, using some or all of this apparatus, which would enable you to determine the resistivity of the material.
(ii) Describe how you would make the necessary measurements, ensuring that you have a range of values.
(iii) Show how a value of the resistivity is determined from your measurements. *(9 marks)*

**(b)** A sheet of carbon-reinforced plastic measuring 80 mm × 80 mm × 1.5 mm has its two large surfaces coated with highly conducting metal film. When a potential difference of 240 V is applied between the metal films, there is a current of 2.0 mA in the plastic. Calculate the resistivity of the plastic. *(3 marks)*

AQA, 2002

7 **(a)** (i) What is a *superconductor*?
(ii) With the aid of a sketch graph, explain the term *transition temperature*. *(3 marks)*

**(b)** Explain why superconductors are very useful for applicatons which require very large electric currents and name *two* such applicatons. *(3 marks)*

AQA, 2006

# 13 Direct current circuits

## 13.1 Circuit rules

**Learning objectives:**

→ State the rules for series and parallel circuits.

→ State the principles behind these rules.

→ Describe how we use the rules in circuits.

*Specification reference: 3.5.1.4*

### Current rules

1 **At any junction in a circuit, the total current leaving the junction is equal to the total current entering the junction.**

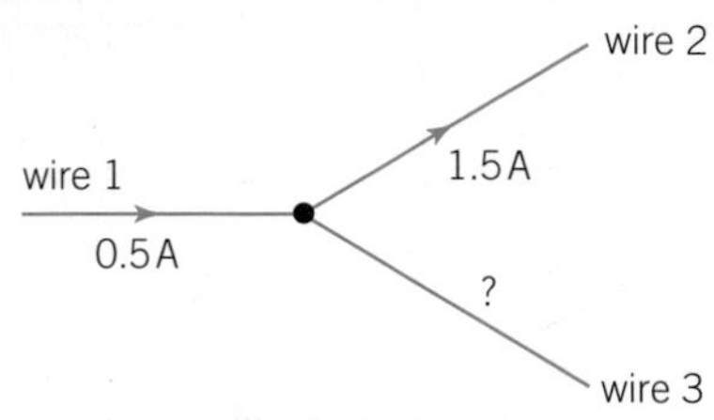


▲ **Figure 1** *At a junction*

For example, Figure 1 shows a junction of three wires where the current in two of the wires (wire 1 and wire 2) is given. The current in wire 3 must be 1.0 A into the junction, because the total current into the junction (= 1.0 A along wire 3 + 0.5 A along wire 1) is the same as the total current out of the junction (= 1.5 A along wire 2).

The junction rule holds because the rates of charge flowing into and out of a junction are always equal. The current along a wire is the charge flow per second. In Figure 1, the charge *entering* the junction each second is 0.5 C along wire 1 and 1.0 C along wire 3. The charge *leaving* the junction each second must therefore be 1.5 C as the junction does not retain charge.

2 **Components in series**

- **The current entering a component is the same as the current leaving the component**. In other words, components do not use up current. The charge per second entering a component is equal to the charge per second leaving it. In Figure 2, ammeters $A_1$ and $A_2$ show the same reading because they are measuring the same current.
- **The current passing through two or more components in series is the same through each component**. This is because the rate of flow of charge through each component is the same at any instant. The same amount of charge passing through any one component each second passes through every other component each second. In Figure 2, the ammeters read the same because the same amount of charge each second passes through each component.

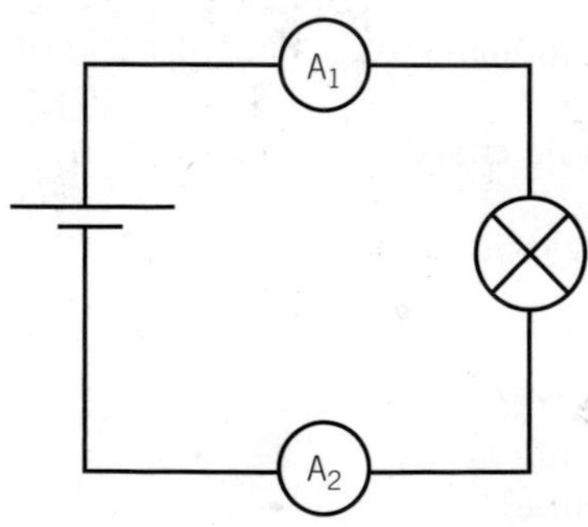


▲ **Figure 2** *Components in series*

### Potential difference rules

The potential difference (abbreviated as pd), or voltage, between any two points in a circuit is defined as the energy transfer per coulomb of charge that flows from one point to the other.

- If the charge carriers lose energy, the potential difference is a potential drop.
- If the charge carriers gain energy, which happens when they pass through a battery or cell, the potential difference is a potential rise equal to the pd across the battery or cell's terminals.

The rules for potential differences are listed below with an explanation of each rule in energy terms.

1 **For two or more components in series, the total pd across all the components is equal to the sum of the potential differences across each component.**

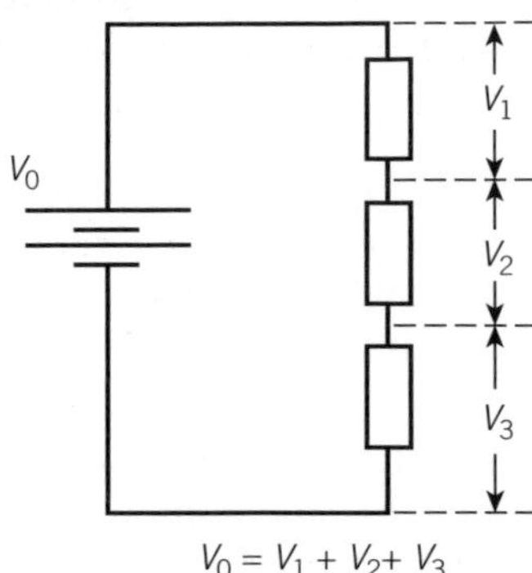


▲ **Figure 3** *Adding potential differences*

In Figure 3, the pd across the battery terminals is equal to the sum of the potential differences across the three resistors. This is because the pd across each resistor is the energy delivered per coulomb of charge to that resistor. So the sum of potential differences across the three resistors is the total energy delivered to the resistors per coulomb of charge passing through them, which is the pd across the battery terminals.

2 **The pd across components in parallel is the same.**

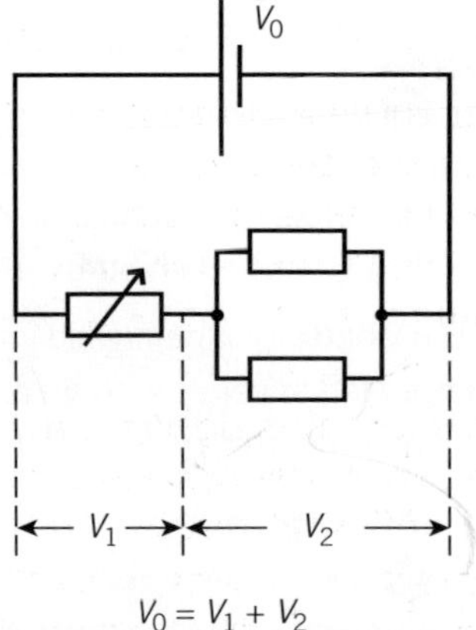


▲ **Figure 4** *Two components in parallel*

In Figure 4, charge carriers can pass through either of the two resistors in parallel. The same amount of energy is delivered by a charge carrier regardless of which of the two resistors it passes through.

Suppose the variable resistor is adjusted so the pd across it is 4 V. If the battery pd is 12 V, the pd across each of the two resistors in parallel is 8 V (= 12 V – 4 V). This is because each coulomb of charge leaves the battery with 12 J of electrical energy and uses 4 J on passing through the variable resistor. Therefore, each coulomb of charge has 8 J of electrical energy to deliver to either of the two parallel resistors.

## Study tip

You should learn the rules in bold within this topic in preparation for the rest of this chapter.

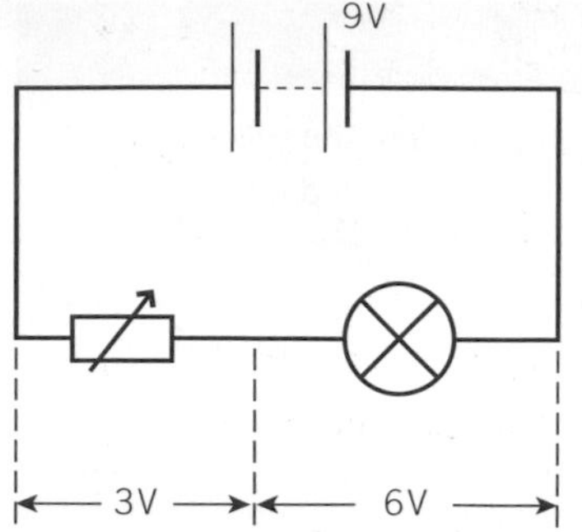

▲ **Figure 5** *The loop rule*

**Hint**

Sources of emf usually possess some internal resistance (see Topic 13.3). This can be discounted if a question states that the internal resistance of a source of emf is negligible.

3 **For any complete loop of a circuit, the sum of the emfs round the loop is equal to the sum of the potential drops around the loop.** This follows from the fact that the total emf in a loop is the total electrical energy per coulomb produced in the loop and the sum of the potential drops is the electrical energy per coulomb delivered round the loop. The above statement follows therefore from the conservation of energy.

For example, in Figure 5, the battery has an emf of 9 V. If the variable resistor is adjusted so that the pd across the light bulb is 6 V, the pd across the variable resistor is 3 V (= 9 V – 6 V). The total emf in the circuit is 9 V due to the battery. This is equal to the sum of the potential differences round the circuit (= 3 V across the variable resistor + 6 V across the light bulb). In other words, every coulomb of charge leaves the battery with 9 J of electrical energy and supplies 3 J to the variable resistor and 6 J to the light bulb.

Each time a charge carrier goes round the circuit

- a certain amount of energy $E$ is transferred to it from the battery
- it transfers energy equal to $\frac{E}{3}$ to the variable resistor, and $\frac{2E}{3}$ to the light bulb.

## Summary questions

1 A battery which has an emf of 6 V and negligible internal resistance is connected to a 6 V 6 W light bulb in parallel with a 6 V 24 W light bulb, as shown in Figure 6.

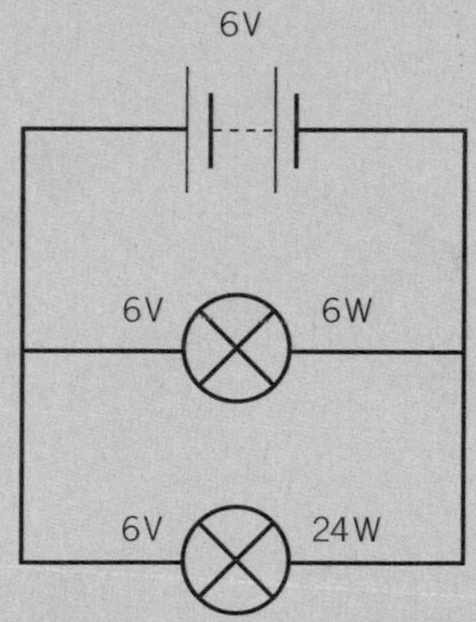

▲ **Figure 6**

Calculate **a** the current through each light bulb, **b** the current from the battery, **c** the power supplied by the battery.

2 A 4.5 V battery of negligible internal resistance is connected in series with a variable resistor and a 2.5 V 0.5 W torch bulb.

**a** Sketch the circuit diagram for this circuit.

**b** The variable resistor is adjusted so that the pd across the torch bulb is 2.5 V. Calculate **i** the pd across the variable resistor, **ii** the current through the torch bulb.

3 A 6.0 V battery of negligible internal resistance is connected in series with an ammeter, a 20 Ω resistor, and an unknown resistor $R$.

**a** Sketch the circuit diagram.

**b** The ammeter reads 0.20 A. Calculate **i** the pd across the 20 Ω resistor, **ii** the pd across $R$, **iii** the resistance of $R$.

4 In **3**, when the unknown resistor is replaced with a torch bulb, the ammeter reads 0.12 A. Calculate **a** the pd across the torch bulb, **b** the resistance of the torch bulb.

# 13.2 More about resistance

**Learning objectives:**

- → Calculate resistances in series and in parallel.
- → Define resistance heating.
- → Calculate the current and pds for each component in a circuit.

*Specification reference: 3.5.1.4*

## Resistors in series

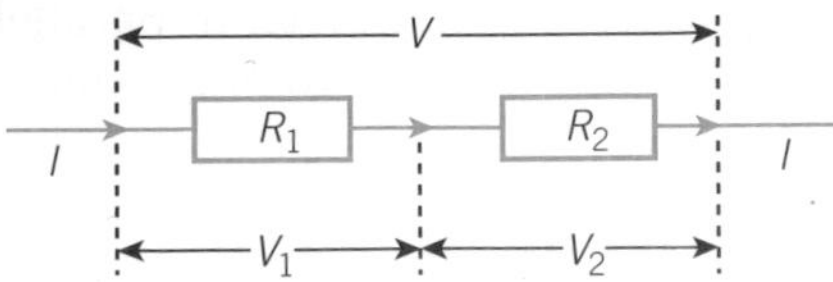


▲ **Figure 1** *Resistors in series*

Resistors in series pass the same current. The total pd is equal to the sum of the individual pds.

- For two resistors $R_1$ and $R_2$ in series, as in Figure 1, when current $I$ passes through the resistors,
  the pd across $R_1$, $V_1 = IR_1$, and
  the pd across $R_2$, $V_2 = IR_2$

The total pd across two resistors, $V = V_1 + V_2 = IR_1 + IR_2$

Therefore, the total resistance $R = \frac{V}{I} = \frac{IR_1 + IR_2}{I} = R_1 + R_2$

- For two or more resistors $R_1$, $R_2$, $R_3$, etc. in series, the theory can easily be extended to show that the **total resistance is equal to the sum of the individual resistances.**

$$\boldsymbol{R = R_1 + R_2 + R_3 + \ldots}$$

## Resistors in parallel

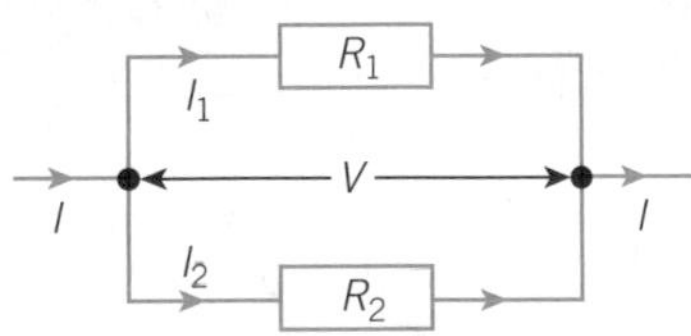


▲ **Figure 2** *Resistors in parallel*

Resistors in parallel have the same pd. The current through a parallel combination of resistors is equal to the sum of the individual currents.

- For two resistors $R_1$ and $R_2$ in parallel, as in Figure 2, when the pd across the combination is $V$,
  the current through resistor $R_1$, $I_1 = \frac{V}{R_1}$
  the current through resistor $R_2$, $I_2 = \frac{V}{R_2}$

The total current through the combination, $I = I_1 + I_2 = \frac{V}{R_1} + \frac{V}{R_2}$

Since the total resistance $R = \frac{V}{I}$, then the total current $I = \frac{V}{R}$

Therefore $\frac{V}{R} = \frac{V}{R_1} + \frac{V}{R_2}$

## Study tip

You need to know how to use the resistor rules correctly.

Notice the resistance of two resistors in parallel is equal to their product divided by their sum. For example, the resistance of a 3 Ω resistor in parallel with a 6 Ω resistor is $\frac{3 \times 6}{(3+6)} = 2\,\Omega$.

## Synoptic link

You learnt about resistance in Topic 12.3, Resistance and Topic 12.4, Components and their characteristics.

Cancelling $V$ from each term gives the following equation, which is used to calculate the total resistance $R$:

$$\frac{1}{R} = \frac{1}{R_1} + \frac{1}{R_2}$$

- For two or more resistors $R_1$, $R_2$, $R_3$, etc. in parallel, the theory can easily be extended to show that the total resistance $R$ is given by

$$\frac{1}{R} = \frac{1}{R_1} + \frac{1}{R_2} + \frac{1}{R_3} + \ldots$$

## Resistance heating

The heating effect of an electric current in any component is due to the resistance of the component. The charge carriers repeatedly collide with the positive ions of the conducting material. There is a net transfer of energy from the charge carriers to the positive ions as a result of these collisions. After a charge carrier loses kinetic energy in such a collision, the force due to the pd across the material accelerates it until it collides with another positive ion.

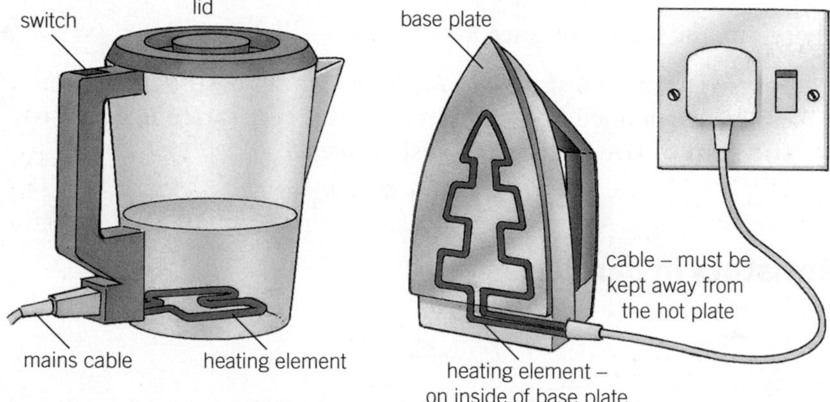


▲ **Figure 3** *Heating elements*

For a component of resistance $R$, when current $I$ passes through it, the pd across the component, $V = IR$.

Therefore the power supplied to the component,

$P = IV = I^2 R \left(= \frac{V^2}{R}\right)$.

Hence the energy per second transferred to the component as thermal energy = $I^2 R$.

If the component is at constant temperature, heat transfer to the surroundings takes place at the same rate. Therefore,

**the rate of heat transfer = $I^2 R$**

- If the component heats up, its temperature rise depends on the power supplied to it ($I^2 R$) and the rate of heat transfer to the surroundings.
- The energy transferred to the object by the electric current in time $t$ = power × time = $I^2 Rt$.
- The energy transfer per second to the component (i.e., the power supplied to it) does not depend on the direction of the current.

## Worked example

The pd across a $1000\,\Omega$ resistor in a circuit was measured at 6.0 V. Calculate the electrical power supplied to the resistor.

*Solution*

Current $I = \frac{V}{R} = \frac{6.0\,\text{V}}{1000\,\Omega} = 6.0\,\text{mA}$

Power $P = I^2R = (6.0 \times 10^{-3})^2 \times 1000 = 0.036\,\text{W}$

### Study tip

The power of the resistor could also have been calculated in one step using the equation $P = \frac{V^2}{R}$. Prove to yourself that you can obtain the same answer using this method.

## Summary questions

1 Calculate the total resistance of each of the resistor combinations in Figure 4.

a

b

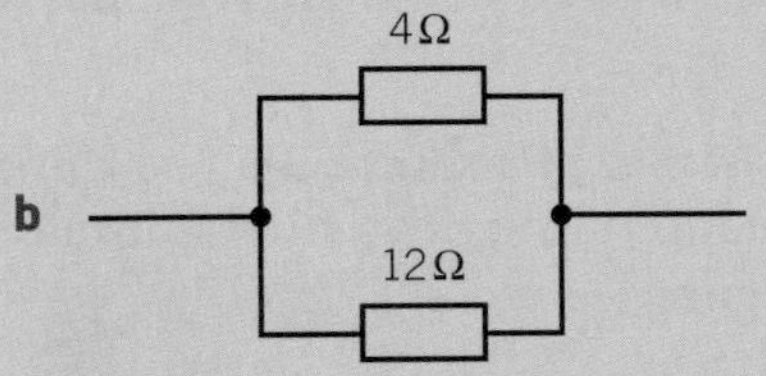


c

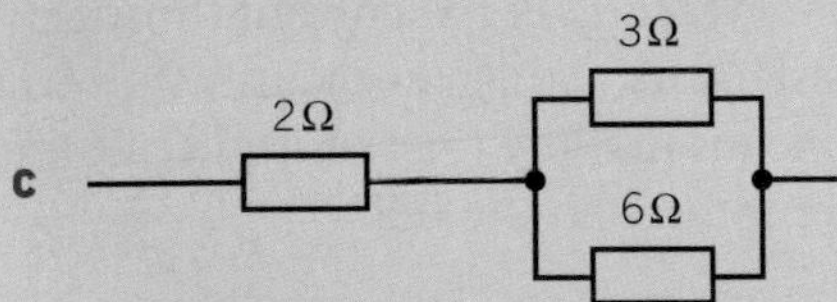


▲ **Figure 4**

2 A $3\,\Omega$ resistor and a $6\,\Omega$ resistor are connected in parallel with each other. The parallel combination is connected in series with a 6.0 V battery and a $4\,\Omega$ resistor, as shown in Figure 5. Assume the battery itself has negligible internal resistance.

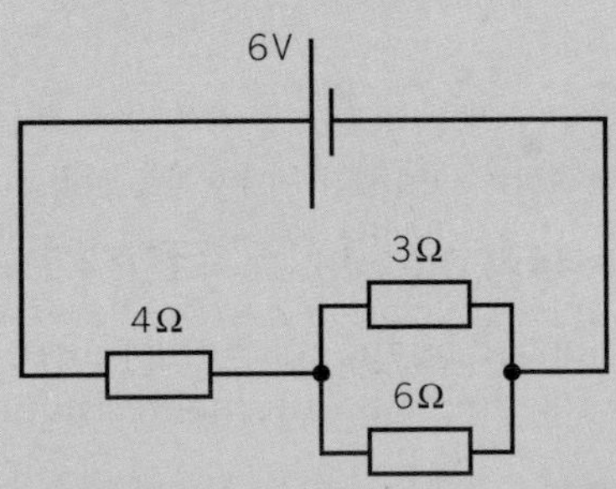


▲ **Figure 5**

Calculate:

**a** the combined resistance of the $3\,\Omega$ resistor and a $6\,\Omega$ resistor in parallel

**b** the total resistance of the circuit

**c** the battery current

**d** the power supplied to the $4\,\Omega$ resistor.

3 A $2\,\Omega$ resistor and a $4\,\Omega$ resistor are connected in series with each other. The series combination is connected in parallel with a $9\,\Omega$ resistor and a 3 V battery of negligible internal resistance, as shown in Figure 6.

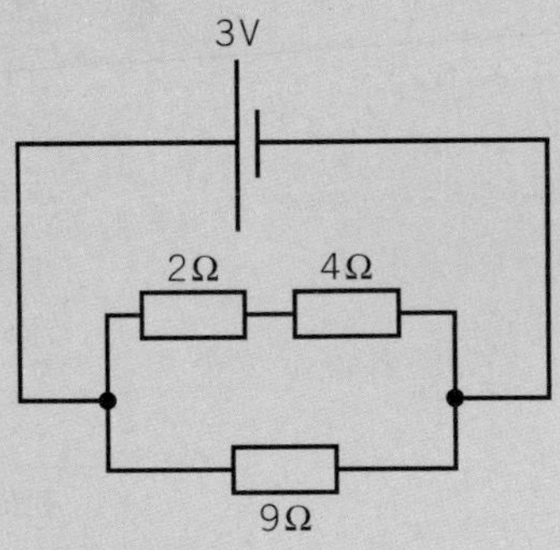


▲ **Figure 6**

Calculate:

**a** the total resistance of the circuit

**b** the battery current

**c** the power supplied to each resistor

**d** the power supplied by the battery.

4 Calculate:

**a** the power supplied to a $10\,\Omega$ resistor when the pd across it is 12 V

**b** the resistance of a heating element designed to operate at 60 W and 12 V.

# 13.3 Electromotive force and internal resistance

## Learning objectives:

- → Explain why the pd of a battery (or cell) in use is less than its emf.
- → Measure the internal resistance of a battery.
- → Describe how much power is wasted in a battery.

*Specification reference: 3.5.1.6*

## Internal resistance

The **internal resistance** of a source of electricity is due to opposition to the flow of charge through the source. This causes electrical energy produced by the source to be dissipated inside the source when charge flows through it.

- The **electromotive force** (emf, symbol $\varepsilon$) of the source is the electrical energy per unit charge produced by the source. If electrical energy $E$ is given to a charge $Q$ in the source,

$$\varepsilon = \frac{E}{Q}$$

- The **pd across the terminals** of the source is the electrical energy per unit charge delivered by the source when it is in a circuit. The terminal pd is less than the emf whenever current passes through the source. The difference is due to the internal resistance of the source.

**The internal resistance of a source is the loss of potential difference per unit current in the source when current passes through the source.**

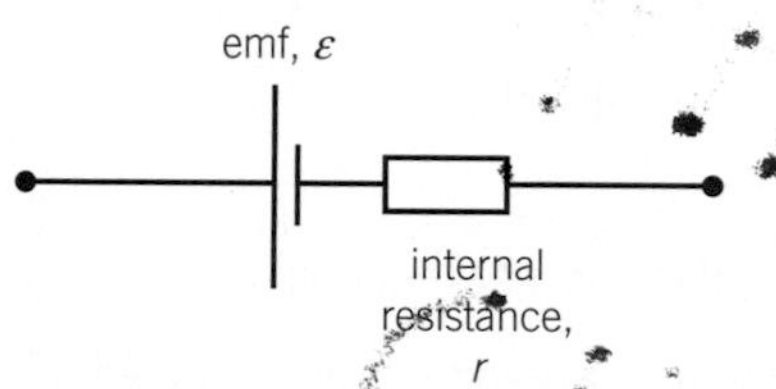


▲ **Figure 1** *Internal resistance*

In circuit diagrams, the internal resistance of a source may be shown as a resistor (labelled 'internal resistance') in series with the usual symbol for a cell or battery, as in Figure 1.

When a cell of emf $\varepsilon$ and internal resistance $r$ is connected to an external resistor of resistance $R$, as shown in Figure 2, all the current through the cell passes through its internal resistance and the external resistor. So the two resistors are in series, which means that the total resistance of the circuit is $r + R$. Therefore, the current through the cell, $I = \dfrac{\varepsilon}{R + r}$.

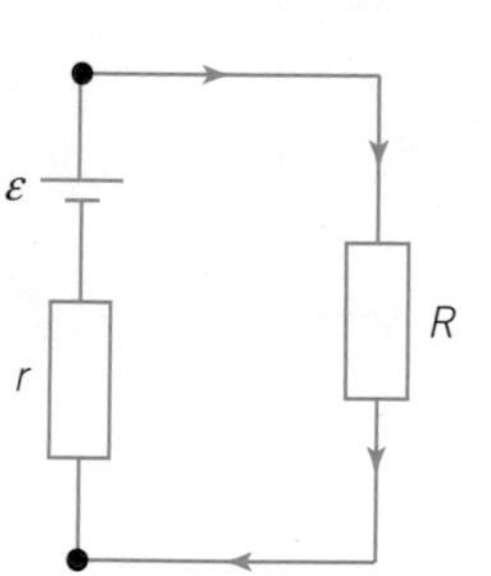


▲ **Figure 2** *Emf and internal resistance*

In other words, the cell emf $\varepsilon = I(R + r) = IR + Ir$ = the terminal pd + the lost, or wasted, pd

$$\varepsilon = IR + Ir$$

The lost pd inside the cell (i.e., the pd across the internal resistance of the cell) is equal to the difference between the cell emf and the pd across its terminals. In energy terms, the lost pd is the energy per coulomb dissipated or wasted inside the cell due to its internal resistance.

## Power

Multiplying each term of the above equation by the cell current $I$ gives

**power supplied by the cell,** $I\varepsilon = I^2 R + I^2 r$

In other words, the power supplied by the cell = the power delivered to $R$ + the power wasted in the cell due to its internal resistance.

The power delivered to $R = I^2 R = \dfrac{\varepsilon^2}{(R + r)^2} R$ since $I = \dfrac{\varepsilon}{R + r}$.

Figure 3 shows how the power delivered to $R$ varies with the value of $R$.

It can be shown that the peak of this power curve is at $R = r$. In other words, when a source delivers power to a load, **maximum power is delivered to the load when the load resistance is equal to the internal resistance of the source**. The load is then said to be matched to the source. (Although you don't need to know this, you might find it useful if you ever need to replace an amplifier or a speaker.)

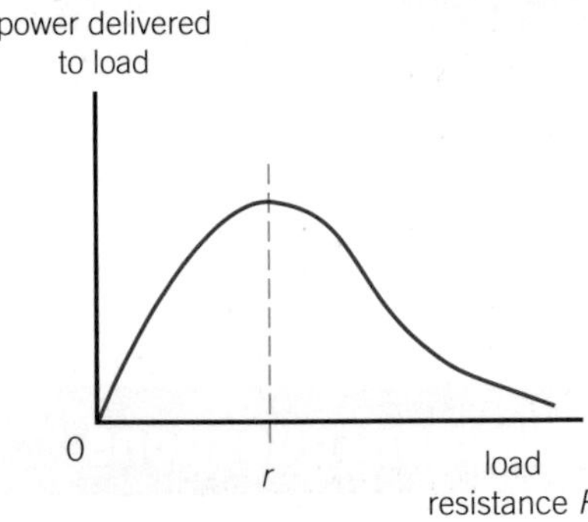


▲ **Figure 3** *Power delivered to a load versus load resistance*

## Measurement of internal resistance

The pd across the terminals of a cell, when the cell is in a circuit, can be measured by connecting a high-resistance voltmeter directly across the terminals of the cell. Figure 4 shows how the terminal pd can be measured for different values of current.

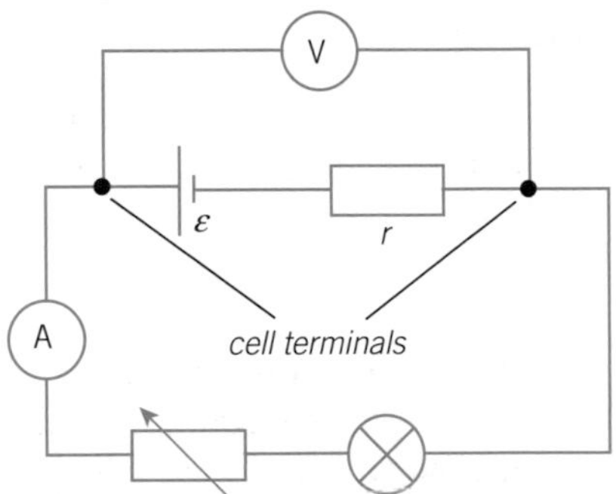


▲ **Figure 4** *Measuring internal resistance*

The current is changed by adjusting the variable resistor. The lamp (or a fixed resistor) limits the maximum current that can pass through the cell. The ammeter is used to measure the cell current. The measurements of terminal pd and current for a given cell may be plotted on a graph, as shown in Figure 5.

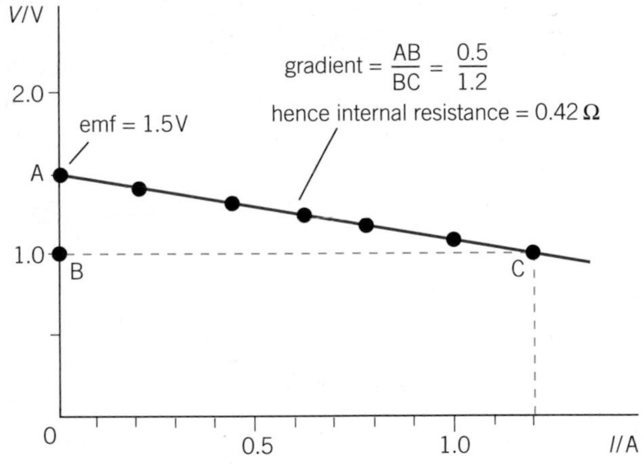


▲ **Figure 5** *A graph of terminal pd versus current*

The terminal pd decreases as the current increases. This is because the lost pd increases as the current increases.

- The terminal pd is equal to the cell emf at zero current. This is because the lost pd is zero at zero current.
- The graph is a straight line with a negative gradient. This can be seen by rearranging the equation $\varepsilon = IR + Ir$ to become $IR = \varepsilon - Ir$. Because $IR$ represents the terminal pd $V$, then

$$V = \varepsilon - Ir$$

By comparison with the standard equation for a straight line, $y = mx + c$, a graph of $V$ on the $y$-axis against $I$ on the $x$-axis gives a straight line with a gradient $-r$ and a $y$-intercept $\varepsilon$.

**Study tip**

Ensure you can relate $V = \varepsilon - Ir$ to $y = mx + c$

Figure 5 shows the gradient triangle ABC in which AB represents the lost pd and BC represents the current. So the gradient AB/BC = lost voltage ÷ current = internal resistance $r$.

### *Note:*

The internal resistance and the emf of a cell can be calculated if the terminal pd is known for two different values of current.

- For current $I_1$, the terminal pd $V_1 = \varepsilon - I_1 r$.
- For current $I_2$, the terminal pd $V_2 = \varepsilon - I_2 r$.

Subtracting the first equation from the second gives:

$$V_1 - V_2 = (\varepsilon - I_1 r) - (\varepsilon - I_2 r) = I_2 r - I_1 r = (I_2 - I_1)r$$

Therefore, $r = \dfrac{V_1 - V_2}{I_2 - I_1}$.

So $r$ can be calculated from the above equation and then substituted into either equation for the cell pd to enable $\varepsilon$ to be calculated.

## Worked example

A cell of unknown emf $\varepsilon$ and internal resistance $r$ was connected in series with an ammeter, a switch, and a 10.0 Ω resistor P, as shown in Figure 6.

When the switch was closed, the ammeter reading was 0.40 A. When P was replaced by a 5.0 Ω resistor Q, the ammeter reading became 0.65 A. Calculate the internal resistance of the cell.

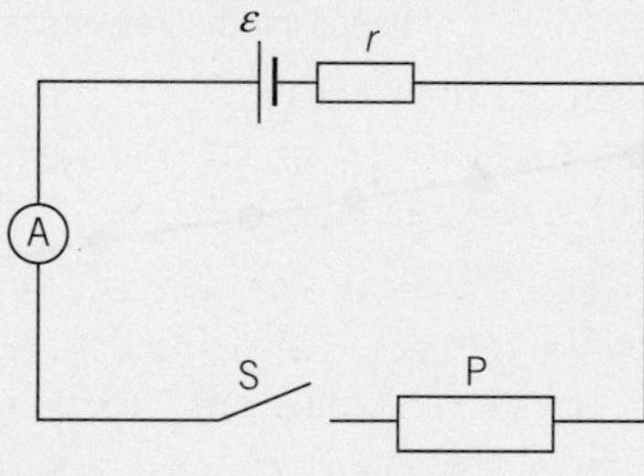


▲ **Figure 6**

### *Solution*

Applying the equation $\varepsilon = IR + Ir$ to the circuit with P,
$\varepsilon = (0.40 \times 10.0) + (0.4r) = 4.0 + 0.4r$

Applying the equation $\varepsilon = IR + Ir$ to the circuit with Q,
$\varepsilon = (0.60 \times 5.0) + (0.6r) = 3.0 + 0.6r$

Therefore

$4.0 + 0.4r = 3.0 + 0.6r$

Rearranging this equation gives

$0.6r - 0.4r = 4.0 - 3.0$

$0.2r = 1.0$

$r = \dfrac{1.0}{0.2} = 5.0\,\Omega$

By substituting the value of $r$ into either of the equations, $\varepsilon = 6.0\,\text{V}$.

## Summary questions

1 A battery of emf 12 V and internal resistance 1.5 Ω is connected to a 4.5 Ω resistor. Calculate:

  a the total resistance of the circuit

  b the current through the battery

  c the lost pd

  d the pd across the cell terminals.

2 A cell of emf 1.5 V and internal resistance 0.5 Ω is connected to a 2.5 Ω resistor. Calculate:

  a the current

  b the terminal pd

  c the power delivered to the 2.5 Ω resistor

  d the power wasted in the cell.

3 The pd across the terminals of a cell was 1.1 V when the current from the cell was 0.20 A, and 1.3 V when the current was 0.10 A. Calculate:

  a the internal resistance of the cell

  b the cell's emf.

4 A battery of unknown emf $\varepsilon$ and internal resistance $r$ is connected in series with an ammeter and a resistance box $R$. The current was 2.0 A when $R$ = 4.0 Ω and 1.5 A when $R$ = 6.0 Ω. Calculate $\varepsilon$ and $r$.

# 13.4 More circuit calculations

## Learning objectives:

→ Calculate currents in circuits with:

- resistors in series and parallel
- more than one cell
- diodes in the circuit.

*Specification reference: 3.5.1.4; 3.5.1.6*

## Circuits with a single cell and one or more resistors

Here are some rules:

1. Sketch the **circuit diagram** if it is not drawn.
2. To calculate the **current** passing through the cell, calculate the total circuit resistance using the resistor combination rules. Don't forget to add on the internal resistance of the cell if that is not negligible.

$$\textbf{cell current} = \frac{\textbf{cell emf}}{\textbf{total circuit resistance}}$$

3. To work out the current and pd for each resistor, start with the **resistors in series with the cell** which therefore pass the same current as the cell current.

**Pd across each resistor in series with the cell = current × the resistance of each resistor.**

4. To work out the current through **parallel resistors**, work out the combined resistance and multiply by the cell current to give the pd across each resistor.

$$\textbf{Current through each resistor} = \frac{\textbf{pd across the parallel combination}}{\textbf{resistor's resistance}}$$

### Study tip

Check the pds round a circuit add up to the battery pd.

## Circuits with cells in series

The same rules as above apply except the current through the cells is calculated by dividing the overall (net) emf by the total resistance.

- If the cells are connected in the same direction in the circuit, as in Figure 1a, the net emf is the sum of the individual emfs. For example, in Figure 1a, the net emf is 3.5 V.
- If the cells are connected in opposite directions to each other in the circuit, as in Figure 1b, the net emf is the difference between the emfs in each direction. For example, in Figure 1b, the net emf is 0.5 V in the direction of the 2.0 V cell.
- The total internal resistance is the sum of the individual internal resistances. This is because the cells, and therefore the internal resistances, are in series.

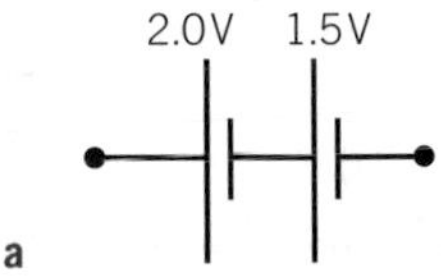


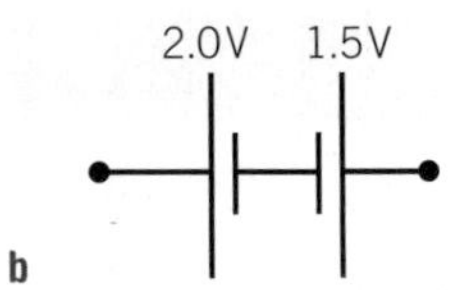


▲ **Figure 1** *Cells in series*

### Worked example

A cell of emf 3.0 V and internal resistance 2.0 Ω and a cell of emf 2.0 V and internal resistance 1.0 Ω are connected in series with each other and with a 7.0 Ω resistor, as in Figure 2. Calculate the pd across the 7.0 Ω resistor.

*Solution*

The net emf of the two batteries = 3.0 + 2.0 = 5.0 V in the direction of the 3.0 V cell.

The total circuit resistance = 1.0 Ω + 2.0 Ω + 7.0 Ω = 10.0 Ω

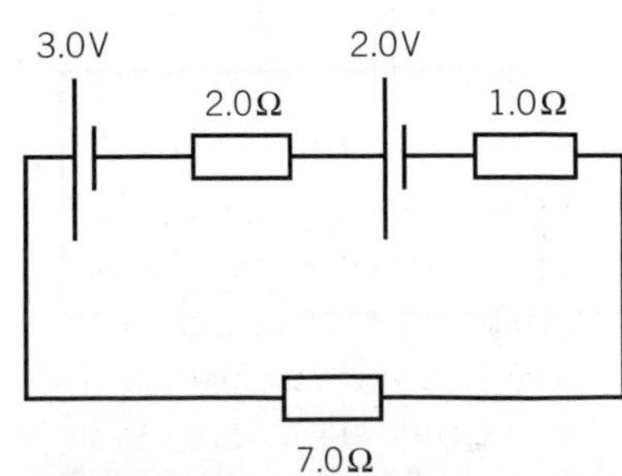


▲ **Figure 2**

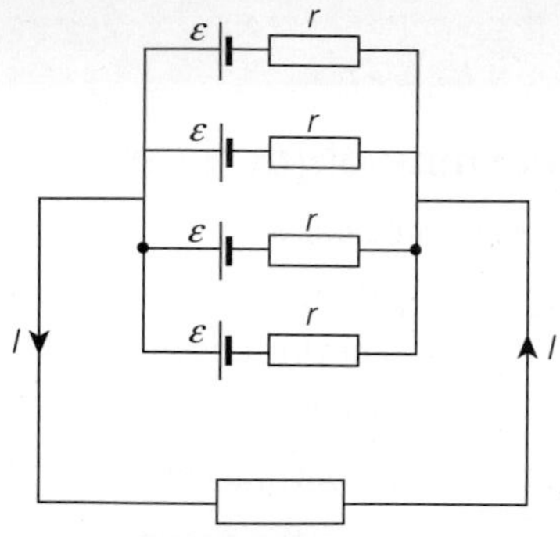


▲ **Figure 3** *Cells in parallel*

Therefore the cell current $= \frac{\text{net emf}}{\text{total circuit resistance}} = \frac{5.0\,\text{V}}{10.0\,\Omega} = 0.50\,\text{A}$

The pd across the 7.0 Ω resistor = current × resistance
$= 0.50\,\text{A} \times 7.0\,\Omega = 3.5\,\text{V}$

## Circuits with identical cells in parallel

For a circuit with $n$ identical cells in parallel, the current through each cell = $I/n$, where $I$ is the total current supplied by the cells.

Therefore, the lost pd in each cell $= \frac{I}{n} r = \frac{Ir}{n}$, where $r$ is the internal resistance of each cell.

Hence the terminal pd across each cell, $V = \varepsilon - \frac{Ir}{n}$.

Each time an electron passes through the cells, it travels through one of the cells only (as the cells are in parallel), therefore the cells act as a source of emf $\varepsilon$ and internal resistance $r/n$.

### Solar panels

A solar panel consists of many parallel rows of identical solar cells in series. For example, using single solar cells with a maximum emf of 0.45 V and an internal resistance of 1.0 Ω, a row of 20 cells in series would have a maximum emf of 9.0 V and an internal resistance of 20 Ω. Forty such rows in parallel would still give a maximum emf of 9.0 V but would have an internal resistance of 0.5 Ω (= 20 Ω/40).

### Synoptic link

For more about solar panels, see 'Renewable energy' in Chapter 10.

## Diodes in circuits

Assume that a silicon diode has:

- a forward pd of 0.6 V whenever a current passes through it
- infinite resistance in the reverse direction or at pds less than 0.6 V in the forward direction.

Therefore, in a circuit with one or more diodes:

- a pd of 0.6 V exists across a diode that is forward-biased and passing a current
- a diode that is reverse-biased has infinite resistance.

For example, suppose a diode is connected in its forward direction in series with a 1.5 V cell of negligible internal resistance and a 1.5 kΩ resistor, as in Figure 4.

The pd across the diode is 0.6 V because it is forward-biased. Therefore, the pd across the resistor is 0.9 V (= 1.5 V – 0.6 V). The current through the resistor is therefore $6.0 \times 10^{-4}$ A (= 0.9 V/1500 Ω).

However, if the diode in Figure 4 was reversed, the circuit current would be zero, so the pd across the resistor would also be zero. The pd across the diode would therefore be 1.5 V (equal to the cell emf).

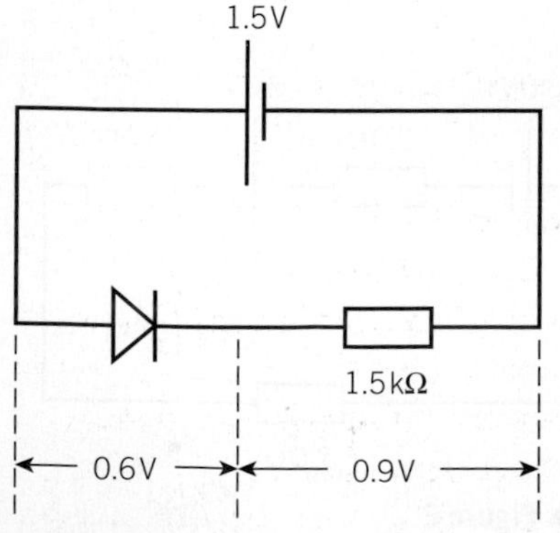


▲ **Figure 4** *Using a diode*

### Kirchhoff's laws

The following two circuit rules, which are called Kirchhoff's laws, can be used to analyse any dc circuit, regardless of how many loops and cells are in the circuit. The example on the following page illustrates their use.

**At any junction in a circuit, the total current entering the junction is equal to the total current leaving the junction.**

**For any complete loop in a circuit, the sum of the emfs around the loop is equal to the sum of the potential drops around the loop.**

*Worked example*

Two cells, P and Q, and a resistor R of resistance 4.0 Ω are connected in parallel with each other, as shown in Figure 5. Cell P has an emf of 2.0 V and an internal resistance of 1.5 Ω. Cell Q has an emf of 1.5 V and an internal resistance of 2.0 Ω. Calculate the current in each cell.

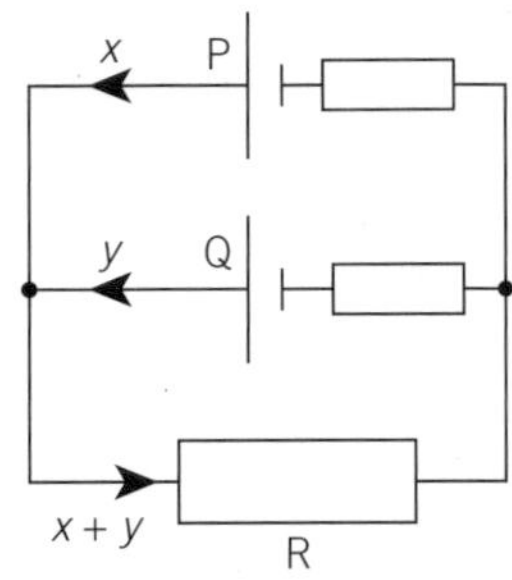


▲ **Figure 5**

*Solution*

Let the current in P be $x$ and the current in Q be $y$.

Using Kirchhoff's first law, the current in R is therefore $x + y$.

**Consider the complete loop consisting of P and R:**

The sum of the emfs in the loop is 2.0 V.

The pd across the internal resistance of P is $1.5x$.

The pd across R is $4.0(x + y)$.

Therefore, the sum of the pds around the loop = $1.5x + 4.0(x + y) = 5.5x + 4.0y$.

Using Kirchhoff's second law, $\mathbf{5.5x + 4.0y = 2.0}$. **(1)**

**Consider the complete loop consisting of Q and R:**

The sum of the emfs in the loop is 1.5 V.

The pd across the internal resistance of Q is $2.0y$.

The pd across R is $4.0(x + y)$.

Therefore, the sum of the pds around the loop = $2.0y + 4.0(x + y) = 4.0x + 6.0y$.

Using Kirchhoff's second law, $\mathbf{4.0x + 6.0y = 1.5}$. **(2)**

Multiplying **(1)** by 3 gives $16.5x + 12.0y = 6.0$.

Multiplying **(2)** by 2 gives $8.0x + 12.0y = 3.0$.

Subtracting **(2)** from **(1)** gives $8.5x = 3.0$.

$x = \frac{3.0}{8.5} - 0.35$ A.

Prove for yourself that substituting this value into either **(1)** or **(2)** gives $y = -0.015$ A.

*Note:*

The negative value of $y$ shows that the current in Q is in the reverse direction to that assumed in Figure 5.

**Q:** If cell Q had been reversed, the total emf in the loop of Q and R would have been −1.5 V. Calculate the current in each cell for this circuit.

**A:** $x = 1.06$ A $y = 0.96$ A

## Summary questions

1 A cell of emf 3.0 V and negligible internal resistance is connected to a 4.0 Ω resistor in series with a parallel combination of a 24.0 Ω resistor and a 12.0 Ω resistor, as shown in Figure 6. Calculate:

**a** the total resistance of the circuit

**b** the cell current

**c** the current and pd for each resistor.

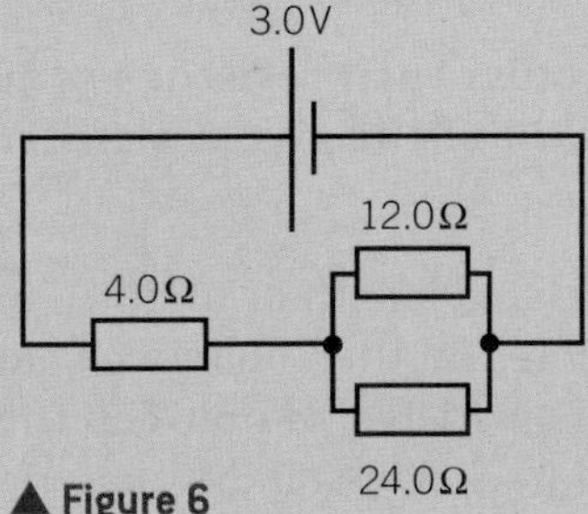


▲ **Figure 6**

2 A 15.0 Ω resistor, a battery of emf 12.0 V with an internal resistance of 3.0 Ω, and a battery of emf 9.0 V with an internal resistance of 2.0 Ω are connected in series. The batteries act in the same direction in the circuit. Sketch the circuit diagram and calculate:

**a** the total resistance of the circuit

**b** the cell current

**c** the current and pd across the 15 Ω resistor.

3 Two 8 Ω resistors and a battery of emf 12.0 V and internal resistance 8 Ω are connected in series with each other. Sketch the circuit diagram and calculate **i** the power delivered to each external resistor, **ii** the power wasted due to internal resistance.

# 13.5 The potential divider

**Learning objectives:**

- → Describe a potential divider.
- → Explain how we can supply a variable pd from a battery.
- → Explain how we can design sensor circuits.

*Specification reference: 3.5.1.5*

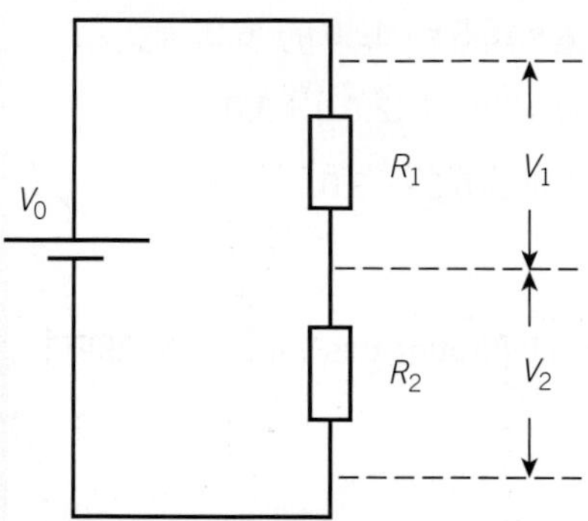


▲ **Figure 1** *A potential divider*

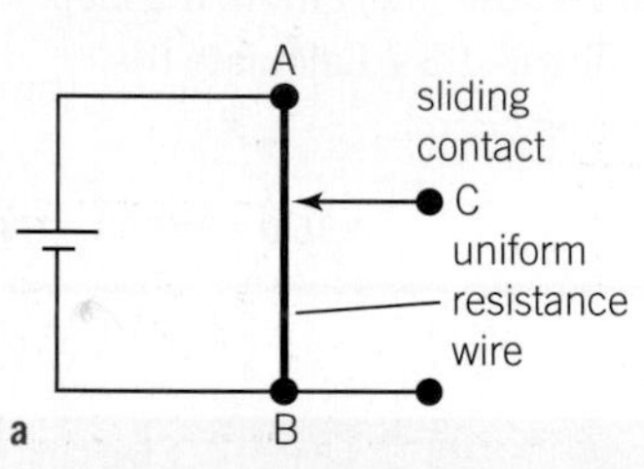


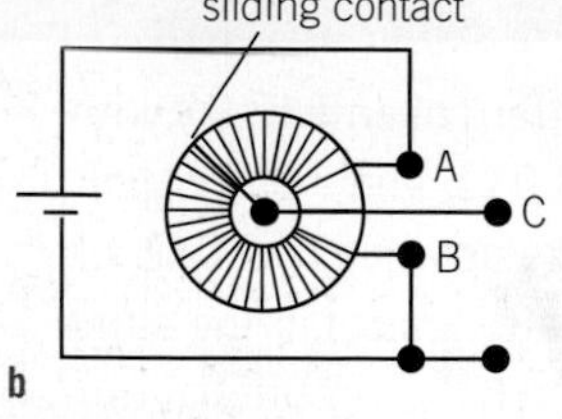


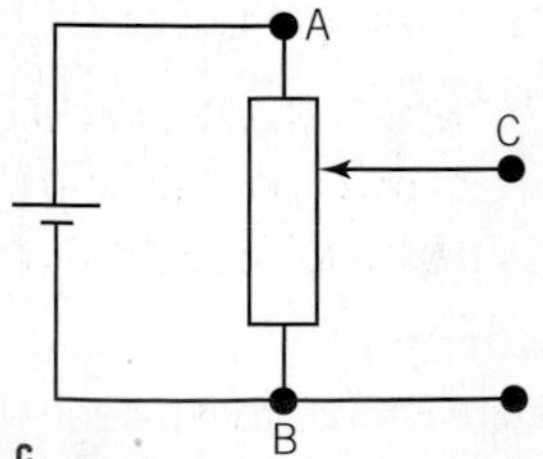


▲ **Figure 2** *Potential dividers used to supply a variable pd*
**a** *A linear track using resistance wire*
**b** *A circular track* **c** *Circuit symbol*

## The theory of the potential divider

A **potential divider** consists of two or more resistors in series with each other and with a source of fixed potential difference. The potential difference of the source is divided between the components in the circuit, as they are in series with each other. By making a suitable choice of components, a potential divider can be used:

- to supply a pd which is fixed at any value between zero and the source pd
- to supply a variable pd
- to supply a pd that varies with a physical condition such as temperature or pressure.

### To supply a fixed pd

Consider two resistors $R_1 + R_2$ in series connected to a source of fixed pd $V_0$, as shown in Figure 1.

The total resistance of the combination = $R_1 + R_2$.

Therefore, the current $I$ through the resistors is given by:

$$I = \frac{\text{pd across the resistors}}{\text{total resistance}} = \frac{V_0}{R_1 + R_2}$$

So the pd $V_1$ across resistor $R_1$ is given by

$$V_1 = IR_1 = \frac{V_0 R_1}{R_1 + R_2}$$

and the pd $V_2$ across resistor $R_2$ is given by

$$V_2 = IR_2 = \frac{V_0 R_2}{R_1 + R_2}$$

These two equations show that the pd across each resistor as a proportion of the source pd is the same as the resistance of the resistor in proportion to the total resistance. In other words, if the resistances are 5 kΩ and 10 kΩ, rcspectively:

- the pd across the 5 kΩ resistor is $\frac{1}{3}\left(=\frac{5}{15}\right)$ of the source pd
- the pd across the 10 kΩ resistor is $\frac{2}{3}\left(=\frac{10}{15}\right)$ of the source pd.

Also, dividing the equation for $V_1$ by the equation for $V_2$ gives

$$\frac{V_1}{V_2} = \frac{R_1}{R_2}$$

This equation shows that

**the ratio of the pds across each resistor is equal to the resistance ratio of the two resistors.**

### To supply a variable pd

The source pd is connected to a fixed length of uniform resistance wire. A sliding contact on the wire can then be moved along the wire, as illustrated in Figure 2, giving a variable pd between the contact and one end of the wire. A uniform track of a suitable material may be used instead of resistance wire. The track may be linear or circular, as in Figure 2. The circuit symbol for a variable potential divider is also shown in Figure 2c.

A variable potential divider can be used:

- As a simple audio volume control to change the loudness of the sound from a loudspeaker. The audio signal pd is supplied to the potential divider in place of a cell or battery. The variable output pd from the potential divider is supplied to the loudspeaker.
- To vary the brightness of a light bulb between zero and normal brightness (Figure 3). In contrast with using a variable resistor in series with the light bulb and the source pd, the use of a potential divider enables the current through the light bulb to be reduced to zero. If a variable resistor in series with the light bulb had been used, there would be a current through the light bulb when the variable resistor is at maximum resistance.

### Study tip

Ensure you understand the basic potential divider rule.

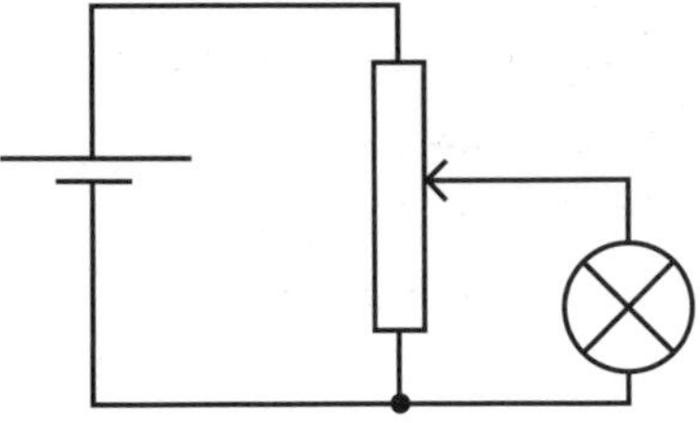

▲ **Figure 3** *Brightness control using a variable potential divider*

## Sensor circuits

A **sensor circuit** produces an output pd which changes as a result of a change of a physical variable such as temperature or light intensity.

1 A **temperature sensor** consists of a potential divider made using a thermistor and a variable resistor, as in Figure 4.

With the temperature of the thermistor constant, the source pd is divided between the thermistor and the variable resistor. By adjusting the variable resistor, the pd across the thermistor can then be set at any desired value. When the temperature of the thermistor changes, its resistance changes so the pd across it changes. For example, suppose the variable resistor is adjusted so that the pd across the thermistor at 20 °C is exactly half the source pd. If the temperature of the thermistor is then raised, its resistance falls, so the pd across it falls.

2 A **light sensor** uses a light-dependent resistor (LDR) and a variable resistor, as in Figure 5. The pd across the LDR changes when the incident light intensity on the LDR changes. If the light intensity increases, the resistance of the LDR falls and the pd across the LDR falls.

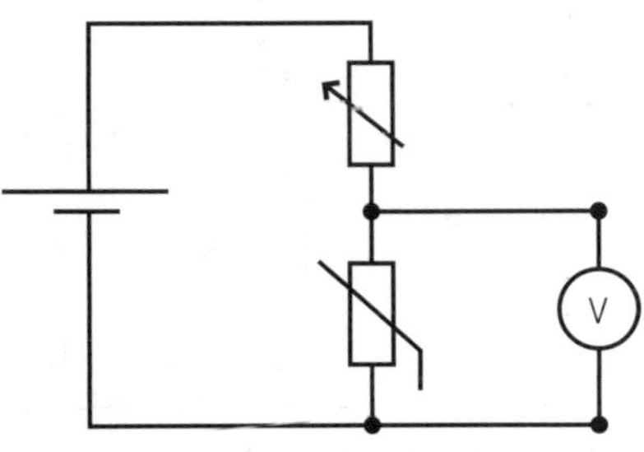


▲ **Figure 4** *A temperature sensor*

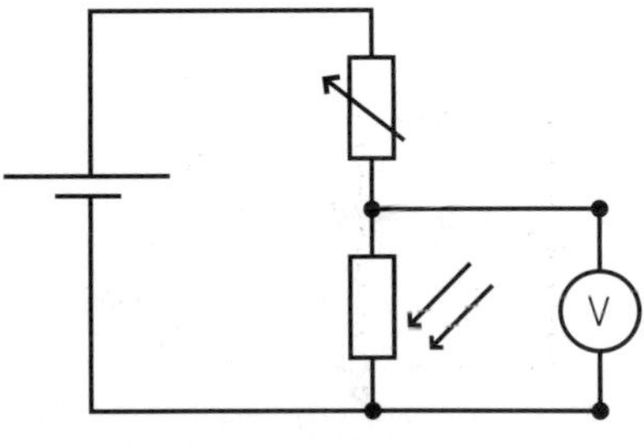


▲ **Figure 5** *A light sensor*

## Summary questions

**1** A 12 V battery of negligible internal resistance is connected to the fixed terminals of a variable potential divider which has a maximum resistance of 50 . A 12 V light bulb is connected between the sliding contact and the negative terminal of the potential divider. Sketch the circuit diagram and describe how the brightness of the light bulb changes when the sliding contact is moved from the negative to the positive terminal of the potential divider.

**2** **a** A potential divider consists of an 8.0 resistor in series with a 4.0 resistor and a 6.0 V battery of negligible internal resistance. Calculate **i** the current, **ii** the pd across each resistor.

**b** In the circuit in **a**, the 4 resistor is replaced by a thermistor with a resistance of 8 at 20 C and a resistance of 4 at 100 C. Calculate the pd across the fixed resistor at **i** 20 C, **ii** 100 C.

**3** A light sensor consists of a 5.0 V cell, an LDR, and a 5.0 k resistor in series with each other. A voltmeter is connected in parallel with the resistor. When the LDR is in darkness, the voltmeter reads 2.2 V.

**a** Calculate **i** the pd across the LDR, **ii** the resistance of the LDR when the voltmeter reads 2.2 V.

**b** Describe and explain how the voltmeter reading would change if the LDR were exposed to daylight with no adjustment made to the variable resistor.

1 A student is given three resistors of resistance 3.0 Ω, 4.0 Ω, and 6.0 Ω, respectively.

**(a)** Draw the arrangement, using all three resistors, which will give the largest resistance.

**(b)** Calculate the resistance of the arrangement you have drawn.

**(c)** Draw the arrangement, using all three resistors, which will give the smallest resistance.

**(d)** Calculate the resistance of the arrangement you have drawn. *(5 marks)*

AQA, 2005

2 In the circuit shown in **Figure 1**, the battery has negligible internal resistance.

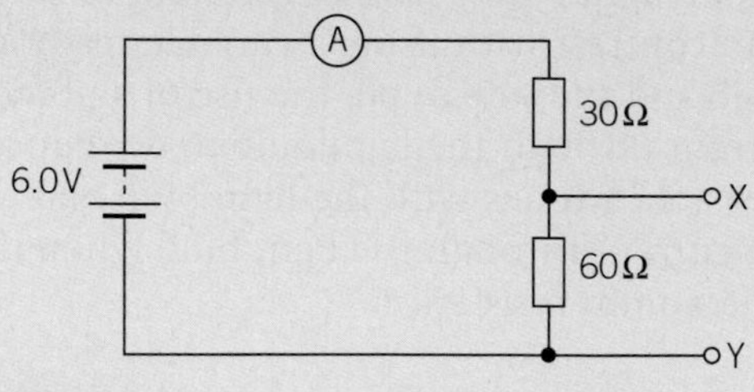


▲ **Figure 1**

Calculate the current in the ammeter when

**(a)** the terminals X and Y are short-circuited, that is, connected together *(2 marks)*

**(b)** the terminals X and Y are connected to a 30 Ω resistor. *(4 marks)*

AQA, 2002

3 **(a)** Define the electrical resistance of a component. *(2 marks)*

**(b)** Calculate the total resistance of the arrangement of resistors in **Figure 2**. *(3 marks)*

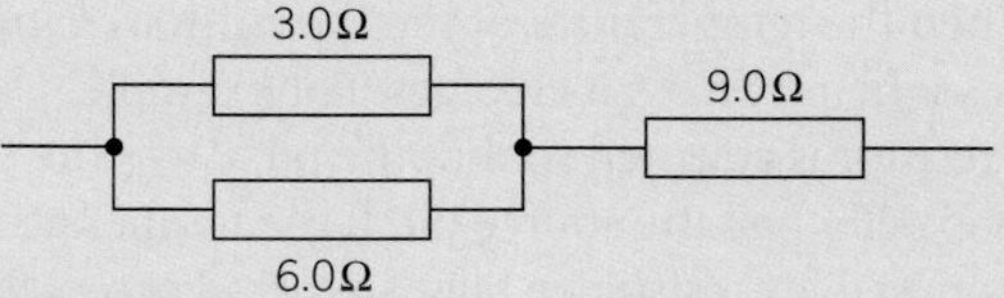


▲ **Figure 2**

**(c)** (i) Calculate the current in the 3.0 Ω resistor in **Figure 2** when the current in the 9.0 Ω resistor is 2.4 A.

(ii) Calculate the total power dissipated by the arrangement of resistors in **Figure 2** when the current in the 9.0 Ω resistor is 2.4 A. *(4 marks)*

AQA, 2006

4 In the circuit shown in **Figure 3** the resistor network between the points P and Q is connected in series to a resistor R, an ammeter, and a battery of negligible internal resistance.

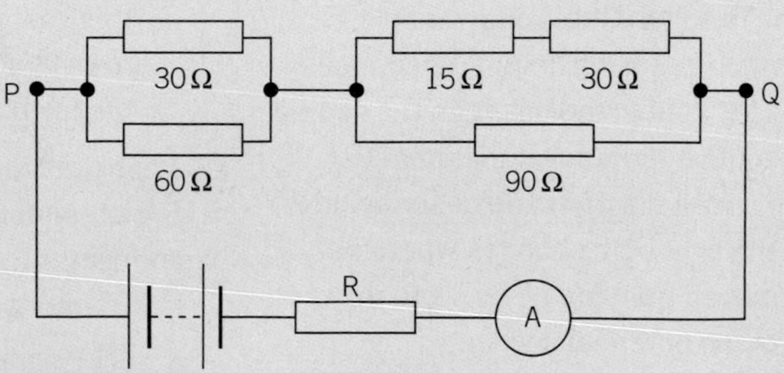


▲ **Figure 3**

**(a)** Determine the equivalent resistance of the network between the points P and Q. *(3 marks)*

**(b)** √x (i) If the current through the ammeter is 50 mA, calculate the total charge that flows through the resistor R in 4 minutes.

(ii) If 18 J of energy are transferred to the resistor R in this time, calculate the potential difference across R.

(iii) Calculate the resistance of R.

(iv) Calculate the emf of the battery. *(6 marks)*

AQA, 2007

**5** √x In the circuit shown in **Figure 4**, the battery has an emf of 12 V and an internal resistance of 2.0 Ω. The resistors A and B each have resistance of 30 Ω.

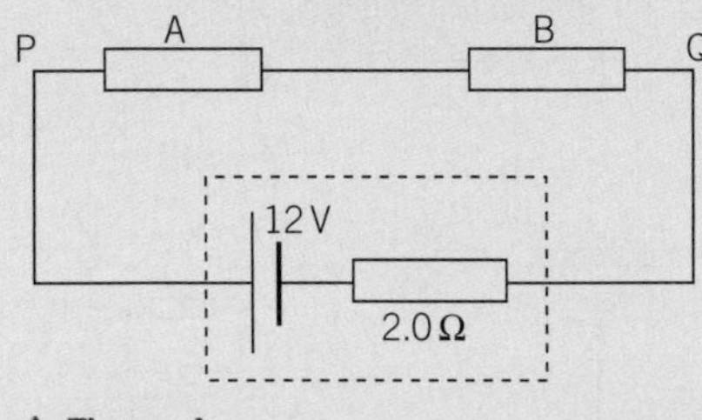


▲ **Figure 4**

Calculate:

(i) the total current in the circuit,

(ii) the voltage between the points P and Q,

(iii) the power dissipated in resistor A,

(iv) the energy dissipated by resistor A in 20 s. *(8 marks)*

AQA, 2003

**6** **Figure 5** shows an EHT supply of emf 5000 V and internal resistance 2 MΩ.

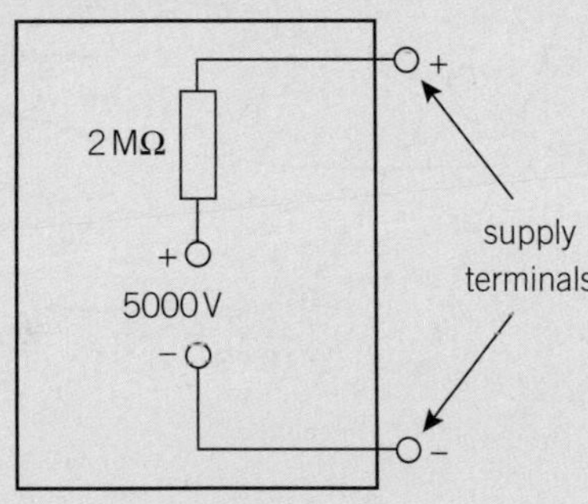


▲ **Figure 5**

**(a)** A lead of negligible resistance is connected between the supply terminals producing a short circuit.

(i) State the magnitude of the terminal potential difference between the supply terminals.

(ii) √x Calculate the current in the circuit.

(iii) √x Calculate the minimum power rating for the resistor used to provide the internal resistance. *(4 marks)*

**(b)** Explain briefly why the supply is designed with such a high internal resistance. *(1 mark)*

AQA, 2005

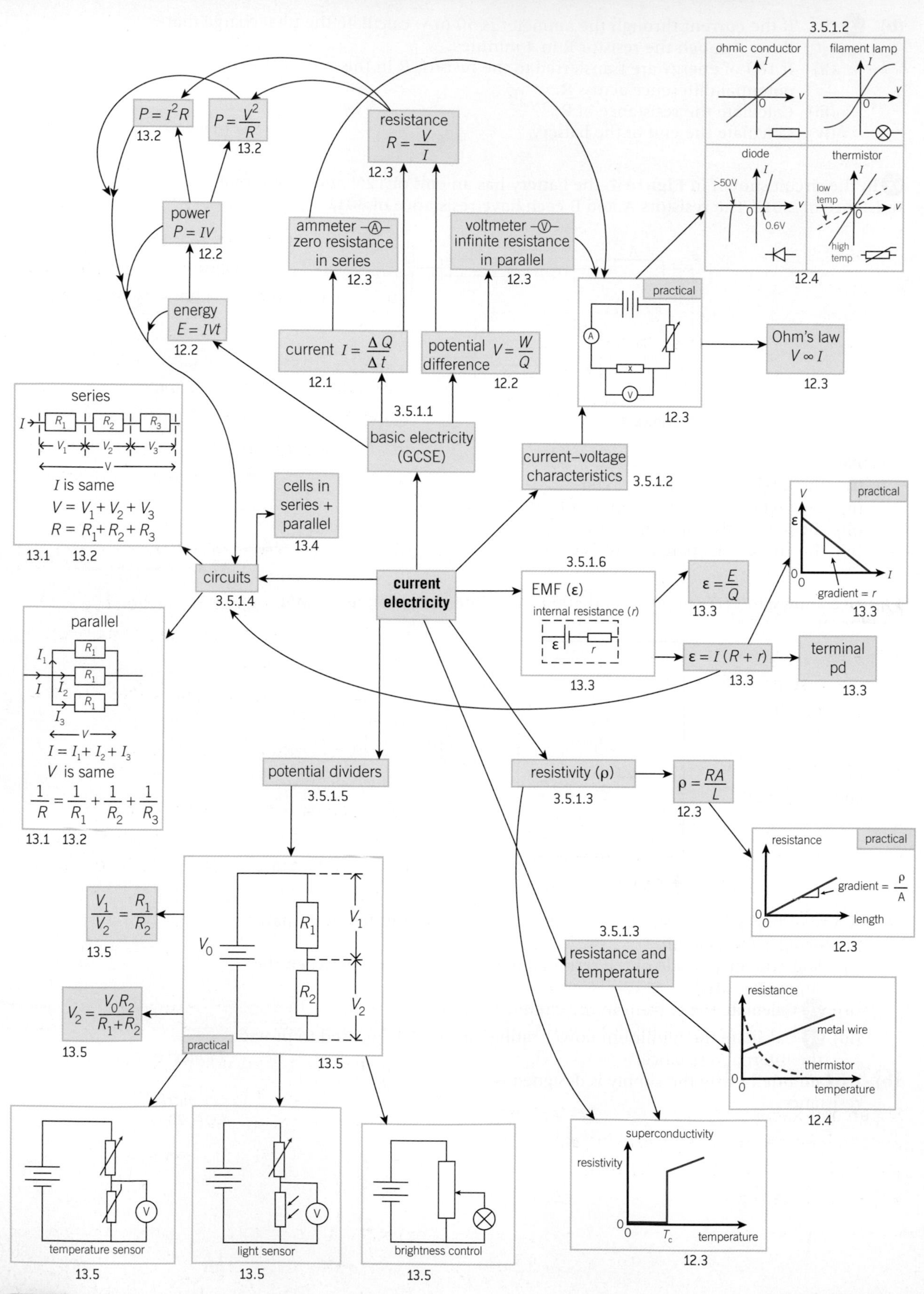

3.5.1.2
ohmic conductor
filament lamp
diode
thermistor
>50V
0.6V
low temp
high temp
12.4
$P = I^2R$
13.2
$P = \frac{V^2}{R}$
13.2
resistance $R = \frac{V}{I}$
12.3
power $P = IV$
12.2
ammeter zero resistance in series
12.3
voltmeter infinite resistance in parallel
12.3
practical
energy $E = IVt$
12.2
current $I = \frac{\Delta Q}{\Delta t}$
12.1
potential difference $V = \frac{W}{Q}$
12.2
Ohm's law $V \propto I$
12.3
12.3
3.5.1.1
basic electricity (GCSE)
current–voltage characteristics
3.5.1.2
series
I is same
$V = V_1 + V_2 + V_3$
$R = R_1 + R_2 + R_3$
13.1 13.2
cells in series + parallel
13.4
circuits
3.5.1.4
current electricity
3.5.1.6
EMF (ε)
internal resistance (r)
$\varepsilon = \frac{E}{Q}$
13.3
practical
gradient = r
13.3
$\varepsilon = I(R + r)$
13.3
terminal pd
13.3
13.3
parallel
$I = I_1 + I_2 + I_3$
V is same
$\frac{1}{R} = \frac{1}{R_1} + \frac{1}{R_2} + \frac{1}{R_3}$
13.1 13.2
potential dividers
3.5.1.5
resistivity (ρ)
3.5.1.3
$\rho = \frac{RA}{L}$
12.3
resistance
practical
gradient = $\frac{\rho}{A}$
length
12.3
$\frac{V_1}{V_2} = \frac{R_1}{R_2}$
13.5
$V_2 = \frac{V_0 R_2}{R_1 + R_2}$
13.5
practical
13.5
3.5.1.3
resistance and temperature
resistance
metal wire
thermistor
temperature
12.4
temperature sensor
13.5
light sensor
13.5
brightness control
13.5
superconductivity
resistivity
$T_c$
temperature
12.3

## Practical skills

In this section you have met the following skills:

- use of digital meters to obtain measurements of current, voltage and resistance (e.g. resistance measurement)
- use of a micrometer to find the diameter of wire (e.g. resistivity experiment)
- use of electrical equipment safely, connecting up circuits correctly using circuit diagrams with ammeters in series and voltmeters in parallel with a range of circuit components (e.g. measurement of internal resistance)
- use of current and voltage sensors with a data logger to collect data (investigating sensor circuits).

In all practical experiments:

- present experimental data in a table with headings and units using significant figures correctly
- repeat results to improve reliability
- record the precision of any instrument used
- plot a graph from experimental and/or processed data, labelling axes correctly with quantity and units and choosing appropriate scales
- consider the accuracy of your results/ conclusion, compared to a known value and consider reasons for and types of error
- identify uncertainties in measurements and combine them to find the uncertainty in the overall experiment.

## Maths skills

In this section you have met the following skills:

- use of standard form and conversion to standard form from units with prefixes (e.g. mA, kV)
- use calculators to solve calculations involving powers of ten
- use of an appropriate number of significant figures in all answers to calculations
- use of appropriate units in all answers to calculations
- calculation of a mean from a set of repeat readings (e.g. diameter of wire)
- calculate the area of cross section of a wire
- use fractions when using the equation for resistors in parallel
- change the subject of an equation in order to solve calculations
- plot a graph from data provided or found experimentally
- relate $y = mx + c$ to a linear graph with physics variables to find the gradient and intercept and understand what the gradient and intercept represent in physics (e.g. finding emf and internal resistance).

## Extension task

Superconductors are introduced in this textbook. Research further into this topic using other books and the internet to find out more about:

- their properties
- which materials are superconductors
- where they are used in industry
- how they could be used in the future.

Present your findings, for example, as a presentation.

# Practice questions: Electricity

**1** **(a)** A filament lamp labelled '12 V, 2.0 A' has a constant resistance of 2.0 Ω for electrical currents up to 0.50 A.

Sketch on graph paper the current–voltage graph for this lamp for positive voltages up to 12 V and currents up to 2.0 A. Show clearly any calculations you made in order to answer the question. *(3 marks)*

**(b)** Sketch on a copy of the axes below the current–voltage characteristic for a semi-conductor diode. *(3 marks)*

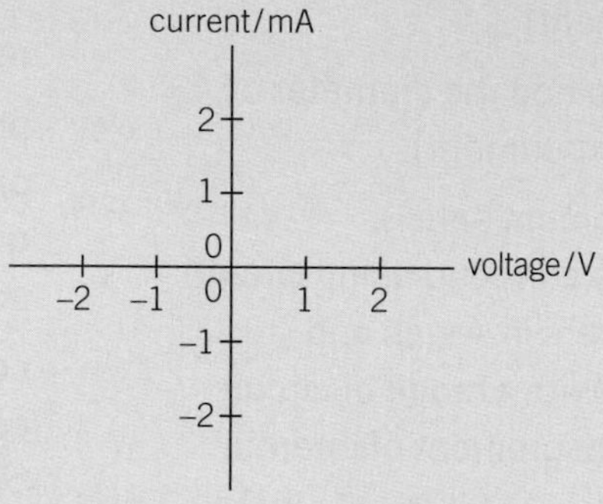


AQA, 2003

**2** Figure 1 shows a circuit including a thermistor T in series with a 2.2 kΩ resistor R. The emf of the power source is 5.0 V with negligible internal resistance.

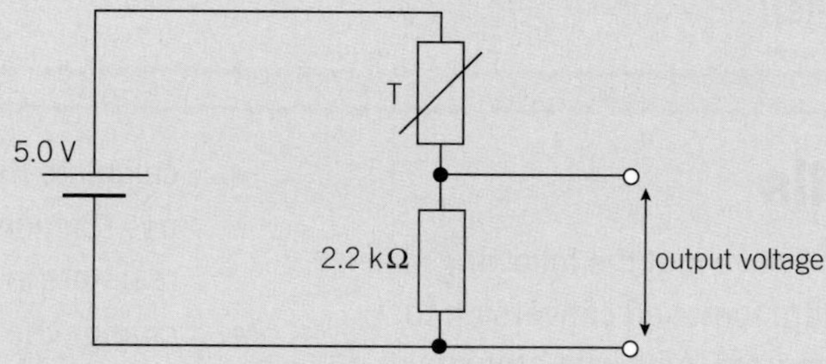


▲ **Figure 1**

**(a)** The resistor and thermistor in Figure 1 make up a potential divider. State what is meant by a potential divider. *(1 mark)*

**(b)** The output voltage is 1.8 V when the thermistor is at 20 °C. Calculate the thermistor resistance at this temperature. *(2 marks)*

**(c)** State and explain what happens to the output voltage when the temperature of the thermistor is raised above 20 °C. *(2 marks)*

**(d)** An alarm is connected across the output voltage terminals. The alarm sounds when the voltage across it exceeds 3.0 V. Describe how you would modify the circuit so that it would switch the alarm on if the temperature of the thermistor exceeds 25 °C. *(2 marks)*

AQA, 2007

**3** The circuits in **Figure 2** and **Figure 3** both contain a 6.0 V supply of negligible internal resistance. Each circuit is designed to operate a 2.5 V, 0.25 A filament lamp **L**.

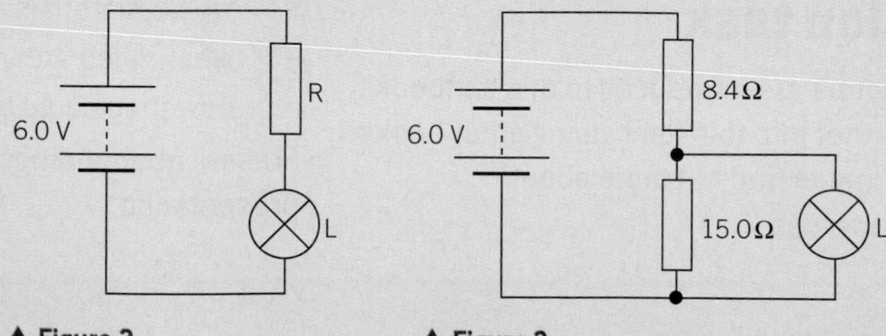


▲ **Figure 2** ▲ **Figure 3**

**(a)** Calculate the resistance of the filament lamp when working normally. *(2 marks)*

**(b)** Calculate the resistance of the resistor that should be used for R in **Figure 2**. *(2 marks)*

**(c)** Calculate the total resistance of the circuit in **Figure 3**. *(3 marks)*

**(d)** Explain which circuit dissipates the lower total power. *(3 marks)*

AQA, 2005

4 The heating circuit of a hairdryer is shown in **Figure 4**. It consists of two heating elements, $R_1$ and $R_2$, connected in parallel. Each element is controlled by its own switch.

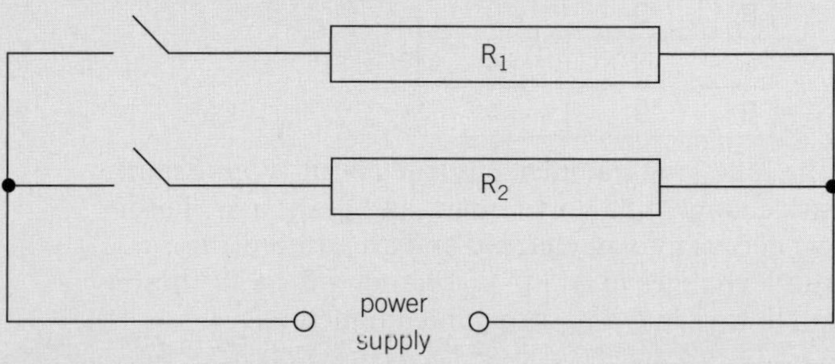

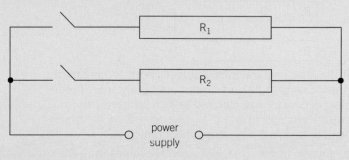

▲ **Figure 4**

The elements are made from the same resistance wire. This wire has a resistivity of $1.1 \times 10^{-6}\,\Omega\,m$ at its working temperature. The cross-sectional area of the wire is $1.7 \times 10^{-8}\,m^2$ and the length of the wire used to make $R_1$ is 3.0 m.

**(a)** Show that the resistance of $R_1$ is about 190 Ω. *(3 marks)*

**(b)** Calculate the power output from the heating circuit with only $R_1$ switched on when it is connected to a 240 V supply. *(2 marks)*

**(c)** With both elements switched on, the total power output is three times that of $R_1$ on its own.

(i) Calculate the length of wire used to make the coil $R_2$.

(ii) Calculate the total current with both elements switched on. *(5 marks)*

AQA, 2004

5 **Figure 5** shows a power supply connected to a car battery in order to charge the battery. The terminals of the same polarity are connected together to achieve this.

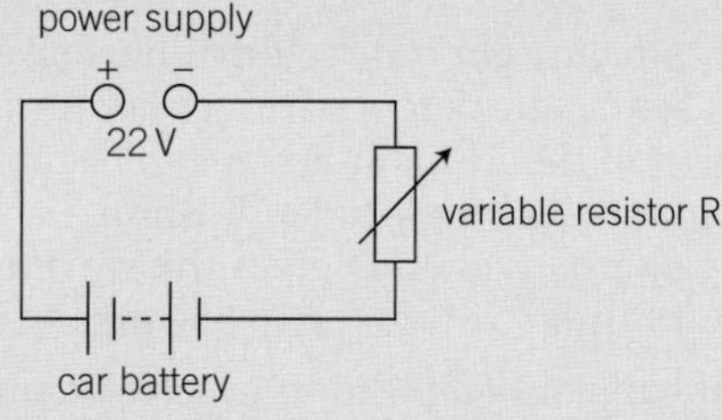


▲ **Figure 5**

**(a)** The power supply has an emf of 22V and internal resistance 0.75 Ω. When charging begins, the car battery has an emf of 10 V and internal resistance of 0.15 Ω. They are connected together via a variable resistor.

(i) Calculate the total emf of the circuit when charging begins.

(ii) The resistor R is adjusted to give an initial charging current of 0.25 A. Calculate the value of R. *(4 marks)*

**(b)** The car battery takes 8.0 hours to charge. Calculate the charge that flows through it in this time assuming that the current remains at 0.25 A. *(1 mark)*

AQA, 2003

# Further practice questions: multiple choice

1 Which one of the following alternatives **A–D** gives the ratio $\frac{\text{specific charge of a carbon } ^{12}_{6}\text{C nucleus}}{\text{specific charge of the proton}}$?

**A** $\frac{1}{3}$ **B** $\frac{1}{2}$ **C** 1 **D** 2

2 How many baryons and mesons are there in an atom of $^{9}_{4}$Be?

| | baryons | mesons |
|---|---|---|
| **A** | 5 | 0 |
| **B** | 9 | 0 |
| **C** | 5 | 4 |
| **D** | 9 | 4 |

3 Which one of the following statements is true about $\beta^{+}$ emission?

**A** An up quark changes to a down quark and emits a $W^{-}$ boson.
**B** A $W^{-}$ boson decays into an electron and an antineutrino.
**C** A down quark changes to an up quark and emits a $W^{+}$ boson.
**D** A $W^{+}$ boson decays into a positron and a neutrino.

4 A conduction electron at the surface of a metal escapes from the surface after absorbing a photon. The work function of the metal is $\phi$ and the metal is at zero potential. Which one of the following inequalities about wavelength $\lambda$ of the photon is true?

**A** $\lambda < \frac{hc}{\phi}$ **B** $\lambda < \frac{h\phi}{c}$ **C** $\lambda < \frac{\phi}{hc}$ **D** $\lambda < \frac{c}{h\phi}$

5 Which one of the following statements about polarisation is true?

**A** Sound waves can be polarised.
**B** Some electromagnetic waves cannot be polarised.
**C** The vibrations of a polarised wave are always in the same plane.
**D** Light waves are always polarised.

6 Which of the following is correct for the first two harmonics of a stationary wave on a string?

**A** The frequency of the first harmonic is twice the frequency of the second harmonic.
**B** Both harmonics have a node at the midpoint of the string.
**C** Between two nodes all parts of the wave vibrate in phase.
**D** Adjacent nodes of the second harmonic are twice as far apart as adjacent nodes of the first harmonic.

7 A parallel beam of monochromatic light is directed at normal incidence at two narrow parallel slits spaced 0.7 mm apart. Interference fringes are formed on a screen which is perpendicular to the direction of the incident beam at a distance of 0.900 m from the slits. The distance across 5 fringe spaces is measured at 4.2 mm.
Which one of the following alternatives **A–D** gives the wavelength of the light?

**A** 130 nm **B** 450 nm **C** 550 nm **D** 650 nm

8 Two particles X and Y at the same initial position accelerate uniformly from rest along a straight line. After 1.0 s, X is 0.20 m ahead of Y. The separation of X and Y after 2.0 s from the start is

**A** 0.40 m **B** 0.80 m **C** 1.20 m **D** 1.60 m

**9** Two trolleys, P and Q, travelling in opposite directions at the same speed, collide and move together after the collision in the direction in which P was originally travelling. Which **one** of the following statements about the collision is true?

**A** The collision is elastic.

**B** The mass of P is greater than the mass of Q.

**C** The force exerted by P on Q is greater than the force exerted by Q on P.

**D** The change of momentum of P is greater than the change of momentum of Q.

**10** A box of mass $m$ slides down a slope at constant velocity as shown in **Figure 1**. The slope is inclined at angle $\theta$ to the horizontal. The diagram shows the frictional force $F$ and the normal reaction force $N$ of the surface on the block.

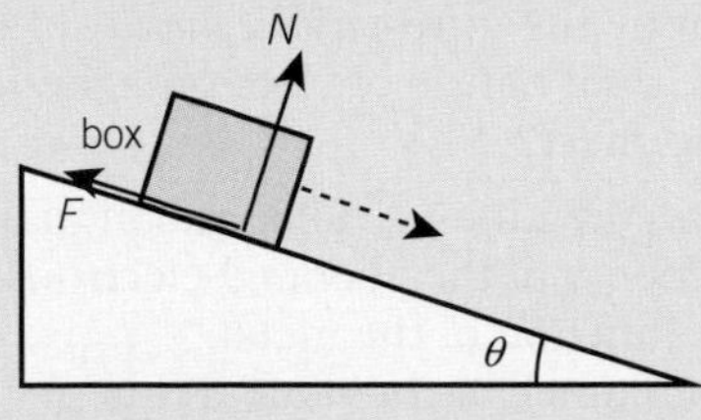


▲ **Figure 1**

Which one of the following statements about forces $F$ and $N$ is true?

**A** $F = mg\cos\theta$ **B** $F = N\sin\theta$ **C** $F = N + mg$ **D** $F^2 = (mg)^2 - N^2$

**11** Two wires P and Q of the same material have lengths $L$ and $2L$, and different diameters $d_p$ and $d_q$ respectively.
When the same force is applied to each wire, the extension of P is 4 times the extension of Q. Which one of the following alternatives **A–D** gives the ratio $\frac{d_p}{d_q}$?

**A** $\frac{1}{2\sqrt{2}}$ **B** $\frac{1}{\sqrt{2}}$ **C** $\sqrt{2}$ **D** $2\sqrt{2}$

**12** Four resistors of resistances 1.0Ω, 2.0Ω, 3.0Ω, and 4.0Ω are connected together as shown in **Figure 2**. A battery is connected across the 1.0Ω resistor.

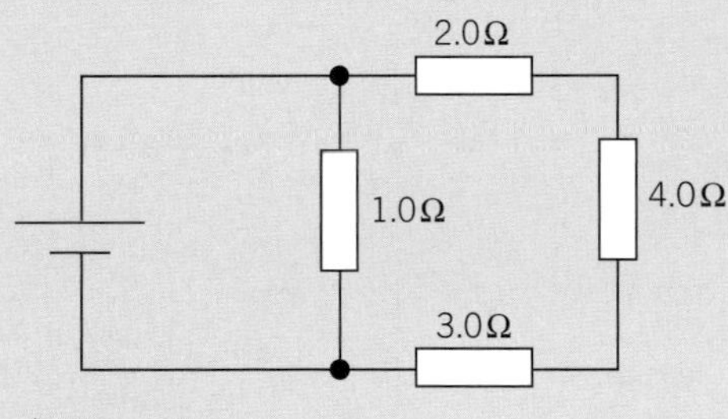


▲ **Figure 2**

If $I_0$ is the battery current and $I_1$ is the current in the 4.0Ω resistor, which one of the following alternatives **A–D** gives the ratio $\frac{I_0}{I_1}$?

**A** 0.1 **B** 0.5 **C** 1.5 **D** 10

# Further practice questions

**1** **(a)** An electron is trapped in a solid between a group of atoms where the potential is +2.8 V.
The de Broglie wavelength of this electron is 1.2 nm. Calculate
(i) its speed
(ii) its kinetic energy
(iii) the sum of its kinetic energy and its potential energy. *(5 marks)*

**(b)** (i) Calculate the energy of a photon of wavelength 650 nm.
(ii) State and explain whether or not the electron in part (a) can escape from this group of atoms as a result of absorbing this photon. *(3 marks)*

AQA, 2005

**2** **(a)** (i) State what is meant by the *wave–particle duality* of electromagnetic radiation.
(ii) Which aspect of the dual nature of electromagnetic radiation is demonstrated by the photoelectric effect? *(2 marks)*

**(b)** A metal plate is illuminated with ultraviolet radiation of frequency $1.67 \times 10^{15}$ Hz. The maximum kinetic energy of the liberated electrons is $3.0 \times 10^{-19}$ J.
(i) Calculate the work function of the metal.
(ii) The radiation is maintained at the same frequency but the intensity is doubled. State what changes, if any, occur to the number of electrons released per second and to the maximum kinetic energy of these electrons.
(iii) The metal plate is replaced by another metal plate of different material. When illuminated by radiation of the same frequency no electrons are liberated. Explain why this happens and what can be deduced about the work function of the new metal. *(7 marks)*

AQA, 2001

**3** **Figure 3** shows the forces acting on a stationary kite. The force *F* is the force that the air exerts on the kite.

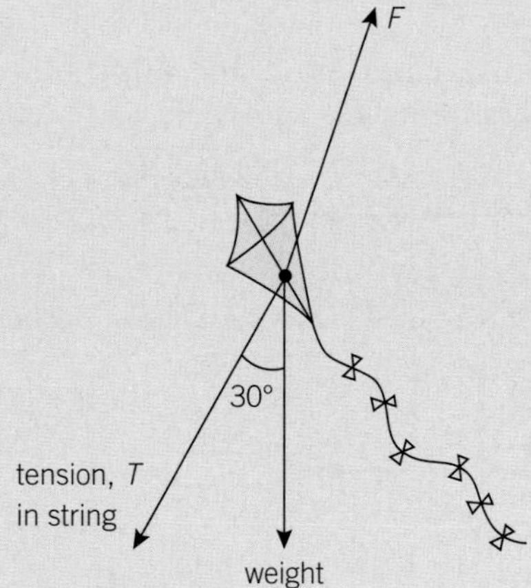


▲ **Figure 3**

**(a)** Show on the diagram how force *F* can be resolved into horizontal and vertical components. *(1 mark)*

**(b)** The magnitude of the tension, *T*, is 25 N.
Calculate:
(i) the horizontal component of the tension
(ii) the vertical component of the tension. *(2 marks)*

**(c)** (i) Calculate the magnitude of the vertical component of *F* when the weight of the kite is 2.5 N.
(ii) State the magnitude of the horizontal component of *F*.
(iii) Hence calculate the magnitude of *F*. *(5 marks)*
AQA, 2002

**4** A sprinter is shown before a race, stationary in the *set* position, as shown in **Figure 4**. Force **F** is the resultant force on the sprinter's finger tips. The reaction force **Y** on her forward foot is 180 N, and her weight **W** is 520 N. **X** is the vertical reaction force on her back foot.

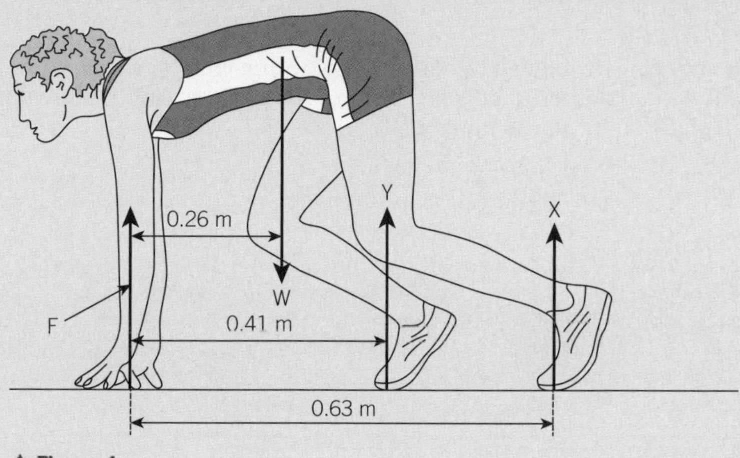


▲ **Figure 4**

**(a)** (i) Calculate the moment of the sprinter's weight **W** about her fingertips. Give an appropriate unit.
(ii) By taking moments about her finger tips, calculate the force on her back foot **X**.
(iii) Calculate the force **F**. *(6 marks)*

**(b)** The sprinter starts running and reaches a horizontal velocity of 9.3 m s$^{-1}$ in a distance of 35 m.
(i) Calculate her average acceleration over this distance.
(ii) Calculate the resultant force necessary to produce this acceleration. *(4 marks)*
AQA, 2012

**5** In a vehicle impact, a car ran into the back of a lorry. The car driver sustained serious injuries, which would have been much less had the car been fitted with a driver's air bag.
**(a)** Explain why the effect of the impact on the driver would have been much less if an air bag had been fitted and had inflated in the crash. *(4 marks)*
**(b)** Calculate the deceleration of the car if it was travelling at a speed of 18 m s$^{-1}$ when the impact occurred and was brought to rest in a distance of 2.5 m. *(2 marks)*
AQA, 2004

**6** A wind turbine, as shown in **Figure 5**, has blades of length 22 m. When the wind speed is 15 m s$^{-1}$ its output power is 1.5 MW.

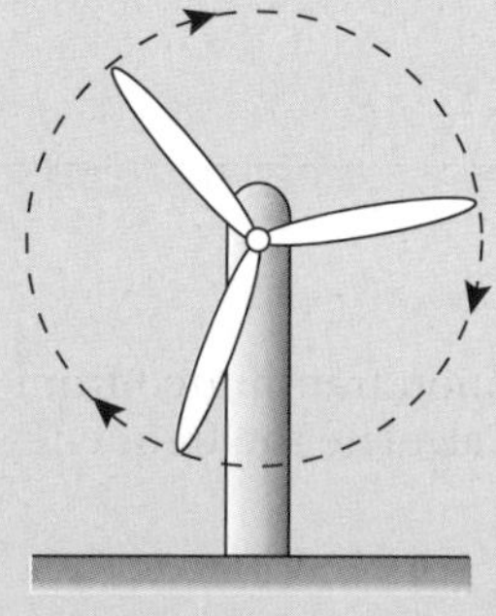

▲ **Figure 5**

(a) The volume of air passing through the blades each second can be calculated by considering a cylinder of radius equal to the length of the blade. Show that $2.3 \times 10^4\,m^3$ of air passes through the blades each second.

(b) Calculate the mass of air that passes through the blades each second.
density of air = $1.2\,kg\,m^{-3}$

(c) Calculate the kinetic energy of the air reaching the blades each second.

(d) Assuming that the power output of the turbine is proportional to the kinetic energy of the air reaching the blades each second, discuss the effect on the power output if the wind speed decreased by half. *(7 marks)*

AQA, 2005

7 Sail systems are being developed to reduce the running costs of cargo ships. The sail and ship's engines work together to power the ship. One of these sails is shown in **Figure 6** pulling at an angle of 40° to the horizontal.

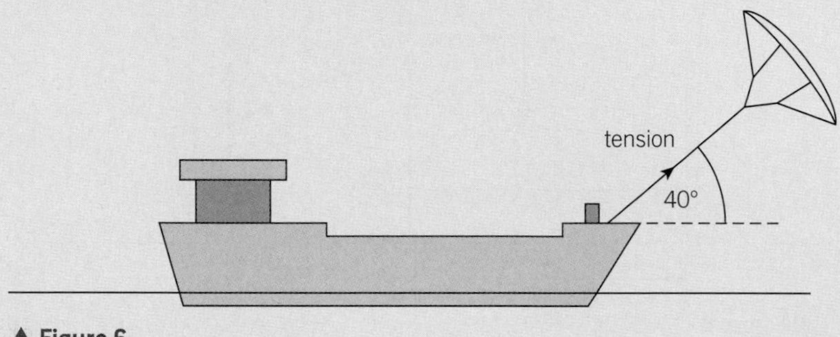


▲ **Figure 6**

(a) The average tension in the cable is 170 kN. Show that, when the ship travels 1.0 km, the work done by the sail on the ship is $1.3 \times 10^8\,J$. *(2 marks)*

(b) With the sail and the engines operating, the ship is travelling at a steady speed of $7.0\,m\,s^{-1}$.

(i) Calculate the power developed by the sail.

(ii) Calculate the percentage of the ship's power requirement that is provided by the wind when the ship is travelling at this speed. The power output of the engines is 2.1 MW. *(4 marks)*

(c) The angle of the cable to the horizontal is one of the factors that affects the horizontal force exerted by the sail on the ship. State *two* other factors that would affect this force. *(2 marks)*

AQA, 2012

8 An aerial system consists of a horizontal copper wire of length 38 m supported between two masts, as shown in **Figure 7**. The wire transmits electromagnetic waves when an alternating potential is applied to it at one end.

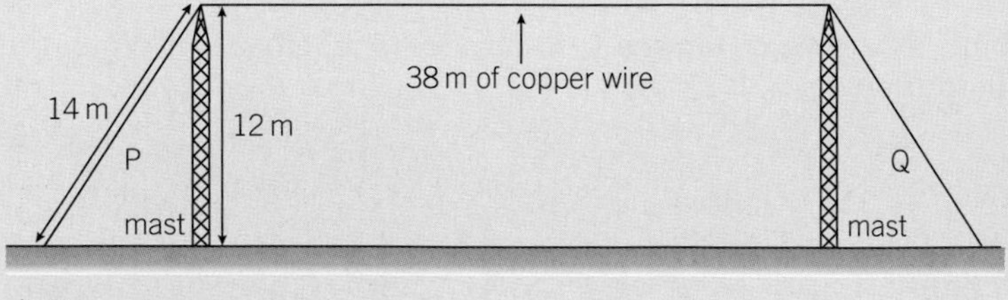


▲ **Figure 7**

(a) The wavelength of the radiation transmitted from the wire is twice the length of the copper wire. Calculate the frequency of the transmitted radiation. *(1 mark)*

**(b)** The ends of the copper wire are fixed to masts of height 12.0 m. The masts are held in a vertical position by cables, labelled P and Q, as shown in **Figure 7**.

(i) P has a length of 14.0 m and the tension in it is 110 N. Calculate the tension in the copper wire.

(ii) The copper wire has a diameter of 4.0 mm. Calculate the stress in the copper wire.

(iii) Discuss whether the wire is in danger of breaking if it is stretched further due to movement of the top of the masts in strong winds.
breaking stress of copper = $3.0 \times 10^8$ Pa *(7 marks)*

AQA, 2006

**9** The filament of a 60 W, 230 V mains lamp is a coil of thin tungsten wire.

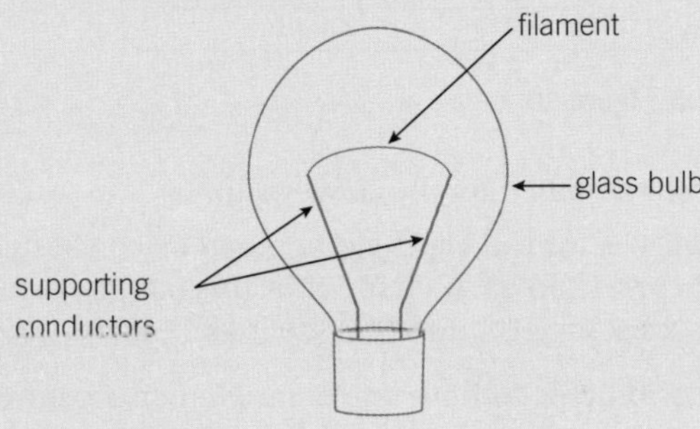


▲ **Figure 8**

**(a)** When the lamp is new, the filament wire has a radius of 80 μm and operates at a temperature of 2500 K.

(i) Calculate the resistance of the filament at 2500 K.

(ii) Show that the length of the wire in the filament is 0.25 m.
resistivity of tungsten = $7.0 \times 10^{-5}\,\Omega$m *(4 marks)*

**(b)** Near the end of its working life, the radius of the filament has decreased to 70 μm and its working temperature is 2300 K. At this temperature, the resistivity of tungsten is $6.4 \times 10^{-5}\,\Omega$m. Discuss whether the lamp consumes electrical energy at a rate of 60 W near the end of its working life. *(3 marks)*

AQA, 2006

**10** **Figure 9** shows how the resistance $R$ of three electrical components varies with temperature $\theta$ in °C.

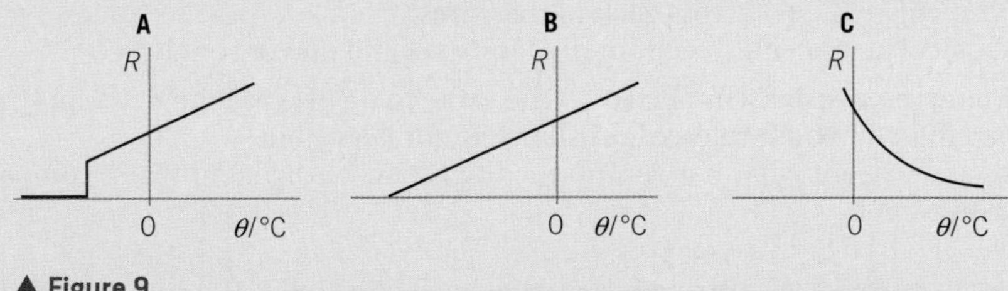


▲ **Figure 9**

**(a)** Indicate below which one of **A**, **B**, or **C** shows the correct graph for

(i) a wire-wound resistor

(ii) a thermistor

(iii) a superconductor. *(2 marks)*

**(b)** The metal wire used to manufacture the wire-wound resistor has a resistance per metre of 26 Ω and a diameter of 0.23 mm. Calculate the resistivity of the material from which the wire is made. *(4 marks)*

AQA, 2007

**11** A manufacturer asks you to design the heating element in a car rear-window demister. The design brief calls for an output of 48 W at a potential difference of 12 V. **Figure 10** shows where the eight elements will be on the car window before electrical connections are made to them.

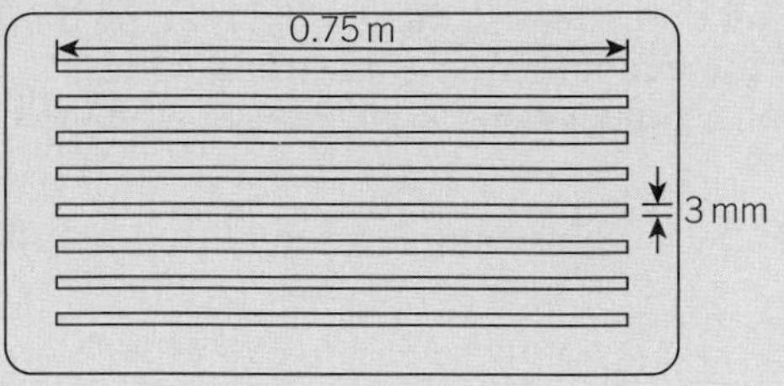


▲ **Figure 10**

**(a)** Calculate the current supplied by the power supply. *(1 mark)*

**(b)** One design possibility is for the eight elements to be connected in parallel.
(i) Calculate the current in each element in this parallel arrangement.
(ii) Calculate the resistance required for each element. *(3 marks)*

**(c)** Another design possibility is to have the eight elements connected in series.
(i) Calculate the current in each element in this series arrangement.
(ii) Calculate the resistance required for each element. *(4 marks)*

**(d)** State *one* disadvantage of the series design compared to the parallel arrangement. *(1 mark)*

**(e)** The series design is adopted. Each element is to have a rectangular cross section of 0.12 mm by 3.0 mm. The length of each element is to be 0.75 m.
(i) State the unit of resistivity.
(ii) Calculate the resistivity of the material from which the element must be made. *(3 marks)*

AQA, 2002

**12** A copper connecting wire is 0.75 m long and has a cross-sectional area of $1.3 \times 10^{-7}\,m^2$.

**(a)** Calculate the resistance of the wire. (Resistivity of copper = $1.7 \times 10^{-7}\,\Omega\,m$.) *(2 marks)*

**(b)** A 12 V 25 W lamp is connected to a power supply of negligible internal resistance using two of the connecting wires. The lamp is operating at its rated power.
(i) Calculate the current flowing in the lamp.
(ii) Calculate the pd across each of the wires.
(iii) Calculate the emf (electromotive force) of the power supply. *(4 marks)*

**(c)** The lamp used in **b** is connected by the same two wires to a power supply of the same emf, but whose internal resistance is not negligible.
State and explain what happens to the brightness of the lamp when compared to its brightness in **b**. *(2 marks)*

AQA, 2013

**13** When a note is played on a string instrument such as a violin, the sound it produces consists of the first harmonic and many higher harmonics. **Figure 11** shows the shape of the string for a stationary wave that corresponds to a higher harmonic. The positions of maximum and zero displacement for this harmonic are shown. The ends A and B of the string are fixed, and P, Q, and R are points on the string.

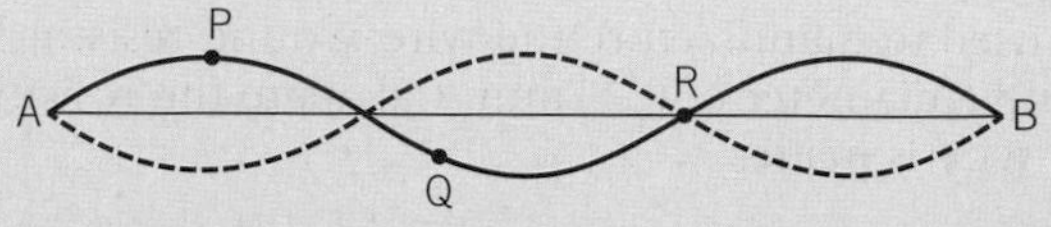


▲ **Figure 11**

**(a)** (i) Describe and compare the motion of points P and Q on the string. *(3 marks)*
(ii) What can you say about the motion of point R on the string? *(1 mark)*
(iii) What is the phase relationship between point Q on the string and the midpoint of the string? *(1 mark)*

**(b)** The string has a length of 0.60 m and is vibrating at a frequency of 510 Hz.
(i) Calculate the wavelength and the speed of the progressive waves on the wire. *(2 marks)*
(ii) The tension in the wire is 10 N. Calculate the mass per unit length of the wire. *(3 marks)*

**14** (a) A laser emits *monochromatic light*. Explain the meaning of the term monochromatic light. *(1 mark)*

**Figure 12** shows a laser emitting blue light directed at a single slit, where the slit width is greater than the wavelength of the light. The intensity graph for the diffracted blue light is shown.

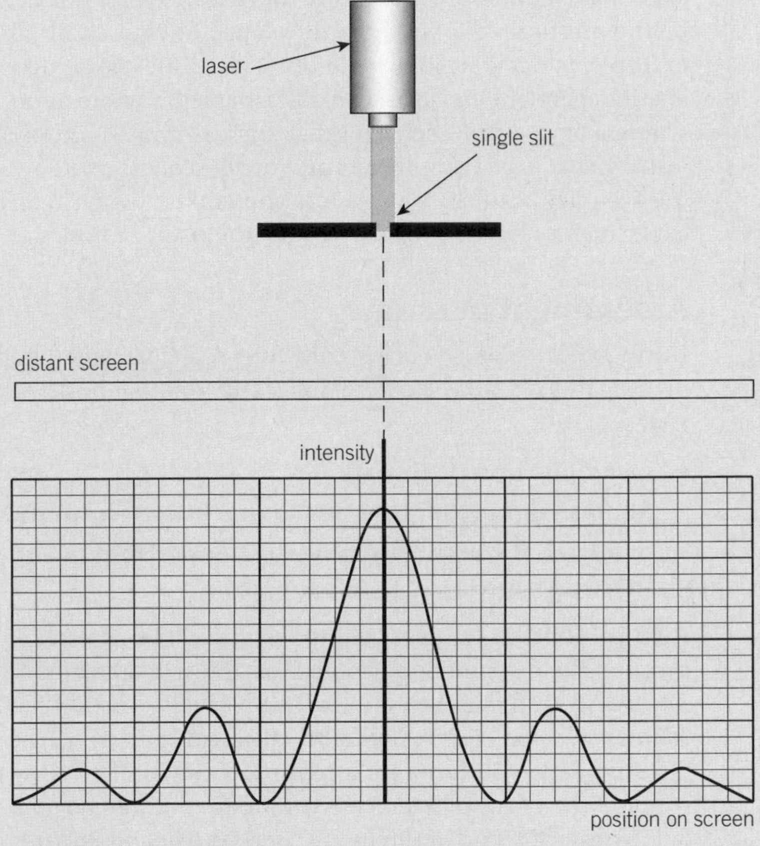


▲ **Figure 12**

**(b)** On the axes shown in **Figure 12**, sketch the intensity graph for a laser emitting red light. *(2 marks)*

**(c)** State and explain **one** precaution that should be taken when using laser light. *(2 marks)*

**(d)** The red laser light is replaced by a non-laser source emitting white light. Describe how the appearance of the pattern would change. *(3 marks)*

AQA, 2013

# Section 5
## Skills in AS Physics

### Chapters in this section:


**14** Practical work in physics

**15** About practical assessment

**16** More on mathematical skills


### Moving on from GCSE

Practical work is an integral feature of your AS physics course as it helps you develop your understanding of important concepts and applications. It also teaches you how scientists work in practice. You will find out how important discoveries are made in the subject. You might even make important discoveries yourself. AS level practical skills build on the practical skills you have developed in your GCSE course. For example, you will need to build on your GCSE knowledge of dependent and independent variables, control variables, and precision. In addition, you will be able to learn new skills, such as how to use measuring instruments (e.g., micrometers and oscilloscopes) and how to determine the accuracy of the measurements you make. Always remember that experiments and investigations are at the heart of how science works.

### Assessment overview

During your course, you will carry out practical experiments and investigations to develop your skills and you will be assessed on how well you can:

- **carry out practical work**
- **analyse data from practical experiments and investigations**
- **evaluate the results of practical experiments and investigations.**

Practical skills such as evaluating results will be assessed indirectly through the examination papers that you will take at the end of your AS course (and at the end of the full A level course). Practical skills such as safe and correct use of practical equipment will be assessed directly when you carry out appropriate practical work as part of your course. However, such direct assessment will only be on a 'pass or fail' basis at the end of the full A level course and will not contribute to your marks at either AS level or A level. See Chapter 15 for more about practical assessment.

# 14 Practical work in physics

## 14.1 Moving on from GCSE

### In the laboratory

The experimental skills you will develop during your course are part of the tools of the trade of every physicist. Data loggers and computers are commonplace in modern physics laboratories, but awareness on the part of the user of precision, reliability, errors, and accuracy are just as important as when measurements are made with much simpler equipment. Let's consider in more detail what you need to be aware of when you are working in the physics laboratory.

#### *Safety and organisation*

Your teacher will give you a set of safety rules and should explain them to you. You must comply with them at all times. You must also use your common sense and organise yourself so that you work safely. For example, if you set up an experiment with pulleys and weights, you need to ensure they are stable and will not topple over.

#### *Working with others*

Most scientists work in teams, each person cooperating with other team members to achieve specific objectives. This is effective because, although each team member may have a designated part to play, the exchange of ideas within the team often gives greater insight and awareness as to how to achieve the objectives.

In your AS level practical activities, you will often work in a small group in which you need to cooperate with the others in the group so everyone understands the objectives of the practical activity and everyone participates in planning and carrying out the activity.

#### *Planning*

At AS level, you may be asked to plan an experiment or investigation. The practical activities you carry out during your course should enable you to prepare a plan. Here are the key steps in drawing up a plan:

1. Decide in detail what you intend to investigate. Note the independent and dependent variables you intend to measure and note the variables that need to be controlled. The other variables need to be controlled to make sure they do not change. A **control variable** that can't be kept constant would cause the **dependent variable** to alter.
2. Select the equipment necessary for the measurements. Specify the range of any electrical meters you need.
3. List the key stages in the method you intend to follow and make some preliminary measurements to check your initial plans. Consider safety issues before you do any preliminary tests. If necessary, modify your plans as a result of your preliminary tests.
4. If the aim of your investigation is to test a hypothesis or theory or to use the measurements to determine a physical quantity (e.g., resistivity), you need to know how to use the measurements you make. See Topic 14.4 Analysis and evaluation for notes about how to process and use measurement data.

## Carrying out instructions and recording your measurements

In some investigations, you will be expected to follow instructions supplied to you either verbally or on a worksheet. You should be able to follow a sequence of instructions without guidance. Part of the direct assessment of your practical work is on how well you follow instructions. However, always remember safety first and, if the instructions are not clear, ask your teacher to clarify them.

When you record your measurements, tabulate them with a column for the independent variable and one or more columns for the dependent variable to allow for repeat readings and average values, if appropriate. The table should have a clear heading for each of the measured variables, with the unit shown after the heading, as below.

Single measurements of other variables (e.g., control variables) should be recorded together, immediately before or after the table. In addition, you should record the precision (i.e., the least detectable reading) of each measurement. This information is important when you come to analyse and evaluate your measurements.

▼ **Table 1** *Tabulating the measurements from an investigation of pd against current for a wire*

| Potential difference / V | Current / A | | | Average current / A |
|---|---|---|---|---|
| | 1st set | 2nd set | 3rd set | |
| | | | | |
| | | | | |

length of wire / m = ______

diameter of wire / mm = ______, ______, ______

average diameter of wire / mm = ______

# 14.2 Making careful measurements

## Measurements and errors

Measurements play a key role in science, so they must be:

1. **valid** – the measurements are of the required data or can be used to give the required data and have been obtained by an acceptable method. For example, a voltmeter connected across a variable resistor in series with a lamp and a battery would not measure the potential difference across the lamp.
2. **repeatable** and **reproducible** – the same results are obtained if the original experimenter repeats the investigation using the same method and equipment, or if the investigation is reproduced by another person or by using different equipment or techniques.

**Errors of measurement** are important in finding out how accurate a measurement is. We need to consider errors in terms of differences from the mean value. Consider the example of measuring the diameter, *d*, of a uniform wire using a micrometer. Suppose the following diameter readings are taken for different positions along the wire from one end to the other

0.34 mm, 0.33 mm, 0.36 mm, 0.33 mm, 0.35 mm

- The **range** of the measurements is given by the maximum and minimum value of the measurements. Here the range is from 0.33 mm to 0.36 mm. We will see later we can use this to estimate the **uncertainty** or probable error of the measurement.
- The **mean value**, <*d*>, is 0.34 mm, calculated by adding the readings together and dividing by the number of readings. If the difference between each reading and <*d*> changed regularly from one end of the wire to the other, it would be reasonable to conclude that the wire was non-uniform. Such differences are called **systematic errors**. If there is no obvious pattern or bias in the differences, the differences are said to be **random errors**.

What causes random errors? In the case of the wire, vibrations in the machine used to make the wire might have caused random variations in its diameter along its length. The experimenter might not use or read the micrometer correctly consistently.

The range of the diameter readings above is from 0.33 mm to 0.36 mm. The readings lie within 0.015 mm (i.e., half the range) of the mean value, which we will round up to 0.02 mm. The diameter can therefore be written as 0.34 ± 0.02 mm. The diameter is accurate to ±0.02 mm. The uncertainty or probable error in the mean value of the diameter is therefore ±0.02 mm.

## Using instruments

Instruments used in the laboratory range from the very basic (e.g., a millimetre scale) to the highly sophisticated (e.g., a multichannel data

**Hint**

**A measurement checklist**

At AS level, you should be able to:

- measure length using a ruler, vernier calipers, and a micrometer
- weigh an object and determine its mass using a spring balance, a lever balance, or a top pan balance
- use a protractor to measure an angle and use a set square
- measure time intervals using clocks, stopwatches, and the time base of an oscilloscope
- measure temperature using a thermometer
- use ammeters and voltmeters with appropriate scales
- use an oscilloscope.

In addition you should be able to:

- distinguish between systematic errors (including zero errors) and random errors
- understand what is meant by accuracy, sensitivity, linearity, reliability, precision, and validity
- read analogue and digital displays.

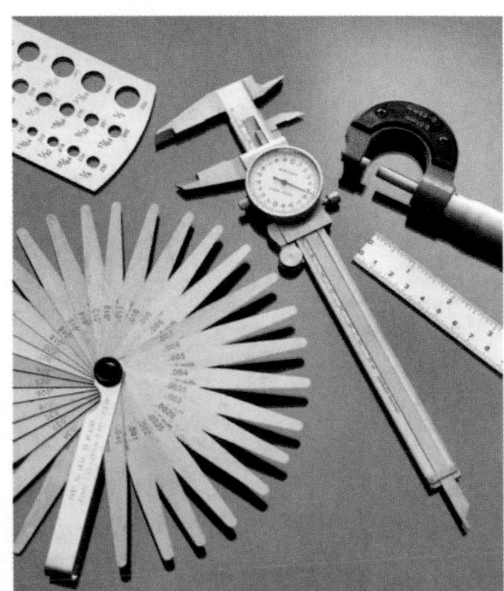

▲ **Figure 1** *Physics instruments*

recorder). Whatever type of instrument you use, you need to know what the following terms mean:

### *Zero error*

Does the instrument read zero when it is supposed to? If not, the zero reading must be taken into account when taking measurements otherwise there will be a systematic error in the measurements.

### *Uncertainty*

The uncertainty is the interval within which the true value can be expected to lie, expressed as a ± value (e.g., $I = 2.6 \pm 0.2$ A). If the readings are the same, the instrument precision should be used as the uncertainty. Uncertainty can be given with a level of confidence or probability (e.g., $I = 2.6 \pm 0.2$ A, at a level of confidence of 90%).

### *Accuracy*

An **accurate** measurement is one that is close to the accepted value. **Accuracy** is a measure of confidence in a measurement and is usually expressed as the uncertainty in the measurement.

### *Precision*

The **precision of a measurement** is the degree of exactness of the measurement. The precision is given by the extent of the random errors – a precise set of measurements will have little spread around the mean.

- If the reading of an instrument fluctuates when it is being taken, take several readings and calculate the mean and range of the measurements. The precision of the measurement is then given by the range of the readings.
- If the reading is constant, estimate the precision of a measurement directly from the instrument (or use its specified precision).

Precise readings are not necessarily accurate readings because systematic errors could make precise readings higher or lower than they ought to be.

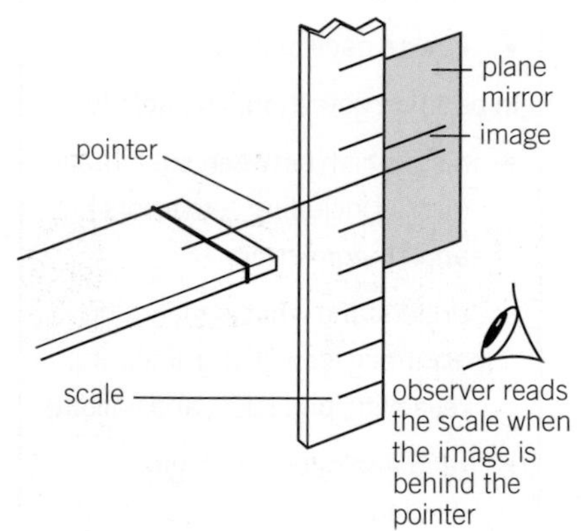


▲ **Figure 2** *Reading a scale*

### *Linearity*

This is a design feature of many instruments – it means the reading is directly proportional to the magnitude of the variable that causes the reading to change. For example, if the scale of a moving coil meter is **linear**, the reading of the pointer against the scale should be proportional to the current.

Measurement errors are caused in analogue instruments if the pointer on the scale is not observed correctly. The observer must be directly in front of the pointer when the reading is made. Figure 3 shows how a plane mirror is used for this purpose. The image of the pointer must be directly behind the pointer to ensure the observer views the scale directly in front of the pointer.

▲ **Figure 3** *A multimeter*

### *Instrument range*

Multirange instruments such as multimeters have a range dial that needs to be set according to the maximum reading to be measured. For example, if the dial can be set at 0–0.10 A, 0–1.00 A, or 0–10.0 A, you would use the 0–1.00 A range to measure the current through a 0.25 A torch bulb as the 0–0.10 A range is too low and the 0–1.00 A range is more sensitive than the 0–10.0 A range.

# 14.3 Everyday physics instruments

## Rulers and scales

Metre rulers are often used as vertical or horizontal scales in mechanics experiments.

To set a metre ruler in a vertical position:

- use a set square perpendicular to the ruler and the bench, if the bench is known to be horizontal, or
- use a plumb line (a small weight on a string) to see if the ruler is vertical. You need to observe the ruler next to the plumb line from two perpendicular directions. If the ruler appears parallel to the plumb line from both directions, then it must be vertical.

To ensure a metre ruler is horizontal, use a set square to align the metre ruler perpendicular to a vertical metre ruler.

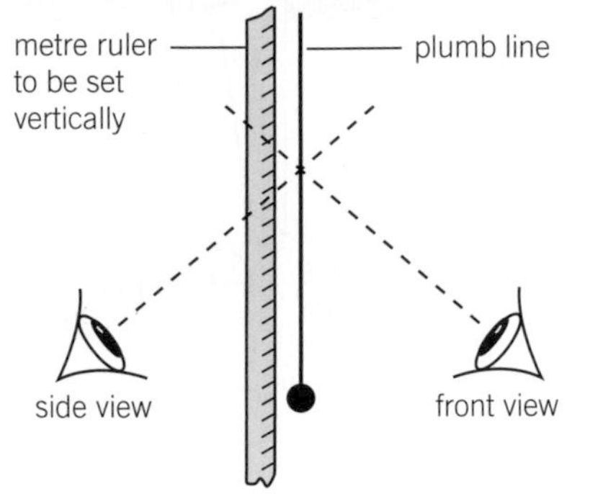

▲ **Figure 1** *Finding the vertical – if the metre ruler appears parallel to the plumb line from the front and the side, the ruler must be vertical*

## Micrometers and verniers

Micrometers give readings to within 0.01 mm. A **digital** micrometer gives a read-out equal to the width of the micrometer gap. An **analogue** micrometer has a barrel on a screw thread with a pitch of 0.5 mm. For such a micrometer:

- the edge of the barrel is marked in 50 equal intervals so each interval corresponds to changing the gap of the micrometer by 0.5/50 mm = 0.01 mm
- the stem of the micrometer is marked with a linear scale graduated in 0.5 mm marks
- the reading of a micrometer is where the linear scale intersects the scale on the barrel.

Figure 2 shows a reading of 4.06 mm. Note that the edge of the barrel is between the 4.0 and 4.5 mm marks on the linear scale. The linear scale intersects the 6th mark after the zero mark on the barrel scale. The reading is therefore 4.00 mm from the linear scale +0.06 mm from the barrel scale.

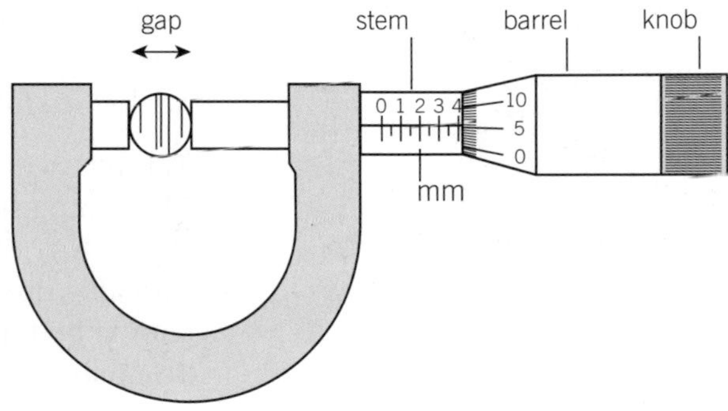


▲ **Figure 2** *Using a micrometer*

To use a micrometer correctly:

1. Check its zero reading and note the zero error if there is one.
2. Open the gap (by turning the barrel if analogue) then close the gap on the object to be measured. Turn the knob until it slips. Don't overtighten the barrel.
3. Take the reading and note the measurement after allowing if necessary for the zero error.
4. Note that the precision of the measurement is ±0.010 mm because the precision of the reading and the zero reading are both ±0.005 mm. So the difference between the two readings (i.e., the measurement) has a precision of 0.010 mm.

**Vernier calipers** are used for measurements of distances up to 100 mm or more. Readings can be made to within 0.1 mm. The sliding

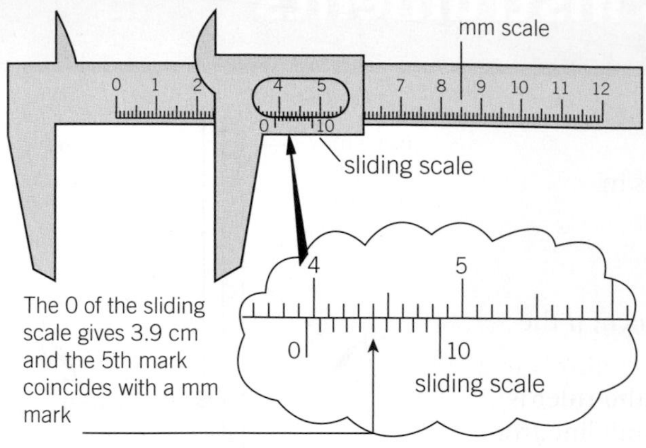


▲ **Figure 3** *Using a vernier*

scale of an analogue vernier has ten equal intervals covering a distance of exactly 9 mm so each interval of this scale is 0.1 mm less than a 1 mm interval. To make a reading:

1. The zero mark on the sliding scale is used to read the main scale to the nearest millimetre. This reading is rounded down to the nearest millimetre.
2. The mark on the sliding scale closest to a mark on the millimetre scale is located and its number noted. Multiplying this number by 0.1 mm gives the distance to be added on to the rounded-down reading.

Figure 3 shows the idea. The zero mark on the sliding scale is between 39 and 40 mm on the mm scale. So the rounded-down reading is 39 mm. The 5th mark after the zero on the sliding scale is nearest to a mark on the millimetre scale. So the extra distance to be added on to 39 mm is 0.5 mm (= 5 × 0.1 mm). Therefore, the reading is 39.5 mm.

## Timers

Stopwatches used for interval timings are subject to human error because reaction time, about 0.2 s, is variable for any individual. With practice, the delays when starting and stopping a stopwatch can be reduced. Even so, the precision of a single timing is unlikely to be better than 0.1 s. Digital stopwatches usually have read-out displays with a resolution of 0.01 s, but human variability makes such precision unrealistic and the precision of a single timing is the same as for an analogue stopwatch.

Timing oscillations requires timing for as many cycles as possible. The timing should be repeated several times to give an average (mean) value. Any timing that is significantly different to the other values is probably due to miscounting the number of oscillations so that timing should be rejected. For accurate timings, a fiducial mark is essential. The mark should be lined up with the centre of the oscillations so it provides a referencc position to count the number of cycles as the object swings past it each cycle.

Electronic timers use automatic switches, or gates, to start and stop the timer. However, just as with a digital stopwatch, a timing should be repeated, if possible several times, to give an average value. Light gates may be connected via an interface unit to a microcomputer. Interrupt signals from the light gates are timed by the microcomputer's internal clock. A software program is used to provide a set of instructions to the microcomputer.

### Synoptic link

You will meet the use of light gates with data loggers in more detail in Topic 15.2, Direct assessment.

## Balances

A balance is used to measure the weight of an object. Spring balances are usually less precise than lever balances. Both types of balance are usually much less precise than an electronic top-pan balance. The scale or read-out of a balance may be calibrated for convenience in kilograms or grams. The accuracy of an electronic top-pan balance can easily be tested using accurately known masses.

# 14.4 Analysis and evaluation

## Data processing

**For a single measurement,** the precision of the measuring instrument determines the precision of the measurement. A micrometer with a precision of 0.01 mm gives readings that each have a precision of 0.01 mm.

**For several readings,** the number of significant figures of the mean value should be the same as the precision of each reading. For example, consider the following measurements of the diameter of a wire: 0.34 mm, 0.33 mm, 0.36 mm, 0.33 mm, 0.35 mm. The mean value of the diameter readings works out at 0.342 mm, but the third significant figure cannot be justified as the precision of each reading is 0.01 mm. Therefore the mean value is rounded down to 0.34 mm.

### *Note:*

The uncertainty in the mean value is ±0.02 mm (i.e., half the range) as explained in Topic 14.2.

## Using error estimates

How confident can you be in your measurements and any results or conclusions you draw from your measurements? If you work out what each uncertainty is, as a percentage of the measurement (the percentage uncertainty), you can then see which measurement is least accurate. You can then think about how that measurement could be made more accurately.

### Worked example

The mass and diameter of a ball bearing were measured and the uncertainty of each measurement was estimated.

The mass, $m$, of a ball bearing = $4.85 \times 10^{-3} \pm 0.02 \times 10^{-3}$ kg

The diameter, $d$, of the ball bearing = $1.05 \times 10^{-2} \pm 0.01 \times 10^{-2}$ m

Calculate and compare the percentage uncertainty of these two measurements.

### *Solution*

The percentage uncertainty of the mass $m = \dfrac{0.02}{4.85} \times 100\% = 0.4\%$

The percentage uncertainty of the diameter $d = \dfrac{0.01}{1.05} \times 100\% = 1.0\%$

The diameter measurement is therefore more than twice as inaccurate as the mass measurement.

### *More about errors*

1 When two measurements are added or subtracted, the uncertainty of the result is the sum of the uncertainties of the measurements. For example, the mass of a beaker is measured when it is empty and then when it contains water:

- the mass of an empty beaker = 65.1 ± 0.1 g
- the mass of the beaker and water = 125.6 ± 0.1 g.

### Study tip

At the end of a calculation, don't give the answer to as many significant figures as shown on your calculator display. Give your answer to the same number of significant figures as the data with the least number of significant figures.

## Hint

**Percentage uncertainty**

To work out the percentage uncertainty of $A$, you could:

- Calculate the area of cross section for $d = 0.34 - 0.01$ mm = 0.33 mm.

  This should give an answer of $8.55 \times 10^{-8}\,\text{m}^2$.

- Calculate the area of cross section for $d = 0.34 + 0.01$ mm = 0.35 mm.

  This should give an answer of $9.62 \times 10^{-8}\,\text{m}^2$.

  Therefore, the area lies between $8.55 \times 10\,\text{m}^2$ and $9.62 \times 10^{-8}\,\text{m}^2$.

  In other words, the area is $(9.08 \pm 0.53) \times 10^{-8}\,\text{m}^2$

  (as $9.08 - 0.53 = 8.55$ and $9.08 + 0.53 = 9.62$).

  The percentage uncertainty of $A$ is $\frac{0.53}{9.08} \times 100\% = 5.8\%$.

  This is twice the percentage uncertainty of $d$.

  It can be shown as a general rule that for a measurement $x$, the percentage uncertainty in $x^n$ is $n$ times the percentage uncertainty in $x$.

## Synoptic link

You met straight line graphs in Topic 7.4, Free fall.

Then the mass of the water could be as much as

$(125.6 + 0.1) - (65.1 - 0.1)\,\text{g} = 60.7\,\text{g}$, or as little as
$(125.6 - 0.1) - (65.1 + 0.1)\,\text{g} = 60.3\,\text{g}$.

The mass of water is therefore $60.5 \pm 0.2\,\text{g}$.

2 When a measurement in a calculation is raised to a power $n$, the percentage uncertainty is increased $n$ times. For example, suppose you need to calculate the area $A$ of cross section of a wire that has a diameter of $0.34 + 0.01$ mm. You will need to use the equation $A = \pi d^2/4$. The calculation should give an answer of $9.08 \times 10^{-8}\,\text{m}^2$. The percentage uncertainty of $d$ is $\frac{0.01}{0.34} \times 100\% = 2.9\%$. So the percentage uncertainty of $A$ is 5.8% (= 2 × 2.9%). The consequence of this rule is that in any calculation where a quantity is raised to a higher power, the uncertainty of that quantity becomes much more significant.

## Graphs and errors

Straight line graphs are important because they are used to establish the relationship between two physical quantities. Consider a set of measurements of the distance fallen by an object released from rest and the time it takes. A graph of distance fallen against (time)$^2$ should be a straight line through the origin. If the line is straight, the theoretical equation $s = \frac{1}{2}gt^2$ (where $s$ is the distance fallen and $t$ is the time taken) is confirmed. The value of $g$ can be calculated, as the gradient of the graph is equal to $\frac{1}{2}g$. If the straight line does not pass through the origin, there is a systematic error in the distance measurement. Even so, the gradient is still $\frac{1}{2}g$.

### A best-fit test

Suppose you have obtained your own measurements for an experiment and you use them to plot a graph that is predicted to be a straight line. The plotted points are unlikely to be exactly straight in line with each other. The next stage is to draw a straight line of best fit so that the points are on average as close as possible to the line. Some problems may occur at this stage:

1 There might be a point much further than any other point from the line of best fit. The point is referred to as an **anomaly**. Methods for dealing with an anomalous point are as follows:

- If possible, the measurements for that point should be repeated and used to replace the anomalous point, if the repeated measurement is much nearer the line.
- If the repeated measurement confirms the anomaly, there could be a fault in the equipment or the way it is used. For example, in an electrical experiment, it could be caused by a change of the range of a meter to make that measurement. If no fault is found, make more measurements near the anomaly to see if these measurements lead towards the anomaly. If they do, it is no longer an anomaly and the measurements are valid.
- If a repeat measurement is not possible, the anomalous point should be ignored and a comment made in your report (or on the graph) about it.

**2** The points might seem to curve away from the straight line of best fit. The uncertainty of each measurement can be used to give a small range or **error bar** for each measurement. Figure 1 shows the idea. The straight line of best fit should pass through all the error bars. If it doesn't, the following notes might be helpful. You could use the error bars to draw a straight line of maximum gradient and a straight line of minimum gradient.

- Suppose the points lie along a straight line over most of the range but curve away further along the line. This would indicate that a straight line relationship between the plotted quantities is valid only over the range of measurements which produced the straight part of the line.
- Only two or three points in Figure 2 seem to lie on a straight line. In this case, it cannot be concluded that there is a linear relationship between the plotted quantities. You might need to plot further graphs to find out if a different type of graph would give a straight line relationship. A data analysis software package on a computer could be used to test different possible relationships (or for the second year of your course, a log graph could be plotted).

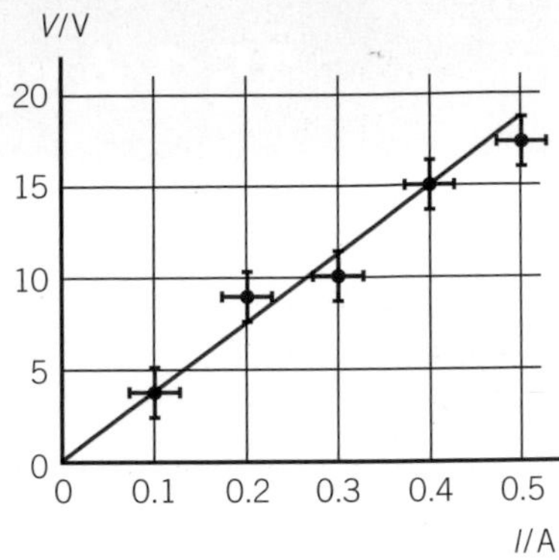


▲ **Figure 1** *Error bars*

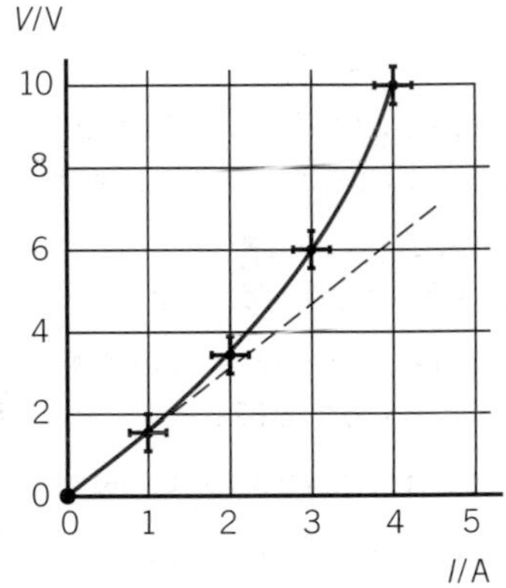


▲ **Figure 2** *Curves*

## Evaluating your results

You should be able to form a conclusion from the results of an investigation. This might be a final calculation of a physical quantity or property (e.g., resistivity) or a statement of the relationship established between two variables. As explained earlier, the degree of accuracy of the measurements could be used as a guide to the number of significant figures in a 'final result' conclusion. Mathematical links established or verified between quantities should be stated in a 'relationship' conclusion.

You always need to evaluate the conclusion(s) of an experiment or investigation to establish its validity. This evaluation could start with a discussion of the strength of the experimental evidence used to draw the conclusions:

- Discuss the reliability of the data and suggest improvements, where appropriate, that would improve the reliability. You may need to consider the effect of the control variables, if the experimental evidence is not as reliable as it should be.
- Discuss the methods taken (or proposed) to eliminate or reduce any random or systematic errors. Describe the steps taken to deal with anomalous results.
- Evaluate the accuracy of the results by considering the percentage uncertainties in the measurements. These can be compared to identify the most significant sources of error in the measurements, which can then lead to a discussion of how to reduce the most significant sources of error.
- Propose improvements to the strategy or experimental procedures, referring to the above discussion on validity as justification for the proposals.
- Suggest further experimental work, based on the strength of the conclusions. Strong conclusions could lead to a prediction and how to test it.

### Study tip

A final result, including its uncertainty, can be compared with the accepted (accurate) value, if known. Such a comparison can be used to evaluate the method used.

For example, for an obtained resistivity value of $4.8 \times 10^{-7} \pm 0.3 \times 10^{-7}\,\Omega\,m$ for a certain metal wire, with an accepted value of $5.2 \times 10^{-7}\,\Omega\,m$, you could review your method to see where improvements can be made (e.g., by making more measurements and taking a mean value).

# 15 About practical assessment

## 15.1 Assessment outline

Practical work is a vital part of any physics course. It helps you to understand new ideas and difficult concepts, and it helps you to appreciate the importance of experiments when testing and developing scientific theories. Practical work also develops the skills that scientists use every day in their work. Such skills involve planning, researching, making and processing measurements, and analysing and evaluating experimental results. The notes in this chapter tell you how your practical skills are assessed at A level.

The AQA A level two-year course is designed so that the first-year course and the AQA AS physics course cover the same content.

- The practical skills that can be assessed as part of a written examination (**indirect** practical skills) are tested in the end-of-course AS and A level examination papers.
- Practical skills that can't be assessed as part of a written examination (**direct** practical skills) will be assessed directly during your course. These skills are practical, technical, and manipulative skills such as following instructions, making measurements, and recording results, which are all essential at A level.
- If you complete the required practical experiments successfully in the two-year A level course, the grade you receive for this practical work will be separate from your grade for the written exam.
- There will be no non-exam assessment of practical skills for the AS course. But you will still need to do practical work and know and understand your practical skills.

***Note:***

Most of you taking physics after GCSE will want to take the full two-year A level course and sit the exam papers at the end of the course. Your school or college may also want you to sit the AS papers in June after studying the common AS/Year 1 A level course. This will give university admissions tutors external evidence of how well you have done in your first year of studying physics after GCSE. Hopefully, your AS results will also give you the confidence to progress to the second year of the A level course. But your AS marks will not contribute to your final A level marks!

Competence in practical skills is acquired through practice. The activities are chosen to give you opportunities to develop your practical skills and to demonstrate competence in each skill, either directly or indirectly. Some skills are easier to get than others, and so you should read the relevant notes in Chapter 14 and Chapter 16 before you start carrying out any practical activity.

Table 1 lists some of the practical activities you may be asked to carry out in your AS/Year 1 A level course. By carrying out these activities, you should become proficient in all of the practical skills assessed directly and indirectly in your AS/Year 1 A level course. For each activity, the table also gives references to the relevant topic in this book. The activities listed in **bold** are required for the practical skills questions in the written papers at both AS and A Level, and may be required for your practical portfolio.

▼ **Table 1**

| Activity | Topic |
|---|---|
| Investigating polarised light | 4.1 |
| Observing wave properties using a ripple tank | 4.3 |
| **Investigating the vibrations of a stretched string** | **4.6** |
| Measuring the refractive index of a transparent substance | 5.2 |
| **Investigating interference using Young's double slit experiment** | **5.4** |
| Using a diffraction grating to measure the wavelength of different colours of light | 5.7 |
| Testing the forces acting on an object in equilibrium | 6.2 |
| Using the principle of moments to weigh an object | 6.3 |
| **Measuring the acceleration of a falling object** | **7.4** |
| Measuring the path of a projectile | 7.7 |
| Investigating the motion of an object falling in a fluid | 8.3 |
| Investigating conservation of momentum | 9.3 |
| Measuring the efficiency of an electric motor used to lift an object | 10.4 |
| Measuring the density of an object | 11.1 |
| Investigating the strength of a spring | 11.2 |
| **Measuring the Young modulus of a wire** | **11.3** |
| **Measuring the resistivity of the material of a wire** | **12.3** |
| Investigating the characteristics of a component such as a resistor or a diode | 12.4 |
| Testing the resistance rules for different resistor combinations | 13.2 |
| Investigating the heating effect of an electric current | 13.2 |
| **Measuring the emf and the internal resistance of a cell or a battery** | **13.3** |
| Using a potential divider to design and test sensor circuits | 13.5 |

# 15.2 Direct assessment

You will need to carry out a sufficient number of practical activities (which may include the required practical experiments) over the A level course to show that you can:

- follow written procedures correctly
- identify variables and how to control them where necessary or take account of them if they cannot readily be controlled
- select appropriate equipment and methods to make accurate measurements
- use a range of practical equipment and materials safely and correctly with little or no assistance
- make and record accurate observations, using appropriate procedures, and making adjustments when necessary
- keep accurate and precise records of experimental data methodically, using appropriate units and conventions
- use appropriate software and tools to process data, carry out research, and report your findings
- use research sources including websites, textbooks, and other printed scientific sources of information
- correctly state sources of information used for research and when planning and drawing conclusions.

You will carry out some of these practical activities when you study topics in the common one-year AS/Year 1 A level course. You will carry out the rest of the activities in the second year of your A level course. The skills listed above are assessed directly by your teacher on the basis of your work in these activities.

**Before you carry out a practical task,** you should eliminate (if possible) or minimise any health and safety hazards. A risk assessment means that you must think about the possible hazards in an activity and plan so that you can eliminate or minimise them. Your teacher should have made a risk assessment in advance of every practical activity to make sure that the practical activities you will undertake are safe. For example, if you are about to stretch springs or wires, you should be provided with eye protection, which you should wear. However, you should also carry out your own risk assessment to make sure that when you use the apparatus you are given, you use it in a safe way.

**When you carry out these activities,** you will be given a set of instructions, and you will be assessed on your ability to:

- demonstrate safe and skilful practical techniques and processes
- obtain information from stated scientific sources of information, including websites and textbooks
- select appropriate methods and make measurements precisely and accurately
- make and record accurate and valid observations and measurements
- use appropriate software to process data and report on your findings
- work with others in experimental activities.

**When carrying out your practical work,** you should know how to:

- use vernier calipers and micrometers to measure small distances
- use appropriate analogue apparatus to measure angles, length and distance, volume, force, temperature, and pressure
- use appropriate digital instruments to measure mass, time, current, and voltage
- use a stopwatch or light gates for timing
- use methods to increase the accuracy of your measurements, such as by repeating and averaging timings, or by using a fiduciary marker, set square, or plumb line
- correctly construct circuits from your own or given circuit diagrams using dc power supplies, cells, and a range of circuit components, including ones where polarity is important
- generate and measure waves by using a microphone and loudspeaker, ripple tank, vibration transducer, or microwave/ radio wave source
- use a suitable light source to investigate characteristics of light, including interference and diffraction
- use ICT such as computer modelling or a data logger with a variety of sensors to collect data, or use software to process data.

The laboratory apparatus you will use to make measurements might include:

- basic apparatus (metre ruler, set square, protractors)
- electrical meters and multimeters (analogue or digital) for measuring current, voltage, and resistance
- a micrometer, vernier calipers
- a top-pan electronic balance, measuring cylinders
- a digital stopwatch or light gates for timing
- thermometers, newtonmeters
- a data logger with sensors
- a pressure gauge.

## About data loggers

A data logger is a device that automatically measures and records a variable at regular intervals. This variable is usually pd. A multi-channel data logger has more than one pair of input terminals and can record multiple inputs simultaneously. To record variations in a physical quantity (e.g., temperature, light intensity, or magnetic-field variations), a suitable sensor needs to be connected to the data logger. The sensor is designed so that its output pd changes as the physical quantity changes. Timing measurements using light gates can also be recorded by connecting the gates to an interface circuit which is connected to a data logger or a computer.

Most data loggers are capable of making thousands of measurements at intervals from milliseconds to hours. The interval value needs to be *keyed in* to the data logger (or set on a dial) before a data logger is used. If the sensor has more than one range, a suitable range must also be selected. The data recorded by a data logger may be displayed in a table or as a graph on a computer screen or on an oscilloscope.

Data logging has many advantages, including

- measuring and recording changes that are too fast to record manually
- making measurements automatically at long intervals and/or in remote locations
- making simultaneous measurements of more than one physical quantity
- processing and analysing recorded data and displaying measured and processed data graphically.

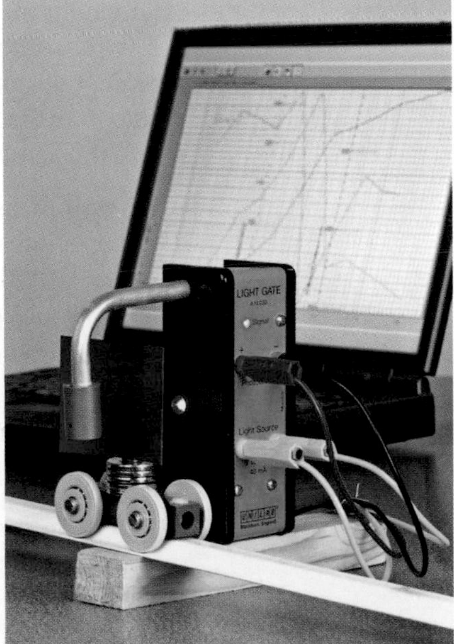


▲ **Figure 1** *A light gate in use*

# 15.3 Indirect assessment

In addition to assessing your knowledge and understanding of all the physics topics you have studied in this course, your written examination papers will include questions that assess your abilities in the following practical skill areas:

**Independent thinking**

- Solve problems set in practical contexts.
- Apply scientific knowledge to practical contexts.

**Use and application of scientific methods and practices**

- Comment on experimental design and evaluate scientific methods.
- Present data in appropriate ways.
- Evaluate results and draw conclusions with reference to measurement uncertainties and errors.
- Identify variables, including ones that must be controlled.

**Numeracy and the application of mathematical concepts in a practical context**

- Plot and interpret graphs.
- Process and analyse data using appropriate mathematical skills.
- Consider margins of error, accuracy, and precision of data.

**Instruments and equipment**

- Know and understand how to use a wide range of experimental and practical instruments, equipment, and techniques.

The practical experiments and investigations you carry out during your course should enable you to develop your competence in all of the direct and indirect practical skill areas that you will be assessed on. For example, in the experiment to measure the resistivity of the material of a wire, you need to appreciate from the start that the resistance of a wire depends on its length and on its diameter. Therefore, you might decide to measure the resistance of wires of different diameters that are all the same length and material. Then you will need to decide:

- how to measure the resistance and diameter of each wire and also their lengths (because this is a control variable and you will need to make sure that it is the same for each wire)
- which measuring instruments to use, how to use the instruments, and how to make sure that your measurements are as accurate as possible
- how to avoid systematic errors (e.g., check zero errors) and random errors (e.g., measure the diameter at several places along each wire and obtain an average value) and how to assess the uncertainty in each of your measurements.

Once you have a set of results, you need to know how to process them and how to use them. You might need to do some research to find the theoretical relationship between the two variables you have measured so that you can plot a graph. In the example, you might plot a graph of the resistance of the wire against the reciprocal of the area of cross section. This graph should be a straight line through the origin, as shown in Figure 1. The theory should tell you that the gradient is equal to the resistivity × the length of the wire. Hence, the resistivity of the wire can be determined. Finally, you might use your uncertainty estimates to determine the overall uncertainty in your value of resistivity.

## Synoptic link

You met resistivity in Topic 12.3, Resistance, and will meet straight line graphs in detail in Topic 16.4, Straight line graphs.

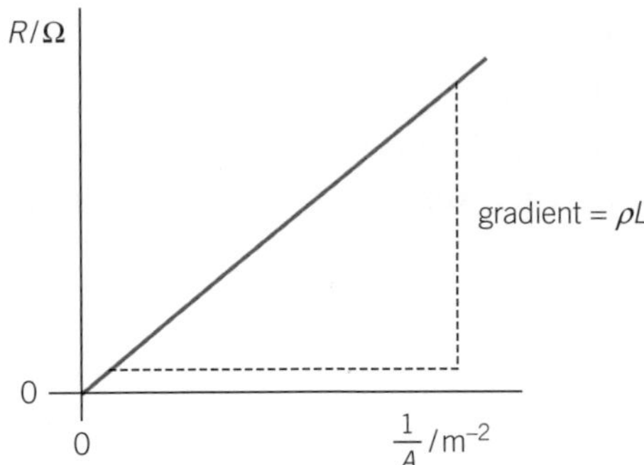


▲ **Figure 1** *Graph of resistance R against 1/area of cross section A for wires of the same material of resistivity ρ and length L*

The practical experiments and investigations you will carry out during this course will help you to develop and practise indirect skills regularly as well as direct skills. Each of the four main sections in this book includes practice questions, which are designed to test your indirect practical skills. So, at the end of this course, you will be well prepared for any questions on indirect practical skills in your written examination papers. You will also be well prepared for physics courses beyond A level.

### Note:

The practical skills questions in the written papers in both AS and A Level assume that you have carried out the practical activities listed in bold on Table 1, page 251. This table also provides information on where you can find more information about these activities within this book.

# 16 More on mathematical skills

## 16.1 Data handling

▼ **Table 1** *SI base units*

| Physical quantity | Unit |
|---|---|
| mass | kilogram (kg) |
| length | metre (m) |
| time | second (s) |
| electric current | ampere (A) |
| temperature | kelvin (K) |

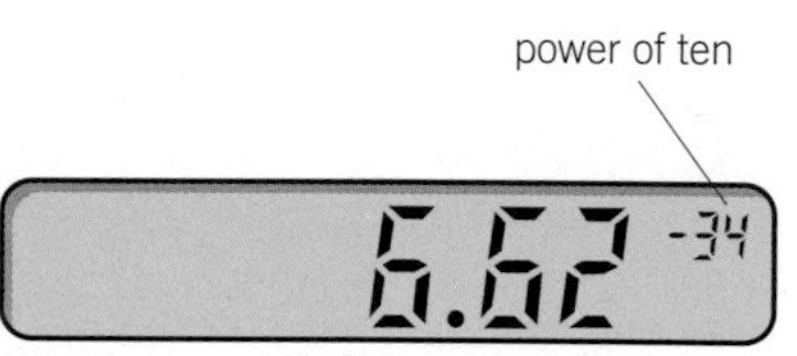


▲ **Figure 1** *Displaying powers of ten*

### Hint

Do not confuse $\log_{10}$ (base 10) with ln or $\log_e$ (base $e$). $\text{Log}_{10}$ allows you to find a number as a power of ten, while ln or $\log_e$ allows you to find a number as a power of $e$, the natural log.

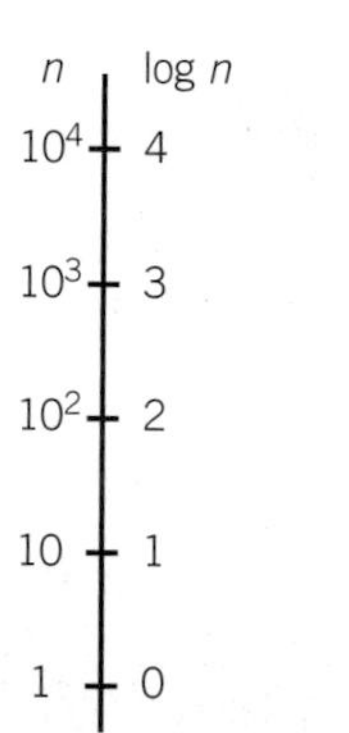


▲ **Figure 2** *A logarithmic scale*

This chapter covers the mathematical skills you will require for your whole AS Physics A course.

### Scientific units

Scientists use a single system of units to avoid unnecessary effort and time converting between different units of the same quantity. This system, the **Système International** (or **SI system**) is based on a defined unit for certain physical quantities including those listed in Table 1. Units of all other quantities are derived from the SI **base units**.

The following examples show how the units of all other physical quantities are derived from the base units.

- The unit of density is the kilogram per cubic metre ($\text{kg}\,\text{m}^{-3}$).
- The unit of speed is the metre per second ($\text{m}\,\text{s}^{-1}$).

### More about using a calculator

1 '**Exp**', '**EE**', or '**× 10^x**', is the calculator button you press to key in a **power of ten**. To key in a number in standard form (e.g., $3.0 \times 10^8$), the steps are as follows:
   - Step 1 Key in the number between 1 and 10 (e.g., 3.0).
   - Step 2 Press the calculator button marked 'Exp' (or 'EE' on some calculators).
   - Step 3 Key in the power of ten (e.g., 8).

   If the display reads '3.0 08' this should be read as $3.0 \times 10^8$ (not $3.0^8$ which means 3.0 multiplied by itself eight times). If the power of ten is a negative number (e.g., $10^{-8}$ not $10^8$), press the calculator button marked '+/–' after step 3 (or before, if you are using a graphic calculator) to change the sign of the power of ten.

2 '**inv**' is the button you press if you want the calculator to give the value of the inverse of a function. For example, if you want to find out the angle, which has a sine of 0.5, you key in 0.5 on the display then press 'inv' then 'sin' to obtain the answer of 30°. Some calculators have a 'second function' or '**shift**' button that you press instead of the 'inv' button.

3 '$\mathbf{log_{10}}$', '**log**',' or '**lg**' is the button you press to find out what a number is as a power of ten. For example, press 'log' then key in 100 and the display will show 2, because $100 = 10^2$. Logarithmic scales have equal intervals for each power of ten.

4 '$\mathbf{x^y}$', '$\mathbf{x^{\square}}$', or '^' allows you to raise a number to any power. For example, if you want to work out the value of $2^8$, key in 2 onto the display then $x^y$ then 8, and press =. The display should then show 256 as the decimal value of $2^8$. Raising a number $N$ to the power $1/n$ gives the $n$th root of $N$. For example, the cube root of 29 791 is $(29\,791)^{1/3}$ which equals 31.

## Significant figures

A calculator display shows a large number of digits. When you use a calculator, you should always round up or round down the final answer of a calculation to the same number of significant figures as the data given. Sometimes, a numerical answer to one part of a question has to be used in a subsequent calculation, in which case the numerical answer to the first part should be carried forward without rounding it up or down. For example, if you need to calculate the value of $d \sin 65°$, where $d = 1.64$, the calculator will show $9.063077870 \times 10^{-1}$ for the sine of 65°. Multiplying this answer by 1.64 then gives 1.486344771, which should then be rounded off to 1.49 so it has the same number of significant figures as 1.64 (i.e., to 3 significant figures).

### Worked example

Calculate the cube root of $2.9 \times 10^6$.

#### Solution

Step 1 Key in $2.9 \times 10^6$ as explained on page 246.

Step 2 Press the $y^x$ button.

Step 3 Key in $(1 \div 3)$.

Step 4 Press =.

The display should show '1.42602' so the answer is 142.6.

## Summary questions

Write your answers to each of the following questions in standard form, where appropriate, and to the same number of significant figures as the data.

**1** Copy and complete the following conversions.

**a** **i** 500 mm = _____ m
**ii** 3.2 m = _____ cm
**iii** 9560 cm = _____ m.

**b** **i** 0.45 kg = _____ g
**ii** 1997 g = _____ kg
**iii** 54 000 kg = _____ g.

**c** **i** 20 $cm^2$ = _____ $m^2$
**ii** 55 $mm^2$ = _____ $m^2$
**iii** 0.050 $cm^2$ = _____ $m^2$.

**2** **a** Write the following values in standard form:
**i** 150 million km in metres
**ii** 365 days in seconds
**iii** 630 nm in metres
**iv** 25.7 μg in kilograms
**v** 150 m in millimetres
**vi** 1.245 μm in metres.

**b** Write the following values with a prefix instead of in standard form.
**i** $3.5 \times 10^4$ m = _____ km
**ii** $6.5 \times 10^{-7}$ m = _____ nm
**iii** $3.4 \times 10^6$ g = _____ kg
**iv** $8.7 \times 10^8$ W = _____ MW = _____ GW.

**3** **a** Use the equation average speed = distance/time to calculate the average speed in m $s^{-1}$ of:
**i** a vehicle that travels a distance of 9000 m in 450 s
**ii** a vehicle that travels a distance of 144 km in 2 h
**iii** a particle that travels a distance of 0.30 nm in a time of $2.0 \times 10^{-18}$ s
**iv** the Earth on its orbit of radius $1.5 \times 10^{11}$ m, given the time taken per orbit is 365.25 days.

**b** Use the equation

$$\text{resistance} = \frac{\text{potential difference}}{\text{current}}$$

to calculate the resistance of a component for the following values of current $I$ and pd $V$:
**i** $V = 15$ V, $I = 2.5$ mA
**ii** $V = 80$ mV, $I = 16$ mA
**iii** $V = 5.2$ kV, $I = 3.0$ mA
**iv** $V = 250$ V, $I = 0.51$ μA
**v** $V = 160$ mV, $I = 53$ mA.

**4** **a** Calculate each of the following:
**i** $6.7^3$
**ii** $(5.3 \times 10^4)^2$
**iii** $(2.1 \times 10^{-6})^4$
**iv** $(0.035)^2$
**v** $(4.2 \times 10^8)^{1/2}$
**vi** $(3.8 \times 10^{-5})^{1/4}$.

**b** Calculate each of the following:
**i** $\dfrac{2.4^2}{3.5 \times 10^3}$
**ii** $\dfrac{3.6 \times 10^{-3}}{6.2 \times 10^2}$
**iii** $\dfrac{8.1 \times 10^4 + 6.5 \times 10^3}{5.3 \times 10^4}$
**iv** $7.2 \times 10^{-3} + \dfrac{6.2 \times 10^4}{2.6 \times 10^6}$.

# 16.2 Trigonometry

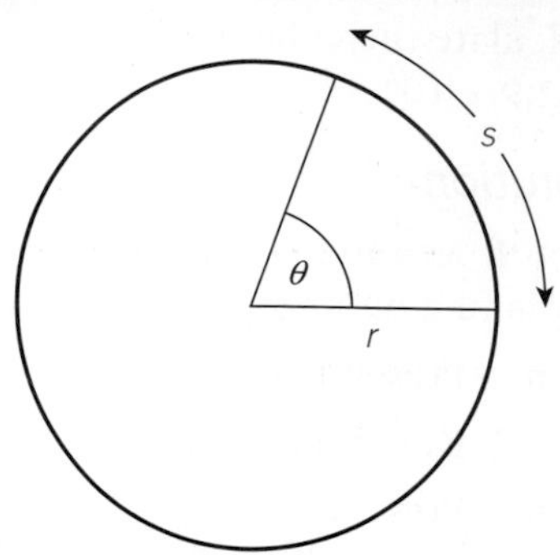


▲ **Figure 1** *Arcs and segments*

## Angles and arcs

- Angles are measured in degrees or radians. The scale for conversion is $360° = 2\pi$ radians. The symbol for the radian is rad, so 1 rad = $360/2\pi = 57.3°$ (to 3 significant figures).
- The circumference of a circle of radius $r = 2\pi r$. So the circumference can be written as the angle in radians ($2\pi$) round the circle × $r$.
- For a segment of a circle, the length of the arc of the segment is in proportion to the angle $\theta$ which the arc makes to the centre of the circle. This is shown in Figure 1. Because the arc length is $2\pi r$ (the circumference) for an angle of 360° (= $2\pi$ radians), then

$$\frac{\textbf{arc length, } s}{2\pi r} = \frac{\theta \textbf{ in degrees}}{360°}$$

## Triangles and trigonometry

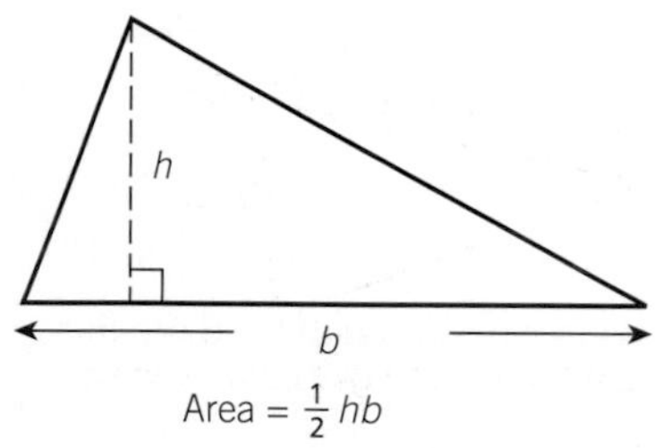


▲ **Figure 2** *The area of a triangle*

### Area rule

As shown in Figure 2,

**the area of any triangle = $\frac{1}{2}$ × its height × its base**

### Trigonometry calculations using a calculator

A scientific calculator has a button you can press to use either degrees or radians. Make sure you know how to switch your calculator from one of these two modes to the other. For example,

- sin 30° = 0.50, whereas sin 30 rad = −0.99
- inv sin 0.17 in degree mode = 9.79°, whereas inv sin 0.17 in rad mode = 0.171.

Also, watch out when you calculate the sine, cosine, or tangent of the product of a number and an angle. For example, sin (2 × 30°) comes out as 1.05 if you forget the brackets, instead of the correct answer of 0.867. The reason for the error is that, unless you insert the brackets, the calculator is programmed to work out sin 2° then multiply the answer by 30.

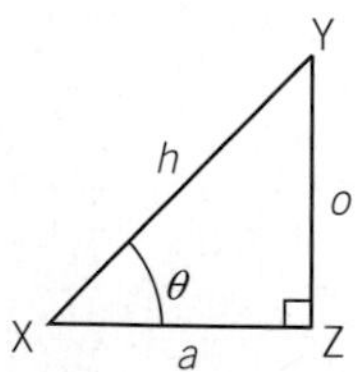


▲ **Figure 3** *A right-angled triangle*

### Trigonometry functions

Consider again the definitions of the sine, cosine, and tangent of an angle, as applied to the right-angled triangle in Figure 3.

$\sin\theta = \frac{o}{h}$ **where** $o$ **= the length of the side opposite angle** $\theta$

$\cos\theta = \frac{a}{h}$ $h$ **= the length of the hypotenuse**

$\tan\theta = \frac{o}{a}$ $a$ **= the length of the side adjacent to angle** $\theta$

### Pythagoras's theorem and trigonometry

Pythagoras's theorem states that for any right-angled triangle, the square of the hypotenuse = the sum of the squares of the other two sides.

Applying Pythagoras's theorem to the right-angled triangle in Figure 3 gives

$$h^2 = o^2 + a^2$$

**Synoptic link**

You met vectors and how to resolve them in Topic 6.1, Vectors and scalars.

Since $o = h \sin\theta$ and $a = h\cos\theta$, then the above equation may be written

$$h^2 = h^2\sin^2\theta + h^2\cos^2\theta$$

Cancelling $h^2$ therefore gives the following useful link between $\sin\theta$ and $\cos\theta$:

$$1 = \sin^2\theta + \cos^2\theta$$

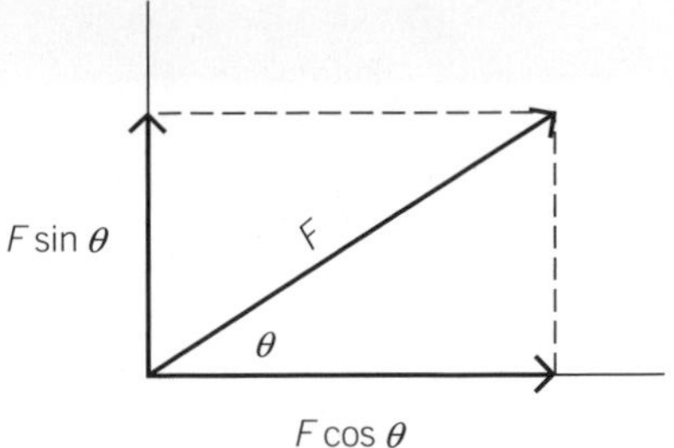


▲ **Figure 4** *Resolving a vector*

## Vector rules

### Resolving a vector

Any vector can be resolved into two perpendicular components in the same plane as the vector, as shown by Figure 4. The force vector $F$ is resolved into a horizontal component $F\cos\theta$ and a vertical component $F\sin\theta$, where $\theta$ is the angle between the line of action of the force and the horizontal line.

### Adding two perpendicular vectors

Figure 5 shows two perpendicular forces $F_1$ and $F_2$ acting on a point object X. The combined effect of these two forces, the resultant force, is given by the vector triangle in Figure 5. This is a right-angled triangle, where the resultant force is represented by the hypotenuse.

- Applying Pythagoras's theorem to the triangle gives $F^2 = F_1^{\,2} + F_2^{\,2}$, where $F$ is the magnitude of the resultant force. Therefore $F = (F_1^{\,2} + F_2^{\,2})^{1/2}$.
- Applying the trigonometry equation $\tan\theta = o/a$, the angle between the resultant force and force $F_1$ is given by $\tan\theta = F_2/F_1$.

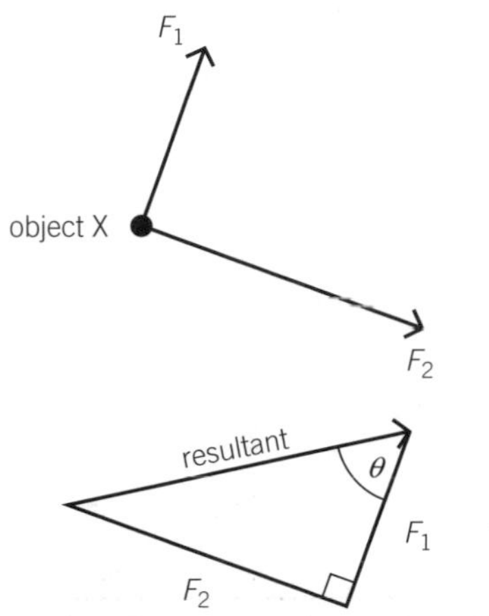


▲ **Figure 5** *Adding two perpendicular vectors*

## Summary questions

**1** **a** Calculate the circumference of a circle of radius 0.250 m.

**b** Calculate the length of the arc of a circle of radius 0.250 m for the following angles between the arc and the centre of the circle:

**i** 360° **ii** 240° **iii** 60°.

**2** For the right-angled triangle XYZ in Figure 6, calculate:

**a** angle YXZ (= $\theta$) if XY = 80 mm and

**i** XZ = 30 mm

**ii** XZ = 60 mm

**iii** YZ = 30 mm

**iv** YZ = 70 mm

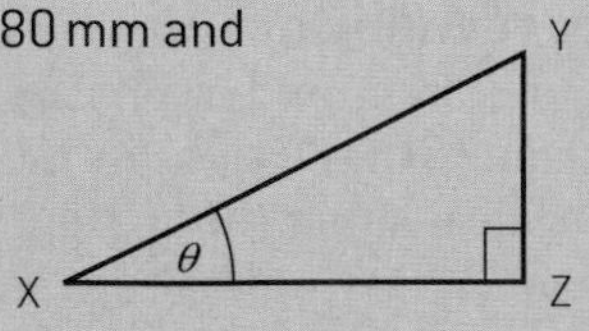

Y
θ
X Z

▲ **Figure 6**

**b** XZ if

**i** XY = 20 cm and $\theta$ = 30°

**ii** XY = 22 m and $\theta$ = 45°

**iii** YZ = 18 mm and $\theta$ = 75°

**iv** YZ = 47 cm and $\theta$ = 25°.

**3** **a** A right-angled triangle XYZ has a hypoteneuse XY of length 55 mm and side XZ of length 25 mm. Calculate the length of the other side.

**b** An aircraft travels a distance of 30 km due north from an airport P to an airport Q. It then travels due east for a distance of 18 km to an airport R. Calculate

**i** the distance from P to R

**ii** the angle QPR.

**4** **a** Calculate the horizontal component A and the vertical component B of:

**i** a 6.0 N force at 40° to the vertical

**ii** a 10.0 N force at 20° to the vertical

**iii** a 7.5 N force at 50° to the horizontal.

**b** Calculate the magnitude and direction of the resultant of a 2.0 N force acting due north and a 3.5 N force acting due east.

# 16.3 More about algebra

## Signs and symbols

If you used symbols in your GCSE course, you might have met the use of *s* for distance and *I* for current. Maybe you wondered why we don't use *d* for distance instead of *s* or *C* for current instead of *I*. The answer is that physics discoveries have taken place in many countries. The first person to discover the key ideas about speed was Galileo, the great Italian scientist, so he used the word for scale from his own language for distance and therefore assigned the symbol *s* to distance. Important discoveries about electricity were made by Ampère, the great French scientist, and he wrote about the intensity of an electric current, so he used the symbol *I* for electric current. The symbols we now use are used in all countries in association with the **SI system of units**.

▼ **Table 1** *Symbols for some physical quantities*

| Physical quantity | Symbol | Unit | Unit symbol |
|---|---|---|---|
| distance | $s$ | metre | m |
| speed or velocity | $v$ | metre per second | $m\,s^{-1}$ |
| acceleration | $a$ | metre per second per second | $m\,s^{-2}$ |
| mass | $m$ | kilogram | kg |
| force | $F$ | newton | N |
| energy or work | $E$ | joule | J |
| power | $P$ | watt | W |
| density | $\rho$ | kilogram per cubic metre | $kg\,m^{-3}$ |
| current | $I$ | ampere | A |
| potential difference or voltage | $V$ | volt | V |
| resistance | $R$ | ohm | Ω |

## Signs you need to recognise

- Inequality signs are often used in physics. You need to be able to recognise the meaning of the signs in Table 2. For example, the inequality $I \geq 3\,A$ means that the current is greater than or equal to 3 A. This is the same as saying that the current is not less than 3 A.

▼ **Table 2** *Signs*

| Sign | Meaning | Sign | Meaning | Sign | Meaning |
|---|---|---|---|---|---|
| > | greater than | >> | much greater than | $\langle x^2 \rangle$ | mean square value |
| < | less than | << | much less than | $\propto$ | is proportional to |
| ≥ | greater than or equal to | ≈ | approximately equals to | Δ | change of |
| ≤ | less than or equal to | $\langle x \rangle$ | mean value | √ | square root |

- The approximation sign is used where an estimate or an order-of-magnitude calculation is made, rather than a precise calculation. For an order-of-magnitude calculation, the final value is written with one significant figure only, or even rounded up or down to the nearest power of ten. Order-of-magnitude calculations are useful as a quick check after using a calculator. For example, if you are asked to calculate the density of a 1.0 kg metal cylinder of height 0.100 m and diameter 0.071 m, you ought to obtain a value of 2530 kg m$^{-3}$ using a calculator. Now let's check the value quickly:

  volume = $\pi$(radius)$^2$ × height
  $= 3 \times (0.04)^2 \times 0.1 = 48 \times 10^{-5}\,\text{m}^3$

  density = mass/volume
  $= 1.0/50 \times 10^{-5} = 2000\,\text{kg}\,\text{m}^{-3}$

  This confirms our 'precise' calculation.

- Proportionality is represented by the $\propto$ sign. A simple example of its use in physics is for Hooke's law – the tension in a spring is directly proportional to its extension:

$$\textit{tension } T \propto \textit{extension } \Delta L$$

By introducing a constant of proportionality $k$, the link above can be made into an equation:

$$T = k\Delta L$$

where $k$ is defined as the spring constant. With any proportionality relationship, if one of the variables is increased by a given factor (e.g., ×3), the other variable is increased by the same factor. So in the above example, if $T$ is trebled, then extension $\Delta L$ is also trebled. A graph of tension $T$ on the $y$-axis against extension $\Delta L$ on the $x$-axis would give a straight line through the origin.

**Synoptic link**

You met Hooke's law and the spring constant in Topic 11.2, Springs.

## More about equations

### Rearranging an equation with several terms

The equation $v = u + at$ is an example of an equation with two terms on the right-hand side. These terms are $u$ and $at$. To make $t$ the subject of the equation:

1 Isolate the term containing $t$ on one side by subtracting $u$ from both sides to give $v - u = at$.

2 Isolate $t$ by dividing both sides of the equation $v - u = at$ by $a$ to give

$$\frac{(v-u)}{a} = \frac{at}{a} = t$$

Note that $a$ cancels out in the expression $\frac{at}{a}$

3 The rearranged equation may now be written

$$t = \frac{(v-u)}{a}$$

## Rearranging an equation containing powers

Suppose a quantity is raised to a power in a term in an equation, and that quantity is to be made the subject of the equation. For example, consider the equation $V = \frac{4}{3}\pi r^3$ where $r$ is to be made the subject of the equation.

1 Isolate $r^3$ from the other factors in the equation by dividing both sides by $4\pi$, then multiplying both sides by 3 to give $\frac{3V}{4\pi} = r^3$.

2 Take the cube root of both sides to give $\left(\frac{3V}{4\pi}\right)^{1/3} = r$.

3 Rewrite the equation with $r$ on the left-hand side if necessary.

### *More about powers*

1 Powers add for identical quantities when two terms are multiplied together. For example, if $y = ax^n$ and $z = bx^m$, then $yz = ax^m\, bx^n = ab\,x^{m+n}$.

2 An equation of the form $y = \frac{k}{z^n}$ may be written in the form $y = kz^{-n}$.

3 The $n$th root of an expression is written as the power $1/n$. For example, the square root of $x$ is $x^{1/2}$. Therefore, rearranging $y = x^n$ to make $x$ the subject gives $x = y^{1/n}$.

## Summary questions

1 Complete each of the following statements:

a if $x > 5$, then $1/x <$

b if $4 < x < 10$, then ____ $< 1/x <$

c if $x$ is positive and $x^2 > 100$ then $1/x$ ____.

2 a Make $t$ the subject of each of the following equations:

i $v = u + at$  iii $y = k(t - t_0)$

ii $s = \frac{1}{2}at^2$  iv $F = \frac{mv}{t}$.

b Solve each of the following equations:

i $2z + 6 = 10$  iii $\frac{2}{z-4} = 8$

ii $2(z + 6) = 10$  iv $\frac{4}{z^2} = 36$.

3 a Make $x$ the subject of each of the following equations:

i $y = 2x^{1/2}$  iii $yx^{1/3} = 1$

ii $2y = x^{-1/2}$  iv $y = \frac{k}{x^2}$.

b Solve each of the following equations:

i $x^{-1/2} = 2$  iii $\frac{8}{x^2} = 32$

ii $3x^2 = 24$  iv $2(x^{1/2} + 4) = 12$.

4 Use the data given with each equation below to calculate:

a the volume $V$ of a wire of radius $r = 0.34$ mm and length $L = 0.840$ m, using the equation $V = \pi r^2 L$

b the radius $r$ of a sphere of volume $V = 1.00 \times 10^{-6}\,\text{m}^3$, using the equation $V = \frac{4}{3}\pi r^3$

c the time period $T$ of a simple pendulum of length $L = 1.50$ m, using the equation $T = 2\pi(L/g)^{0.5}$, where $g = 9.8\,\text{m s}^{-2}$

d the speed $v$ of an object of mass $m = 0.20$ kg and kinetic energy $E_k = 28$ J, using the equation $E_k = \frac{1}{2}mv^2$.

# 16.4 Straight line graphs

## The general equation for a straight line graph

Links between two physical quantities can be established most easily by plotting a graph. One of the physical quantities is represented by the vertical scale (the ordinate, often called the $y$-axis) and the other quantity by the horizontal scale (the abscissa, often called the $x$-axis). The coordinates of a point on a graph are the $x$- and $y$-values, usually written $(x, y)$ of the point.

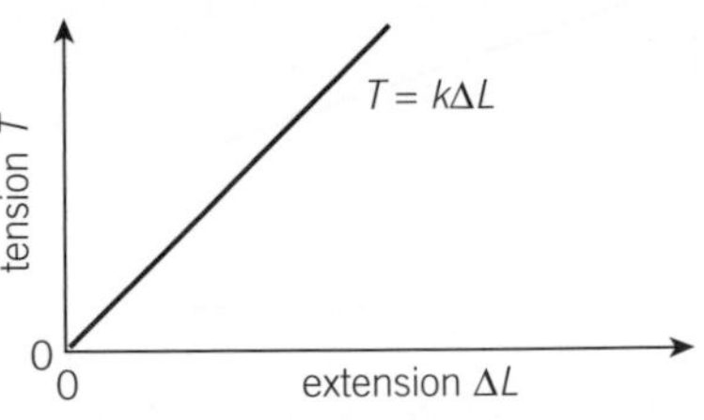


▲ **Figure 1** *Straight line graph*

The simplest link between two physical variables is where the plotted points define a straight line. For example, Figure 1 shows the link between the tension in a spring and the extension of the spring – the gradient of the line is constant and the line passes through the origin. Any situation where the $y$-variable is directly proportional to the $x$-variable gives a straight line through the origin. For Figure 1, the gradient of the line is the spring constant $k$. The relationship between the tension $T$ and the extension $\Delta L$ may therefore be written as $T = k\,\Delta L$.

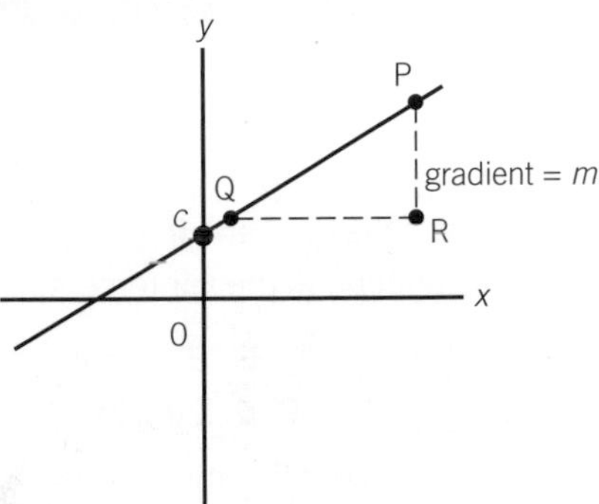


▲ **Figure 2** $y = mx + c$

The general equation for a straight line graph is usually written in the form $\mathbf{y = mx + c}$, where $m$ = the gradient of the line and $c$ = the $y$-intercept.

- The gradient $m$ can be measured by marking two points P and Q far apart on the line. The triangle PQR, as shown in Figure 2, is then used to find the gradient. If $(x_P, y_P)$ and $(x_Q, y_Q)$ represent the $x$- and $y$-coordinates of points P and Q, respectively, then

$$\textbf{gradient } m = \frac{y_P - y_Q}{x_P - x_Q}$$

- The $y$-intercept, $c$, is the point at $x = 0$ where the line crosses the $y$-axis. To find the $y$-intercept of a line on a graph that does not show $x = 0$, measure the gradient as above, then use the coordinates of any point on the line with the equation $y = mx + c$ to calculate $c$. For example, rearranging $y = mx + c$ gives $c = y - mx$. Therefore, using the coordinates of point Q in Figure 2, the $y$-intercept $c = y_Q - mx_Q$.

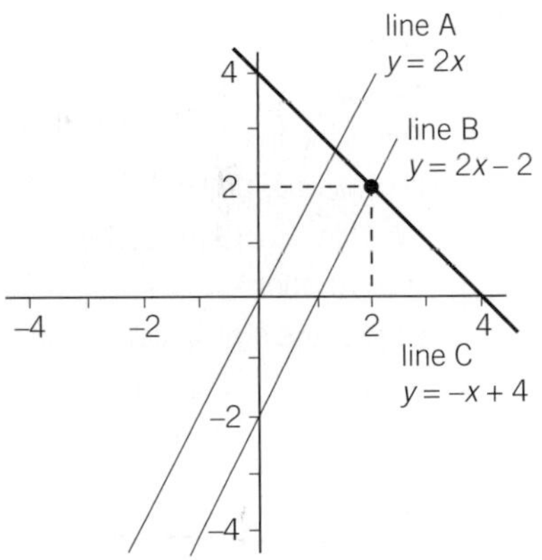


▲ **Figure 3** *Straight line graphs*

### *Examples of straight line graphs*

- Line A $c = 0$, so the line passes through the origin. Its equation is $y = 2x$.
- Line B $m > 0$, so the line has a positive gradient. Its equation is $y = 2x - 2$.
- Line C $m < 0$, so the line has a negative gradient. Its equation is $y = -x + 4$.

## Straight line graphs and physics equations

You need to be able to work out gradients and intercepts for equations you meet in physics that generate straight line graphs. Some further examples in addition to Figure 1 are described below.

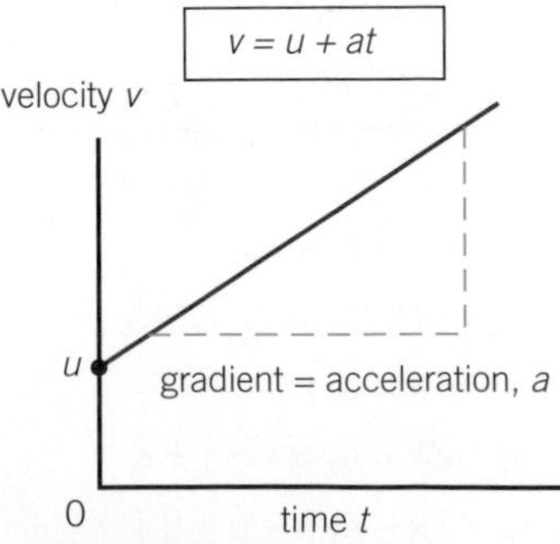


▲ **Figure 4** *Motion at constant acceleration*

1 **The velocity $v$ of an object moving at constant acceleration $a$ at time $t$** is given by the equation $v = u + at$, where $u$ is its velocity at time $t = 0$. Figure 4 shows the corresponding graph of velocity $v$ on the $y$-axis against time $t$ on the $x$-axis.

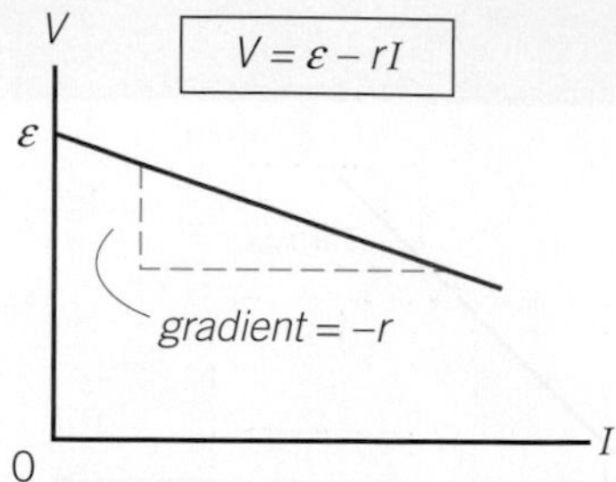


▲ **Figure 5** *pd versus I for a battery*

Rearranging the equation as $v = at + u$ and comparing this with $y = mx + c$ shows that

- the gradient $m$ = acceleration $a$
- the $y$-intercept $c$ = the initial velocity $u$.

2 **The pd, $V$, across the terminals of a battery of emf $\varepsilon$ and internal resistance $r$** varies with current in accordance with the equation $V = \varepsilon - Ir$. Figure 5 shows the corresponding graph of pd, $V$, on the $y$-axis against current, $I$, on the $x$-axis.
Rearranging the equation as $V = -rI + \varepsilon$ and comparing this with $y = mx + c$ shows that
   - the gradient $m = -r$
   - the $y$-intercept $c = \varepsilon$ so the intercept on the $y$-axis gives the emf $\varepsilon$ of the battery.

3 **The maximum kinetic energy $E_{Kmax}$ of a photoelectron** emitted from a metal surface of work function $\phi$ varies with frequency $f$ of the incident radiation, in accordance with the equation $E_{Kmax} = hf - \phi$. Figure 6 shows the corresponding graph of $E_{Kmax}$ on the $y$-axis against $f$ on the $x$-axis.

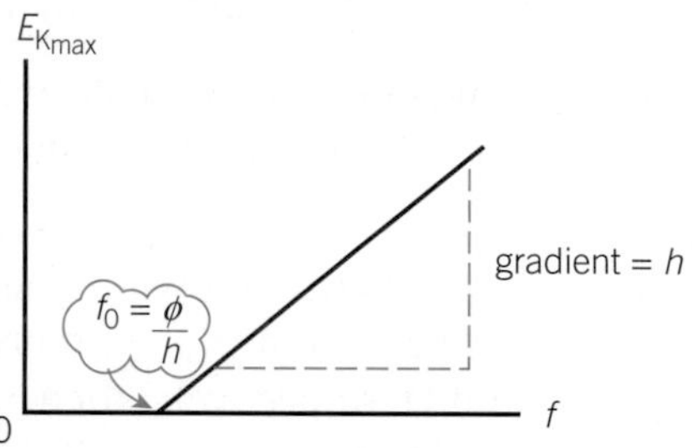


▲ **Figure 6** *Photoelectric emission*

Comparing the equation $E_{Kmax} = hf - \phi$ with $y = mx + c$ shows that
- the gradient $m = h$
- the $y$-intercept $c = -\phi$.

Note that the $x$-intercept is where $y = 0$ on the line. Let the coordinates of the $x$-intercept be $(x_0, 0)$. Therefore $mx_0 + c = 0$ so $x_0 = -c/m$. Applied to Figure 6, the $x$-intercept is therefore $\phi/h$. Since the $x$-intercept is the threshold frequency $f_0$, then $f_0 = \phi/h$.

### Simultaneous equations

In physics, simultaneous equations can be solved graphically by plotting the line for each equation. The solution of the equations is given by the coordinates of the points where the lines meet. For example, lines B and C in Figure 3 meet at the point (2, 2) so $x = 2$, $y = 2$ are the only values of $x$ and $y$ that fit both equations.

Solving simultaneous equations doesn't require graph plotting if the equations can be arranged to fit one of the variables. Start by rearranging to make $y$ the subject of each equation, if necessary. Considering the example above:

equation of line B is $y = 2x - 2$

equation of line C is $y = -x + 4$.

At the point where they meet, their coordinates are the same, so solving $2x - 2 = -x + 4$ gives the $x$-coordinate. Rearranging this equation gives $3x = 6$ so $x = 2$.

Since $y = 2x - 2$, then $y = (2 \times 2) - 2 = 2$.

## Summary questions

1 For each of the following equations, which represent straight line graphs, write down **i** the gradient, **ii** the $y$-intercept, **iii** the $x$-intercept:

**a** $y = 3x - 3$ **c** $y + x = 5$

**b** $y = -4x + 8$ **d** $2y + 3x = 6$.

2 **a** A straight line on a graph has a gradient $m = 2$ and passes through the point $(2, -4)$. Work out **i** the equation for this line, **ii** its $y$-intercept.

**b** The velocity $v$ (in m s$^{-1}$) of an object varies with time $t$ (in s) in accordance with the equation $v = 5 + 3t$. Determine **i** the acceleration of the object, **ii** the initial velocity of the object.

3 **a** Plot the equations $y = x + 3$ and $y = -2x + 6$ over the range from $x = -3$ to $x = +3$. Write down the coordinates of the point P where the two lines cross.

**b** Write down the equation for the line OP, where O is the origin of the graph.

4 Solve the following pairs of simultaneous equations:

**a** $y = 2x - 4$, $y = -x + 2$

**b** $y = 3x - 4$, $x + y = 8$

**c** $2x + 3y = 4$, $x + 2y = 2$.

# 16.5 More on graphs

## Curves and equations

Graphs with curves occur in physics in two ways.

1 In practical work, where one physical variable is plotted against another and the plotted points do not lie along a straight line. For example, a graph of pd on the $y$-axis against current on the $x$-axis for a filament lamp is a curve that passes through the origin.
2 In theory work, where an equation containing two physical variables is not in the form of the equation for a straight line ($y = mx + c$). For example, for an object released from rest, a graph of distance fallen, $s$, on the $y$-axis against time, $t$, on the $x$-axis is a curve, because $s = \frac{1}{2}gt^2$. Figure 1 shows this equation represented on a graph.

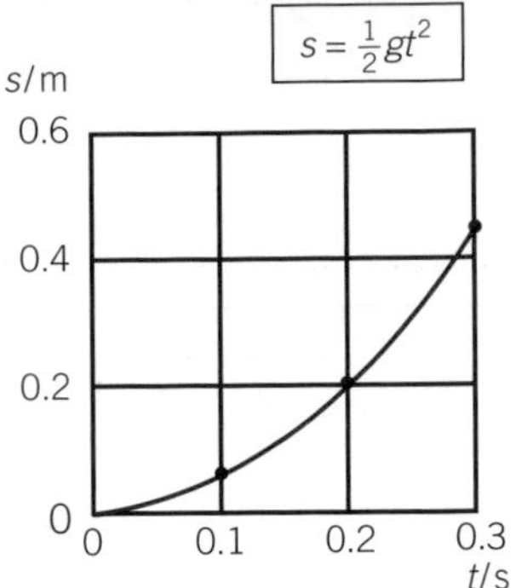


▲ **Figure 1** *s against t*

Knowledge of the general equations for some common curves is an essential part of physics. When a curve is produced as a result of plotting a set of measurements in a practical experiment, few conclusions can be drawn as the exact shape of a curve is difficult to test. In comparison, if the measurements produce a straight line, it can then be concluded that the two physical variables plotted on the axes are related by an equation of the form $y = mx + c$.

If a set of measurements produces a curve rather than a straight line, knowledge of the theory could enable the measurements to be processed in order to give a straight line graph, which would then be a confirmation of the theory. For example, the distance and time measurements that produced the curve in Figure 1 could be plotted as distance fallen, $s$, on the $y$-axis against $t^2$ on the $x$-axis (where $t$ is the time taken). Figure 2 shows the idea.

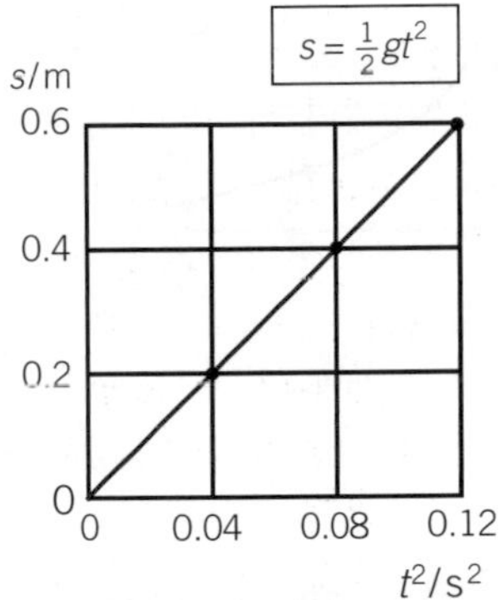


▲ **Figure 2** *s against $t^2$*

If a graph of $s$ against $t^2$ gives a straight line, this would confirm that the relationship between $s$ and $t$ is of the form $s = kt^2$, where $k$ is a constant. Because theory gives $s = \frac{1}{2}gt^2$, it can then be concluded that $k = \frac{1}{2}g$.

## From curves to straight lines

### Parabolic curves

These curves describe the flight paths of projectiles or other objects acted on by a constant force that is not in the same direction as the initial velocity of the object. In addition, parabolic curves occur where the energy of an object depends on some physical variable.

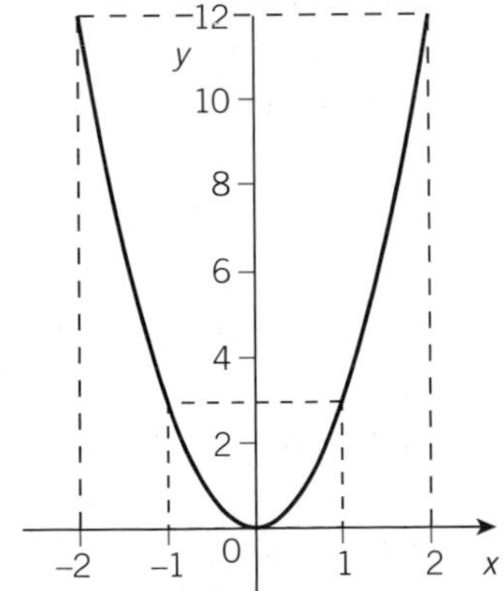


▲ **Figure 3** $y = 3x^2$

The general equation for a parabola is $y = kx^2$. Figure 3 shows the shape of the parabola $y = 3x^2$. Equations of the form $y = kx^2$ pass through the origin and they are symmetrical about the $y$-axis. This is because equal positive and negative values of $x$ always give the same $y$-value.

The **flight path for a projectile** projected horizontally at speed $U$ has coordinates $x = Ut$, $y = \frac{1}{2}gt^2$, where $x$ = horizontal distance travelled, $y$ = vertical distance fallen, and $t$ is the time from initial projection.

Combining these equations gives the flight path equation $y = \frac{gx^2}{2u^2}$

**Synoptic link**

You met projectiles in Topic 7.7, Projectile motion 1, and Topic 7.8, Projectile motion 2.

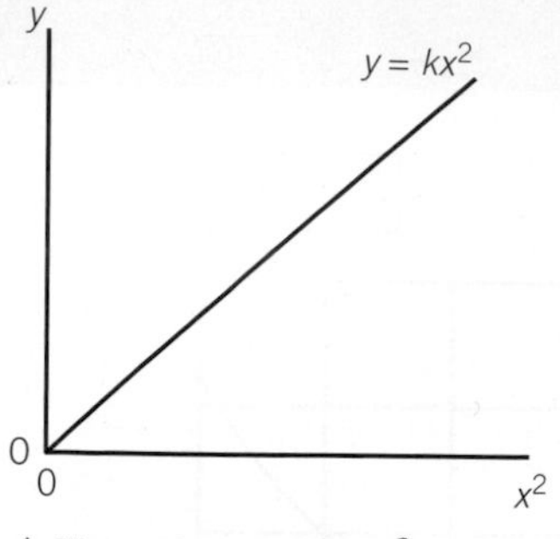


▲ **Figure 4** *y against $x^2$*

which is the same as the parabola equation $y = kx^2$ where $\frac{g}{2u^2}$ is represented by $k$ in the equation.

A set of measurements plotted as a graph of vertical distance fallen, $y$, against horizontal distance travelled, $x$, would be a parabolic curve as shown in Figure 3. However, a graph of $y$ against $x^2$ should give a straight line (of gradient $k$) through the origin (Figure 4), because $y = kx^2$.

### Inverse curves

An inverse relationship between two variables $x$ and $y$ is of the form $y = \frac{k}{x}$, where $k$ is a constant. The variable $y$ is said to be inversely proportional to variable $x$. For example, if $x$ is doubled, $y$ is halved.

Figure 5 shows the curve for $y = \frac{10}{x}$.

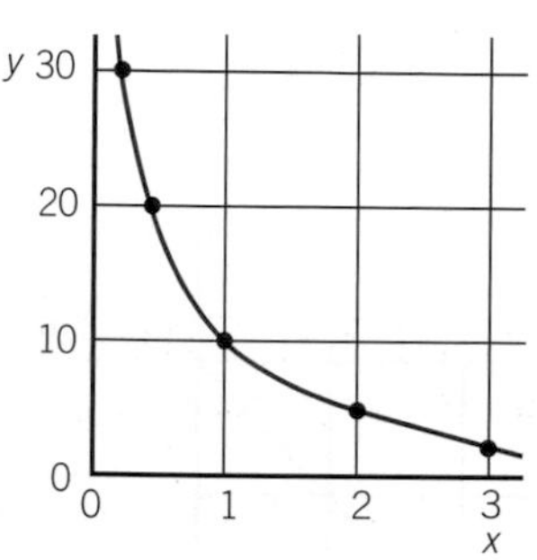


▲ **Figure 5** $y = \frac{10}{x}$

The curve tends towards either axis but never actually meets the axes. The correct mathematical word for 'tending towards but never meeting is asymptotic. Consider the following example.

The resistance $R$ of a wire of constant length $L$ varies with the wire's area of cross section, $A$, in accordance with the equation $R = \frac{\rho L}{A}$, where $\rho$ is the resistivity of the wire.

Therefore, $R$ is inversely proportional to $A$. A graph of $R$ (on the vertical axis) against $A$ would therefore be a curve like Figure 5. However, a graph of $R$ (on the vertical axis) against $1/A$ is a straight line through the origin (Figure 6). The gradient of this straight line is $\rho L$.

**Synoptic link**

You met resistivity in Topic 12.3, Resistance.

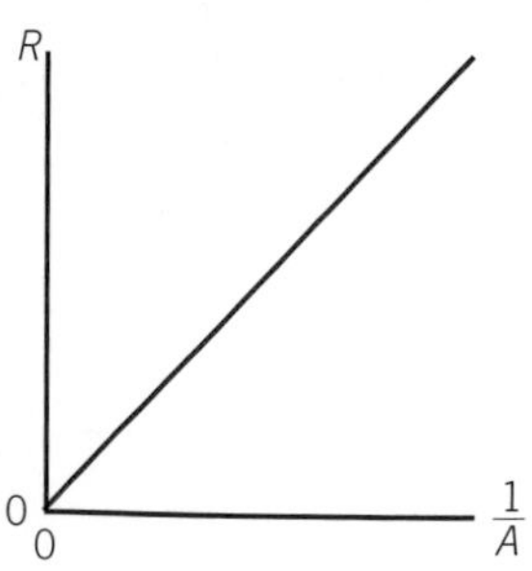


▲ **Figure 6** *R against* $\frac{1}{A}$

## Summary questions

**1** The potential energy, $E_p$, stored in a stretched spring varies with the extension $\Delta L$ of the spring, in accordance with the equation $E_p = \frac{1}{2}k\,\Delta L^2$. Sketch a graph of $E_p$ against **a** $\Delta L$, **b** $\Delta L^2$.

**2** The energy $E_{ph}$ of a photon varies with its wavelength $\lambda$ in accordance with the equation $E_{ph} = hc/\lambda$, where $h$ is the Planck constant and $c$ is the speed of light. Sketch a graph of $E_{ph}$ against **a** $\lambda$, **b** $1/\lambda$.

**3** The current $I$ through a wire of resistivity $\rho$ varies with the length $L$, area of cross section $A$, and pd $V$, in accordance with the equation $I = \frac{VA}{\rho L}$.

- **a** Sketch a graph of $I$ against **i** $V$, **ii** $L$, **iii** $1/L$.
- **b** Explain how you would determine the resistivity from the graph of $I$ against **i** $V$, **ii** $1/L$.

**4** An object released from rest falls at constant acceleration $a$ and passes through a horizontal beam at speed $u$. The distance it falls in time $t$ after passing through the light beam is given by the equation $s = ut + \frac{1}{2}at^2$.

- **a** Show that $\frac{s}{t} = u + \frac{1}{2}at$.
- **b** **i** Sketch a graph of $s/t$ on the vertical axis against $t$ on the horizontal axis.
  - **ii** Explain how $u$ and $a$ can be determined from the graph.

# 16.6 Graphs, gradients, and areas

## Gradients

1 The gradient of a straight line = $\Delta y/\Delta x$, where $\Delta y$ is the change of the quantity plotted on the $y$-axis and $\Delta x$ is the change of the quantity plotted on the $x$-axis. As shown in Figure 1, the gradient of a straight line is obtained by drawing as large a gradient triangle as possible and measuring the height $\Delta y$ and the base $\Delta x$ of this triangle, using the scale on each axis.

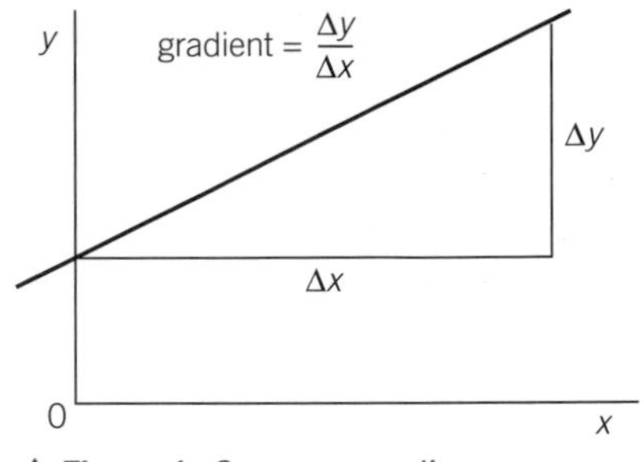


▲ **Figure 1** *Constant gradient*

***Note:***

As a rule, when you plot a straight line graph, always choose a scale for each axis that covers at least half the length of each axis. This will enable you to draw the line of best fit as accurately as possible. The measurement of the gradient of the line will therefore be more accurate. If the $y$-intercept is required and it cannot be read directly from the graph, it can be calculated by substituting the value of the gradient and the coordinates of a point on the line into the equation $y = mx + c$.

**Synoptic link**

You have met lines of best fit in Topic 14.4, Analysis and evaluation.

2 The gradient at a point on a curve = the gradient of the tangent to the curve at that point.

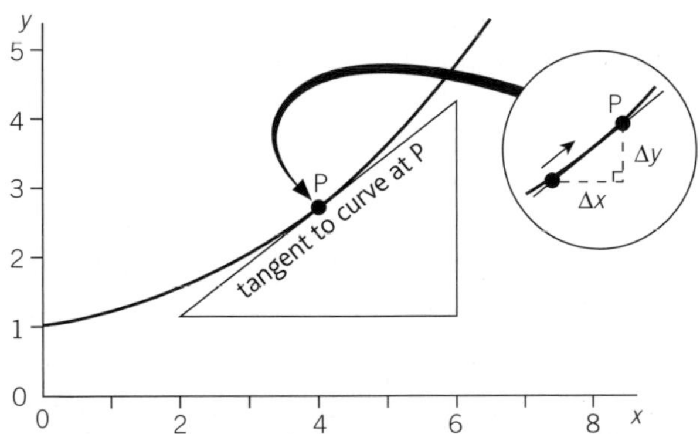


▲ **Figure 2** *Tangent to a curve at a point*

The tangent to the curve at a point is a straight line that touches the curve at that point, without cutting across it. To see why, mark any two points on a curve and join them by a straight line. The gradient of the line is $\Delta y/\Delta x$, where $\Delta y$ is the vertical separation of the two points and $\Delta x$ is the horizontal separation. Now repeat with one of the points closer to the other – the straight line is now closer in direction to the curve. If the points are very close, the straight line between them is almost along the curve. The gradient of the line is then virtually the same as the gradient of the curve at that position. Figure 2 shows this idea. In other words, the gradient of the straight line $\Delta y/\Delta x$ becomes equal to the gradient of the curve as $\Delta x \rightarrow 0$. The curve gradient is written as $\frac{dy}{dx}$ where $\frac{d}{dx}$ means rate of change.

The gradient of the tangent is a straight line and is obtained as explained above. Drawing the tangent to a curve requires practice. This skill is often needed in practical work. The **normal** at the point where the tangent touches the curve is the straight line perpendicular to the tangent at that point. An accurate technique for drawing the normal to a curve using a plane mirror is shown in Figure 3. At the point where the normal intersects the curve, the curve and its mirror image should join smoothly without an abrupt change of gradient where they join. After positioning the mirror surface correctly, the normal can then be drawn and then used to draw the tangent to the curve.

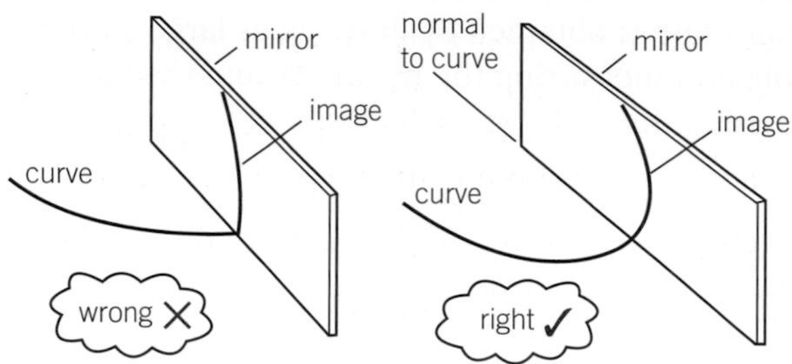


▲ **Figure 3** *Drawing the normal to a curve*

## Turning points

A turning point on a curve is where the gradient of the curve is zero. This happens where a curve reaches a peak with a fall either side (i.e., a maximum) or where it reaches a trough with a rise either side (i.e., a minimum). Where the gradient represents a physical quantity, a turning point is where that physical quantity is zero. Figure 4 shows an example of a curve with a turning point. This is a graph of the vertical height against time for a projectile that reaches a maximum height, then descends as it travels horizontally. The gradient represents the vertical component of velocity. At maximum height, the gradient of the curve is zero, so the vertical component of velocity is zero at that point.

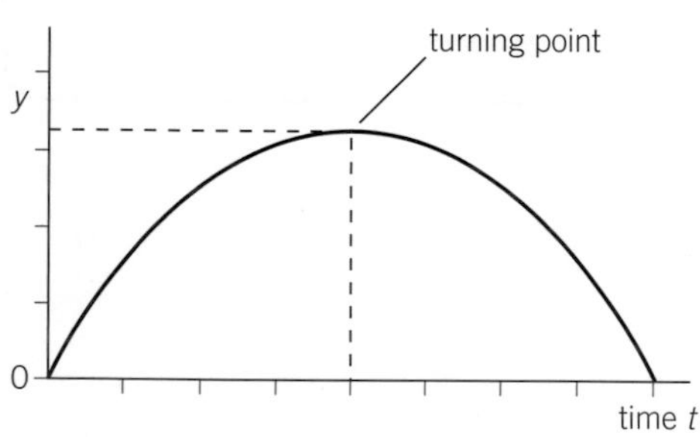


▲ **Figure 4** *Turning points*

### *Note:*

If the equation of a curve is known, the gradient can be determined by the process of **differentiation**. This mathematical process is not needed for AS level physics. The essential feature of the process is that, for a function of the form $y = kx^n$, the gradient (written as $dy/dx$) $= nkx^{n-1}$.

For example, if $y = \frac{1}{2}gt^2$, then $\frac{dy}{dt} = gt$.

## Areas and graphs

The area under a line on a graph can give useful information if the product of the $y$-variable and the $x$-variable represents another physical variable. For example, consider Figure 5, which is a graph of the tension in a spring against its extension. Since tension × extension is force × distance, which equals work done, then the area under the line represents the work done to stretch the spring.

Figure 5b shows a tension against extension graph for a rubber band. Unlike Figure 5a, the area under the curve is not a triangle, but it still represents work done, in this case the work done to stretch the rubber band.

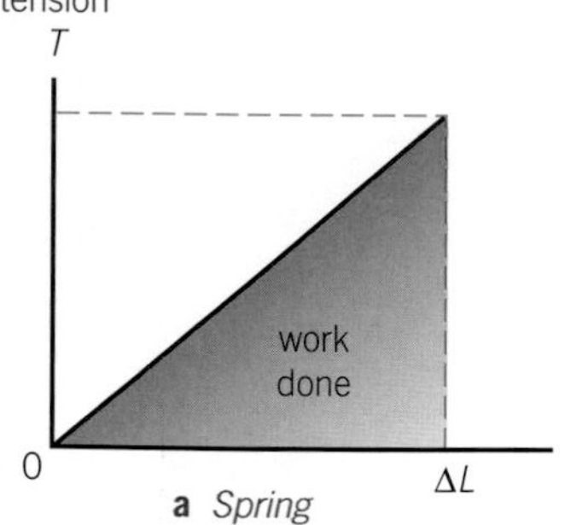


**a** *Spring*

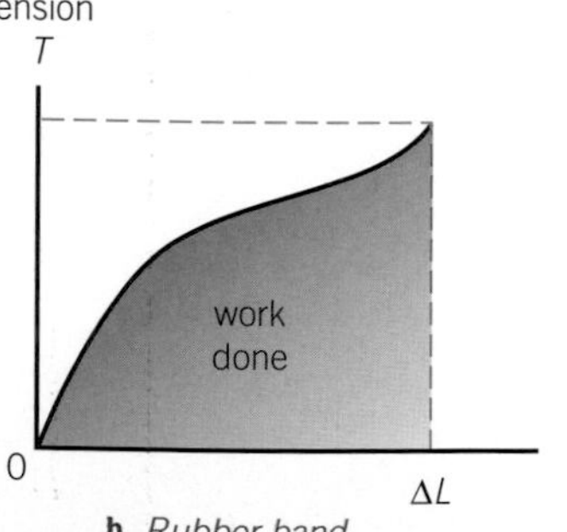


**b** *Rubber band*

▲ **Figure 5** *Tension versus extension*

The product of the $y$-variable and the $x$-variable must represent a physical variable with a physical meaning if the area is to be of use. A graph of mass against volume for different sizes of the same material gives a straight line through the origin. The mass is directly proportional to the volume, and the gradient gives the density. But the area under the line has no physical significance since mass × volume does not represent a physical variable.

Note that even where the area does represent a physical variable, it may not have any physical meaning. For example, for a graph of pd against current, the product of pd and current represents power, but this physical quantity has no meaning in this situation.

More examples of curves where the area is useful include:

- velocity against time, where the area between the line and the time axis represents displacement
- acceleration against time, where the area between the line and the time axis represents change of velocity
- power against time, where the area between the curve and the time axis represents energy
- potential difference against charge, where the area between the curve and the charge axis represents energy.

### Synoptic link

You have met work done as the area under a graph in Topic 11.2, Springs.

## Summary questions $\sqrt{x}$

**1** **a** Sketch a velocity against time graph (with time on the $x$-axis) to represent the equation $v = u + at$, where $v$ is the velocity at time $t$.

**b** What feature of the graph represents **i** the acceleration, **ii** the displacement?

**2** **a** Sketch a graph of current (on the $y$-axis) against pd (on the $x$-axis) to show how the current through an ohmic conductor varies with pd.

**b** How can the resistance of the conductor be determined from the graph?

**3** An electric motor is supplied with energy at a constant rate.

**a** Sketch a graph to show how the energy supplied to the motor increases with time.

**b** Explain how the power supplied to the motor can be determined from the graph.

**4** A steel ball bearing is released in a tube of oil and falls to the bottom of the tube.

**a** Sketch graphs to show how **i** the velocity, **ii** the acceleration of the ball changed with time from the instant of release to the point of impact at the bottom of the tube.

**b** What is represented on graph **a i** by **i** the gradient, **ii** the area under the line?

**c** What is represented on graph **a ii** by the area under the line?

# Useful data for AS Physics

## Data

### Fundamental constants and values

| quantity | symbol | value | units |
|---|---|---|---|
| speed of light in vacuo | $c$ | $3.00 \times 10^{8}$ | $\mathrm{m\,s^{-1}}$ |
| charge of electron | $e$ | $-1.60 \times 10^{-19}$ | C |
| the Planck constant | $h$ | $6.63 \times 10^{-34}$ | J s |
| the Avogadro constant | $N_A$ | $6.02 \times 10^{23}$ | $\mathrm{mol^{-1}}$ |
| molar gas constant | $R$ | 8.31 | $\mathrm{J\,K^{-1}\,mol^{-1}}$ |
| the Boltzmann constant | $k$ | $1.38 \times 10^{-23}$ | $\mathrm{J\,K^{-1}}$ |
| electron rest mass | $m_e$ | $9.11 \times 10^{-31}$ | kg |
| electron charge/mass ratio | $e/m_e$ | $1.76 \times 10^{11}$ | $\mathrm{C\,kg^{-1}}$ |
| proton rest mass | $m_p$ | $1.67(3) \times 10^{-27}$ | kg |
| proton charge/mass ratio | $e/m_p$ | $9.58 \times 10^{7}$ | $\mathrm{C\,kg^{-1}}$ |
| neutron rest mass | $m_n$ | $1.67(5) \times 10^{-27}$ | kg |
| gravitational field strength | $g$ | 9.81 | $\mathrm{N\,kg^{-1}}$ |
| acceleration due to gravity | $g$ | 9.81 | $\mathrm{m\,s^{-2}}$ |
| atomic mass unit | $u$ | $1.661 \times 10^{-27}$ | kg |

# AS equations

## Particle physics

### Fundamental particles

| class | name | symbol | rest energy / MeV |
|---|---|---|---|
| photon | photon | $\gamma$ | 0 |
| lepton | neutrino | $\nu_e$ | 0 |
| | | $\nu_\mu$ | 0 |
| | electron | $e^\pm$ | 0.510999 |
| | muon | $\mu^\pm$ | 105.659 |
| mesons | $\pi$ meson | $\pi^\pm$ | 139.576 |
| | | $\pi^0$ | 134.972 |
| | K meson | $K^\pm$ | 493.821 |
| | | $K^0$ | 497.762 |
| baryons | proton | p | 938.257 |
| | neutron | n | 939.551 |

### Properties of quarks

Antiparticles have opposite signs

| type | charge | baryon number | strangeness |
|---|---|---|---|
| u | $+\frac{2}{3}e$ | $+\frac{1}{3}$ | 0 |
| d | $-\frac{1}{3}e$ | $+\frac{1}{3}$ | 0 |
| s | $-\frac{1}{3}e$ | $+\frac{1}{3}$ | −1 |

### Properties of leptons

| lepton | lepton number |
|---|---|
| particles: $e^-, \nu_e; \mu^-, \nu_\mu; \tau^-, \nu_\tau$ | +1 |
| antiparticles: $e^+, \bar{\nu}_e; \mu^+, \bar{\nu}_\mu; \tau^+, \bar{\nu}_\tau$ | −1 |

## Geometrical equations

arc length = $r\theta$

circumference of circle = $2\pi r$

area of circle = $\pi r^2$

surface area of cylinder = $2\pi rh$

volume of cylinder = $\pi r^2 h$

area of sphere = $4\pi r^2$

volume of sphere = $\frac{4}{3}\pi r^3$

## Photons and energy levels

| | |
|---|---|
| photon energy | $E = hf = \frac{hc}{\lambda}$ |
| photoelectric effect | $hf = \phi + E_{Kmax}$ |
| energy levels | $hf = E_1 - E_2$ |
| de Broglie wavelength | $\lambda = \frac{h}{p} = \frac{h}{mv}$ |

## Electricity

| | |
|---|---|
| current and pd | $I = \frac{\Delta Q}{\Delta t}$ |
| | $V = \frac{W}{Q}$ |
| | $R = \frac{V}{I}$ |
| emf | $\varepsilon = \frac{W}{Q}$ |
| | $\varepsilon = IR + Ir$ |
| resistors in series | $R = R_1 + R_2 + R_3 + \ldots$ |
| resistors in parallel | $\frac{1}{R} = \frac{1}{R_1} + \frac{1}{R_2} + \frac{1}{R_3} + \ldots$ |
| resistivity | $\rho = \frac{RA}{L}$ |
| power | $P = VI = I^2R = \frac{V^2}{R}$ |

## Mechanics

| | |
|---|---|
| Moments | moment = $Fd$ |
| velocity and acceleration | $v = \frac{\Delta s}{\Delta t}$; $a = \frac{\Delta v}{\Delta t}$ |
| equations of motion | $v = u + at$ |
| | $s = \frac{(u + v)t}{2}$ |
| | $v^2 = u^2 + 2as$ |
| | $s = ut + \frac{1}{2}at^2$ |
| force | $F = ma$ |
| | $F = -\frac{\Delta(mv)}{\Delta t}$ |
| work, energy and power | $W = Fs\cos\theta$ |
| | $E_K = \frac{1}{2}mv^2$ |
| | $\Delta E_P = mg\Delta h$ |
| | $P = \frac{\Delta W}{\Delta t}$ $P = Fv$ |

efficiency of a machine = $\frac{\text{useful output power}}{\text{input power}}$

## Materials

| | |
|---|---|
| density | $\rho = \frac{m}{V}$ |
| Hooke's law | $F = k\Delta L$ |
| Young modulus | $= \frac{\text{tensile stress}}{\text{tensile strain}} = \frac{FL}{A\,\Delta L}$ |
| energy stored | $E = \frac{1}{2}k\,\Delta L^2 = \frac{1}{2}F\Delta L$ |

## Waves

| | |
|---|---|
| wave speed | $c = f\lambda$ |
| period | $T = \frac{1}{f}$ |
| fringe spacing | $w = \frac{\lambda D}{s}$ |
| diffraction grating | $d\sin\theta = n\lambda$ |

refractive index of a substance $s$, $n_s = \frac{c}{c_s}$

For two different substances of refractive indices $n_1$ and $n_2$,

| | |
|---|---|
| law of refraction | $n_1\sin\theta_1 = n_2\sin\theta_2$ |
| critical angle | $\sin\theta_c = \frac{n_2}{n_1}$ for $n_1 > n_2$ |

For a stretched string, first harmonic frequency

$$f_0 = \frac{1}{2l}\sqrt{\frac{T}{m}}$$

# Answers to summary questions

## 1.1

1 a i 6p, 6n ii 8p, 8n iii 92p, 143n

iv 11p, 13n v 29p, 34n

b i $^{235}_{92}U$ ii $^{12}_{6}C$ and $^{16}_{8}O$

2 a neutron b electron c neutron

3 a $+3.2 \times 10^{-19}$ C b 63 c $3.04 \times 10^{6}$ C kg$^{-1}$

4 a $2.67 \times 10^{-26}$ kg b 8 neutrons and 10 electrons

## 1.2

1 a electrostatic force b strong nuclear force

c strong nuclear force d electrostatic force

2 a $^{225}_{88}U$ Ra, $^{4}_{2}\alpha$ b $^{65}_{29}Cu$, $^{0}_{-1}\beta$

3 a Each beta emission increases the proton number by 1 and the alpha emission decreases it by 2, so the number of protons at the end is unchanged, which means the final nuclide is a bismuth nucleus.

b i 209 ii 82p, 127n

4 a A hypothesis is an untested idea or theory.

b i It is uncharged; it hardly interacts with matter.

ii The Sun, a nuclear reactor, a beta-emitting isotope

## 1.3

1 b i $5.1 \times 10^{14}$ Hz ii $1.5 \times 10^{6}$ Hz

3 a $7.0 \times 10^{14}$ Hz b $4.6 \times 10^{-19}$ J

4 a $4.7 \times 10^{14}$ Hz, $3.1 \times 10^{-19}$ J b $4.8 \times 10^{15}$

## 1.4

1 a 939 MeV (5 sf values for *c*, *e* and *h* give 938.26 MeV)

2 The rest energy of an electron and a positron is less than 2 MeV but not for a proton–antiproton pair, so pair production can happen for an electron and positron.

3 a 0.511 MeV b i 1.180 MeV

4 a A proton in the nucleus changes into a neutron, and a positron and a neutrino are created and emitted from the nucleus.

b No photon is involved in positron emission; no neutrino is emitted in pair production.

## 1.5

1 See 1.5 Fig 1

2 See 1.5 Fig 4

3 a See 1.5 Fig 3 b 0.001 fm, $10^{-27}$ s

4 a The W boson is charged and the photon is uncharged; the W boson has a non-zero rest mass and the photon has zero rest mass; the W boson is the exchange particle of the weak nuclear force whereas the virtual photon is the exchange particle of the electromagnetic force.

b In Fig 4b, a $W^+$ boson interacts with a neutrino, which turns into a $\beta^-$ particle. In Fig 5, the $W^+$ boson decays into a $\beta^-$ particle and a neutrino.

## 2.1

1 a the electron, the muon, the $\pi^+$ meson ,the $K^0$ meson, the proton, the neutron

b the $K^0$ meson, the neutron

2 a weak b weak c strong

3 a A muon b They both decay; the $\pi$ meson has a greater rest mass than a muon.

4 a i They decay into $\pi$ mesons, muons and antineutrinos and antimuons and neutrinos.

ii They decay into muons and antineutrinos and antimuons and neutrinos.

b electrons, neutrinos and antineutrinos

## 2.2

1 a i A particle that experiences the strong interaction

ii A particle that does not experience the strong interaction

b i a lepton ii a baryon iii a meson

2 a Both carry negative charge; the muon does not interact through the strong nuclear force, the $\pi^-$ meson does (or they have different rest masses).

b Both carry positive charge; the K meson has more rest mass (or is strange).

c Both are uncharged; a neutron is a baryon whereas a $K^0$ meson is a meson (or they have different rest masses and strangeness).

3 a lepton b baryons, mesons c mesons

4 a $\pi^- \pi^+ \pi^+$ b 74 MeV

## 2.3

1 a Both are negatively charged and do not interact through the strong nuclear force; they have different rest masses.

b Both are uncharged and do not interact through the strong nuclear force; the electron neutrino does not interact with the muon and the muon neutrino does not interact with the electron.

2 a i A strong interaction **ii** a weak interaction

**iii** a weak interaction

b i $\mu^- \rightarrow e^- + \bar{\nu}_e + \nu_\mu$ **ii** 105.5 MeV

3 a i 0 **ii** +1 **iii** +1 **iv** 0

b i +1 **ii** −1 **iii** −1 **iv** −1

4 a No, charge is not conserved. **b** Yes

c No, the (electron) lepton number is not conserved.

d No, neither the electron lepton number nor the muon lepton number is conserved.

## 2.4

1 a meson = quark + antiquark, baryon = 3 quarks

b proton = uud, neutron = udd

2 a See Figure 3 **b** a u quark in a proton changes to a d quark and emits a $W^+$ boson, which decays into a positron and a neutrino.

3 a dds has a charge of −1 ($= 3 \times -\frac{1}{3}$) and a strangeness of −1 due to the s quark

b $d\bar{u}$ + uud → $u\bar{s}$ + sdd; an up antiquark and an up quark annihilate each other and create a strange quark and a strange antiquark, which form a meson, consisting of an up quark and a strange antiquark, and a baryon, consisting of a strange quark and two down quarks.

## 2.5

1 a ii baryon number

c ii proton = uud, neutron = udd

**ii** uud + uud → uud + udd + $u\bar{d}$

2 a i $+\frac{1}{3}$ **ii** −1 **iii** 0 **b i** +1 **ii** −1 **iii** 0

3 a sss

b i charge = 0, strangeness = −2

**ii** the weak interaction

**iii** sss → ssu + $d\bar{u}$; an s quark changes to a u quark, and a d quark and a u antiquark are created.

## 3.1

2 a i $6.7 \times 10^{14}$ Hz, $4.4 \times 10^{-19}$ J

**ii** $2.0 \times 10^{14}$ Hz, $1.3 \times 10^{-19}$ J

3 a $1.7 \times 10^{14}$ Hz **b** $2.7 \times 10^{-19}$ J

4 a $3.1 \times 10^{-19}$ J **b** $1.6 \times 10^{-19}$ J **c** $2.5 \times 10^{14}$ Hz

## 3.2

2 a $1.6 \times 10^{12}$

3 a $3.4 \times 10^{-19}$ J **b** $1.5 \times 10^{15}$ **c** $2.5 \times 10^{12}$

4 a i $4.0 \times 10^{14}$ Hz **ii** $2.7 \times 10^{-19}$ J,

b $2.7 \times 10^{-19}$ J

## 3.3

1 Similarity; energy is absorbed by the atom; Difference; an electron stays in the atom when excitation occurs but leaves the atom when ionisation occurs.

2 a $1.66 \times 10^{-18}$ J, **b** 1.6 eV

3 a less **b** more, increases, decreases

4 a An electron inside the atom moves to an outer shell.

b The electron has insufficient kinetic energy to excite or ionise the atom and so cannot be absorbed.

## 3.4

1 a 9.0 eV

b There are 3 energy levels below the 5.7 eV level. There are 3 possible transitions to the ground state, 2 to the first excited state and 1 to the second excited state. The energy changes for these transitions are all different.
Hence there are 6 possible photon energies.

2 a The energy of an electron in the atom increases in excitation and decreases in de-excitation. Excitation can occur through photon absorption or electron collision. De-excitation only occurs through photon emission.

b i There are two excited levels at 1.8 eV and 4.6 eV above the ground state;

**ii** photon energies 1.8 eV, 2.8 eV and 4.6 eV.

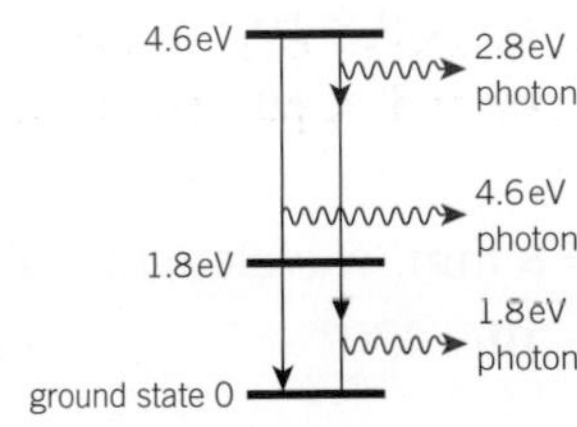

**3 a** The intermediate level could be at 0.6 eV or at 3.2 eV. The diagram shows one of these alternatives.

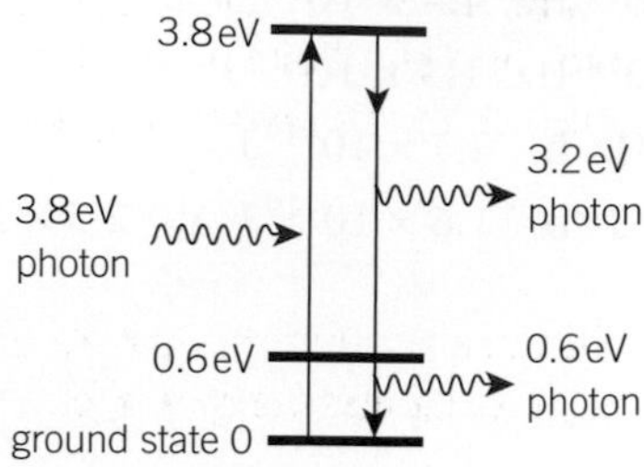


**b** An electron in the atom absorbs the 3.8 eV photon and moves to an outer shell. It moves to an inner shell and emits a 3.2 eV photon. An electron moves from this shell to the ground state, emitting a 0.6 eV photon.

**4** When the electricity supply is switched off, excitation by collision of the gas atoms in the tube ceases because the supply of electrons to the tube is cut off. Therefore the mercury atoms no longer emit ultraviolet radiation and so the coating atoms can no longer emit light.

## 3.5

**1** A line spectrum consists of discrete coloured lines whereas a continuous spectrum has a continuous spread of colours. A line spectrum is due to photons of certain energies only, whereas a continuous spectrum is due to a continuous spread of photon energies.

**2 a** $1.28 \times 10^{-19}$ J **b** $1.55 \times 10^{-6}$ m

**3 a** $3.21 \times 10^{-19}$ J **b** 2.0 eV

**4** The energy levels of the atoms of an element are unique to those atoms. The photons emitted by them have energies equal to the energy differences between the levels. Therefore the photon energies are characteristic of the atoms and can be used to identify the element.

## 3.6

**3 a** $3.6 \times 10^{-11}$ m **b** $2.0 \times 10^{-14}$ m

**4 a** $1.3 \times 10^{-27}$ kg m s$^{-1}$, $1.5 \times 10^{3}$ m s$^{-1}$

**b** $1.3 \times 10^{-27}$ kg m s$^{-1}$, 0.78 m s$^{-1}$

## 4.2

**1 a** 0.10 m **b** $1.9 \times 10^{-2}$ m

**2 a** 10 GHz **b** $5.0 \times 10^{14}$ Hz

**3** 1.0 V, 1.0 kHz

**4 a i** amplitude = 8 mm, wavelength = 47 mm

**ii** 180° **iii** 270°

**b** +8 mm

## 4.5

**1 b** 8.0 m

**2 a** 2.0 m **b i** 180° **ii** 225° **iii** 0

**4 b** 30 mm

## 4.6

**1 a** 1.6 m **b** 410 m s$^{-1}$

**2 a** 0.4 m **b** 0.53 m

**3 a** $2.4 \times 10^{-4}$ kg m$^{-1}$

**b** 0.20 mm

## 4.7

**1 a i** 5.0 V cm$^{-1}$ **ii** 2.4 cm **b i** 1.63 V **ii** 1.15 V

**2 a** 22 ms **b** 45.5 Hz

**3 a** 12.5 V, 8.8 V **b** 10 ms, 100 Hz

**4** A straight horizontal line 2.0 cm above the centre line would be seen.

The waveform seen on the screen has a peak height above the centre of 2.83 cm and 1.5 cycles would appear on the screen.

## 5.1

**1 a i** 14.9° **ii** 28.9° **iii** 40.6°

**b i** 58.7° **ii** See diagram below

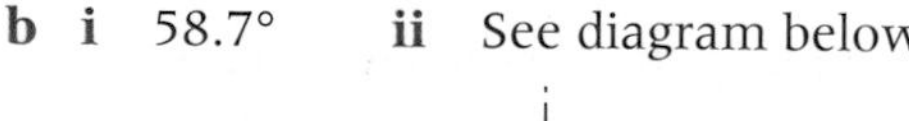

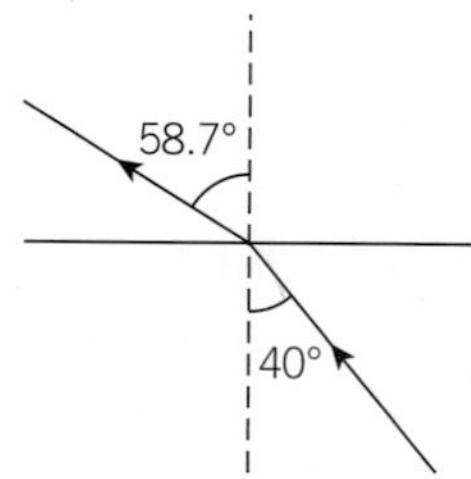


**2 a i** 19.5° **ii** 35.2° **b i** 59.4° **ii** 74.6°

**3 a** 1.53 **b** 34.5°

**4 b ii** 72.6°

## 5.2

**1 a i** 0.040 m **ii** 0.030 m, **b** 18.5°

**2 a i** $1.97 \times 10^{8}$ m s$^{-1}$ **ii** $2.26 \times 10^{8}$ m s$^{-1}$

**3 a** 25.4° **ii** 35°

**4 a i** red 49.5° **ii** blue 50.8°

## 5.3

**1 b i** 41° **ii** 49°

**2 b i** 34° **ii** 34°

**3 a** 65° **b i** 30° **ii** 45°

## 5.4

2 550 nm

3 0.9 mm

4 0.75 m

## 5.5

3 1.1 mm

## 5.6

2 a 6 mm

## 5.7

1 a 10.9°, 22.2° b 5

2 a 2 b 0.58 (= 35′)

3 a 1090 b 69.9°

4 a 599 $mm^{-1}$ b 3

## 6.1

1 a 3.7 N at 33° to 3.1 N b 17.1 N at 21° to 16 N

c 1.4 N at 45° to 3 N and 1 N

2 a 14.0 N in the same direction

b 6.0 N in the direction of the 10 N force

c 10.8 N at 22° to the 10 N force

3 6.1 kN vertically up, 2.2 kN horizontal

4 a 268 N b 225 N

## 6.2

1 a 7.3 N b 7.3 N at 31.5° to the vertical

2 b i 2.7 N ii 4.7 N

3 a 139 N b 95 N

4 a 73° b 6.8 N

## 6.3

1 300 N

2 b 6.2 N

3 0.27 m

4 6.75 N

## 6.4

1 0.51 N at 100 mm mark, 0.69 N at the 800 mm mark

2 a 122 N at 1.0 m end and 108 N at the other end, both vertically upwards

b 122 N at 1.0 m end and 108 N at the other end, both vertically downwards

3 620 kN, 640 kN

4 a 100 N, 50 N b 150 N

## 6.5

1 The centre of mass is higher if the upper shelves are filled instead of the lower shelves. If tilted, it will topple over at a smaller angle with the upper shelves full, than if they were empty.

2 89 N

3 a 48°

b Yes, they will raise the overall centre of mass so it will topple on a less steep slope.

## 6.6

1 a 50 N b 250 N

2 a 1800 N m b 1800 N

3 a 6.0 kN b 10.8 kN

4 1.5 N in the cord at 40° to the vertical, 1.9 N in the other cord

## 6.7

1 a 15 N b 3.0 N c 10.8 N

2 7 N

3 a 6.8 N b 52°

4 18.0 N

5 a 16.9 kN b 16.9 kN

6 a 6.2 N b 12.2 N

7 Move it further 50 mm away from the pivot.

8 a 6.8 N b 9.8 N

9 a 2200 N b 3100 N

10 b 950 N at X, 750 N at Y

11 a 2820 kN b 1660 kN and 1540 kN (to 3sf)

12 a 8.0 N and 16.0 N b 38 N and 76 N

13 b 11 kN, 11 kN

14 b 28.4 N

## 7.1

1 a 80 $km\,h^{-1}$ b 22 $m\,s^{-1}$

2 a 2.5 $m\,s^{-1}$ b 3.0 $m\,s^{-1}$

3 a $2.5 \times 10^4$ $km\,h^{-1}$ b $7.0 \times 10^3$ $m\,s^{-1}$

4 a 45 000 m b 28.3 $m\,s^{-1}$

5 b i 4.0 km

ii 30 $m\,s^{-1}$ then 25 $m\,s^{-1}$ in the opposite direction

## 7.2

**1 a** $1.5\,m\,s^{-2}$ **b** $-2.5\,m\,s^{-2}$

**2 a** $0.45\,m\,s^{-2}$ **b** $7.9\,m\,s^{-1}$

**3 b** $0.60\,m\,s^{-2}$, 0, $-0.40\,m\,s^{-2}$

## 7.3

**1 a** $2.0\,m\,s^{-2}$ **b** 221 m

**2 a** 43 s **b** $-0.93\,m\,s^{-2}$

**3 a i** $0.2\,m\,s^{-2}$ **ii** 90 m

**b i** $-0.75\,m\,s^{-2}$ **ii** 8.0 s

**d** $3.0\,m\,s^{-1}$

**4 a** $5.0\,m\,s^{-2}$ **b** 7.5 m **c** 18 m **d** $6.4\,m\,s^{-1}$

## 7.4

**1 a** 4.0 m **b** $8.8\,m\,s^{-1}$

**2 a** 3.2 s **b** $31\,m\,s^{-1}$

**3 a i** 3.9 s **ii** $38\,m\,s^{-1}$

**4 a** $1.6\,m\,s^{-2}$ **b** $3.6\,m\,s^{-1}$ **c** 0.64 m

## 7.5

**1 a i** 83 s **ii** 127 s

**2 a** 600 s

**3 a** velocity / $m\,s^{-1}$

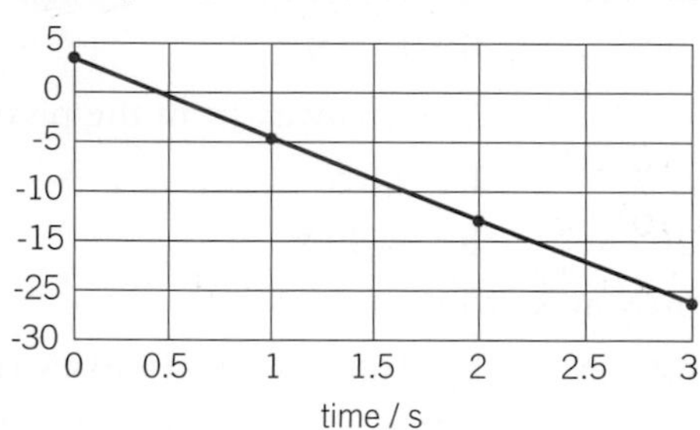


**4 a i** 0.61 s **ii** $5.9\,m\,s^{-1}$ **iii** 0.43 s **iv** $4.2\,m\,s^{-1}$

## 7.6

**1 a i** 52 s **ii** $0.49\,m\,s^{-2}$

**b i** 406 m **ii** $-1.04\,m\,s^{-2}$

**2 a** 15 m **b** $-0.13\,m\,s^{-2}$

**c** $0.67\,m\,s^{-1}$ downwards, 13.4 m from the start

**3 a i** 80 m **ii** $8.0\,m\,s^{-1}$

**b i** 65 s **ii** $-0.12\,m\,s^{-2}$

**4 a i** $180\,m\,s^{-1}$ **ii** 2.7 km

**b** 4.4 km **c** $290\,m\,s^{-1}$

## 7.7

**1 a** $32\,m\,s^{-1}$ **b** 2.8 s **c** 39 m

**2 a** 3.0 s **b** 49 m

**c** $34\,m\,s^{-1}$ (at 62° to the horizontal)

**3 a** 0.20 s **b** $11.7\,m\,s^{-1}$

**4 a** 354 m **b i** 1020 m **ii** 1020 m

**c** $146\,m\,s^{-1}$

## 7.8

**1 a** 470 mm **b** $3.0\,m\,s^{-2}$ **c** $2.7\,m\,s^{-1}$

**2 a** 2.02 s **b** 50.5 m

**3 a** $3.5\,m\,s^{-1}$, $3.0\,m\,s^{-1}$ **b i** 150 m **ii** 20 m

## 8.1

**1 a** $0.24\,m\,s^{-2}$ **b** 190 N **c** 0.024

**2 a** $2.4\,m\,s^{-2}$ **b** 12 000 N

**3 a** 360 N **b** 23 s

**4 a** $-1.3 \times 10^5\,m\,s^{-2}$ **b** 260 N

## 8.2

**1 a** 5400 N **b** 7700 N

**2 a** 60 N **b** 270 N

**3 a** 11.8 kN **b** 11.8 kN **c** 12.3 kN **d** 12.3 kN

**4 a** $1.0\,m\,s^{-2}$ **b** 12.5 N

## 8.3

**1 a i** $0.04\,m\,s^{-1}$ **ii** 1.5 N

**3 a** $0.14\,m\,s^{-2}$ **b** 520 m

## 8.4

**1 a i** 7.2 m **ii** 33.7 m **b** 4.1 m

**2 a** $6.75\,m\,s^{-2}$ **b** 6750 N

**4 b** 76 m

## 8.5

**1 a** $2.0g$ **b** 24 kN

**2 a** 80 ms **b** 375 kN

**3 a** $7.5\,m\,s^{-2}$ **b** 6750 N

**4 b i** $6.4g$ **ii** 4250 N

## 9.1

**1 a i** $1.2 \times 10^{-18}\,kg\,m\,s^{-1}$ **ii** $0.050\,kg\,m\,s^{-1}$

**iii** $14\,kg\,m\,s^{-1}$

**b i** 6.0 kg **ii** $20\,m\,s^{-1}$

**2 a** $3.6 \times 10^5\,kg\,m\,s^{-1}$ **b** 60 s

**3 a** $5.4 \times 10^6\,kg\,m\,s^{-1}$ **b** 45 s

**4 a** $9.0 \times 10^3\,kg\,m\,s^{-1}$

**b i** $-8.4 \times 10^3\,kg\,m\,s^{-1}$ **ii** $1.0\,m\,s^{-1}$

## 9.2

1 a $1600\,kg\,m\,s^{-1}$ b $3200\,N$

2 a $3000\,kg\,m\,s^{-1}$ b $7.5\,kN$

3 a $-4.2 \times 10^{-23}\,kg\,m\,s^{-1}$ b $-1.9 \times 10^{-13}\,N$

4 a $-2.1 \times 10^{-23}\,kg\,m\,s^{-1}$ b $-9.5 \times 10^{-14}\,N$

## 9.3

1 $0.72\,m\,s^{-1}$

2 $0.7\,m\,s^{-1}$ in the same direction

3 $0.05\,m\,s^{-1}$ in the direction the 1.0 kg trolley was moving in

4 $-0.63\,m\,s^{-1}$ in the opposite direction to its initial direction

## 9.4

2 a $9.0\,m\,s^{-1}$ in the same direction b 24 kJ

3 a $1.1\,m\,s^{-1}$ in the reverse direction b 20 J

4 a i $1.0\,m\,s^{-1}$ b The driver of the 250 kg car experiences a force from his/her car which slows him/her down; the other driver experiences a force from his/her car that accelerates him/her.

## 9.5

1 $0.35\,m\,s^{-1}$

2 a $0.25\,m\,s^{-1}$; the mass of A and X was greater than the mass of B, so B moved away faster.

3 a i $0.10\,m\,s^{-1}$ ii 15 mJ b $0.19\,m\,s^{-1}$

4 a $9.0\,m\,s^{-1}$ b i 1.1 J ii 81 J

## 10.1

1 a 200 J b 4.5 J

2 a 48 J b 24 J c 0

3 a 1000 J b 600 J c 400 J

4 a 2.4 N b 0.12 J

## 10.2

1 a 9.0 J b 9.0 J c 1.8 m

2 a 15.7 kJ b 5.8 kJ c 9.9 kJ d 20 N

3 a 590 kJ b 2.4 kJ c 470 kJ d 122 kJ e 1.6 kN

## 10.3

1 a 1.1 kJ b $62.5\,J\,s^{-1}$

2 500 MW

3 a 156 MJ b 140 MJ c 12 MW

4 122 m

## 10.4

1 a $450\,J\,s^{-1}$ b $1800\,J\,s^{-1}$

2 a 480 J b 50 J c 10%

3 a $570\,MJ\,s^{-1}$ b $6.2 \times 10^{5}\,kg$

4 a 600 b 3.7 MJ c 8%

### Renewable energy

1 $1.0 \times 10^{7}\,m^{2}$

2 4 MW

3 $125\,m^{3}\,s^{-1}$

4 a $6.3 \times 10^{11}\,kg$ b 430 MW

## 11.1

1 a $8.0 \times 10^{-4}\,m^{3}$ b $3.1 \times 10^{3}\,kg\,m^{-3}$

2 a 6.3 kg b $2.0 \times 10^{-3}\,m^{3}$ c $3.1 \times 10^{3}\,kg\,m^{-3}$

3 a $9.6 \times 10^{-6}\,m^{3}$ b $7.5 \times 10^{-2}\,kg$

4 a i 0.29 kg ii 0.12 kg b $2.3 \times 10^{3}\,kg\,m^{-3}$

## 11.2

1 a 0.40 m b 12.5 N

2 a 20 N b 100 mm c $200\,N\,m^{-1}$

3 a 40 N b 200 mm

4 a $12.3\,N\,m^{-1}$ b $8.8 \times 10^{-2}\,J$ c 2.2 N

## 11.3

1 $1.0 \times 10^{9}\,Pa$

2 $1.3 \times 10^{11}\,Pa$

3 a $9.4 \times 10^{8}\,Pa$ b $1.1 \times 10^{-2}\,m$

## 11.4

1 a $3.3 \times 10^{6}\,Pa$ b $2.8 \times 10^{-4}\,m$ c 0.21 J

2 a 2.3 mm b $1.7 \times 10^{-2}\,J$

3 a 470 kN b 47 J

4 a 10.0 J b 4.2 J

## 12.1

1 a i 3.5 C ii 210 C b i 3.0 A ii 0.15 A

2 a $3.8 \times 10^{15}$ b $1.9 \times 10^{22}$

3 a 72 mC b $4.5 \times 10^{17}$

4 a 1600 s b 8000 s

## 12.2

1 a 29 kJ b 720 J

2 a 2.0 A b 22 kJ

3 i 48 kJ ii 3.5 A

4 a 12 kJ b 4.5 W c 2700 s

## 12.3

1 a 6.0 Ω, 10 V, 0.125 mA, 160 Ω, 2.5 mA
 b 7.5 Ω

2 31 Ω

3 0.11 mΩ

4 a $1.8 \times 10^{-6}$ Ωm  b 33 mm

## 12.4

1 a 0.25 A, 12 Ω
 b The filament would become brighter and hotter until it melts and breaks as a result.

2 a 0.03 mA  b 0.38 mA

3 a

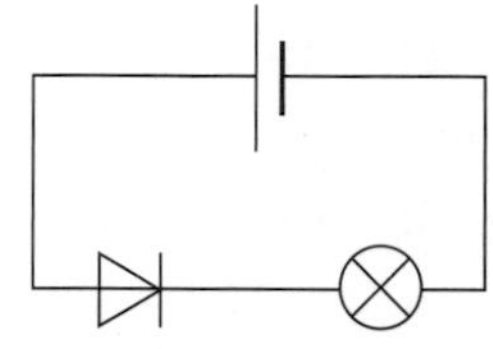

 b The diode would then be 'reverse-biased' so the current in the circuit would be negligible.

4 a 30.4 Ω  b 46°C

## 13.1

1 a 1.0 A, 4.0 A  b 5.0 A  c 30 W

2 b i 2.0 V  ii 0.20 A

3 b i 4.0 V  ii 2.0 V  iii 10 Ω

4 a 3.6 V  b 30 Ω

## 13.2

1 a 16 Ω  b 3.0 Ω  c 4 Ω

2 a 2 Ω  b 6 Ω  c 1.0 A  d 4.0 W

3 a 3.6 Ω  b 0.83 A
 c 2 Ω: 0.5 W, 4 Ω: 1.0 W, 9 Ω: 1.0 W  d 2.5 W

4 a 14.4 W  b 2.4 Ω

## 13.3

1 a 6.0 Ω  b 2.0 A  c 3.0 V  d 9.0 V

2 a 0.5 A  b 1.25 V  c 0.63 W  d 0.13 W

3 a 2.0 Ω  b 1.5 V

4 a 12 V, 2 Ω

## 13.4

1 a 12.0 Ω  b 0.25 A
 c 4 Ω: 0.25 A, 1.0 V, 24 Ω: 0.08 A, 2.0 V; 12 Ω: 0.17 A, 2.0 V

2 a 20.0 Ω  b 1.05 A  c 1.05 A, 15.8 V

3 i 2.0 W  ii 2.0 W

## 13.5

1 The brightness of the light bulb increases gradually from zero to maximum brightness.

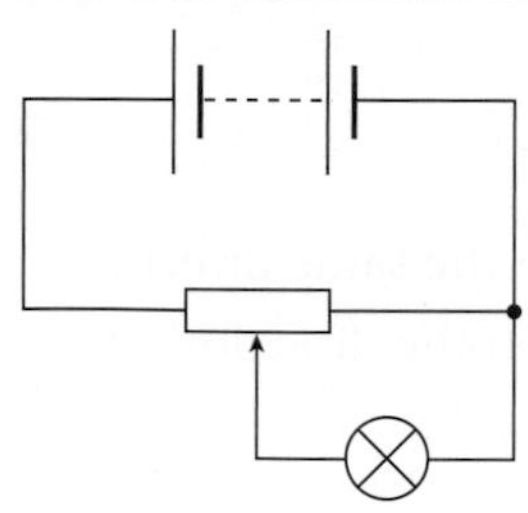

2 a i 0.5 A  ii 8.0 Ω: 4.0 V; 4.0 Ω: 2.0 V
 b i 3.0 V,  ii 4.0 V

3 a i 2.8 V  iii 6.4 kΩ

## 16.1

1 a i 0.500 m  ii 320 cm  iii 95.6 m
 b i 450 g  ii 1.997 kg  iii $5.4 \times 107$ g
 c i $2.0 \times 10^{-3}$ m$^2$  ii $5.5 \times 10^{-5}$ m$^2$
  iii $5.0 \times 10^{-6}$ m$^2$

2 a i $1.50 \times 10^{11}$ m  ii $3.15 \times 10^{7}$ s
  iii $6.3 \times 10^{-7}$ m  iv $2.57 \times 10^{-8}$ kg
  v $1.50 \times 10^{5}$ mm  vi $1.245 \times 10^{-6}$ m
 b i 35 km  ii 650 nm  iii $3.4 \times 10^{3}$ kg
  iv 870 MW (= 0.87 GW)

3 a i 20 m s$^{-1}$  ii 20 m s$^{-1}$  iii $1.5 \times 10^{8}$ m s$^{-1}$
  iv $3.0 \times 10^{4}$ m s$^{-1}$
 b i $6.0 \times 10^{3}$ Ω  ii 5.0 Ω  iii $1.7 \times 10^{6}$ Ω
  iv $4.9 \times 10^{8}$ Ω  v 3.0 Ω

4 a i 301  ii $2.8 \times 10^{9}$  iii $1.9 \times 10^{-23}$
  iv $1.2 \times 10^{-3}$  v $2.0 \times 10^{4}$  vi $7.9 \times 10^{-2}$
 b i $1.6 \times 10^{-3}$  ii $5.8 \times 10^{-6}$
  iii 1.7  iv $3.1 \times 10^{-2}$

## 16.2

1 a 1.57 m  b i 1.57 m  ii 1.05 m  iii 0.26 m

2 a i 68°  ii 41°  iii 22°  iv 61°
 b i 17 cm  ii 16 m (15.6 m to 3sf)  iii 4.8 mm
  iv 101 cm

3 a 49 mm  b i 35 km  ii 31°

4 a i 3.9 N, 4.6 N  ii 3.4 N, 9.4 N  iii 4.8 N, 5.7 N
 b 4.0 N, 30° to 3.5 N

## 16.3

**1** **a** 0.2 **b** 0.1, 0.25 **c** < 0.1

**2** **b** **i** 2 **ii** −1 **iii** 4.25 **iv** $\frac{1}{3}$

**3** **b** **i** 0.25 **ii** ± 2.8 **iii** ± 0.5 **iv** 4

**4** **a** $3.1 \times 10^{-7}\,\mathrm{m}^3$

**b** $6.2 \times 10^{-3}\,\mathrm{m}$

**c** 2.5 s

**d** $17\,\mathrm{m\,s^{-1}}$

## 16.4

**1** **a** **i** 3 **ii** −3 **iii** 1

**b** **i** −4 **ii** 8 **iii** 2

**c** **i** −1 **ii** 5 **iii** 5

**d** **i** −1.5 **ii** 3 **iii** 2

**2** **a** **i** $y = 2x - 8$ **ii** −8

**b** **i** $3\,\mathrm{m\,s^{-2}}$ **ii** $5\,\mathrm{m\,s^{-1}}$

**3** **a** (1, 4) **b** $y = 4x$

**4** **a** $x = 2, y = 0$ **b** $x = 3, y = 5$ **c** $x = 2, y = 0$

## 16.5

**3** **b** **i** use gradient = $\frac{A}{\rho L}$ **ii** use gradient = $\frac{VA}{\rho}$

**4** **b** **ii** $u$ = $y$-intercept, $\frac{1}{2}a$ = gradient

## 16.6

**1** **b** **i** gradient **ii** area under line

**2** **b** Resistance = $\frac{1}{\text{gradient}}$

**3** **b** The power is constant and is represented by the gradient of the line.

**4** **b** **i** acceleration **ii** distance fallen **c** velocity

# Glossary

## A

**acceleration** change of velocity per unit time.

**acceleration of free fall** acceleration of an object acted on only by the force of gravity.

**accurate** a measurement that is obtained, using calibrated instruments correctly, is said to be accurate.

**accuracy** how close a measurement or answer is to the true value.

**alpha radiation** particles that each consist of two protons and two neutrons.

**amplitude** maximum displacement of a vibrating particle; for a transverse wave, it is the distance from the middle to the peak of the wave.

**annihilation** when a particle and its antiparticle meet, they destroy each other and become radiation.

**antibaryon** a hadron consisting of three antiquarks.

**antimatter** *antiparticles* that each have the same rest mass and, if charged, have equal and opposite charge to the corresponding particle. See *annihilation* and *pair production*.

**antimuon** antiparticle of the muon; see *muon*.

**antineutrino** the antiparticle of the *neutrino*.

**antinode** fixed point in a stationary wave pattern where the amplitude is a maximum.

**antiparticle** There is an antiparticle for every type of particle. A particle and its corresponding antiparticle have equal rest mass and, if charged, equal and opposite charge.

**antiquark** antiparticle of a quark.

**atomic number** $Z$ the number of protons in the nucleus of an atom.

## B

**baryon** a hadron consisting of three quarks.

**base units** the units that define the SI system (e.g., the metre, the kilogram, the second, the ampere).

**beta radiation** $\beta^-$ particles are fast moving electrons emitted by unstable neutron-rich nuclei or by free neutrons when they decay; $\beta^+$ particles are fast moving *positrons* emitted by unstable proton-rich nuclei.

**braking distance** the distance travelled by a vehicle in the time taken to stop it.

**breaking stress** see *ultimate tensile stress*.

**brittle** snaps without stretching or bending when subject to stress.

## C

**centre of mass** the centre of mass of a body is the point through which a single force on the body has no turning effect.

**charge carriers** charged particles that move through a substance when a pd is applied across it.

**circuit rule for current (Kirchhoff's first law)**
1. The current passing through two or more components in series is the same through each component.
2. At a junction, the total current in = the total current out.

**circuit rules for pd (Kirchhoff's second law)**
1. For two or more components in series, the total pd across all the components is equal to the sum of the pds across each component.
2. The sum of the emfs round a complete loop in a circuit = the sum of the pds round the loop.

**coherent** two sources of waves are coherent if they emit waves with a constant phase difference.

**conservation rules** conservation of energy, momentum, charge, *baryon number* and *lepton numbers* applies to all particle interactions. Conservation of *strangeness* applies to strong interactions only.

**couple** pair of equal and opposite forces acting on a body but not along the same line.

**critical angle** the angle of incidence of a light ray must exceed the critical angle for *total internal reflection* to occur.

**critical temperature of a superconducting material** temperature at and below which its resistivity is zero.

**cycle** interval for a vibrating particle (or a wave) from a certain displacement and velocity to the next time the particle (or the next particle) that has the same displacement and velocity.

## D

**de Broglie hypothesis** matter particles have a wave-like nature characterised by the de Broglie wavelength.

**de Broglie wavelength** the wavelength of a matter particle = $\frac{h}{p}$, where $p$ is the momentum of the particle.

**de-excitation** process in which an atom loses energy by photon emission, as a result of an electron inside an atom moving from an outer shell to an inner shell.

**density of a substance** mass per unit volume of the substance

**diffraction** spreading of waves on passing through a gap or near an edge.

**diffraction grating** a plate with many closely-ruled parallel slits on it.

**dispersion** splitting of a beam of white light by a glass prism into colours.

**displacement** distance in a given direction.

**drag force** the force of fluid resistance on an object moving through the fluid.

**ductile** stretches easily without breaking.

## E

**efficiency** the ratio of useful energy transferred (or the useful work done) by a machine or device to the energy supplied to it.
The ratio of the machine's output power to its input power.

**effort** the force applied to a machine to make it move.

**elastic limit** point beyond which a wire is permanently stretched.

**elasticity** property of a solid that enables it to regain its shape after it has been deformed or distorted.

**electrolysis** process of electrical conduction in a solution or molten compound due to ions moving to the oppositely charged electrode.

**electrolyte** a solution or molten compound that conducts electricity.

**electromagnetic interaction (or force)** interaction (or force) between two charged objects.

**electromagnetic radiation** see *electromagnetic wave*.

**electromagnetic wave** a wavepacket or photon consisting of transverse electric and magnetic waves in phase and at right angles to each other.

**electromotive force (emf)** the amount of electrical energy per unit charge produced inside a source of electrical energy.

**electron capture** process in which an inner-shell electron of an atom is captured by the nucleus.

**electron volt** amount of energy equal to $1.6 \times 10^{-19}$ J defined as the work done when an electron is moved through a pd of 1 V.

**endoscope** optical fibre device used to see inside cavities.

**energy** the capacity to do work; see *work*.

**energy levels** the energy of an electron in an electron shell of an atom.

**equilibrium** state of an object when at rest or in uniform motion.

**error bar** representation of an uncertainty on a graph.

**error of measurement** uncertainty of a measurement. Errors can include *systematic* (including *zero error*) and *random*.

**excitation** process in which an atom absorbs energy without becoming ionised as a result of an electron inside an atom moving from an inner shell to an outer shell.

## F

**First harmonic** Pattern of stationary waves on a string when it vibrates at its lowest possible frequency.

**fluorescence** glow of light from a substance exposed to ultraviolet radiation; the atoms de-excite in stages and emit visible photons in the process.

**force** any interaction that can change the velocity of an object.

**free body force diagram** a diagram of an object showing only the forces acting on the object.

**frequency** the number of cycles of a wave that pass a point per second.

**friction** force opposing the motion of a surface that moves or tries to move across another surface.

**fundamental mode of vibration** see *first harmonic*.

## G

**gamma radiation** high-energy photons emitted by unstable nuclei or produced in particle annihilations.

**gravitational field strength** force of gravity per unit mass on a small object.

**ground state** lowest energy state of an atom.

## H

**hadron** particles and antiparticles that can interact through the strong interaction.

**Hooke's law** the extension of a spring is proportional to the force needed to extend it.

## I

**inertia** resistance of an object to change of its motion.

**interference** formation of points of cancellation and reinforcement where two coherent waves pass through each other.

**internal resistance** resistance inside a source of electrical energy; the loss of pd per unit current in the source when current passes through it.

**ion** a charged atom.

**ionisation** process of creating ions.

**isotopes** atoms of an element with different numbers of neutrons and the same number of protons.

## K

**kaon (or K meson)** a *meson* that consists of a *strange quark* or *antiquark* and another quark or antiquark.

**kinetic energy** the energy of an object due to its motion.

## L

**laser** device that produces a parallel coherent beam of monochromatic light.

**lepton** electrons, positrons, muons and antimuons, neutrinos and their antiparticles are classified as leptons because they cannot interact through the strong interaction. They interact through the weak interaction and, in the case of electrons, positrons, muons and antimuons, through the electromagnetic interaction.

**lepton number** a lepton number is assigned to every lepton (+1) and antilepton (−1), on the basis that the total lepton number for each branch of the lepton family is always conserved.

**light-dependent resistor** resistor which is designed to have a resistance that changes with light intensity.

**limit of proportionality** the limit beyond which, when a wire or a spring is stretched, its extension is no longer proportional to the force that stretches it.

**linear** two quantities are said to have a linear relationship if the change of one quantity is proportional to the change of the other.

**load** the force to be overcome by a machine when it shifts or raises an object.

**longitudinal waves** waves with a direction of vibration parallel to the direction of propagation of the waves.

## M

**mass number** see *nucleon number*

**mass** measure of the inertia or resistance to change of motion of an object.

**matter waves** the wave-like behaviour of particles of matter.

**meson** a hadron consisting of a quark and an antiquark.

**modal dispersion** the lengthening of a light pulse as it travels along an optical fibre, due to rays that repeatedly undergo total internal reflection having to travel a longer distance than rays that undergo fewer total internal reflections.

**moment of a force about a point** force × perpendicular distance from the line of action of the force to the point.

**momentum** mass × velocity.

**motive force** the force that drives a vehicle.

**muon** a *lepton* that is negatively charged and has a greater rest mass than the electron.

## N

**negative temperature coefficient** the resistance of a semiconductor decreases when its temperature is increased.

**neutrino** uncharged *lepton* with a very low rest mass compared with the electron.

**neutrino types (or 'branches')** there are three types of neutrinos, the electron neutrino, the muon neutrino, and the tau neutrino. This specification only requires knowledge of the electron neutrino and the muon neutrino branches (and their antiparticles) of the lepton 'family'.

**Newton's first law of motion** an object remains at rest or in uniform motion unless acted on by a resultant force.

**Newton's second law of motion** the rate of change of momentum of an object is proportional to the resultant force ($F$) on it. Newton's 2nd law may be written as $F = (\Delta mv)/\Delta t$. For constant mass, this equation becomes $F = ma$ where acceleration $a = (\Delta v/\Delta t)$

**node** fixed point in a stationary wave pattern where the amplitude is zero.

**nucleon** a neutron or proton in the nucleus.

**nucleon number $A$** the number of neutrons and protons in a nucleus; also referred to as *mass number*.

**nuclide** a type of nucleus with a particular number of protons and neutrons.

## O

**Ohm's law** The pd across a metallic conductor is proportional to the current, provided the physical conditions do not change.

**optical fibre** a thin flexible transparent fibre used to carry light pulses from one end to the other.

## P

**pair production** when a gamma photon changes into a particle and an antiparticle.

**pascal** unit of pressure or stress equal to 1 N m$^{-2}$.

**path difference** the difference in distances from two coherent sources to an interference fringe.

**period** time for one complete cycle of a wave to pass a point.

**phase difference** the fraction of a cycle between the vibrations of two vibrating particles, measured either in radians or degrees.

**photoelectric effect** emission of electrons from a metal surface when the surface is illuminated by light of frequency greater than a minimum value known as the *threshold frequency*.

**photon** packet or 'quantum' of electromagnetic waves.

**pion (or π meson)** a *meson* that consists of an up or down *quark* and an up or down *antiquark*.

**plane-polarised waves** transverse waves that vibrate in one plane only.

**plastic deformation** deformation of a solid beyond its elastic limit.

**positive temperature coefficient** the resistance of a metal increases when its temperature is increased.

**positron** *antiparticle* of the electron.

**potential difference** work done or energy transfer per unit charge between two points when charge moves from one point to the other.

**potential divider** two or more resistors in series connected to a source of pd.

**potential energy** the energy of an object due to its position.

**power** rate of transfer of energy.

**precision of a measurement** precise measurements are ones in which there is very little spread about the mean value. Precision depends only on the extent of random error and gives no indication of how close the results are to the true value.

**precision of an instrument** the smallest non-zero reading that can be measured, also sometimes referred to as the instrument sensitivity or resolution.

**pressure** force per unit area acting on a surface perpendicular to the surface.

**principle of conservation of energy** energy cannot be created or destroyed.

**principle of moments** for an object in equilibrium, the sum of the clockwise moments about any point = the sum of the anticlockwise moments about that point.

**probable error** estimate of the *uncertainty* of a measurement.

**progressive waves** waves which travel through a substance or through space if electromagnetic.

**projectile** a projected object in motion acted on only by the force of gravity.

**proton number** see atomic number

## Q

**quark** protons and neutrons and other hadrons consist of quarks. There are six types of quarks, the up quark, the down quark, the strange quark, the charmed quark, the top quark and the bottom quark. This specification only requires knowledge of the up, down, and strange quarks and their antiquarks.

**quark model (or standard model)** a quark can join with an antiquark to form a *meson* or with two other quarks to form a *baryon*. An antiquark can join with two other antiquarks to form an *antibaryon*.

## R

**radian** measure of an angle defined such that 2π radians = 360°.

**random error** error of measurement due to readings that vary randomly with no recognisable pattern or trend or bias.

**range of a set of a readings** the range of a set of readings of the same measurement is given by the minimum and the maximum reading.

**range of an instrument** the minimum and the maximum reading that can be obtained using the instrument.

**refraction** change of direction of a wave when it crosses a boundary where its speed changes.

**refractive index** speed of light in free space / speed of light in the substance.

**resistance** pd/current.

**resistivity** resistance per unit length × area of cross-section.

**rest energy** energy due to rest mass $m_0$, equal to $m_0c^2$, where $c$ is the speed of light in free space.

## S

**scalar** a physical quantity with magnitude only.

**semiconductor** a substance in which the number of charge carriers increases when the temperature is raised.

**sensitivity of an instrument** the output reading per unit input quantity.

**SI system** the scientific system of units.

**specific charge** charge/mass value of a charged particle.

**spectrometer** instrument used to measure light wavelengths very accurately.

**speed** change of distance per unit time.

**stationary waves** wave pattern with nodes and antinodes formed when two or more progressive waves of the same frequency and amplitude pass through each other.

**stiffness constant** the force per unit extension needed to extend a wire or a spring.

**stopping distance** *thinking distance + braking distance*.

**stopping potential** the minimum potential that needs to be applied to a metal plate to attract all the photoelectrons emitted from its surface back to the surface.

**strain** extension per unit length of a solid when deformed.

**strangeness number** a strangeness number is assigned to every particle and antiparticle on the basis that strangeness is always conserved in the strong interaction, but not in a weak interaction or a decay involving a strange quark or antiquark.

**stress** force per unit area of cross-section in a solid perpendicular to the cross-section.

**strong interaction** interaction between two hadrons.

**strong nuclear force** attractive force between nucleons that holds the nucleons in the nucleus.

**superconductor** a material that has zero electrical resistance.

**superposition** the effect of two waves adding together when they meet.

**systematic error** causes readings to differ from the true value by a consistent amount each time a measurement is made. Sources of systematic errors can include the environment, methods of observation, or instruments used.

## T

**terminal pd** the potential difference across the terminals of a power supply.

**terminal speed** the maximum speed reached by an object when the drag force on it is equal and opposite to the force causing the motion of the object.

**thermistor** resistor which is designed to have a resistance that changes with temperature.

**thinking distance** the distance travelled by a vehicle in the time it takes the driver to react.

**threshold frequency** minimum frequency of light that can cause *photoelectric effect*.

**total internal reflection** A light ray travelling in a substance is totally internally reflected at a boundary with a substance of lower refractive index, if the angle of incidence is greater than a certain value known as the *critical angle*.

**transverse waves** waves with a direction of vibration perpendicular to the direction of propagation of the waves.

**types of light spectra**

continuous spectrum
– continuous range of colours corresponding to a continuous range of wavelengths,

line emission spectrum
– characteristic coloured vertical lines, each corresponding to a certain wavelength,

line absorption spectrum
– dark vertical lines against a continuous range of colours, each line corresponding to a certain wavelength.

## U

**ultimate tensile stress** tensile stress needed to break a solid material.

**uncertainty** the interval within which the true value can be expected to lie.

**useful energy** energy transferred to where it is wanted when it is wanted.

## V

**vector** a physical quantity with magnitude and direction.

**velocity** change of displacement per unit time.

**virtual photon** carrier of the electromagnetic force; a photon exchanged between two charged particles when they interact.

## W

**W boson** carrier of the weak nuclear force; W bosons have non-zero rest mass and may be positive or negative.

**wave – particle duality** matter particles have a wave-like nature as well as a particle-like nature; photons have a particle-like nature as well as a wave-like nature.

**wavefronts** lines of constant phase (e.g., wavecrests).

**wavelength** the least distance between two adjacent vibrating particles with the same displacement and velocity at the same time (e.g., distance between two adjacent wave peaks).

**weak interaction** interaction between two leptons and between a lepton and a hadron (e.g., electron capture).

**weak nuclear force** force responsible for beta decay.

**weight** the force of gravity acting on an object.

**work** force × distance moved in the direction of the force.

**work function** minimum amount of energy needed by an electron to escape from a metal surface.

## Y

**yield point** point at which the stress in a wire suddenly drops when the wire is subjected to increasing strain.

**Young's fringes** parallel bright and dark fringes observed when light from a narrow slit passes through two closely spaced slits.

**Young modulus** stress/strain (assuming the limit of propotionality has not been exceeded).

## Z

**zero error** any indication that a measuring system gives a false reading when the true value of a measured quanity is zero. A zero error may result in a systematic uncertainty.

# Index


absorption, photons 30–33, 37
acceleration 96, 120–127, 144–150, 186
- contact/impact times 150
- dynamics 122–124
- gravity 125–127
- non-uniform 121, 123
- performance tests 120
- terminal speed 144–145
- two-stage problems 130–131
- uniform 120–127

accuracy 245–246
addition
- potential differences 215
- vectors 96–98, 261

airbags 151
air resistance, projectiles 134–135
algebra 263–264
alpha radiation 6–7, 166
ammeters 209–210
amplitude 54, 60–61
analysis 249–251
annihilation, matter 10–11
anodes, photocells 32–33
anomalies 249–251
antibaryons 25
antimatter 10–12
antimuons 19, 26
antineutrino–proton interactions 14
antineutrinos 7, 14, 23
antinodes 60–63
antiparticles 10–12, 19, 26–27
antiquarks 21, 22, 24–25
argon, starter switches 38
atomic numbers 4–5
atoms
- excitation energies 35–40, 43
- fluorescence 37–38
- radioactive decay 6–7
- spectra 39–40
- structure 4

average speed 118–119

balanced forces 100–102
balances 248
baryons 21, 25–27
batteries 209
beam power, lasers 9
best-fit tests 250–251
beta radiation/decay 7, 14–15, 25, 166
Bohr atom 40
bosons 14–15
boundaries 71–72, 74
braking distances 146–148
breaking stress 191
bright fringes 77–81
brightness control 227
brittle materials 191
de Broglie wavelengths 42–43

cable tension 142
calipers 247–548
cancellation, double slits 78
capture, electrons 15
cathode ray oscilloscopes (CRO) 54, 64–65
cathodes, photocells 32–33
cells 209, 223–225
centre of mass 104
changing velocity 118–119
charge carriers 202–203
charge conservation 26
circuit diagrams, symbols 209, 226–227
circular track variable resistors 226
circular waves 58–59
coherence, interference 79
coherent bundles 74
coherent sources 58–59, 76–86
collapsible steering wheels 151
collisions
- elastic/inelastic 164–165
- head-on 163
- ionisation 34–35
- particles 26–27
- rebounding 159–160

colour 79–81
combinations of quarks 25
communication, optical fibres 73–74
complete waves 54
components
- electrical 209–211, 224, 226–227
- vectors 98–99

compression, longitudinal waves 52
conduction, electricity 32–33, 202–203, 208
conductors 203, 208
conservation 23, 24, 26–27, 161–3, 170
constant acceleration 120, 122–137, 265–266
constant velocity 119
contact time, impacts 150
continuous spectra 87
control variables 243
cosmic rays 18
couples 106
critical angles 73
critical temperatures 208
CRO *see* cathode ray oscilloscopes
crumple zones 151
current
- direct current circuits 214, 223–225
- electrical 202–203
- photoelectric 32–33
- power 205

dark fringes 77–81
data loggers 255
data processing 249
decays, particles and antiparticles 26–27
deceleration 146–148, 150
- *see also* acceleration

de-excitation 36–40
deformation of solids 189–193
density 184–185
dependent variables 243
diamonds 73
differentiation 270
diffraction 41–43, 57, 82–87
diodes 209, 210, 224–225
direct current circuits 214–229
- cells 223–225
- current rules 214
- internal resistance 220–222, 266
- parallel 215
- potential difference 214–216
- potential dividers 226–227
- resistors 217–219
- series 214–215

discharge tubes 80
dispersion 72, 74
displacement 54, 96, 118
- forces 171–173
- longitudinal waves 52
- motion graphs 128–129
- phase 55
- superpositions 58–63
- transverse waves 53
- two-stage problems 130–131
- velocity–time graphs 123

displacement–time graphs 119
distance–time graphs 118–119
double slit interference 76–81
drag 134–135, 144–145
duality, quantum phenomena 41–43
ductile materials 191
DVDs 80

dynamics, constant
acceleration 122–124

efficiency, energy 177–178
Einstein, A. 30–33
elastic collisions 164–165
elastic energy 193
elasticity
materials 186–193
solids 189–193
springs 186–188
elastic limits 186, 190–191
elastic potential 188
electrical circuits
cells 223–225
components 209–211, 224, 226–227
diodes 209, 210, 224–225
internal resistance 220–222, 266
Kirchhoff's laws 224–225
loop rule 216
parallel 215, 217–218
potential dividers 226–227
resistors 206–208, 211, 217–219
sensing 227
series 214–215, 217
electricity 200–229
conduction 202–203
current and charge 202–203
direct current circuits 214–229
efficiency 178
potential difference 204–205
potential dividers 226–227
power 175, 205
renewable energy 179–181
resistance 206–208, 211, 217–222
superconductivity 208
electromagnetic force 13, 20–21
electromagnetic radiation
double slit interference 76–78
duality 41
optics 68–91
photons 30–33, 36–41
refraction 68–73
electromagnetic waves 8
electromotive force (emf) 204–205, 216, 220–222, 266
electron neutrinos 22
electrons 20
antiparticles 10, 12
de Broglie wavelengths 42–43
capture 15
conduction 32–33
decay 26
de-excitation 36–40
diffraction 42–43
emission 30–35, 41–43
excitation 35–40, 43
hydrogen atoms 40
photon interactions 30–33, 36
shells 34–40, 43
as waves 41–43
work functions 31
electron volts 34–35
electroscopes 30
elements 4–7, 39–40
emf *see* electromotive force
emission
electrons 30–35, 41–43
photoelectrons 30–33
photons 36–40
energy
conservation 26
efficiency 177–178
elastic 193
ionisation 34–35
kinetic 173–174
photons 9, 30–33
potential 173–174
potential difference 204–205
power 175–176
renewable 179–181
stored in springs 188
strain 192–193
work 170–172
energy changes 173–174
energy levels 35–40, 43
engines, power 176
equilibria
conditions 111
couples 106
forces 96–117
rules 110–113
stability 107–109
supports 105
three forces 101–102
errors 245–246, 249–250
evaluation 249–251
excitation energies 35–40, 43
ionisation 35–36
photons 36
spectra 39–40
explosions 166–167

filament lamps 80
filters, polarising 53
first harmonics 62
fixed resistors 226
fluids, drag 144
fluorescence 37–38, 178
force carriers 13–15
force–distance graphs 171–172
forces
acceleration 120–127, 144–150, 186
air resistance 134–135
balanced 100–102
conservation of momentum 161–163
couples 106
displacement 171–172
drag 134–135, 144–145
electromotive 204–205, 216, 220–222
in equilibrium 96–117
free body diagrams 110
friction 138, 146–147, 177
gravity 140
impacts 158–160
impulse 156–157
lifting 112
moments 103–104
momentum 154–169
nuclear 6–7, 14, 20–21, 23
opposing 141–142
pulleys 143
rebounding impacts 159–160
scalars/vectors 96–99
slopes 108–111, 143
springs 186–188
stable/unstable equilibria 107–109
statics calculations 114–115
supports 105
vectors 96
weight 140
work 170–172, 175–181
force-time graphs 156–160
free body force diagrams 110
free fall 125–127
frequency 54
harmonics 62–63
photoelectric effect 30–33
stationary waves 60–63
friction 138, 146–147, 177
fringe separation 77–81, 82–84
fundamental particles 10–15

Galileo 126
gamma radiation 7
glass, light refraction 68–69
gold leaf electroscopes 30
gravity 125–127, 140
ground states 36

hadrons 20–23
harmonics 62–63
head-on collisions 163
heat, rate of transfer 218–219
heating, resistance 218–219

helium 40
Hooke's law 172, 186
horizontal projection 132–133
hydroelectric power 180–181
hydrofoils 145
hydrogen atoms 40

identical cells 224
impacts
  forces 149–150, 158–160
  head-on collisions 163
  rebounds 159–160
  times 150
impulse, momentum 156–157
inclined plane tests 125
indicators 209
indirect assessment 256–257
inelastic collisions 164–165
inertia 140
insulators 203
intensity
  photoelectric effects 32–33
  Young's fringes 83–84
interactions, particles 13–15
interference 58–59, 76–86
  coherence 79
  colour 79–81
  diffraction gratings 85–86
  double slits 76–81
  single slits 82–84
  superposition 58–59
  wavelengths 79–81
internal resistance 220–222, 266
ionisation energies 34–35
isotopes 4–7

joules 170–171
junctions, current 214

kaons 18–20, 24–25
kinetic energy 173–174
  de Broglie wavelengths 42
  conduction electrons 32–33
  definition 173
  photoelectrons 31, 33, 266
  potential 173
Kirchhoff's laws 224–225

laboratory practice 243–244
lamps 37–38, 178, 80
lasers 9, 80
LDR *see* light-dependent resistors
LED *see* light-emitting diodes
leptons 18–29
  classification 20–21
  collisions 22–23
  conservation 26
  rules 23
lift, air 134
lifting 112
light
  coherent sources 76, 80
  diffraction 41–43, 57, 82–87
  double slits 76–81
  photons 30–33
  refraction 68–73
  sensor circuits 227
  single slits 82–84
  sources 37–38, 76, 80, 209, 227
  speed 72, 74
  wave-particle duality 41–43
light bulbs 37–38, 178, 227
light-dependent resistors (LDR) 209, 227
light-emitting diodes (LED) 209
limit of proportionality 190–193
line absorption spectra 87
linear track variable resistors 226
line emission spectra 87
line spectra 39–40
liquids, density measurement 184
locality, maps of 96
longitudinal waves 52
loop rule 216
low-energy light bulbs 38

magnetic resonance 43
maps of locality 96
mass
  centre of 104
  numbers 5
  *see also* point masses
material dispersion 74
materials 184–197
  conduction 203
  deformation 189–193
  density 184–185
  elasticity 186–193
  insulation 203
  loading/unloading 192–193
  resistivity 207–208
  semiconductors 203, 210–211
  springs 186–188
matter 4–47
  annihilation 10–11
  force carriers 13–15
  leptons 18–29
  pair production 10–11
  particles 4–17
  quantum phenomena 30–47
  quark combinations 24–25
  strong nuclear force 6–7, 20–21
  waves 41–42
  weak nuclear force 14, 20, 23
medical endoscopes 74
mercury 35, 36–38
mesons 18, 21, 25–27
metals 31–33, 191, 208
micrometers 247–248
microscopes 83, 43
microwaves 59, 61
modal dispersion 74
model explosions 167
moments 103–106
momentum 13, 154–169
  de Broglie wavelengths 42
  conservation 161–163
  definition 154
  force-time graphs 156–157
  head-on collisions 163
  impulse 156–157
  radioactive decay 166
monochromatic light 74, 80
MR (magnetic resonance) 43
muon neutrinos 22
muons 18, 20, 23, 26
muscles, power 175

negative temperature coefficients 211
neutrinos 7, 14, 20–23
neutrons 20
the newton 155
newtonmeters 140
Newton's laws of motion 138–156, 161–167
  first law 138–139, 154–155
  second law 139, 141–156
  third law 161–167
nodes, waves 58, 60–63
non-uniform acceleration 121, 123
normal 68–69
nucleons 2, 4

Ohm's law 206
optical fibres 73–74
optics 68–91
  diffraction 82–86
  double slit interference 76–81
  polarisation 53
  refraction 68–73
  single-slit diffraction 82–84
  spectra 87
  total internal reflection 73–75
oscilloscopes 54, 65–66

pair production 10–11
parabolic curves/paths 134–135, 267–268
parallel cells 224
parallel circuits 215, 217–218, 224

parallel springs 187
partial reflection 68
particle accelerators 19
particle-like nature 41
pascals 189–193
path differences 77–78
pendulums 174
percentage efficiency 178
performance tests 120
period, definition 54
permanent extension 192–193
perpendicular components, vectors 98–99
phase 55, 60–63, 76–78
photoelectric currents 32–33
photoelectric effect 8, 30–33
photoelectrons 30–33, 266
photons 8–9
  emission 36–40
  energy 9, 30–33
  excitation 36
  pair production 11
  spectra 39–40
  virtual 13–14
photovoltaic cells 180–181
pions 18, 20, 24–25
pipes 61
pitch 63
pivots 103–104
plane-polarisation 53
plastic deformation 190–193
point masses 100–101, 108–111, 143
polarisation, transverse waves 53
polymers 192–193
polythene 192–193
positive temperature coefficients 211
positrons 10, 12, 26
potential difference 34–35, 64–65, 204–205, 214–216, 220–222, 266
potential dividers 209–210, 226–227
potential energy 173–174
power 175–176, 179–181, 220–221, 266
powered vehicles, drag 145
practical assessments 252–257
principle of conservation of energy 170
principle of conservation of momentum 161–163
principle of moments 103–104
progressive waves 58, 60–63
projectile-like motion 134
projectiles 132–135, 267–268
proton–antineutrino interactions 14
proton numbers 4–5
protons 20
pulleys 143, 177
pulse dispersion 74
Pythagoras's theorem 260–261

quantised, definition 32
quantum phenomena 30–47
  de Broglie wavelengths 42–43
  de-excitation 36–40
  excitation energies 35–40
  fluorescence 37–38
  ionisation 34–35
  matter waves 41–42
  photoelectric effects 30–33
  photons 30–33
  spectra 39–40
  technology 43
  wave-particle duality 41–43
quarks 18–29
  baryons and mesons 21
  combinations 25
  conservation 26–27
  hadron collisions 22
  properties 25
  strangeness 24–25

radiation 6–9, 25, 30–33, 36–43, 166
radio waves 53
range, measurement 245–246
rarefaction 52
rate of heat transfer 218–219
ratio of potential differences 226
rebounding impacts 159–160
recoil 166–167
reflection 56, 68, 73–75
refraction 56
  boundaries 71–72
  glass 68–69
  optics 68–73
  total internal reflection 73
  triangular prisms 69
refractive index 68–69, 73
regular solids, density measurement 184
reinforcement 77–78
renewable energy 179–181
resistance
  direct current circuits 217–219
  electrical 206–208
  heat 211, 218–219
  internal 220–222, 266
  semiconductors 210–211
  temperature 208, 211
resistive forces 144–145
  *see also* drag; friction
resistivity 207–208
resistors 207–211, 217–219, 226–227
  direct current circuits 217–219
  fixed 226
  heat 211, 218–219
  in parallel 217–218
  in series 217
  variable 226–227
rest energy 10, 21
resultant 97–99, 261
ripple tanks 56
rockets, thrust 142
rotation 103–104
rubber bands 192–193
rules 247

safety 149–151, 243
satellite dish design 57
scalars 96
scale diagrams 96–99, 111
scales 247
scanning tunneling microscopes (STM) 43
scientific units 258
Searle's apparatus 190
seat belts 151
second harmonics 62
semiconductors 203, 210–211
sensor circuits 227
series cells 223–224
series circuits 214–215, 217
series springs 187
shells 34–40, 43
sigma particles 24
single cells 223
single slit diffraction 82–84
SI system 258, 262–263
skidding 146
slopes 108–111, 143
Snell's law 68
solar heating 180–181
solar photovoltaic power 180–181
solids
  deformation 189–193
  regular/irregular 184
sound 52, 61, 63, 65
sources
  coherent light 76, 80
  electromotive force 204
  light 37–38, 76, 80, 209, 227
specific charge 5
spectra 39–40, 72, 87
spectrum analysers 63
speed 118–137
  acceleration 120–137
  definition 118
  light 72, 74
  oscilloscope readout 64–65
  ultrasound 65

waves 54–55
speed–time graphs 128–129
springs 186–188
energy stored 188
Hooke's law 172
series/parallel 187
SQUIDs 43
stable equilibria 107–109
starter switches 38
static forces 114–115
stationary waves 58, 60–63
steel wire 193
stiffness 191
STM (scanning tunneling microscopes) 43
stopping distances 146–148
stopping potential 31–32
straight lines
graphs 265–266
motion 122–131
strain 189–193
strain energy 192–193
strangeness 19, 24–26
strange particles 19
stress 189–193
strings 62–63
strong nuclear force 6–7, 20–21
structure of atoms 4
substances, densities 185
sunlight 80
superconducting quantum interference devices (SQUIDs) 43
superconductivity 43, 208
supercrests 58
superfluids 40
superposition, waves 58–63
supertroughs 58
support forces 105
symbols, electrical components 209, 226–227
Système International (SI system) 258, 262–263

tape charts 144
temperature 208, 211, 227
tensile strain 142, 189–193
tensile stress 189–193
tension, cables 142
terminal speed 144–145
thermistors 209, 227
thinking distance 146
third harmonics 62
three forces in equilibrium 101–102
threshold frequencies 30–33
thrust 142
tidal power 180–181
tilting 107–109
time base circuits 64
timers 248
toppling 107–109
torque 103–104
total internal reflection 73–75
total resistance 217–218
towing trailers 141
transmission electron microscopes (TEM) 43
transparent substances 68–72
transverse waves 52–63
triangle of forces 110–111
triangular prisms 69
turning forces 103–104
turning points, curves 270
two-support problems, moments 105

ultimate tensile stress (UTS) 191
ultrasound, speed 65
uncertainty 245–246
uniform acceleration 120–127, 265–266
uniform objects, mass 104
units 258
unloading materials 192–193
unpolarised waves 53
unstable equilibria 107–109
useful energy 178
UTS (ultimate tensile stress) 191

vacuum photocells 32–33
vapour lamps 80
variable resistors 209–210, 226–227
vectors 96–99, 110, 261
vehicle bumpers 151
vehicles
drag 145
safety 149–151
slopes 108–111, 143
stopping distances 146–148
velocity 119
acceleration 120–137
de Broglie wavelengths 42
definition 118
non-uniform acceleration 121, 123
terminal 144–145
two-stage problems 130–131
uniform acceleration 120–127, 265–266
vectors 96
velocity–time graphs 123, 128–129
vernier calipers 247–248
vertical projection 132
vibrations 52–53, 58, 60–63
virtual photons 13–14
voltage *see* potential difference
voltmeters 209–210

water power 180–181
wavefronts 56, 85–86
wavelengths
de Broglie 42–43
colour 79–81
definition 54
harmonics 62–63
interference 79–81
stationary waves 60–63
wave-like nature 41
wave-particle duality 41–43
waves 50–91
circular 58–59
coherent sources 58–59
diffraction 57, 82–87
double slit interference 76–81
electromagnetic 52–91
harmonics 62–63
interference 58–59
light 30–40
longitudinal 52
matter 41–42
optics 68–91
phase 55, 60–63
plane-polarisation 53
progressive 58, 60–63
properties 56–59
reflection 56
refraction 56, 68–72
speed 54–55, 65
stationary 58, 60–63
superposition 58–63
transverse 52–63
vibrations 52–53, 58, 60–63
W bosons 14–15
weak nuclear force 14, 20, 23
weight 125, 140
white light 72, 81
wind power 179, 181
wires, elastic energy 193
work 170–172, 175–181
definition 170–171
efficiency 177–178
elasticity 193
electrical power 205
energy 170–172
potential difference 204–205
work functions, metals 31–33

yield point 190
Young modulus 190
Young's fringes 76–81, 82–84

**Acknowledgements**

The author wishes to acknowledge the active support, advice and contributions he has received from Marie Breithaupt, Janet Custard and Patrick Organ and from Sze-Kie Ho and Alison Schrecker and their colleagues at OUP.

The publishers would like to thank the following for permission to reproduce photographs.

p.4: Drs A. Yazdani & D.J. Hornbaker/Science Photo Library; p.9: Lightpoet/Shutterstock; p.10: National Institute on Aging/Science Photo Library; p11: American Institute Of Physics/Science Photo Library; p.12: Lawrence Berkeley National Laboratory/ Science Photo Library; p.18: Prof. G. Piragino/Science Photo Library; p.19: David Parker/Science Photo Library; p.21: Xenotar/ iStockphoto; p.22: Jade Experiment, Desy & University Of Manchester/Science Photo Library; p.27: ESA Hubble Space Telescope /NASA; p.32: 360b/Shutterstock; p.38: Sauletas/Shutterstock; p.39(T): SteveUnit4/Shutterstock; p.39(B): Dept. Of Physics, Imperial College/Science Photo Library; p.43: Hank Morgan/Science Photo Library; 1.CO: Bruce Rolff/Shutterstock; p.54: Tomas Mikula/ Shutterstock; p.57(TR): Andrew Lambert Photography/ Science Photo Library; p.57(TL): Andrew Lambert Photography/Science Photo Library; p.57(B): Kati1313/ iStockphoto; p.59: Berenice Abbott/Science Photo Library; p.60: Tommaso lizzul/Shutterstock; p.63: Eliks/ Shutterstock; p.72: SteveUnit4/Shutterstock; p.76: Sheila Terry/Science Photo Library; p.82: RichLegg/iStockphoto; p.83: Lafayette-Picture/Shutterstock; p.86: Sciencephotos/ Alamy; p.87(T): Giphotostock/Science Photo Library; p.87(C): Dept. Of Physics, Imperial College/Science Photo Library; p.87(B): Physics Dept., Imperial College/Science Photo Library; 4.CO: Andy_R/iStockphoto; p.118: Brian A Jackson/Shutterstock; p.120: Natursports/Shutterstock; p.125: Georgios Kollidas/Shutterstock; p.130: 3Dsculptor/ Shutterstock; p.132: Loren Winters/Visuals Unlimited, Inc. /Science Photo Library; p.144: Joggie Botma/ Shutterstock; p.147: photosync/Shutterstock; p.149(T): Trl Ltd./Science Photo Library; p.149(B): Trl Ltd./Science Photo Library; p.158(L): Stephen Dalton/Science Photo Library; p.158(C): Stephen Dalton/Science Photo Library; p.158(R): Stephen Dalton/Science Photo Library; p.166: ChameleonsEye/Shutterstock; : Taina Sohlman/ Shutterstock; p.179: CJimenez/Shutterstock; p.180(L): Y.Derome/Publiphoto Diffusion/Science Photo Library; p.180(TR): Pavel Vakhrushev/Shutterstock; p.180(BR): FooTToo/Shutterstock; 6.CO: Kamil Martinovsky/ Shutterstock; p.203: The Biochemist Artist/Shutterstock; p.204: Oleksiy Mark/Shutterstock; p.205(L): Maxx-Studio/Shutterstock; p.205(R ): Marcio Jose Bastos Silva/ Shutterstock; 12.CO: Chones/Shutterstock; : Andrew Lambert Photography/Science Photo Library; 14.CO: Drbouz/iStockphoto; p.245: Rafe Swan/Cultura/Science Photo Library; p.246: Volodymyr Krasyuk/Shutterstock; : Dutourdumonde Photography/Shutterstock; : Alan Jeffery/Shutterstock; p.255: Martyn F. Chillmaid/Science Photo Library;